T0340091

Piperidine-Based Drug Discovery

Heterocyclic Drug Discovery Series

Piperidine-Based Drug Discovery

Ruben Vardanyan
University of Arizona Tucson, AZ, USA

Elsevier
Radarweg 29, PO Box 211, 1000 AE Amsterdam, Netherlands
The Boulevard, Langford Lane, Kidlington, Oxford OX5 1GB, United Kingdom
50 Hampshire Street, 5th Floor, Cambridge, MA 02139, United States

Notices
Knowledge and best practice in this field are constantly changing. As new research and experience broaden
our understanding, changes in research methods, professional practices, or medical treatment may become
necessary.

Practitioners and researchers must always rely on their own experience and knowledge in evaluating and
using any information, methods, compounds, or experiments described herein. In using such information or
methods they should be mindful of their own safety and the safety of others, including parties for whom they
have a professional responsibility.

To the fullest extent of the law, neither the Publisher nor the authors, contributors, or editors, assume any
liability for any injury and/or damage to persons or property as a matter of products liability, negligence or
otherwise, or from any use or operation of any methods, products, instructions, or ideas contained in the
material herein.

British Library Cataloguing-in-Publication Data
A catalogue record for this book is available from the British Library

Library of Congress Cataloging-in-Publication Data
A catalog record for this book is available from the Library of Congress

ISBN: 978-0-12-805157-3

For Information on all Elsevier publications
visit our website at https://www.elsevier.com/books-and-journals

  Working together
to grow libraries in
developing countries

www.elsevier.com • www.bookaid.org

Publisher: John Fedor
Acquisition Editor: Anneka Hess
Editorial Project Manager: Anneka Hess
Production Project Manager: Anitha Sivaraj
Cover Designer: Mark Rogers

Typeset by MPS Limited, Chennai, India

Contents

3. 2-Substituted and 1,2-Disubstituted Piperidines

4. 3-Substituted and 1,3-Disubstituted Piperidines

5. 4-Substituted and 1,4-Disubstituted Piperidines

6.   Piperidin-4-Ylidene Substituted Tricyclic Compounds

7.   Piperidine-Based Nonfused Biheterocycles With C−N and C−C Coupling

Chapter 1

Introduction

1.1 THE SCOPE OF THE MATERIAL UNDER CONSIDERATION

Heterocyclic compounds constitute the largest and most varied family of organic chemistry that is gaining enormous importance in chemical and pharmaceutical industry. Numerous agrochemicals, information storages, electronics, plastics and optics modifiers and stabilizers, cosmetics additives, etc., are heterocyclic in nature.

Piperidine and its functionalized derivatives are increasingly popular building blocks in a vast array of synthetic protocols. The piperidine ring can be recognized in the structure of many synthetic compounds of practical interest and in the structure of many alkaloids and other natural or synthetic compounds with various biological activities known today. Piperidine is the compound which gives black pepper its spicy taste and gave the name of the compound in question.

Today it is possible to assert unequivocally that the mainstream of pharmaceuticals is heterocyclic and the leading heterocycle in the structure of pharmaceuticals is piperidine, which is the most encountered heterocycle found in pharmaceutical agents [1].

A search of the chemical and patent literature reveals thousands of references to this simple ring system, which is present in the structures of potential drugs in clinical and preclinical research.

As of October 8, 2015, the day of beginning the work on this monograph, 93,984 references containing the concept "piperidine" were found in SciFinder—the world's largest and most reliable collection of chemistry and related science, which, of course, is not an exhaustive number of publications on the subject under consideration. By January 10, 2017, the date on which work on this monograph was completed, that number grew to 97,972, which constitutes the appearance of roughly 4000 additional publications within a year and a half. Information about piperidine-containing compounds exists in publications and patents that do not contain the word piperidine as well as in other sources of scientific information, which we did not use. For example, by analyzing the scaffold content of the CAS Registry from more than 24 million organic compounds it has been found that the most frequently occurring ring references on heterocycles concern piperidine (191.803) [2].

Piperidine-Based Drug Discovery. DOI: http://dx.doi.org/10.1016/B978-0-12-805157-3.00001-6

The huge contribution in the development of piperidine-derived series of drugs belongs to Otto Eisleb, Anton Ebnother, Solomon Snyder, Samuel McElvain, Ivan Nazarov, Miroslav Protiva, and, of course, the great Dr. Paul Janssen, the most prolific drug inventor of all time: his team has produced more than 60 new therapeutics, most of which belong to the piperidine series [3–5].

This book was conceived as an attempt to show the panorama of drugs, the structure of which contains a piperidine ring, and to show methods for their synthesis. It is necessary to emphasize here the word "drugs."

Any attempt to create a more or less complete picture of a biologically active compound containing a piperidine ring in its structure was doomed to fail due to the inability to "grasp the immensity" and to show the entire existing material in reasonable frames.

But even the relatively limited list of piperidine drugs also creates a number of problems, the first of which is the mode of their classification and, consequently, the presentation.

One alternative classification is the attempt to build material based on their pharmacological properties thereby putting them into a traditional order according to, e.g., a generally accepted narration in pharmacology textbooks, just as we did in our two previous books [6,7].

However, to put the emphasis on the structure and methods of the synthesis of drugs requires another way of presentation that is more acceptable from the standpoint of an organic chemist. It was decided that the order and degree of substitution of the piperidine ring would be the best way to sort existing piperidine drugs.

So, the collected factual material in general was divided and distributed into chapters and subchapters according to the following generalization: derivatives of 1-substituted piperidines (such as trihexyphenidyl (Artane) (**1.1.1**)); derivatives of 2-substituted piperidines or 1,2-disubstituted piperidines (such as bupivacaine (Marcaine)) (**1.1.2**)); derivatives of 3-substituted piperidines and 1,3-substituted piperidines (such as troxipide (Aplace)) (**1.1.3**)); and tofacitinib (Xeljanz) (**1.1.4**) (Fig. 1.1).

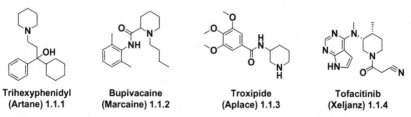

| Trihexyphenidyl | Bupivacaine | Troxipide | Tofacitinib |
| (Artane) 1.1.1 | (Marcaine) 1.1.2 | (Aplace) 1.1.3 | (Xeljanz) 1.1.4 |

FIGURE 1.1   Construction of subchapters in the book.

More diverse groups are represented by derivatives of 4-substituted and 1,4-disubstituted piperidines — fexofenadine (Allegra) (**1.1.4**), ebastine (Evastin) (**1.1.5**), astemizole (Hismanal) (**1.1.6**), indoramin (Baratol) (**1.1.7**), fentanyl (Sublimaze) (**1.1.8**), metopimazine (Vogalene) (**1.1.9**), ketanserin (Sufrexal) (**1.1.10**) (Fig. 1.2).

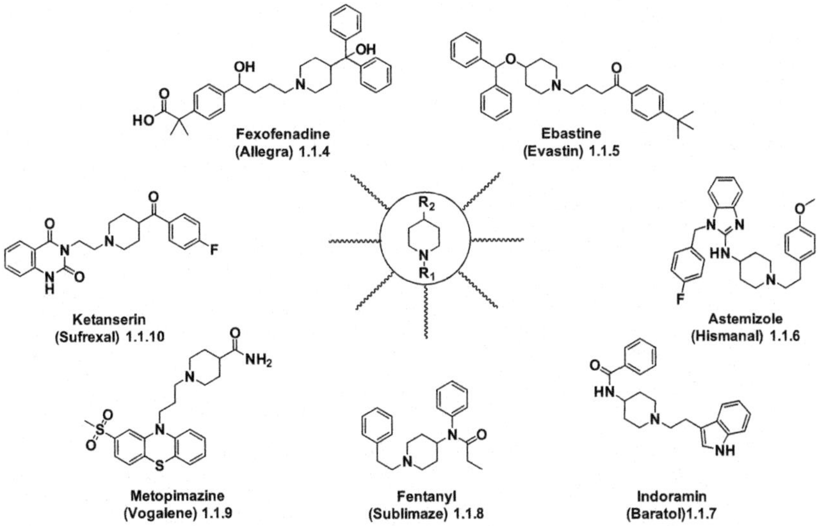

FIGURE 1.2 Structures of some of 1,4-disubstituted piperidine-based drugs.

No less common in the market are 1,4,4-trisubstituted piperidines exemplified by haloperidol (Haldol) (**1.1.11**), loperamide (Imodium) (**1.1.12**), trimeperidine (Promedol) (**1.1.13**), pethidine (Meperidine) (**1.1.14**), ketobemidon (Cliradon) (**1.1.15**), pipamperone (Dipiperon) (**1.1.16**), piritramide (Dipidolor) (**1.1.17**), alfentanil (Alfenta) (**1.1.18**), and remifentanil (Ultiva) (**1.1.19**), which in turn can be subdivided to alcohols, amines, ketones, amides, etc. Examples of these compounds are presented on the (Fig. 1.3).

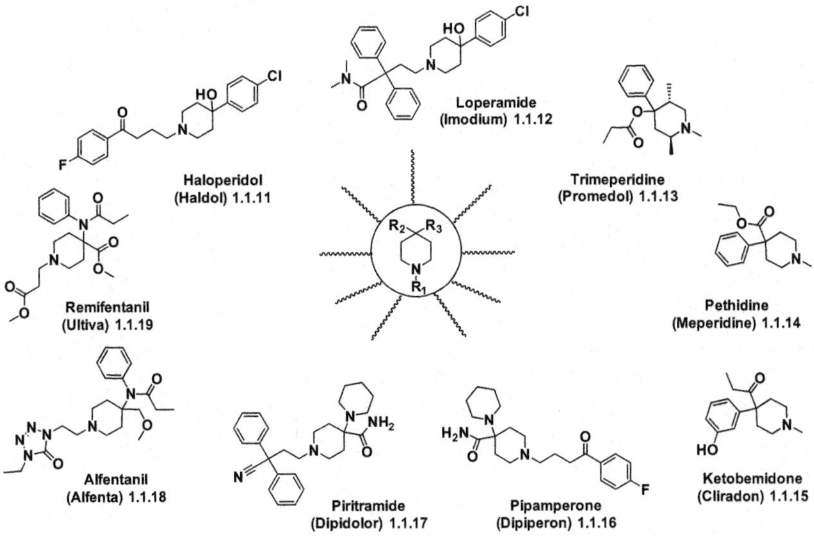

FIGURE 1.3 Structures of some of 1,4,4-trisubstituted piperidine-based drugs.

Another group of piperidine drugs could be represented as a combination of a piperidine ring with another heterocycle, and they can be classified as biheterocyclic compounds where the piperidine ring is not fused with another heterocycle and, in turn, is subdivided to those where: it is just bonded to another heterocycle by single C—C bond (such as naratriptan (Amerge) (**1.1.20**)); it is bonded to another heterocycle by double C=C bond (such as ketotifen (Zaditor) (**1.1.21**)); or by single C—N bond (such as pimozide (Orap)) (**1.1.22**). Another group could be represented as derivatives of piperidines fused with another heterocycle at 3,4-positions (such as clopidogrel (Plavix) (**1.1.23**)). The third group is described as derivatives of piperidines with spirofusion with another heterocycle at 4,4-positions (such as spiperone, (Spiropitan) (**1.1.24**)) (Fig. 1.4).

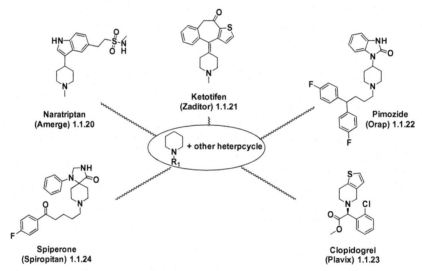

FIGURE 1.4 Structures of some of drugs represented as a combination of piperidine ring with another heterocycle.

In this book we have described the synthesis of about 150 piperidine drugs, but we did not know how to determine their relative importance.

Building a panorama of about 150 drugs, derivatives of piperidine, we did not know how to determine their relative importance and the relevance of a drug in the entire arsenal of piperidine drugs. The same general question also relates to the entire list of existing drugs in medicinal practice. So we had to come up with a way to determine some Comparative Drug Significance Index (DSI) or Comparative Drug Impact Factor (CDIF). In our opinion, one of the simplest possible solutions of the problem could be a determination of a number of publications on the drug that definitely shows a real interest in the scientific community of the drug.

Of course, this parameter depends on the influence of Pharma at every time interval, which, of course, also reflects a real significance of a drug. Therefore, after mentioning the name of a medicine described in this book, we decided to mention and bring in the number of publications devoted to it in SciFinder.

It seems that this approach to some extent could reflect the Comparative DSI.

Perhaps it will be much more correct to determine Comparative DSI as a sum of certain indicators. For instance, as a sum of some conditional figures derived from the number of publications divided by one thousand and sales, divided by one billion, e.g., as:

$$\sum = \frac{\text{Citations}}{10^3} + \frac{\text{Sales(\$)}}{10^9}$$

1.2 GENERAL METHODS OF THE SYNTHESIS OF PIPERIDINE COMPOUNDS

Piperidine (**1.2.1**) itself and most of its derivatives are easily produced by catalytic hydrogenation of the corresponding pyridine (**1.2.2**) derivatives over nickel, palladium, or ruthenium catalysts at $170-200°C$ [8–11]. Other reducing agents are sodium in ethanol or tin in hydrochloric acid [12].

Pyridine itself was first synthesized from acetylene and hydrogen cyanide [13,14]. More affordable sources of pyridine are coal tar, light-oil, and middle-oil fractions. Pyridine has been produced commercially from coal-tar sources contains only about 0.1% pyridine since the 1920s. Nowadays, most pyridine is produced synthetically via a number of synthetic processes. Reaction of acetaldehyde and formaldehyde with ammonia is the most widely used industrial method for pyridine production (Chichibabin pyridine synthesis) [14–17]. This thermal cyclocondensation reaction that occurs by passing on heating aldehydes and ammonia over a contact catalyst such as alumina is a low-yield process. Besides aldehydes, ammonia gas also reacts with acetylene or acetonitrile to give pyridine derivatives. The transformation is regarded as aldol condensations in conjunction with a Michael-type reaction and ring closures with ammonia.

Pyridine can also be prepared from furfuryl alcohol or furfural by passing a mixture of tetrahydrofurfuryl alcohol, H_2, and NH_3 over a Ni catalyst at about 200°C under high pressure [18,19] from glutaric acid, its anhydride or alkyl glutarates with ammonia, and H_2 at high temperatures and under high pressure in the presence of Ru-C or Co catalysts [20,21]. Pyridine can be prepared by oxidative dealkylation of alkylated pyridines, which are obtained as by-products via the synthesis of other pyridines using air over vanadium-, nickel-, silver- or platinum-based catalysts [22–24] (Fig. 1.5).

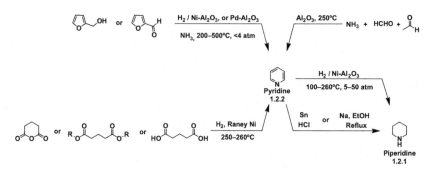

FIGURE 1.5 The major methods for the synthesis of pyridine followed by hydrogenation piperidine.

The major methods for the synthesis of pyridine derivatives, which in principle could be hydrogenated to corresponding substituted piperidine compounds include Hantzsch, Bönnemann, Kröhnke, Boger, Gattermann-Skita pyridine synthesis, and Ciamician-Dennstedt rearrangement which are well reviewed [25].

Onwards, in this introduction, it was decided to describe briefly the basic approaches for the synthesis of piperidine compounds, mainly functionalized piperidines of practical meaning, which can serve as basic starting materials and are the most ubiquitous heterocyclic building blocks for the synthesis of piperidine-based drugs. A huge amount of synthetic effort spent on the preparation of these compounds is reviewed [26−56].

A plethora of methodologies are available for the synthesis of piperidine derivatives and includes nucleophilic substitution reactions, intermolecular Michael, hydroamination, Diels−Alder reactions, ring-closing, metathesis, aldol reactions, Dieckmann condensations ene reactions, and others.

Nucleophilic Substitution Reactions

One of the methods of piperidine ring creation is based on the nucleophilic substitution of two leaving groups in a linear chain that contains the desired substitution pattern such as halides-, acetyl-, tosyl- or mesyl- groups, even hydroxyl- groups separated by five carbon atoms.

Several examples of reaction of amines with dihalides (**1.2.3**) to give substituted piperidines (**1.2.4, 1.2.5**) are described [57−59]. The reaction conditions require a large excess of amine, long-term reflux. Reaction with volatile amines need to be carried out in a sealed tube. The main difficulty of this type of cyclization is the correct choice of reaction conditions. Dilute reaction mixture gives a slow reaction that is too concentrated and thereby fraught with the intramolecular nucleophilic substitution or polymerization reaction.

The intramolecular displacement of a halide by a nitrogen nucleophile followed by deprotection of the functional groups, if necessary, is an analog and a well-established method for forming piperidine rings.

Phosphonamides (**1.2.6**) that have a chlorine group on the distance of five carbon atoms easily form piperidine ring when t-BuOK in dimethylformamide (DMF) is used as a base to give substituted diphenyl(piperidin-1-yl) phosphine oxides (**1.2.7**), which, after oxidative cleavage, give substituted piperidines (**1.2.5**) [60]. Compounds with N-tosyl or N-nosyl groups disposed on the distance of five carbon atoms (**1.2.8**) have been converted into piperidines (**1.2.9**) using 4-dimethylaminopyridine (DMAP) to effect NH-deprotonation and intramolecular iodo- of bromo- substitution. Cleavage of the N-tosyl group resulted in poor yields of the free amine. However, deprotection of the N-nosyl compounds was achieved without difficulty. The N-nosyl group was removed by treatment with thiophenol and K_2CO_3 in acetonitrile/dimethyl sulfoxide (DMSO) to give substituted piperidines (**1.2.5**) [61].

A simple, one-pot preparation of piperidine ring via efficient chlorination of amino alcohols (**1.2.10**) with use of $SOCl_2$ in dimethoxyethane was proposed recently and obviates the need for the N-protection-cyclization-deprotection sequence commonly employed for this type of transformation [62] (Fig. 1.6).

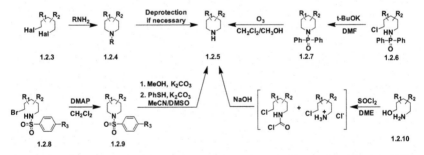

FIGURE 1.6 Methods for the synthesis of substituted piperidines by nucleophilic substitution reactions.

The intramolecular nucleophilic substitution of an alcohol moiety is a less commonly used method in piperidine derivatives synthesis, but intramolecular substitution of an activated alcohol moiety (tosylate, mesylate, triflate, acetate) is another more common method for the synthesis of piperidine ring.

The N-tosylamino alcohols (**1.2.11**) are easily transformed to the desired piperidines under Mitsunobu conditions (triphenylphosphine (PPh$_3$), azodicarboxylate (DEAD), tetrahydrofuran (THF)) to give high yields of 1-tosylpiperidines (**1.2.12**). Reduction of obtained tosylate with sodium bis

(2-methoxyethoxy)aluminumhydride (Red-Al) yields desired piperidines (**1.2.5**) [63].

The case of *N*-Boc-amino alcohols (**1.2.13**) for the transformation to piperidine (**1.2.5**) needs preliminary conversion of the alcohols (**1.2.13**) into the triflate (**1.2.14**) followed by removal of the Boc-protecting group with HCl, and cylization in dioxane/saturated NaHCO₃ solution at 50°C (high dilution) [64].

Reaction of bis-tosylates [65] and bis-mesylates [66] (**1.2.15**) which takes place with large excess of primary amines is also described (Fig. 1.7).

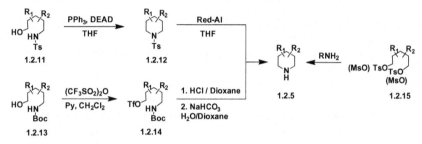

FIGURE 1.7 Methods for the synthesis of substituted piperidines by nucleophilic substitution reactions.

Among numerous synthetic strategies a rather different approach was demonstrated implementing reduction of cyclic imines (**1.2.17**) which can be prepared in different ways.

One-pot reaction occurred when substituted 5-bromopentanenitrile (**1.2.16**) was reacted with Grignard reagent followed with sodium borohydride reduction to give 2,3,4,5-tetrahydropy-ridine derivative − imine (**1.2.17**), which was reduced to substituted piperidine (**1.2.5**) using sodium borohydride in methanol [67].

Another method implements a highly enantioselective catalytic method for the hydrosilylation of imines (**1.2.17**) in the presence of the ethylenebis (η⁵-1,2,3,4-tetrahydroindenyl)titanium difluoride precatalyst ((EBTHI)TiF₂). The method was proposed for the preparation of alkaloid (*S*)-coniine and a constituent of fire-ant venom (2*R*,6*R*)-*trans*-solenopsin A. For this purpose chloro ketones (**1.2.18**) were converted into azido ketones (**1.2.19**) via solid/liquid phase-transfer catalysis, followed by aza-Wittig cyclization to give intermediate (**1.2.20**), which afforded the desired imines. Obtained imines underwent catalytic asymmetric reduction of the imine group with phenylsilane in the presence of (EBTHI)TiF₂ (which had been activated by treatment with pyrrolidine and methanol, followed by acidic hydrolysis of the initially formed aminosilane [68] (Fig. 1.8).

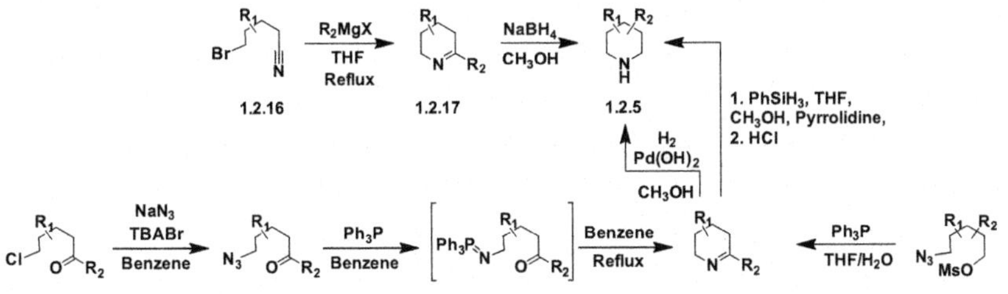

FIGURE 1.8 Methods for the synthesis of substituted piperidines by nucleophilic substitution reactions.

The Staudinger reaction, which is a very mild azide reduction, makes it possible to use azides as an NH_2-group synthons. This reaction was employed and for the piperidine ring synthesis. The azide (**1.2.21**) was reacted with triphenylphosphine resulting in intermediate iminophosphoranes, which in turn, released the free amines in aqueous tetrahydrofuran. The obtained amines immediately cyclized to imines (**1.2.17**). Hydrogenation in the presence of palladium hydroxide in methanol furnished requested piperidine (**1.2.5**) [69,70] (Fig. 1.8).

Intermolecular Michael Reactions

Different approaches for construction of functionalized piperidine ring was demonstrated implementing Michael addition reactions as a key step.

A successful example of highly enantioselective tandem intermolecular Michael reaction and application of obtained functionalized piperidine derivative (**1.2.27**) for the syntheses of several alkaloids.

The synthesis started from N-benzyl-2,2,2-trifluoroacetamide (**1.2.22**), which was alkylated with 2-(2-bromoethyl)-1,3-dioxane using sodium hydride as a base followed by partial hydrolysis dioxolane protecting group with oxalic acid which resulted aldehyde (**1.2.23**). Wittig-type reaction of (**1.2.23**) with (ethoxycarbonylmethylene)triphenylphosphorane gave (**1.2.24**), which, by the hydrolysis of trifluoroacetamide group, resulted in the amine product (**1.2.25**). The obtained amine (**1.2.25**), was used in the first Michael reaction via treating with methyl vinyl ketone to furnish key compound (**1.2.26**).

The last (**1.2.26**) was then treated with 1 equivalent of (R)-1-phenylethyl-1-amine as a chiral base in THF to give the desired optically active ethyl 2-((3R,4R)-3-acetyl-1-benzylpiperidin-4-yl)acetate (**1.2.27**) [71] (Scheme 1.1).

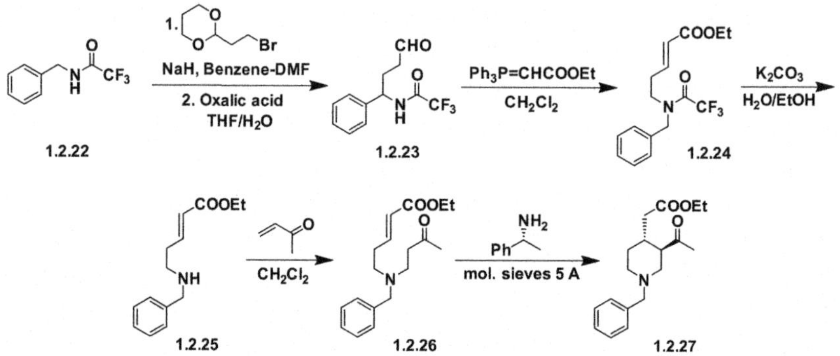

SCHEME 1.1 Synthesis of substituted piperidines implementing Michael addition reactions as key steps.

A convergent one-pot construction of piperidine framework has been accomplished through conjugate addition of a N-nucleophile to an electrophilic olefin followed by intramolecular trapping of the generated enolate by a built-in α,β-unsaturated acceptor. The well-known versatility of the nitroethylene and its propensity to polymerization has been overcome by its generation in situ from 1-benzoyloxy-2-nitroethane. Thus, treatment of equimolecular amounts of the latter with 6-(benzylamino)hex-3-en-2-one (**1.2.28**) at room temperature led to the direct formation of the desired result (**1.2.29**) [41] (Scheme 1.2).

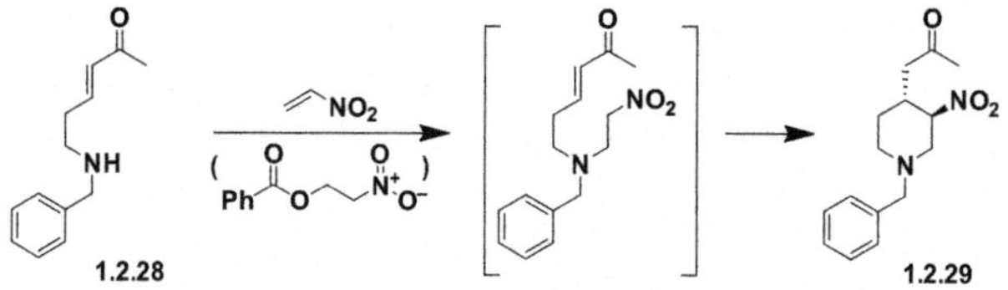

SCHEME 1.2 Synthesis of substituted piperidines implementing Michael addition reaction as a key step.

Enantiopure piperidines were synthesized starting from the 7-oxo-2-enimide (**1.2.30**), which on treatment with benzylamine in toluene at $-15°C$ in presence of MgSO$_4$, give intermediate imine, which then attaches intramolecularly to the conjugate double bond. The obtained product was hydrogenated at 1 atm. In presence of Pd/C to afford the piperidine (**1.2.31**). The N-benzyl group in (**1.2.31**) may be readily removed with the concomitant esterification by hydrogenation under rather forced conditions (3 atm hydrogen pressure, 24 hours) to give (**1.2.32**). Esterification, while retaining the N-benzyl group, can also been achieved by a reaction with MgClOMe solution at 0°C to give (**1.2.33**). (MgClOMe was prepared from MeMgCl in diethyl ether by reaction with methanol at 0°C (Scheme 1.3).

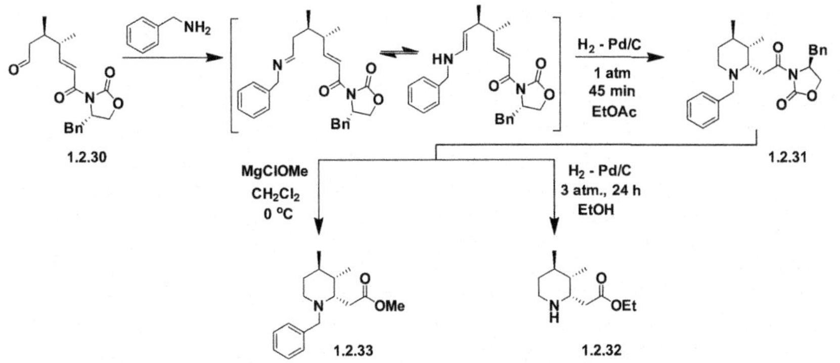

1.2.30 1.2.31 1.2.33 1.2.32

SCHEME 1.3 Synthesis of substituted piperidines implementing Michael addition reaction as a key step.

Another strategy implementing intramolecular Michael addition, which is the crucial step of the proposed synthesis, consists of intramolecular cyclization enone (**1.2.34**), which takes place smoothly under reflux in benzene in the presence of $BF_3 \cdot Et_2O$ to give hexafluoroacetone protected 4-oxo-L-pipecolic acid (**1.2.35**). Simultaneous deprotection of the vicinal amino and carboxylic functions proceeds under very mild conditions (*i*-PrOH/H_2O) at room temperature gives 4-oxo-L-pipecolic acid (**1.2.36**) [72] (Scheme 1.4).

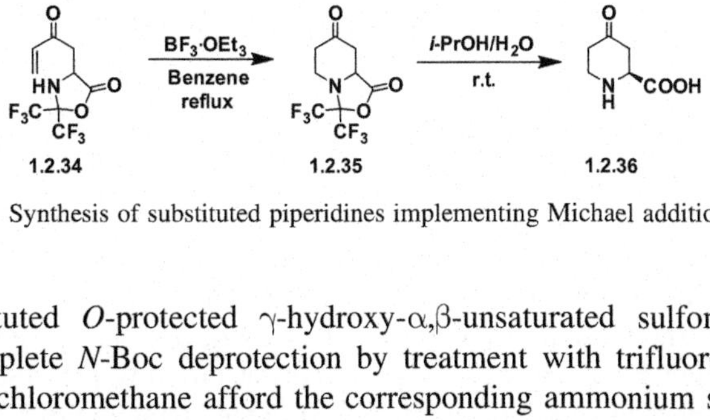

1.2.34 1.2.35 1.2.36

SCHEME 1.4 Synthesis of substituted piperidines implementing Michael addition reaction as a key step.

N-substituted *O*-protected γ-hydroxy-α,β-unsaturated sulfones (**1.2.37**) after a complete *N*-Boc deprotection by treatment with trifluoroacetic acid (TFA) in dichloromethane afford the corresponding ammonium salts, which, after isolation, redissolution in THF, cooling at −78°C, and a workup with an excess of Et_3N, gave piperidines (**1.2.38a, 1.2.38b**). Regardless of the substitution at nitrogen atom, the cyclizations were complete in less than 30 minutes, giving mixtures of isomeric cis/trans piperidines (**1.2.38a, 1.2.38b**) [73] (Scheme 1.5).

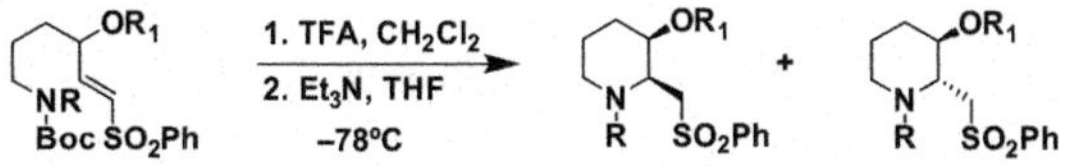

1.2.37 1.2.38a 1.2.38b

SCHEME 1.5 Synthesis of substituted piperidines implementing Michael addition reaction as a key step.

Catalyzed Hydroamination Reactions

The employment of hydroaminations to the intramolecular cyclization of amino olefins to piperidine derivatives is of big synthetic utility and could be mediated by different catalysts.

For example, N-Boc protected 5-phenylpent-4-en-1-amine (**1.2.39**) in dichloromethane in the presence of 2.5% methanol was converted to N-Boc protected 2-phenyl-3-(phenylselanyl) piperidine (**1.2.40**) using chiral 2,6-bis [1-(R)-ethoxyethyl]phenylselenenyl trifluoromethanesulfonate (ArCSeOTf) prepared in situ to give a diastereoisomeric mixture in a 25:1 ratio. The absolute stereochemistry of the major product was assessed by removal of the chiral organoselenium moiety by reduction with triphenyltin hydride in the presence of catalytic amount of azobisisobutyronitrile in refluxing toluene and comparison of the optical rotation of the resulting product (**1.2.41**) with literature data [74] (Scheme 1.6).

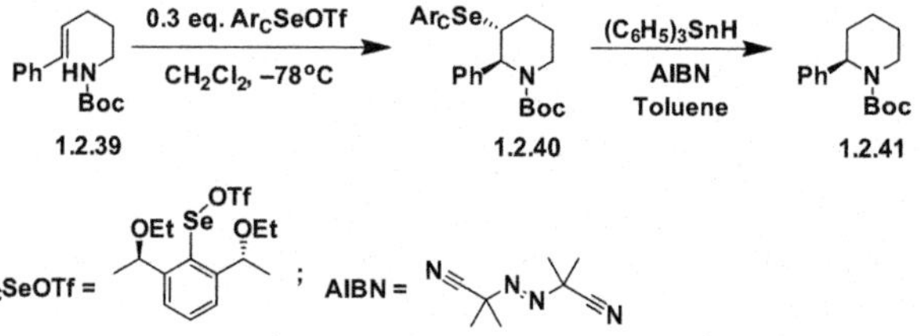

SCHEME 1.6 Synthesis of substituted piperidines implementing catalyzed hydroamination reaction.

Another example of intramolecular reaction of amines with double bond of is lanthanocene-catalyzed, hydroamination. It was shown that the complex $Cp_2NdCH(TMS)_2$ converted 2-substituted 8-nonen-4-amines such as (2R,4S)-2-((tert-butyldiphenylsilyl)oxy)non-8-en-4-amine (**1.2.42**) to 2,6-disubstituted piperidine (**1.2.43**) with greater than 100:1 selectivity for the formation of the cis isomer. A short synthesis of pinidinol (**1.2.44**), an alkaloid isolated from various pine and spruce species, was carried out this way [75] (Scheme 1.7). Other examples of this reaction are described in the literature [76,77].

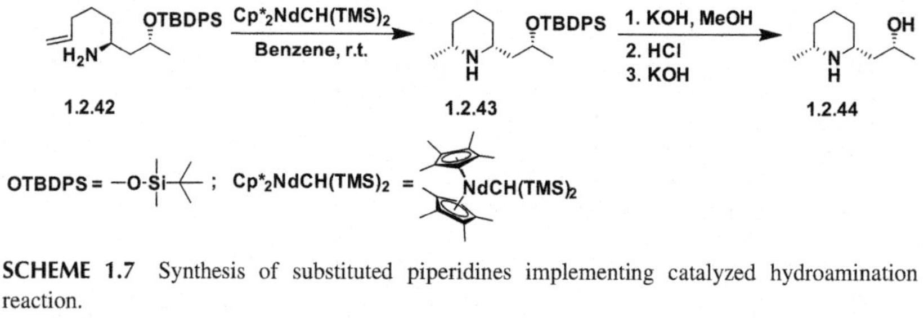

SCHEME 1.7 Synthesis of substituted piperidines implementing catalyzed hydroamination reaction.

Allyl alcohol (**1.2.45**) was stereoselectively cyclized in the presence of PdCl$_2$(MeCN)$_2$ to form piperidine (**1.2.46a, 1.2.46b**) with a cis:trans diastereomeric ratio of 8:1 [78] (Scheme 1.8).

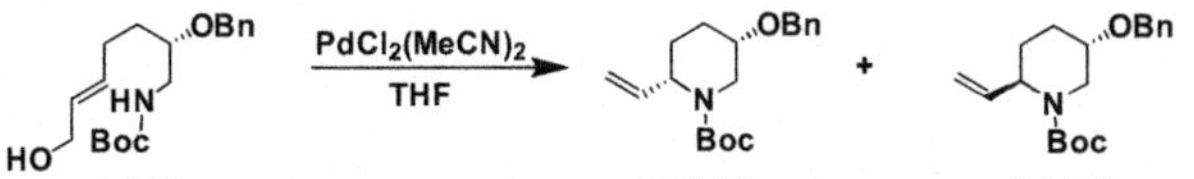

SCHEME 1.8 Synthesis of substituted piperidines implementing catalyzed hydroamination reaction.

There are many other examples of intramolecular hydroaminations reactions with double bond, triple bond, and allenic systems that are well reviewed in [30,32,34,77].

Aza-Diels–Alder Reactions

The aza-Diels–Alder reaction is an important tool for the preparation of substituted piperidines, which allows the construction of functionalized piperidine derivatives with regio-, diastereo- and enantio-selectivity [79]. There could be three ways to carry out these [4 + 2] cycloaddition reactions for creation of piperidine ring:

The reaction of imines (**1.2.47**) or iminium salts with carbon dienes is the most implemented way reported to date.

1-Azadienes (**1.2.48**) in Diels–Alder reaction are rarely used and because of low conversion and competitive imine addition often thwarting and has not proved to be one of the major routes to piperidine synthesis.

2-Azadienes (**1.2.49**) have been very poorly studied as starting materials for the synthesis of piperidine derivatives via the Diels–Alder reaction (Fig. 1.9).

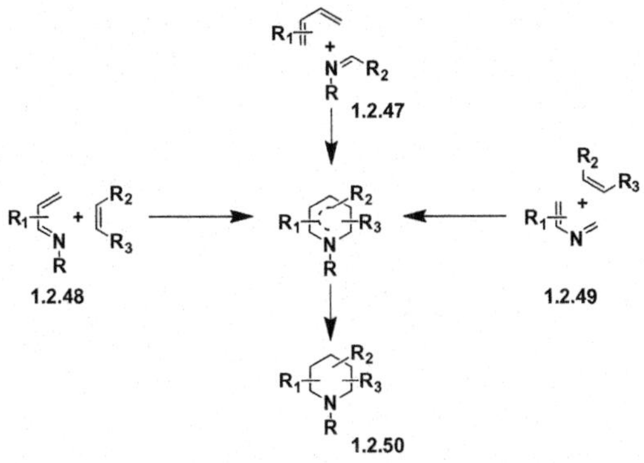

FIGURE 1.9 Synthesis of substituted piperidines implementing aza-Diels–Alder reactions.

All of obtained intermediate tetrahydropyridine compounds have been hydrogenated to final piperidine compounds (**1.2.50**).

For the reaction of imines (**1.2.47**) with carbon dienes it is necessary to have electron-poor imine and an electron-rich diene the most implemented way reported to date.

In general, it has been considered necessary to use of *N*-acyl and *N*-tosyl imines, or imines protected with silylamine groups [80−83]. But benzylimines also have been implemented in this reaction [84,85]. In the cases of *N*-acyl and *N*-tosyl imines (4 + 2) cycloaddition reactions are typically catalyzed by Lewis acids, which vary with the structure of reagents and solvents. The reaction of benzylimines have to be carried out in DMF, using 1 equivalent of TFA and catalytic amount of water, or in trifluoroethanol containing catalytic amount of (TFA) and molecular sieves [85].

One of the classical examples of this reaction is the reaction of the diene (**1.2.51**) with ethyl 2-(tosylimino)acetate (**1.2.52**) at room temperature in toluene to give *cis*-2,6-disubstituted piperidine (**1.2.53**) [86] (Scheme 1.9).

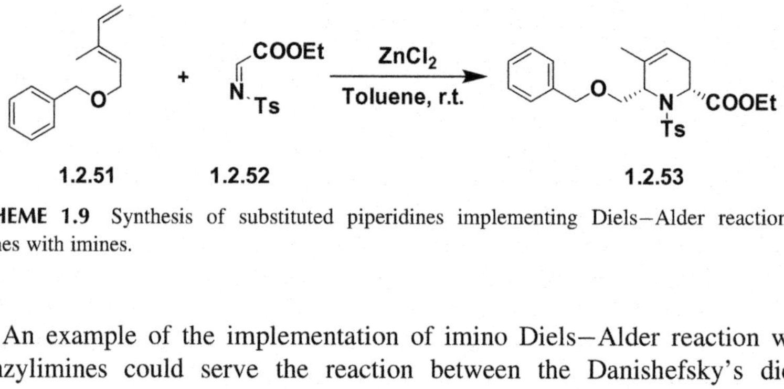

SCHEME 1.9 Synthesis of substituted piperidines implementing Diels−Alder reaction of dienes with imines.

An example of the implementation of imino Diels−Alder reaction with benzylimines could serve the reaction between the Danishefsky's diene (**1.3.54**) and imine (**1.2.55**) obtained from benzylamine and (*R*)-2,3-diisopropylideneglyceraldehyde. The reaction was carded out in acetonitrile with 0.2 eq. of ZnI$_2$ at −40°C with excellent yield but with low stereoselectivity to give a mixture of diastereomeric enaminones−dihydropyridones (**1.2.56a**, **1.2.56b**) [84] (Scheme 1.10).

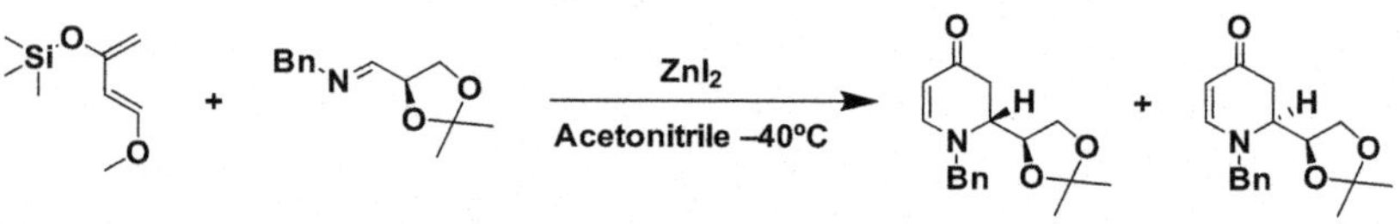

SCHEME 1.10 Synthesis of substituted piperidines implementing Diels−Alder reaction of dienes with imines.

The reaction between Danishefsky's diene (**1.2.54**) and enantiomerically pure imine (**1.2.57**) obtained from (*R*)-2,3-di-*O*-benzylglyceraldehyde (*S*)-*N*-a-(methylbenzyl)imine was carried out in the same conditions, but using 1.1 eq. ZnI$_2$ give cyclic only a single diastereomer – enaminone (**1.2.58**). The double bond was reduced at $-78°C$ with L-Selectride (lithium tri-*sec*-butyl (hydrido)borate) to give 4-piperidinone (**1.2.59**). Keto group in obtained 4-piperidone (**1.2.59**) was protected by ketalization with ethyleneglycol to give dioxolane (**1.2.60**). The obtained compound was *N*-debenzylated under H$_2$ at 1 atm. in ethanol, for 3 hours at room temperature using 20% Pd(OH)$_2$/C as a catalyst and then, without separation of intermediate secondary amine, was acylated with di-tertbutyl dicarbonate in THF in the presence of diisopropylethylamine (DIPEA) giving Boc-derivative (**1.2.61**). For *O*-debenzylation the product (**1.2.61**) was again hydrogenated at room temperature using the same catalyst – 20% Pd(OH)$_2$/C under H$_2$ at 1 atm. in ethanol, but now for 24 hours, to produce the desired polyfunctionalized piperidine derivative (**1.2.62**) [84] (Scheme 1.11).

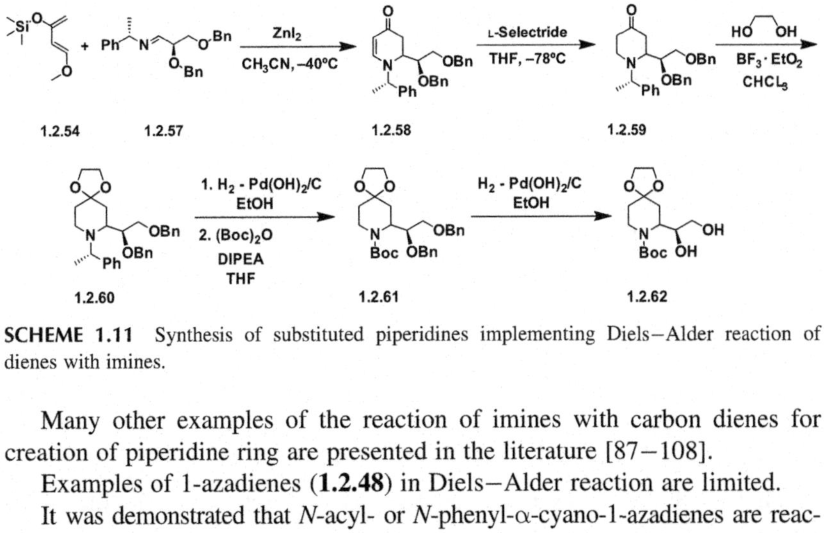

SCHEME 1.11 Synthesis of substituted piperidines implementing Diels–Alder reaction of dienes with imines.

Many other examples of the reaction of imines with carbon dienes for creation of piperidine ring are presented in the literature [87–108].

Examples of 1-azadienes (**1.2.48**) in Diels–Alder reaction are limited.

It was demonstrated that *N*-acyl- or *N*-phenyl-α-cyano-1-azadienes are reactive substrates in the Diels–Alder reaction with a range of dienophiles such as ethyl vinyl ether, styrene, 1-hexene, and alkyl acrylates, and this process can provide an efficient method for the preparation of synthetically useful piperidine derivatives [109–113]. For example, *N*-acyl- or *N*-phenyl-2-cyano-1-aza-1,3-butadienes (**1.2.63**) undergo efficient Diels–Alder cycloaddition with ethyl acrylate and vinyl ether giving 1,2,3,4-tetrahydropyridines (**1.2.64**) and (**1.2.65**).

The only regioisomer observed for the reaction of both (*N*-acyl- and *N*-phenyl) azadienes (**1.2.63**) with dienophile that is activated with an electron-withdrawing group (methyl acrylate) has the carboxyl group β- to the nitrogen atom, and the only regioisomer observed for the reaction electron-donating group (vinyl ethers) has the ethoxy- group α- to the nitrogen atom. The Alder endo rule is fully applicable to the stereochemical pathway in the reactions of azadienes with mentioned dienofiles [113] (Scheme 1.12).

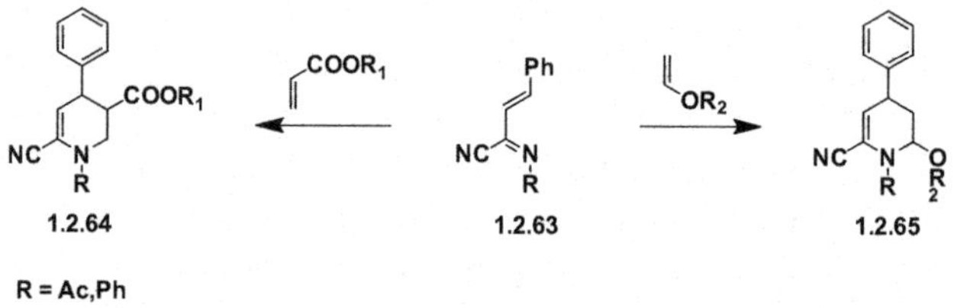

R = Ac,Ph

SCHEME 1.12 Synthesis of substituted piperidines implementing 1-azadienes in Diels−Alder reaction.

Another approach to overcome the reluctance of 1-azadienes to undergo in Diels−Alder reaction was demonstrated with the implementation of the diene (**1.2.66**) − an α,β-unsaturated hydrazone derived from methacrolein and 1,1-dimethylhydrazine. This compound reacted regioselectively with a variety of dienophiles (acrylonitrile, methylvinylketone, dimethylmaleate, and dimethylfumarate) to give the corresponding adducts (**1.2.67, 1.2.68**). Reductive cleavage of the N−N bond with zinc in acetic acid along with a simultaneous reduction of the carbon-carbon double bond gave tetrahydro-pyridines (**1.2.69, 1.2.70**).

Interestingly, the dimethylamine substituent had reversed the "normal" regiochemistry of the Diels−Alder reaction [114] (Scheme 1.13).

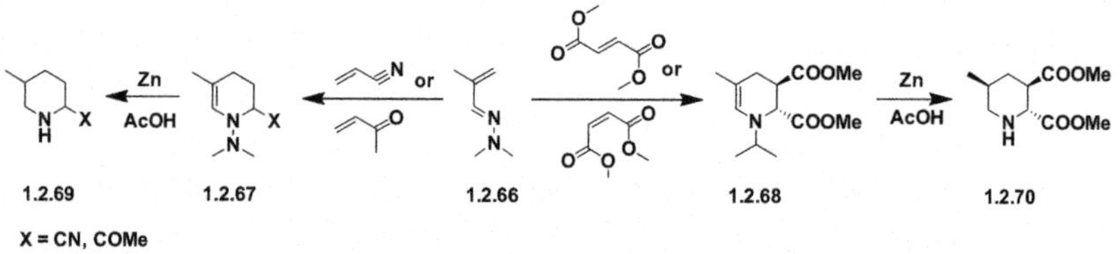

X = CN, COMe

SCHEME 1.13 Synthesis of substituted piperidines implementing 1-azadienes in Diels−Alder reaction.

1-Azadiene (**1.2.71**) derived from α,β-unsaturated aldehyde and Enders' hydrazines cycloaddition to cyclic dienophiles with high facial selectivities gave (**1.2.72**). Esterification followed by the cleavage of the N−N bond readily affected with zinc in acetic acid with the reduction of the double bond, which leads to a mixture of epimers at C_3 (**1.2.73a, 1.2.73b**) [115] (Scheme 1.14).

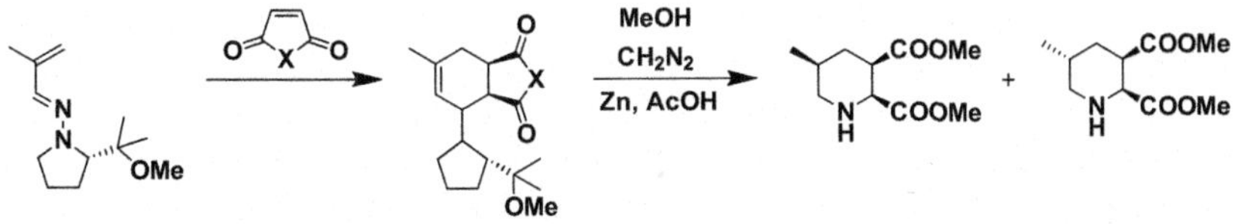

1.2.71 1.2.72 1.2.73a 1.2.73b

X = O, NPh

SCHEME 1.14 Synthesis of substituted piperidines implementing 1-azadienes in Diels–Alder reaction.

Reactions of the *N*-(phenylsulfonyl)- (**1.2.74**) or *N*-(methylsulfonyl)-2-(ethoxycarbonyl)-1-aza-1,3-butadienes with series of vinyl ethers were found to proceed 6-methoxy-1-(phenylsulfonyl)-1,4,5,6-tetrahydropyridine-2-carboxylates (**1.2.75–1.2.78**) with full preservation of the dienophile olefin geometry in the reaction products [116] (Scheme 1.15).

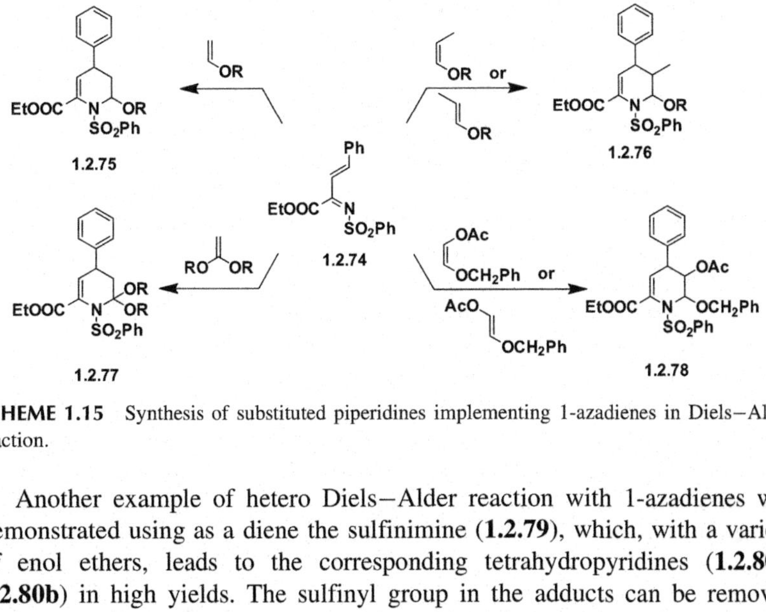

SCHEME 1.15 Synthesis of substituted piperidines implementing 1-azadienes in Diels–Alder reaction.

Another example of hetero Diels–Alder reaction with 1-azadienes was demonstrated using as a diene the sulfinimine (**1.2.79**), which, with a variety of enol ethers, leads to the corresponding tetrahydropyridines (**1.2.80a, 1.2.80b**) in high yields. The sulfinyl group in the adducts can be removed with MeLi followed by treatment with AcCl to give corresponding *N*-acetyl-tetrahydropyridine (**1.2.81**) [117] (Scheme 1.16).

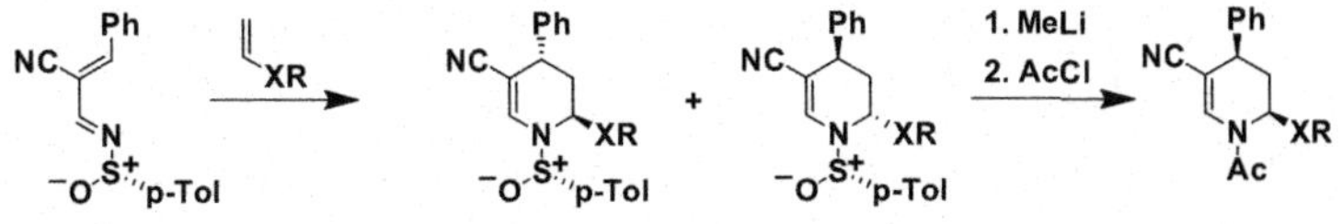

1.2.79 1.2.80a 1.2.80 b 1.2.81

XR = OMe, OEt, Ot-Bu, SMe, SPh

SCHEME 1.16 Synthesis of substituted piperidines implementing 1-azadienes in Diels–Alder reaction.

The few available studies on 2-aza-1,3-dienes have indeed shown that they are able to undergo Diels—Alder reaction with dienophiles. The utility dienes such as vinylformimidate (**1.2.82**) for the synthesis of functionalized piperidines (**1.2.83**) and (**1.2.84**) was demonstrated by their facile reaction with methyl acrylate and maleic anhydride [118] (Scheme 1.17).

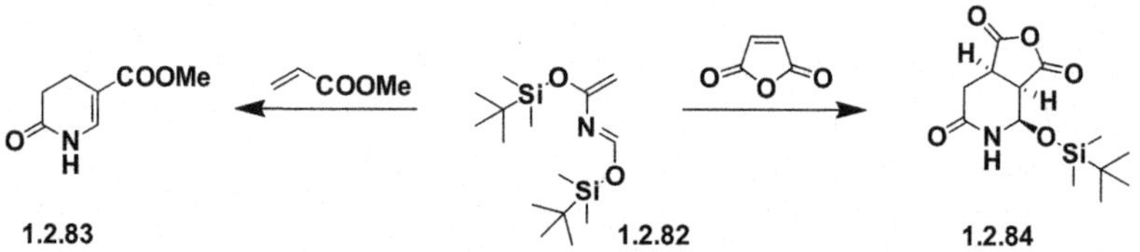

SCHEME 1.17 Synthesis of substituted piperidines implementing 2-azadienes in Diels—Alder reaction.

Intramolecular Ene Reactions

The ene reaction is defined as the addition of a compound with a multiple bond (an enophile) to an olefin with an allylic hydrogen (an ene) [119–126]. The ene reaction is mechanistically related to the much better known Diels—Alder reaction since both reactions proceed through concerted six electrons cyclic transition states. Initially by intramolecular ene reactions were classified to three types [125]. Type four reactions, which are a variation of type one are occasionally observed were added to this classification system later [126].

There are few examples of implementation of ene reactions for the synthesis of functionalized piperidines.

Zirconium-catalyzed cyclization using Grignard reagent was reported. Treatment of the N-allylbut-3-en-1-amine (**1.2.85**) with zirconocene catalyst (S)-(EBTHI)ZrBINOL gave intermediate zirconacycle (**1.2.86**) which addition of n-BuMgCl gave another intermediate (**1.2.87**) which on reaction with an electrophile (O$_2$ in this case) followed by quenching with 10% HCl afforded the piperidine (**1.2.88**) [127] (Scheme 1.18).

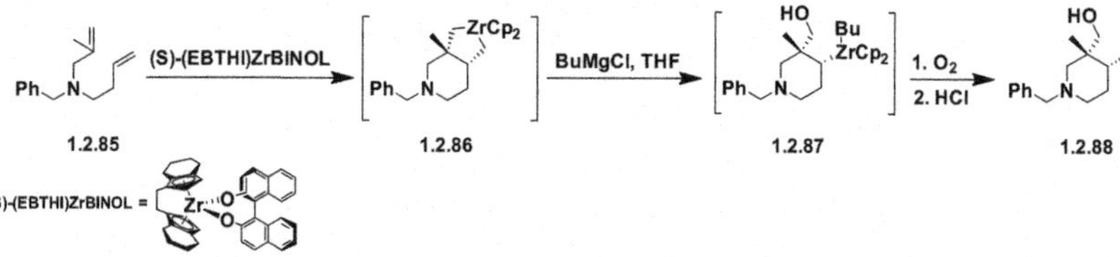

SCHEME 1.18 Synthesis of substituted piperidines implementing intramolecular ene reaction.

Tetraene (**1.2.89**) on treatment with Pd(OAc)$_2$ and PPh$_3$ afforded the piperidine (**1.2.90**) in 86% yield [128] (Scheme 1.19).

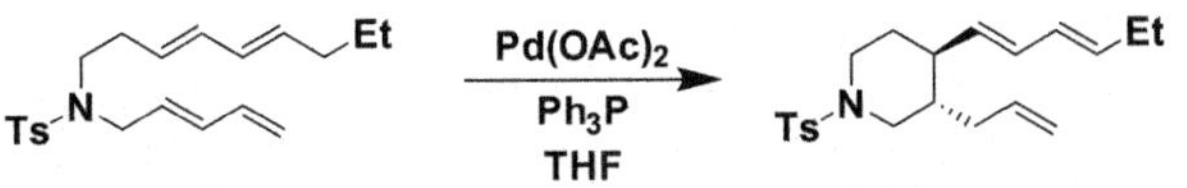

1.2.89 1.2.90

SCHEME 1.19 Synthesis of substituted piperidines implementing intramolecular ene reaction.

Another example of the ene carbocyclisation to the synthesis of piperidines was demonstrated via treatment of the triene (**1.2.91**) in toluene with bis (2,2′bipyridine)iron(0), (bpy-Fe(0)) as a the catalyst, followed by acetalization, which afforded the piperidine (**1.2.92**) in 85% yield [129] (Scheme 1.20).

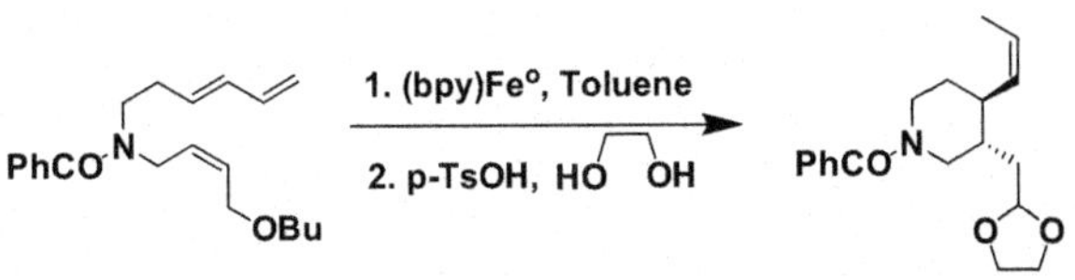

1.2.91 1.2.92

SCHEME 1.20 Synthesis of substituted piperidines implementing intramolecular ene reaction.

Imino-ene reactions are relatively rare implemented in organic synthesis [130]. The use of imines as enophiles in type one ene reaction for piperidine ring synthesis which is involved in complex heterocyclic systems is described by several groups [131–134]. Intramolecular formal ene reactions for the synthesis of relatively simple 3-amino-piperidines or 3-hydroxy-piperidines are described on the Schemes 1.21–1.23.

Lewis acid catalyzed cyclization of *N*-benzylimines (**1.2.93**) depending on used Lewis acid ($FeCl_3$ or $TiCl_4$) in CH_2Cl_2 at room temperature gave two *cis*-configurated 2-alkyl-3-(*N*-benzylamino)-4-isopropenylpiperidines (**1.2.94a, 1.2.94b**) and two *cis*-configurated 2-alkyl-3-(*N*-benzylidene)amino-4-isopropylpiperidines (**1.2.95a, 1.2.95b**). Use of $FeCl_3$ favors the formation of products (**1.2.94**) whereas $TiCl_4$ favors the formation of (**1.2.95**) [135] (Scheme 1.21).

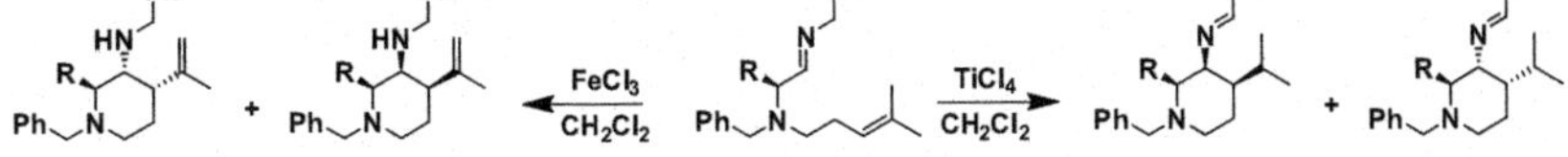

1.2.94b 1.2.94a 1.2.93 1.2.95a 1.2.95b

SCHEME 1.21 Synthesis of substituted piperidines implementing intramolecular ene reaction.

When *N*-benzylimines (**1.2.93**) *N*-tosylimines (**1.2.96**) were instead treated with Lewis acids, a mixture of 3-aminopiperidine (**1.2.97**) and the 3-iminopiperidine (**1.2.98**) was obtained [135] (Scheme 1.22).

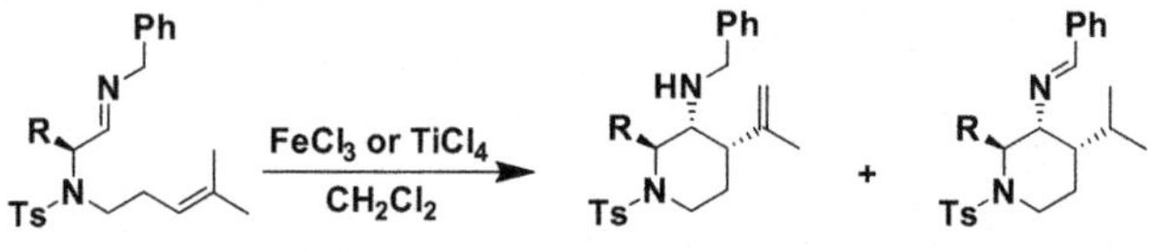

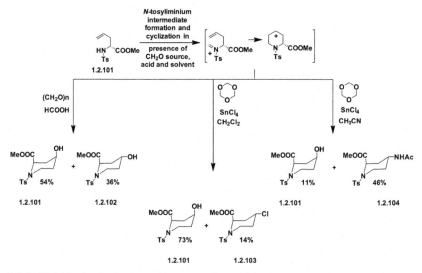

SCHEME 1.22 Synthesis of substituted piperidines implementing intramolecular ene reaction.

Related approach (carbonyl-ene reactions) as was shown above, but starting from 2-(but-3-en-1-ylamino)acetaldehyde derivative (**1.2.99**) was demonstrated for the synthesis of 3-hydroxy-piperidines (**1.2.100**) [136] (Scheme 1.23).

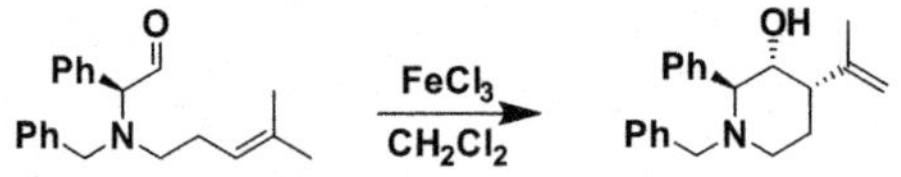

SCHEME 1.23 Synthesis of substituted piperidines implementing intramolecular ene reaction.

Besides imino-ene reactions interesting examples of iminium ion ene reactions for functionalized piperidine ring synthesis are described in the literature.

The carbamate-protected (*N*-COOMe, *N*-Ts, *N*-Fmoc) amino esters such as (**1.2.101**) readily cyclized with paraformaldehyde or trioxan and depending of implemented acid and solvent could give different products through formation of intermediate iminium ion, which readily cylizes to intermediate cyclic cation to give final compounds (**1.2.102, 1.2.103, or 1.2.104**) [137] (Scheme 1.24).

SCHEME 1.24 Synthesis of substituted piperidines implementing intramolecular iminium ion ene reaction.

Intermediate iminium ion (**1.2.106**) induced by the treatment of an α-cyanoamine (**1.2.105**) with titanium (IV) chloride was implemented for the synthesis of 1,2,5,6-tetrahydropyridine (**1.2.107**). Reduction of the double bond afforded (±)-coniine (**1.2.108**) in 90% yield [138] (Scheme 1.25).

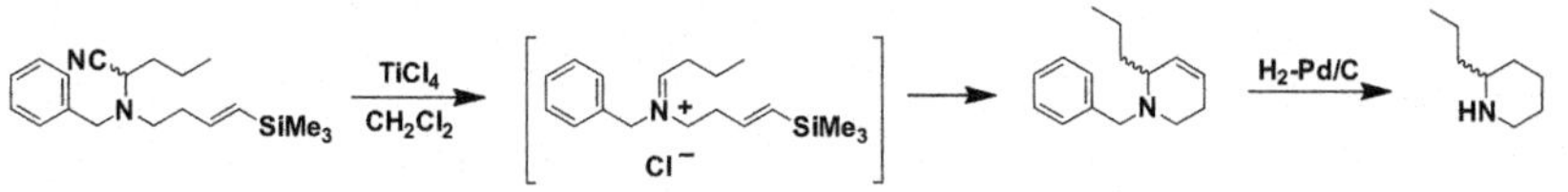

1.2.105. 1.2.106 1.2.107 1.2.108

SCHEME 1.25 Synthesis of substituted piperidines implementing intramolecular iminium ion ene reaction.

A similar cyclization methodology with *N*-methoxymethyl alkyne (**1.2.109**) using as a nucleophile trimethylsilyl chloride gave 3-(1-chloroethylidene)-1-(phenylmethyl)piperidine (**1.2.110**) [139] (Scheme 1.26).

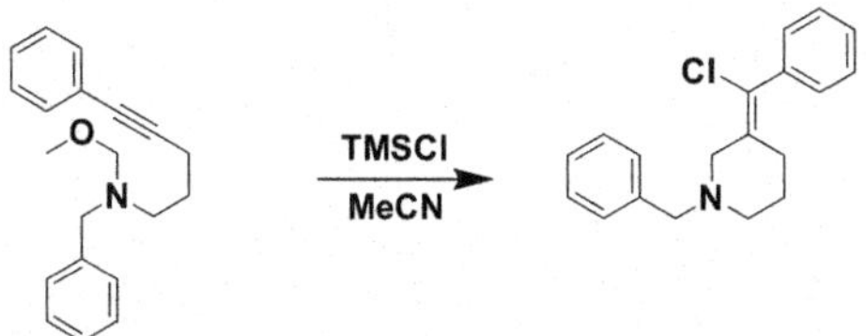

1.2.109 1.2.110

SCHEME 1.26 Synthesis of substituted piperidines implementing cyclization with *N*-methoxymethyl alkyne.

Another example of ene reaction occurred with an allylboronate (**1.2.111**) when it was dissolved in acetonitrile and $LiBF_4$ and water were added to the solution to give *N*-methoxycarbonyl-*cis*-3-vinyl-4-piperidin-ol (**1.2.112**) [140] (Scheme 1.27).

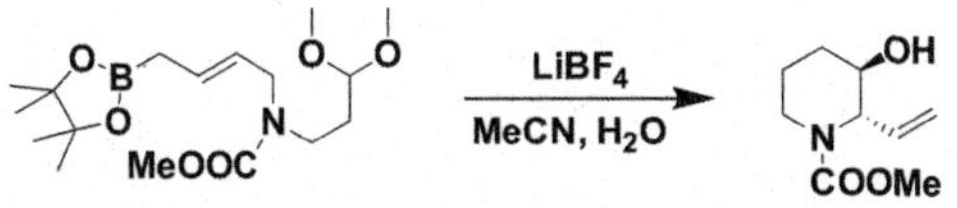

1.2.111 1.2.112

SCHEME 1.27 Synthesis of substituted piperidines implementing intramolecular cyclization of methyl (*E*)-(3,3-dimethoxypropyl)(4-(4,4,5,5-tetramethyl-1,3,2-dioxaborolan-2-yl)but-2-en-1-yl) carbamate.

Palladium catalyzed ene-halogenocyclization is also described. When α-iodoester (**1.2.113**) was heated with $Pd(PPh_3)_4$ in hexamethylphosphoramide in the presence of proton sponge (*N,N,N′,N′*- 1,8-dimetylaminonaphthalene) as a scavenger of hydrogen iodide, piperidine derivative (**1.2.114**) was obtained [141] (Scheme 1.28).

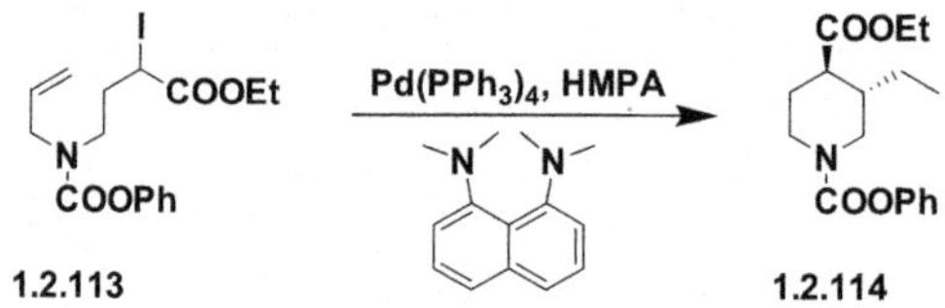

1.2.113 1.2.114

SCHEME 1.28 Synthesis of substituted piperidines implementing ene-halogenocyclization reaction.

Many other ene-cyclization reactions for piperidine ring synthesis are also described in literature. Some of the are summarized in [30,34,130].

Ring-Closing Metathesis Reactions

Ring-closing metathesis reactions have been developed by Grubbs for creating cyclic systems from acyclic diolefins [142,143] and more than once have been employed for the synthesis of tetrahydropyridine systems.

The use of various amino acid–derived diolefins as substrates for the ring-closing olefin metathesis reaction has been shown to be an efficient transformation for the synthesis of functionalized piperidines.

For example, the D-allylglycine-derived substrates (**1.2.115**) were cyclized using the phenyl-substituted ruthenium alkylidene catalyst $(PCy_3)_2(Cl)_2$-Ru=CHPh, refluxing in CH_2Cl_2 to give tetrahydropyridines (**1.2.116**). Employment of Boc-protecting technic gave the best yields. Without a nitrogen protecting group, ring closure was not observed. Similarly, cyclization of the acrylic amides (**1.2.117**) give corresponding α,β-unsaturated lactams (**1.2.118**) with excellent yields [144] (Scheme 1.29).

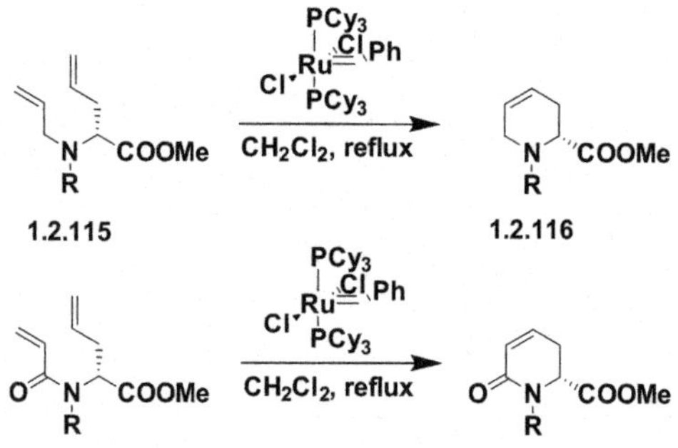

1.2.115 1.2.116

1.2.117 1.2.118

R = Boc, PMB (4-Methoxybenzyl ether), Fem (ferrocenylmethyl)

SCHEME 1.29 Synthesis of substituted piperidines implementing ring-closing metathesis reaction.

Another example of ring-closing metathesis reaction is its implementation for the synthesis of the naturally occurring (−)-halosaline (**1.2.122**), a piperidine alkaloid isolated from saxauls of the genus Haloxylon.

A domino metathesis of (**1.2.119**) using the same ruthenium alkylidene catalyst gave the cyclic silyl ether (**1.2.120**) in high yield. Treatment of (**1.2.120**) with tetrabutylammonium fluoride (TBAF) gave (**1.2.121**) in high yield. The last was quantitatively hydrogenated over Pd/C catalyst under atmospheric pressure, and the tosyl-group was cleaved using Na/Hg in methanolic phosphate buffer to give pure (−)-halosaline (**1.2.122**) an 82% yield [145] (Scheme 1.30).

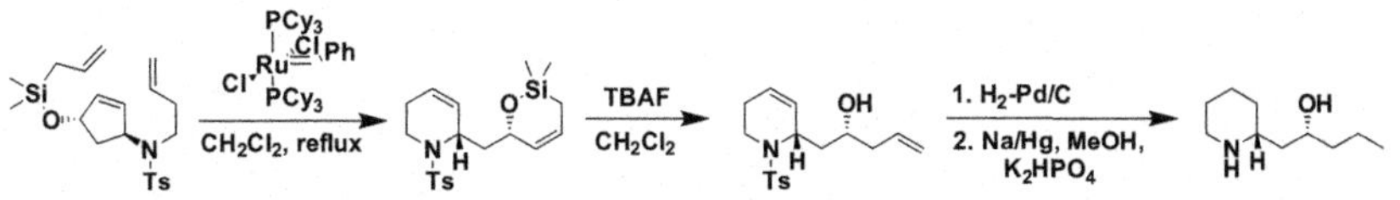

SCHEME 1.30 Synthesis of substituted piperidines implementing ring-closing metathesis reaction.

Synthesis of Piperidin-4-Ones

The piperidin-4-ones are probably the most versatile starting material that can be suitably modified in order to synthesize biologically active compounds with excellent receptor interactions. They have been synthesized using a variety of methods.

The Synthesis of 2,2,6,6-Tetramethyl-4-Oxopiperidine

The synthesis of 2,2,6,6-tetramethyl-4-oxopiperidine − triacetonamine (**1.2.124**) − has been known since 1874, when acetone was converted to phorone (**1.2.123**), which, reacting with ammonia, gave triacetonamine [146]. The first preparative method goes back to 1927 and still used in practice [147]. Many modifications of this reaction [148−152], including methods for industrial scale production catalyzed by nitrate and ammonium hydroxide, calcium-containing zeolites [153−159] have been proposed, but the most convenient method for the laboratory scale is condensation of acetone with ammonia in presence of ammonium chloride [160] (Scheme 1.31).

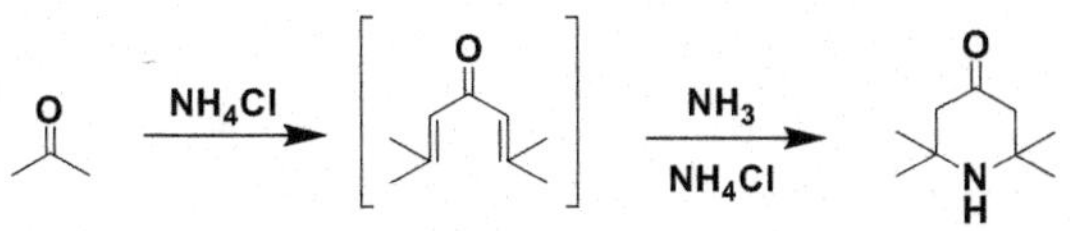

SCHEME 1.31 The synthesis of 2,2,6,6-tetramethyl-4-oxopiperidine—triacetonamine.

Triacetonamine is primarily used as a light stabilizer for plastics, vulcanization accelerators, finds use as a chemical feedstock for different synthesis, mainly for the synthesis of hindered amine 2,2,6,6-tetramethylpiperidine and as well as the radical oxidizer — 4-hydroxy-2,2,6,6-tetramethylpiperidin-1-oxyl — an agent for detoxifying reactive oxygen species in medicine [161]. Chemistry of triacetonamine is well reviewed [162].

Petrenko-Kritschenko Reaction for the Synthesis of 2,6-Disubstituted Piperidin-4-Ones

The preparation of 2,6-diaryl-4-oxopiperidine-3,5-dicarboxylates (**1.2.125**) is a classic multicomponent reaction included into the list of name reactions as Petrenko-Kritschenko reaction. Piperidin-4-ones have been synthesized by the condensation of acetonedicarboxylic ester with two equivalents of aromatic aldehyde in the presence of ammonia or primary amine [163,164].

The reaction was later extended to aliphatic aldehydes by Mannich [165,166]. It has been shown also that 3,5-dialkyl-4-oxopiperidinedicarboxylic esters (**1.2.126**) have been obtained when α,α'-dialkylacetonedicarboxylic esters are condensed with formaldehyde and methylamine. Other aldehydes cannot be substituted for formaldehyde [166] (Scheme 1.32).

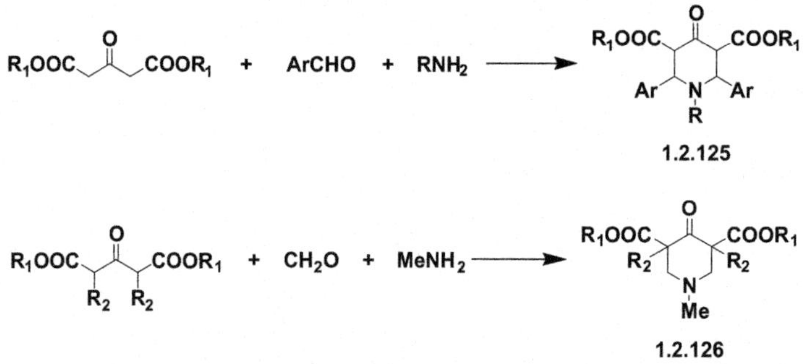

SCHEME 1.32 Synthesis of 2,6-disubstituted piperidines implementing Petrenko-Kritschenko reaction.

Acetonedicarboxylic acid esters can be replaced by acetone or other aliphatic ketones. Treatment of ethyl acetoacetate with benzaldehyde and aniline in ethanol in the presence of malonic acid gives ethyl 1,2,6-triphenyl-4-oxo-piperidine-3-carboxylate (**1.2.127**), which is on hydrolysis with HCl 1,2,6-triphenylpiperidin-4-one (**1.2.128**) [165] (Scheme 1.33).

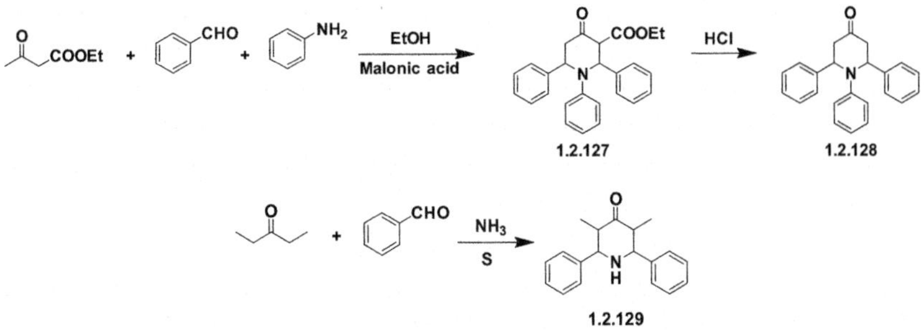

1.2.127   1.2.128

1.2.129

SCHEME 1.33 Synthesis of 2,6-disubstituted piperidines implementing variations of Petrenko-Kritschenko reaction.

The mode of carrying out of this reaction is very simple. In general, the amine or of ammonium acetate are dissolved in glacial acetic acid, and aldehyde and ketone are added to this solution. After a short heating time yields are high with practically no side reactions [167,168].

When ammonia was passed into a mixture of benzaldehyde and diethylketone containing a little amount of sulfur, the vigorous exothermic reaction yielded 2,6-diphenyl-3,5-dimethyl-4-piperidone (**1.2.129**) [169].

A review on the synthesis of 2,6-disubstituted piperidines and their transformations was published [170] (Scheme 1.33).

Synthesis of Piperidin-4-Ones by Dieckmann Condensation

The Dieckmann condensation of aminodicarboxylate esters followed by hydrolysis and decarboxylation of the resulting cyclic product is a well known and probably the most widely used method for the synthesis of piperidin-4-ones. It seems that the method was proposed by Ruzicka [171,172] and thoughtfully investigated and developed by McElvain [173−177] and reviewed in [178]. The reaction conditions are still being optimized by different investigators [179,180]. The Fig. 1.10 schematically demonstrates Dieckmann cyclization for the synthesis of 4-piperidones, which is influenced by multiple parameters. In general a primary amine is condensed with two equivalents of acrylate to give bis-ester (**1.2.130**). The anion formed by base catalyzed proton abstraction at the carbon at one carboxyl group of obtained bis-ester attacks the other carboxyl to form a 4-oxopiperidine-3-carboxylate (**1.2.131**), which after hydrolysis and decarboxylation gives the desired product (**1.2.132**) (Fig. 1.10).

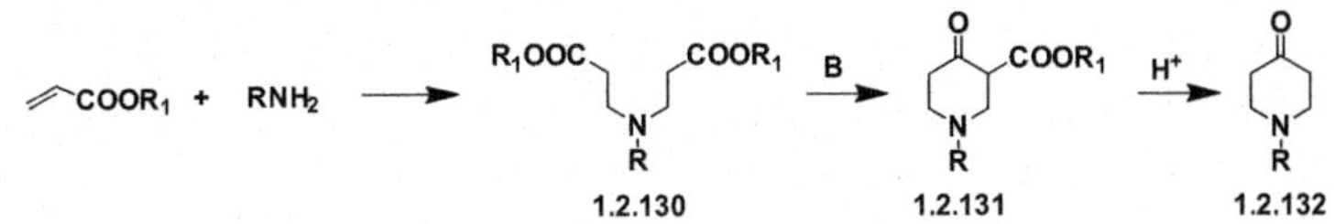

1.2.130   1.2.131   1.2.132

FIGURE 1.10 Synthesis of piperidin-4-ones implementing Dieckmann condensation.

Dieckmann condensation of (**1.2.133**) mediated by sodium/methanol provided the cyclization products as two regioisomers, which were treated with tert-butyldimethylsilyl chloride in presence of 4-DMAP to give a mixture of enol ethers (**1.2.134a, 1.2.134b**), which could not be separated by chromatography; their ratio, 1:4, was determined by 1H NMR [181] (Scheme 1.34).

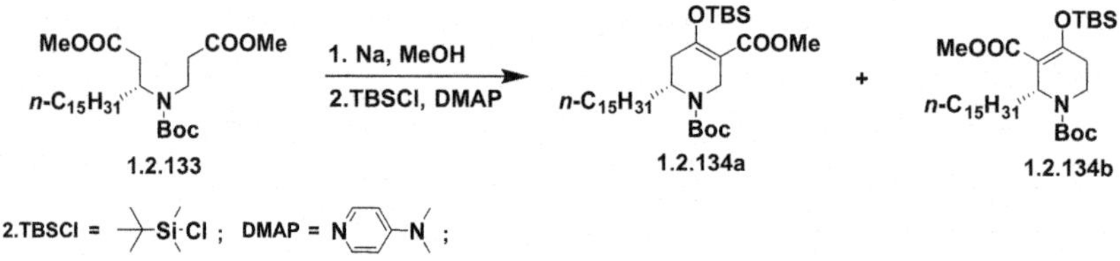

SCHEME 1.34 Synthesis of regioisomers of 2-substituted piperidin-4-ones implementing Dieckmann condensation.

Dieckmann condensation was implemented for the synthesis of piperidin-4-ones with the spiro-attached cyclopropyl ring at the 3-rd position of hetercycle. For that purpose ethyl cyanoacetate (**1.2.135**) was alkylated with 1,2-dibromoethane to afford the functionalized cyclopropane derivative (**1.2.136**). Selective reduction of the cyano group of the resulting cyanoester gave the corresponding amine (**1.2.137**), which upon the Michael addition to methyl acrylate gave product (**1.2.138**) N-protection of which furnished the diester (**1.2.139**). A facile NaH mediated intramolecular Dieckmann cyclization provided the key piperidin-4-one ester (**1.2.140**), which upon decarboxylation under Krapcho conditions (NaCl-H$_2$O-DMSO) [182], afforded the desired compound (**1.2.141**) [183] (Scheme 1.35).

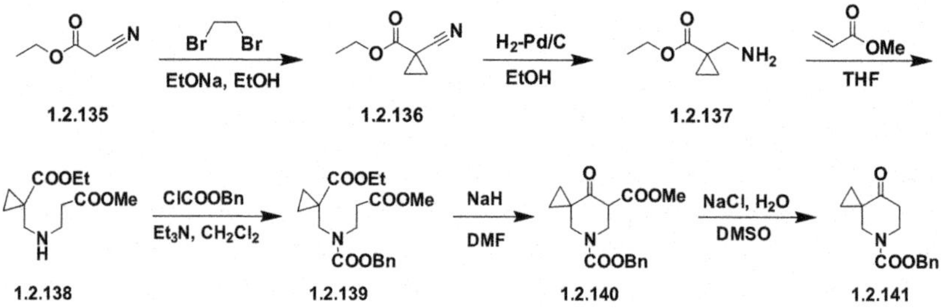

SCHEME 1.35 Synthesis of 3,3-disubstituted piperidin-4-ones implementing Dieckmann condensation.

Synthesis of Piperidin-4-Ones by Miscellaneous Methods

3-Alkylidene-piperidin-4-ones (**1.2.145**) with diverse C-5 substitution patterns have been synthesized via a one-pot cascade sequence. The phosphonium salt (**1.2.143**) was synthesized in excellent yields by reacting methyl

3-(benzyl(2-bromoethyl)amino)-2,2-dimethylpropanoate (**1.2.142**) with tribu-tylphosphine in acetonitrile under reflux. The obtained salt (**1.2.143**) was smoothly deprotonated and the resulting nonstablized ylides cyclized to afford the corresponding β-keto-ylides (**1.2.144**) using *t*-BuOK as the base in toluene. During the Wittig olefination of aliphatic or aromatic aldehydes using a resealable pressure vessel and different solvents series of 3-alkyli-dene-piperidin-4-ones (**1.2.145**) have been synthesized [184] (Scheme 1.36).

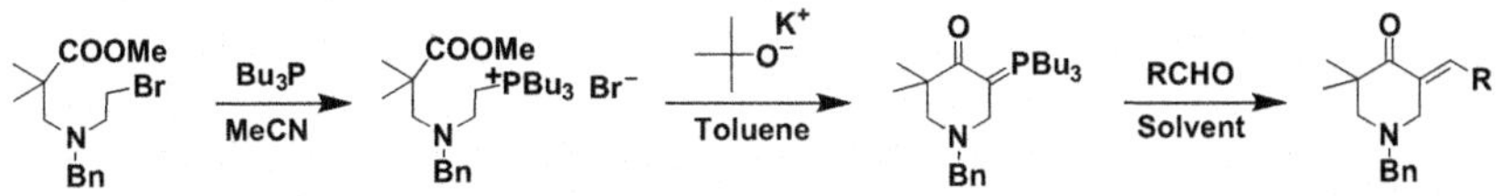

R = Alkyl, Aryl; Solvent = MeCN, Toluene, DMF

SCHEME 1.36 Synthesis of 3,3-alkylidene-piperidin-4-ones via a one-pot cascade reactions sequence.

Special methods were developed for the synthesis of substituted piperidin-4-ones unavailable with the methods described above. For instance, 1-substituted-2,5-dimethylpiperidin-4-ones (**1.2.150**) (Nazarov's piperidones) was proposed to prepare by condensing vinylacetylene (**1.2.146**) with acetone in the presence of powdered KOH (Favorsky reaction), which results in alcohol (**1.2.147**). Dehydration of obtained alcohol (**1.2.147**) to vinyl-isopropenyl-acetylene (2-methylhexa-1,5-dien-3-yne) (**1.2.148**) followed by hydration of a triple bond in diluted H_2SO_4 in MeOH using mercury sulfate as a catalyst (Kucherov reaction) gave an intermediate 2-methylhexa-1,5-dien-3-one (**1.2.149**), which simultaneously isomerizated to propenyl-isopropenylketone (2-methylhexa-1,4-dien-3-one) (**1.2.150**). Obtained 1,4-dien-3-one (**1.2.150**) easily reacts with aliphatic amines to give desired 1-substituted-2,5-dimethylpi-peridin-4-ones (**1.2.151**). 1,2,5-Trimethylpiperidin-4-one is synthesized by this method in industrial scale [185] (Scheme 1.37).

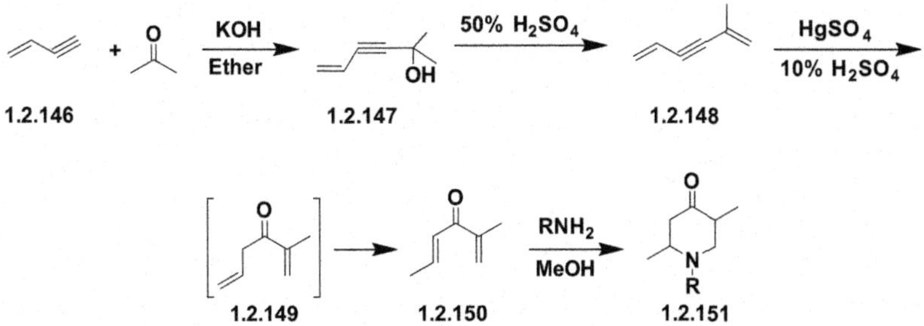

SCHEME 1.37 Synthesis of 1-substituted-2,5-dimethylpiperidin-4-ones (Nazarov's piperidones).

According to another method proposed for the synthesis of the same com-pound, methyl isopropenyl ketone (**1.2.152**) was silylated by trimethylsilyl chloride to give 2-methyl-3-trimethylsilyloxy-buta-1,3-diene (**1.2.153**), which on reaction with 1,1-diethoxyethane, gave 5-hydroxy-2-methylhex-1-en-3-one (**1.2.154**). The latter was cyclocondensed with methylamine to give desired 1,2,5-trimethylpiperidin-4-one [186] (Scheme 1.38).

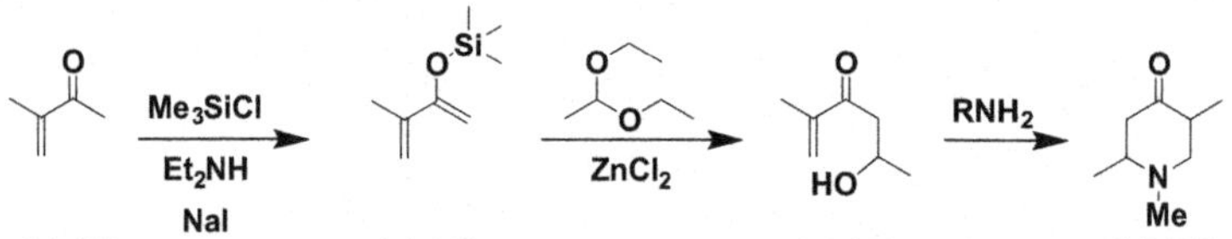

SCHEME 1.38 Synthesis of 1-methyl-2,5-dimethylpiperidin-4-one via cyclocondensation of 5-hydroxy-2-methylhex-1-en-3-one with methylamine.

An efficient synthesis of piperidin-4-ones from secondary amines in two steps has been proposed via a gold catalyst. Authors descried one-pot reaction of secondary amines (**1.2.155**) with 3-butynyl tosylate (**1.2.156**) to give tertiary amines (**1.2.157**) followed by a sequential *m*-chloroperoxybenzoic acid (*m*-CPBA) oxidation to intermediate (**1.2.158**) and cyclization in the presence of gold catalyst Ph$_3$PAuNTf$_2$, at 0°C, which gave piperidin-4-ones (**1.2.159**) in good yields. It was supposed that in situ−generated tertiary amine *N*-oxides can oxidize terminal triple bonds in the presence of gold catalyst, generating oxo-gold-carbene intermediates and subsequently affording piperidin-4-ones [187] (Scheme 1.39). This two-step construction of piperidin-4-ones could become a powerful method if it could be implemented in a large scale synthesis (Fig. 1.11).

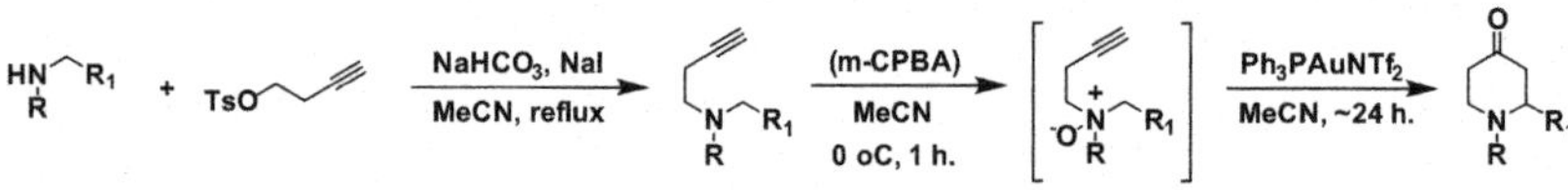

FIGURE 1.11 Synthesis of 1,2-disubstituted piperidin-4-ones implementing reaction of secondary amines with 3-butynyl tosylate.

To assemble 2-substituted-piperidin-4-one (**1.2.165**), the nucleophilic addition to activated 4-methoxypyridines, have been explored. This method has proven to be a powerful way to synthesize substituted piperidines [188].

The *N*-Ts benzimidazole (**1.2.160**) was deprotonated with lithium diisopropylamide (LDA) at −15°C and underwent addition to 4-methoxy-*N*-methyl pyridinium triflate (**1.2.161**) to give the corresponding dihydropyridine (**1.2.162**). Following mild acidic hydrolysis of the obtained enol ether the dihydropyridone (**1.2.163**) was prepared.

Reduction of double bond of enone (**1.2.163**) to the corresponding ketone with lithium aluminum-tri-tert-butoxyhydride (LiAl(*O-t*-Bu)$_3$H) in the presence of copper (I) bromide (CuBr) gave excellent results furnishing (**1.2.164**). Removal of the Ts group with THF, conc. HCl cleanly afforded 4-piperidone (**1.2.163**) [189] (Scheme 1.39).

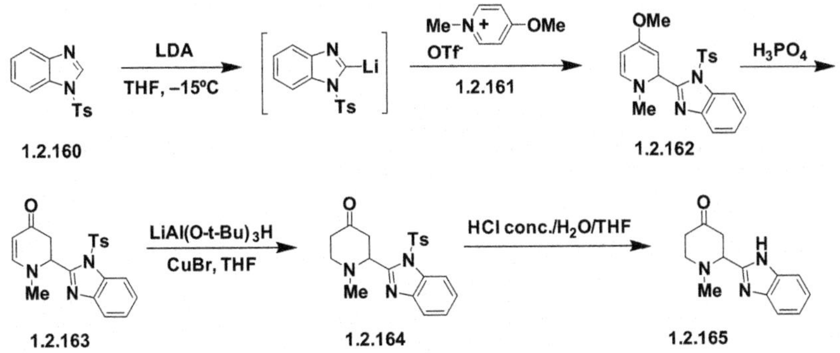

SCHEME 1.39   Synthesis of 2-(1*H*-benzo[*d*]imidazol-2-yl)-1-methylpiperidin-4-one.

A multigram synthesis of 4-oxopipecolates (**1.2.172a, 1.2.172b**) was achieved using a 1,3-dipolar cycloaddition reaction of nitrone (**1.2.166**) with 3-butenol (**1.2.167**) in chloroform. Obtained equimolar mixture of four diastereomers of isoxazoline (**1.2.168**) was mesylated with mesyl chloride in pyridine and the resulting mesylates (**1.2.169**) spontaneously underwent intramolecular nucleophilic displacement to give intermediate salt (**1.2.170**). Reaction of the last with 1,4-diazabicyclo[2.2.2]octane (DABCO) in reflux-ing acetonitrile allowed to cleave of N-O bond getting a mixture of epimeric at C-2 ethyl 4-oxopipecolates (**1.2.171**), which are readily separated chromatographically. Deprotection of the separated epimers led to (*R*)- and (*S*)-4-oxopipecolates (**1.2.172a, 1.2.172b**) [190] (Scheme 1.40).

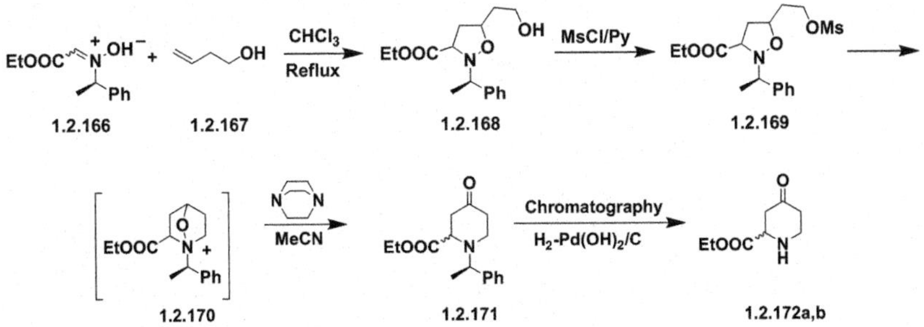

SCHEME 1.40   Synthesis of ethyl 4-oxopiperidine-2-carboxylate.

The intramolecular Mannich reaction is another tool for the assembly substituted piperidin-4-ones.

As an example could serve reaction sequence on the on Fig. 1.12. Sulfinyl amine (**1.2.173**) was treated with five- to sixfold excess of TFA in MeOH, which allowed the release of amine as trifluoroacetic salt (**1.2.173**). The obtained salt (**1.2.174**) was then reacted with an aldehyde in DCM, giv-ing 3-caroxymethylpiperidin-4-one (**1.2.175**). Decarboxylation of the last to (**1.2.176**) was best affected with 48% HBr in methanol. The major isomer of obtained piperidin-4-one (**1.2.176**) was shown to have the C-2 and C-6 substituents in a *cis*-orientation [191] (Fig. 1.12).

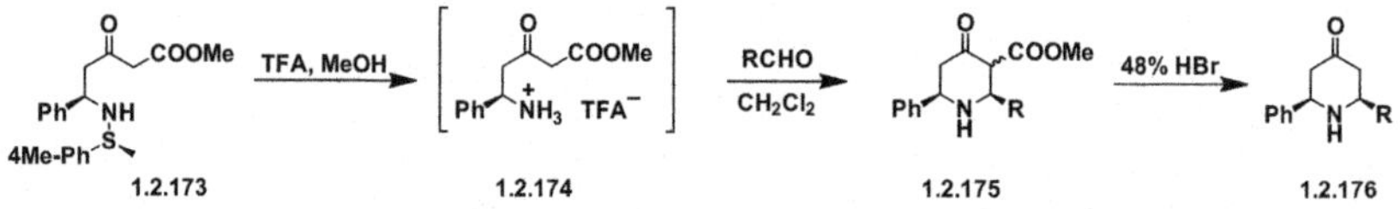

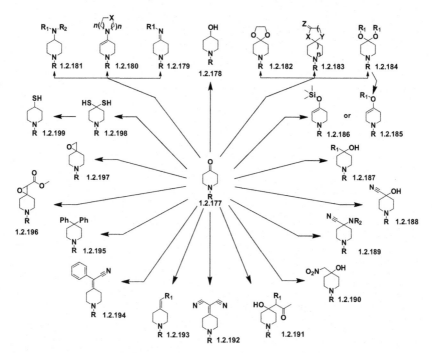

FIGURE 1.12 Synthesis of piperidin-4-ones implementing intramolecular Mannich reaction.

Other approaches for the synthesis of piperidin-4-ones implementing Mannich reaction are also demonstrated in the literature [192–194].

Nucleophilic Addition Reactions to the Carbonyl Group of Piperidin-4-Ones

The piperidin-4-ones have been reported as versatile starting materials for preparation of many drugs and just biologically active compounds [32,39,195–197].

The Fig. 1.13 represents the major transformations of carbonyl group in a nominal piperidin-4-one (**1.2.177**).

FIGURE 1.13 Major transformations of carbonyl group in piperidin-4-ones.

N-Substituted piperidin-4-ones (**1.2.177**) can be reduced to the corresponding piperidinols (**1.2.178**) with H_2 implementing Raney-Ni, Pt, Pd or Ru catalysts [39,198−202], Li, Na, or K in alcohols or amine solvents [203−205], diborane and complex metal hydrides [205−209], which are preferred. Some sophisticated catalysts such as a heterogeneous system composed of silica-supported, caged, compact phosphine silica-SMAP and rhodium(I)-alkoxo complex silica-SMAP-Rh(OMe)(cod), where SMAP is the abbreviation of a silicon-constrained monodentate trialkylphosphine [210], or enzymic "in vitro" reductions, have been implemented [211].

Formation of imines, oximes, hydrazones (**1.2.179**), enamines (**1.2.180**), and related compounds such as amines and hydrazines (**1.2.181**) are another widely used reaction series of carbonyl group transformations in piperidin-4-one series.

Piperidin-4-ones (**1.2.177**) with hydroxylamine hydrochloride in the presence of a base easily produce oximes (**1.2.179**) (R_1=OH) [212−218], which can be hydrogenated with sodium in boiling alcohols [212], or with litium aluminum hydride [214] to give primary amines (**1.2.200**). Reaction of piperidin-4-one oximes (**1.2.179**) with acetylenes in the presence of catalytic pair KOH/DMSO (Trofimov reaction [219]), results in the formation of 4,5,6,7-tetrahydro-1*H*-pyrrolo[3,2-*c*]pyridine derivatives (**1.2.201**) [215].

A Beckmann rearrangement of oximes (**1.2.179**) produced lactams (**1.2.202**) [216,220,221].

A transformation of the oximes (**1.2.179**) to tosylates (**1.2.202**) and employing them into the Neber rearrangement (treatment with 2 equivalents of KOEt in the presence of a desiccating agent) led to 3-amino- 4,4-diethoxypiperidines (**1.2.204**) [213,217].

Several methods for preparation of *N*-alkyl-4-piperidylhydrazines (**1.2.207**) are described. The method comprising formation of hydrazones (**1.2.205**) from acylhydrazines and piperidin-4-ones followed by reduction of the C=N double bond to give hydrazines (**1.2.206**) followed by hydrolysis of protecting acyl group, probably can be considered as an optimal method [216,220].

The conversion of piperidin-4-ones to imines (**1.2.208**), particularly to 4-anilino-piperidines, usually is carried out in two steps, first involving the reaction of (**1.2.177**) with aniline or another primary to give corresponding imine, which is then reduced, usually with sodium cyanoborohydride to give secondary amines (**1.2.209**) [221−227].

The prepared secondary amines were widely used in acylation reactions to give a variety of drugs represented as structure (**1.2.210**) [224−230] and in arylation reactions [230] to give compounds of the type (**1.2.211**), as well as in transformation to corresponding phenyl hydrazines (**1.2.213**) through the procedure of nitrosation to prepare nitroso-compounds (**1.2.212**) followed by reduction to desired phenyl hydrazines (**1.2.213**) [224,225] used for further cyclizations to indole derivatives.

The reaction of piperidin-4-ones (**1.2.177**) with secondary amines give enamines (**1.2.180**) [231−235]. In reductive amination conditions [236−238] piperidin-4-ones give amines (**1.2.181**), which could also be easily prepared via reduction of synthesized enamines [234]. Enamines obtained from piperidin-4-ones easily undergo alkylation and acylation reactions to give 3-substituted derivatives (**1.2.214**) [235,239]. The chemistry of imines and enamines obtained from piperidin-4-ones is well reviewed [197] (Fig. 1.14).

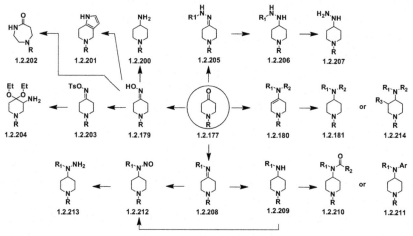

FIGURE 1.14 Transformations of carbonyl group in piperidin-4-ones to corresponding imines, amines, hydrazones, and hydrazines.

Piperidin-4-ones (**1.2.177**) easily form acetals (**1.2.184**). The ethyl and methyl acetals were prepared by treating the appropriate ketone in ethanol or methanol with an excess of hydrogen chloride [240−243]. They easily convert to spirocyclic compounds such as 1,3-dioxolanes (**1.2.182**), 1,3-dioxanes, on reaction with ethyleneglycol and propylene glycol, and more complex spirocycles such as (**1.2.183**) on reaction with 2-aminoethanol, cysteamine, 3-aminopropanol, 3-mercaptopropanol, 2-hydroxyacetamide, 2-aminoacetamide [244], salicylamide [245], etc.

The most frequently used method for enol ether synthesis is the acid-catalyzed thermolysis of acetals − 4,4-dialkoxypiperidines (**1.2.184**) obtained by reaction of piperidin-4-ones with trimethyl or triethyl orthoformate via eliminating of one alcohol moiety (**1.2.184**) → (**1.2.185**) which usually occurs usually under a reduced pressure producing the desired enol ethers [246−248].

Important intermediates in the synthesis of piperidine derivatives are trimethylsilyl enol ethers (**1.2.186**) which can be prepared from piperidin-4-ones (**1.2.177**) on reaction with trimethylsilyl chloride [249,250] or trimethylsilyl triflate [251] in presence of a base.

An infinite number of publications examples of Grignard and alkyl- and aryllithium reagents addition to piperidin-4-ones (**1.2.177**) to give tertiary alcohols (**1.2.187**), (**1.2.215**) and (**1.2.216**) is published in literature [252−263]. Allyl [264−266] and propargyl magnesium bromides [267] give corresponding alcohols (**1.2.217**) and (**1.2.218**). Reaction with vinyl magnesium bromide gives alcohols (**1.2.219**) [268,269], which could be prepared also by partial hydrogenation of acetylenic alcohols (**1.2.220**) [270,271]. Acetylenic alcohols in turn can be prepared or in Favorsky reaction conditions by condensing acetylene or its derivatives with piperidin-4-ones in presence of KOH [270,271], or by reaction of ethynylsodium [272] or ethynyl magnesium bromide with piperidin-4-ones (**1.2.177**) [253,272]. Reformatsky reaction, another reliable, practical reaction for synthetic preparations on an industrial scale, was implemented to piperidin-4-ones. α-Halo esters with in the presence of Zn metal easily react with piperidin-4-ones (**1.2.177**) to give β-hydroxy esters (**1.2.221**) [240,273] (Fig. 1.15).

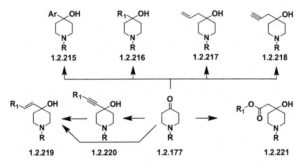

FIGURE 1.15 Transformations of piperidin-4-ones to corresponding alcohols.

Under Strecker reaction conditions promoted by acid, HCN piperidin-4-ones (**1.2.177**) are converted to α-amino nitriles (**1.2.189**), leading to quite interesting bifunctional compounds. Obtained α-amino nitriles could serve as very important precursors of α-amino acids, 1,2-amino alcohols, 1,2-diamines, and intermediates for several transformations, mainly to heterocycles [274,275], giving opportunity to synthesize a variety of heterocycles spirofused to piperidine ring.

Zelinsky and Stadnikoff modification of Strecker reaction is the use of a mixture of an alkali cyanide and ammonium chloride as reagents for the preparation of α-aminonitriles. Reaction of piperidin-4-ones (**1.2.177**) with ammonium chloride and potassium (sodium) cyanide in water, water/alcohol or water ammonia media gives α-amino-nitriles (**1.2.222**) [276–281].

The synthesis of 4-cyano- (4-alkylamino) or 4-dialkylamino-piperidine derivatives (**1.2.189**) takes place with primary and secondary amines and cyanide ion source under a variety of reaction conditions, which are referred to as a Tieman–Strecker synthesis, when the piperidin-4-one is treated with hydrogen cyanide and a primary or secondary amine, either in the Knoevenagel synthesis, when the adduct of the ketone and sodium bisulfite is formed first and then allowed to react with the amine and cyanide ion [282] (Fig. 1.16).

FIGURE 1.16   Dehydrogenation of piperidin-4-ones.

Strecker reaction followed by sulfuric acid mediated hydrolysis, delivered α-dialkylamino amides (**1.2.227**) [280,282,283]. The single publication on transformation of α-amino amides (**1.2.223**) to α-amino acids (**1.2.224**), which takes place with lithium hydroxide in a mixture of THF and water at room temperature), is described in patent [284]. α-amino acids (**1.2.224**) and (**1.2.228**) can be converted to esters (**1.2.225**) and (**1.2.229**) using thionyl chloride in alcohol [285].

When piperidin-4-ones are added to a suspension of ammonium carbonate and sodium cyanide in a water ethanol system and heated at 60°C, spiro-hydantoins (**1.2.226**) are formed [285,286], which easily hydrolyzed to amino acids (**1.2.224**). Direct hydrogenation of (**1.2.222**) was not successful. Instead, the aminonitrile (**1.2.222**) was acylated with trifluoroacetic anhydride in pyridine and generated compound (**1.2.230**) was smoothly hydrogenated on Raney-Ni giving (**1.2.231**), which after deprotection gives a desired diamine (**1.2.232**) [276] (Scheme 1.41).

Analog reactions, which give cyanohydrins instead of α-aminonitriles, was observed by Urech in 1872, 22 years after Strecker's discovery in 1850.

Treatment of the piperidin-4-ones (**1.2.177**) with cyanide salt in acid represents a simple and relatively safe procedure for the synthesis of cyanohydrins (**1.2.188**) [281,287−289]. Another very convenient laboratory method is the sodium bisulfite modification proposed by Pape. Reaction of the piperidin-4-ones (**1.2.177**) with trialkylsilyl cyanide reagents may be used to prepare the *O*-silylated cyanohydrin derivatives (**1.2.233**) [290].

Cyanohydrins obtained from piperidin-4-ones are versatile intermediates that can be easily converted into various multifunctional molecules. They easily undergo hydrolysis to give hydroxyacids (**1.2.234**) upon heating with concentric HCl. Reaction of obtained acids with alcohols in the presence of *p*-TsOH allows it to prepare corresponding esters (**1.2.235**) [287−289].

Hydroxypiperidinecarboxylates (**1.2.235**) and *O*-silylated cyanohydrin (**1.2.233**) were reduced with LiAlH$_4$ to give diol (**1.2.236**) [290] and 2-aminoethanol derivatives (**1.2.237**) [291], respectively (Scheme 1.42).

The described above α-aminonitriles (**1.2.189**) were pointed out to have some peculiar properties, and one of them is the fact that reaction of these α-aminonitriles with Grignard reagents (aromatic as well as aliphatic) results in replacement of the nitrile group (**1.2.239**), whereas with the

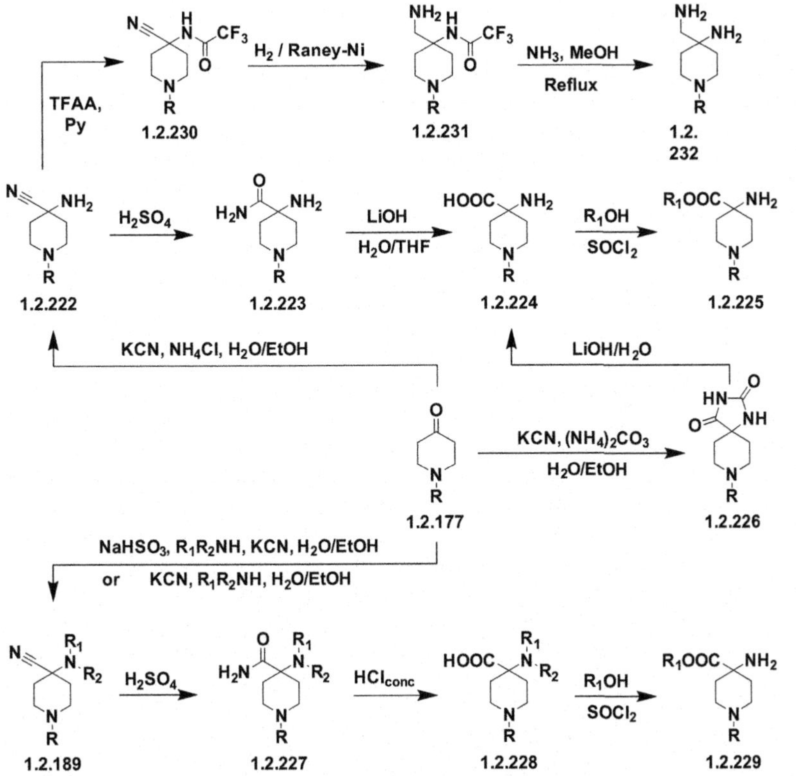

SCHEME 1.41 Transformations of piperidin-4-ones in Strecker reaction and its modifications.

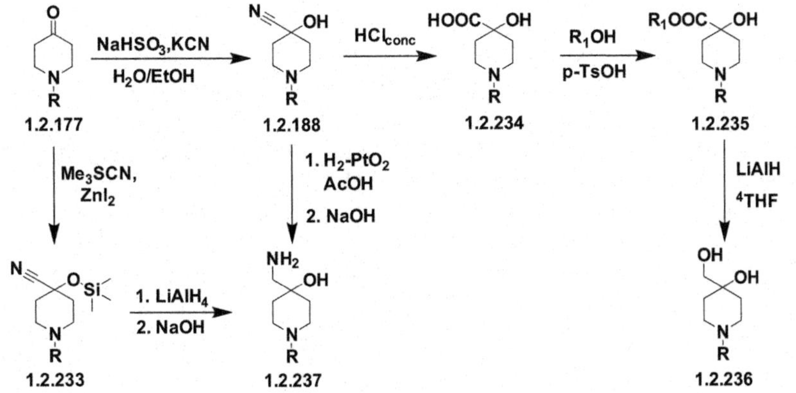

SCHEME 1.42 Transformations of piperidin-4-ones to cyanohydrins.

organolithinm compounds normal ketone formation (**1.2.238**) takes place [292−294].

The reduction of α-aminonitriles (**1.2.189**) with LiAlH$_4$ results in denitrilation (**1.2.240**), whereas the same reduction of the corresponding carboxamides (**1.2.189**) yields the expected diamines (**1.2.241**) [295] (Scheme 1.43).

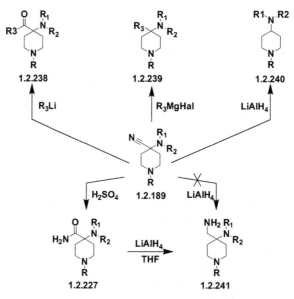

SCHEME 1.43   Some transformations of 4-aminopiperidine-4-carbonitriles.

The piperidin-4-ones (**1.2.177**) that were treated with sodium ethoxide and MeNO$_2$ nitromethylpiperidin-ols (**1.2.190**) were synthesized and consequently reduced to corresponding, aminomethylpiperidinols (**1.2.237**) [296−299].

The piperidin-4-ones (**1.2.177**) in an acid medium (Amberlite IR 120 in water) react with ketones to give products (**1.2.191**) [300].

Series of Knoevenagel condensation of piperidin-4-ones with malononitrile and cyanoacetic ester that correspond to the piperidylidene compounds (**1.2.177**) → (**1.2.192**) have been carried out. The reaction is usually carried out in an inert organic solvent (benzene, toluene, xylene) in the presence of a catalyst mixture, e.g., glacial acetic acid and ammonium acetate under the continuous water separation [240,301−305].

The Wittig reaction and its later variants, the Horner and Wadsworth−Emmons reactions, are the most versatile synthetic routes for the preparation of alkenes from carbonyl compounds (**1.2.177**) → (**1.2.193**) and were implemented to piperidin-4-ones (**1.2.177**).

Reaction of benzyl bromide with triphenylphosphine in ether at room temperature afforded the corresponding phosphonium bromide, treatment of which with sodium methylsulfinyl carbanion in DMSO at 80°C gave the corresponding ylide, which reacted with piperidin-4-ones (**1.2.177**) to form 4-benzylidenepiperidine (**1.2.242**) [306].

Wittig reaction of indole 2-methyltriphenylphosphonium iodide with piperidin-4-ones in ethanol in the presence of lithium, potassium or sodium ethoxides give good yields of the adducts (**1.2.243**) [307].

Wittig reaction between piperidin-4-ones (**1.2.177**) and (methoxymethyl) triphenylphosphonium chloride followed by acidic hydrolysis of the resulting vinyl ether (**1.2.244**) give aldehydes (**1.2.245**) [308—311].

Wittig reaction of piperidin-4-ones with ethoxycarbonyltriphenylphosphonium methylide in ethanol in the presence of sodium ethylate give desired α,β-unsaturated esters (**1.2.246**) [312]; the Wadsworth—Emmons reaction of this piperidin-4-ones with triethyl phosphonoacetate in THF in the presence of *LDA* give the same products (**1.2.246**). The same reaction with diethyl (2-(methoxy(methyl)amino)-2-oxoethyl)phosphonate in presence of LiCl, 1,8-diazabicyclo[5.4.0]undec-7-ene (DBU) in acetonitrile afford compound (**1.2.247**), which on reaction with organolithium reagents give α,β-unsaturated ketones (**1.2.248**) [313] (Scheme 1.44).

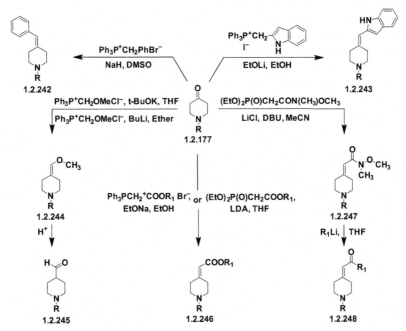

SCHEME   1.44 Transformations   of   piperidin-4-ones   in   Wittig   and   Horner   and Wadsworth—Emmons reactions.

Other types of piperidylidene derivatives − 2-(piperidin-4-ylidene)-2-phenylacetonitriles (**1.2.194**) have been synthesized by the condensation of piperidin-4-ones (**1.2.177**) with phenylacetonitrile in the presence of sodium ethoxide [303,314−316]. A proposed method is the use of a phase-transfer catalysis method condensing reagents in benzene-aqueous media, NaOH, and benzyltrimetyl-ammonioum chloride as a catalyst [317,318].

4,4-Diphenylpiperidines (**1.2.195**) were easily prepared with attractive yields by Friedel−Crafts reaction of arenes with 4-piperidinones, 4-phenyltetrahydropyridines, or 4-phenylpiperidinols in AlCl$_3$-benzene solution [319,320].

Electrophilic aromatic substitution involving piperidin-4-ones (**1.2.177**) → (**1.2.195**) was also accomplished with strong Bronsted acid such as trifluoromethanesulfonic acid at room temperature. It is proposed that the condensation reactions occur through reactive dicationic intermediates [250,321] (Fig. 1.13).

The Darzens condensation was employed for the synthesis of glycidic esters (**1.2.196**). Piperidin-4-ones (**1.2.177**) were reacted with alkyl chloroacetates in the presence of sodium hydride, sodium amide, sodium ethoxide, or potassium tertiary butoxide in toluene. Obtained glycidic esters (**1.2.196**) were hydrolyzed in hydrochloric acid to give 4-formylpiperidines (**1.2.245**) [322−325].

The Corey and Chaykovsky reaction of dimethyloxosulfonium methylide (prepared by a reaction of trimethyloxosulfonium iodide and sodium hydride or potassium tert-butoxide in DMSO) is a very efficient process for the multikilogram scale preparation of epoxides from aldehydes and ketones.

It was implemented to piperidin-4-ones (**1.2.177**) to give oxiranes (**1.2.197**) in high yield [326−331]. A rearrangement of obtained oxiranes (**1.2.197**) in the presence of magnesium bromide etherate give 4-formylpiperidines (**1.2.245**) [331].

Piperidin-4-ones (**1.2.177**) in i-PrOH on passage of H$_2$S give 4,4-dimercaptopiperidines (**1.2.198**), which on reaction with sodium borohydride easily transformed to 4-mercapto-piperidines (**1.2.199**) [332−334] (Fig. 1.13).

Corresponding piperidin-4-ols (**1.2.178**) as it was mentioned above [39,198−211] were prepared by reduction of carbonyl group of piperidin-4-ones (**1.2.177**) to hydroxyl group using practically entire group of methods known for that type of reaction such as catalytic hydrogenation with Pt oxide [198] or Raney Ni [335], Meerwein−Ponndorf−Verley reduction [336] and series of hydride reagents such as lithium aluminum hydride [203,337]. Mainly part of publications devoted to reduction with sodium borohydride [338,339]. Piperidin-4-ols on reaction with SOCl$_2$ or HCl easily transforms them to 4-chloropiperidines [336,340,341], which easily have been converted to Grignard reagents − (piperidin-4-yl)magnesium chlorides(bromides) [341−344].

Nucleophilic Substitution Reactions Involving α-Carbon of Piperidin-4-Ones

Piperidin-4-ones undergo many nucleophilic substitution reactions at α-carbon. The halogenation is one of those reactions that have proceeded α-Halogen ketones; α-bromo ketones mainly belong to an important class of compounds in in medicinal chemistry. A range of pharmaceutically important heterocycles such as thiazoles, imidazole, indoles, etc., were synthesized using α-halogen ketones as starting materials. There are few publications and some patents on the α-halogenation of piperidin-4-ones (**1.2.177**) in different conditions.

Direct bromination of piperidin-4-ones (**1.2.177**) using molecular bromine in AcOH [345], CHCl$_3$ [346], CH$_3$OH [347], THF [348], or using tetra-*N*-butylammonium tribromide in THF [349] provide a good yield of 3-bromo-piperidin-4-ones (**1.2.249**).

α-Chlorination of silyl enol ethers (**1.2.186**) using *N*-chlorosuccinimide (NCS) followed by spontaneous cleavage of volatile intermediate afford the desired 3-chloro-piperidin-4-ones (**1.2.250**) [350–352]. Analog reactions with Selectfluor (1-(chloromethyl)-4-fluoro-1,4-diazoniabicyclo[2.2.2]octane ditetra-fluoroborate) give 3-fluoro-piperidin-4-ones (**1.2.251**) [250] (Scheme 1.45).

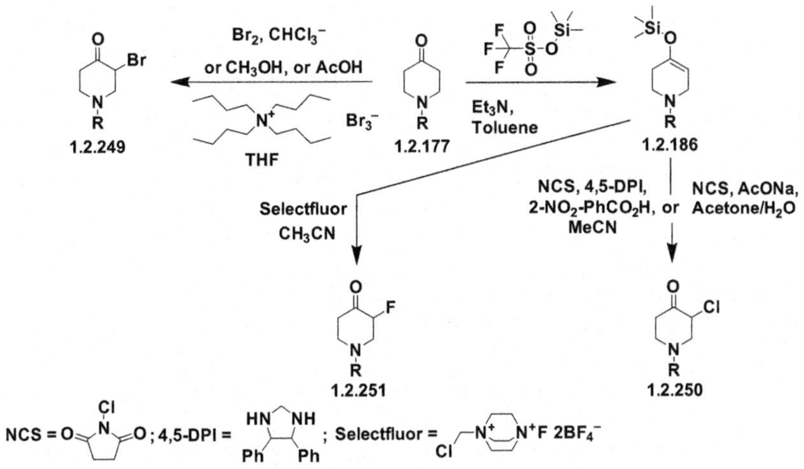

SCHEME 1.45 Halogenation of piperidin-4-ones.

Substitutions at the 3-position in the 4-phenylpiperidines.

Substitutions at the 3-position piperidine-4-ones usually are prepared from enamines derived from piperidine-4-ones (**1.2.177**) and act as nucleophiles in alkylation, acylation (Stork's alkylation or acylation), Michael addition, annulation, Diels–Alder and other reactions typical for enamines in general. Usually enamines (**1.2.180**) are prepared by a reaction of piperidine-4-ones (**1.2.177**) with a secondary amine—pyrrolidine, morpholine,or piperidine in dry toluene on reflux in the presence of catalytic amount of *p*-TsOH and removal of the water formed (Dean—Stark trap).

As mentioned above, the reaction of piperidin-4-ones (**1.2.177**) with secondary amines give enamines (**1.2.180**) [231—235,353—355] available and versatile instruments in the creation of a plethora of functionalized piperidin-4-ones.

Alkylation experiments of enamines derived from 1-benzoyl and 1-alkyl-4-piperidones allowed to synthesize a series of 3-substituted-piperidine-4-ones such as prop-2-ynyl bromide (**1.2.252**), with bromoacetone (**1.2.253**), and γ-bromoacetoacetate (**1.2.254**) [355—357].

Acylation with acetic anhydride [358] and trifluoroacetic anhydride [359] give compounds (**1.2.255**), acylation with cyclopropanecarbonyl chloride [360] compound (**1.2.256**), and with benzoyl chloride [361] compound (**1.2.257**).

Michael addition with ethyl acrylate in MeCN gave compound (**1.2.258**) [362], with acrylonitrile in dioxane compound (**1.2.259**) [353] (Scheme 1.46).

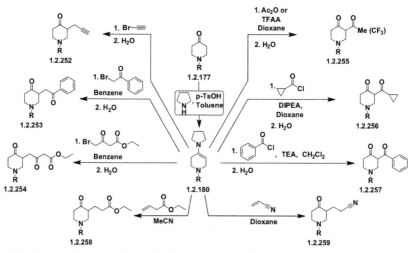

TFAA = trifluoroacetic anhydride; DIPEA = N,N-Diisopropylethylamine; TEA = triethylamine

SCHEME 1.46   Transformations of enamines derived from piperidin-4-ones.

Substitution at the 3-position piperidine-4-ones using enamine chemistry allows to realize some annulation reactions. One of the examples is the synthesis of methylhexahydroisoquinolone (**1.2.260**) prepared from the enamine of benzoylpiperidone (**1.2.180**) and methyl isopropenyl ketone on reflux in dioxane [363]. Another example is reaction of enamine (**1.2.180**) methyl vinylacrylate which on heating in benzene under reflux give a mixture of three intermediate compounds. The compound (**1.2.261**) was one of the three components present in the original mixture, the other two presumably being isomeric hexahydro derivatives which underwent monodehydrogenation. Treatment of the mixture with 5% Pd-C in methanol solution under reflux afforded a 93% yield of the 2-methyl-5-carbomethoxy − 1,2,3,4 − tetrahydroisoquinoline (**1.2.262**) [364].

The 1,3-dipolar cycloaddition of nitriloxide, generated in situ by treating the methyl chlorooxime acetate with triethylamine in dichloromethane to the enamines (**1.2.180**), has been exploited for the synthesis of isoxazole-fused piperidine derivatives (**1.2.263**) [365].

The reactions of troponimines with enamines (**1.2.180**) lead to formal [8 + 2] cycloadducts and subsequent aromatization under the reaction conditions gave *N*-substituted 1,2,3,4-tetrahydrocyclohepta[4,5]pyrrolo[3,2-c] pyridines (**1.2.264**) in modest-to-good yields [366]. Piperidin-4-one−derived enamines (**1.2.180**) were found to undergo an inverse electron demand [4 + 2]cycloaddition Diels−Alder reactions with electron-rich dienophiles. Reaction of enamines (**1.2.180**) with ethyl 1,2,4-triazine-3-carboxylate gave tetrahydronaphthyridine derivative (**1.2.265**) [367]. Reaction with1,4-diaryl-pyridazinopyridazines gave 6,7,8,9-tetrahydropyrido [3,4-g]phthalazines (**1.2.266**) [368]. Many other examples of this inverse electron demand Diels−Alder reaction are described [369−372] (Scheme 1.47).

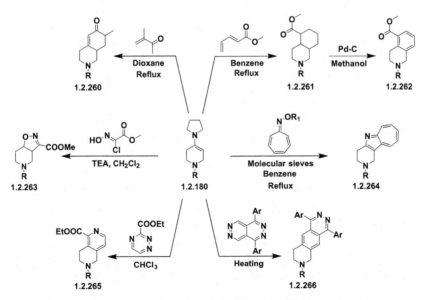

SCHEME 1.47 Transformations of enamines derived from piperidin-4-ones.

Aldol condensations are yet another example of electrophilic substitution at the alpha carbon in piperidin-4-ones chemistry.

The condensation of a variety aromatic aldehydes, particularly benzaldehyde with the piperidin-4-ones (**1.2.177**), was first described about 70 years ago [373] and then was repeated and modified several times. Many references are given in reviews [27,196] and protocols which slightly differ from each other can be found in [374—381].

The vast majority of the synthesis of the cross-conjugated dienones (**1.2.267**) were carried out by the reactions of piperidin-4-ones (**1.2.177**) with two equivalents of aromatic aldehydes under basic conditions. The condensation of benzaldehyde with the *N*-methylpiperidin-4-one may also be affected by acid catalysis [373]. The "mother" paper [373] also showed the possibility of the synthesis of 3-benzylidenepiperidin-4-ones (**1.2.268**) on implementing an equimolar ratio of reagents in presence of potassium hydroxide in ethanol. A special protocol was developed recently for this purpose [382], which involves the use of an equimolar solvent-free mixture of the two reactants *N*-trimethylsilyldimethylamine and magnesium bromide ethyl etherate leading to the formation of the respective product (**1.2.268**) in high yield.

Another special protocol for the synthesis of 3-(hydroxy(aryl)methyl) piperidin-4-one (**1.2.269**) was proposed. It consists of a reaction of the lithium enolate of piperidin-4-ones (**1.2.177**) prepared by its lithiation with LDA in THF followed by treatment with aldehyde and acidic workup [383] (Scheme 1.48).

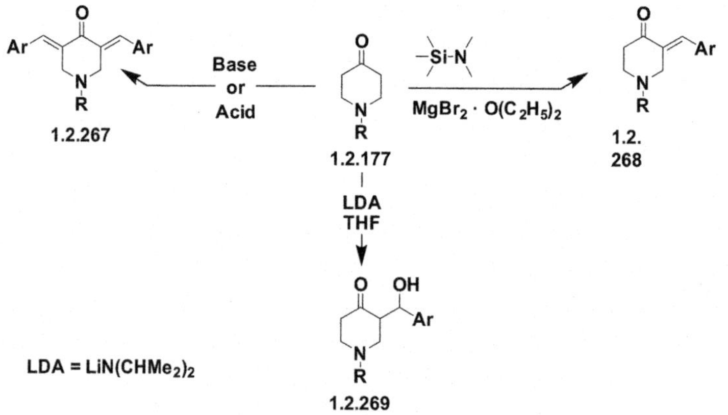

SCHEME 1.48   The condensation of piperidin-4-ones with aromatic aldehydes.

Another challenging series of compounds formally arising as a result of electrophilic substitution at the alpha carbon are β-dicarbonyl representatives of piperidin-4-ones that were widely used in series of heterocylizations resulted in the creation of new fused heterocyclic compounds.

4-Oxopiperidine-3-carbaldehydes (**1.2.270**) were prepared by condensing of piperidin-4-ones (**1.2.177**) with ethyl formate using sodium ethoxide in dry ether [384–386]. Preparation of 3-acyl-piperidin-4-ones (**1.2.255**) via acylation of enamines (**1.2.180**) with acetic anhydride [358] and trifluoroacetic anhydride [359] was described above. Direct acylation of ketones (**1.2.177**) with acetyl cyanide was carried out in THF using as a base LDA [387]. Aroyl substitutions usually are carried out through enamine step [361,388], but a direct acylation with benzoyl chloride implemented as a base lithium bis(trimethylsilyl)amide (LiHMDS) in THF gave a perfect yield of product (**1.2.257**) [389]. β-Ketoesters (**1.2.131**) usually are prepared in bulk with excellent yields on Dieckmann condensation of aminodicarboxylate esters [171–176,178], but a reverse case, when piperidin-4-one (**1.2.177**) was converted to β-ketoester (**1.2.131**) by condensation with dimethyl carbonate on reflux in toluene and used as a base sodium hydride, is also described [390]. All synthesized 1,3-dicarbonyl compounds were widely implemented in series of diverse heterocyclization reactions resulting in the creation of new fused heterocyclic compounds (Scheme 1.49).

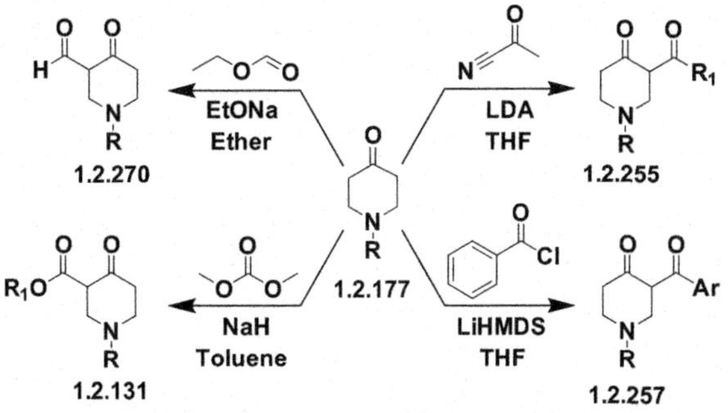

SCHEME 1.49 Transformations of piperidin-4-ones to β-dicarbonyl compounds.

There are described reactions with very important practical implementations, where both reactive centers in piperidin-4-ones (**1.2.177**) — carbonyl and α-carbon atoms — simultaneously participate in some reactions. One example is the Gewald reaction, a method used for the synthesis of substituted-4,5,6,7-tetrahydrothieno[2,3-c]pyridin-2-amines (**1.2.272**) via a multicomponent condensation between sulfur and α-methylene dicarbonyl compound. The first step of the process is a Knoevenagel reaction, which results in a product (**1.2.271**); the further sequence is not known in detail. For example, thyl cyanoacetate reacted with the appropriate piperidin-4-one and elemental sulfur in ethanol in the presence of elemental sulfur providing 2-aminothiophenes (**1.2.272**) [391–398], which underwent multiple transformations (Scheme 1.50).

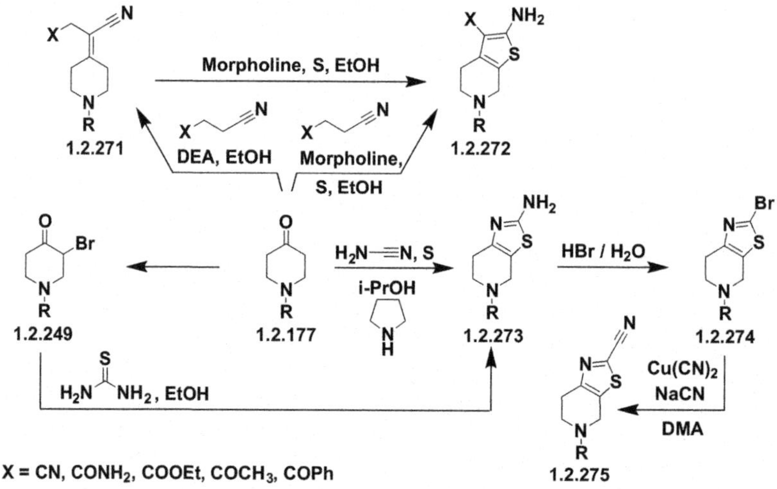

SCHEME 1.50 Transformations of piperidin-4-ones using both reactive centers—carbonyl group and α-carbon atoms—simultaneously.

Various analog transformations take place when (**1.2.177**) instead of α-methylene dicarbonyl compound is used with cyanamide, which in the presence of sulfur powder and a catalytic amount of pyrrolidine in i-propanol on heating gives 2-aminothiazole derivatives (**1.2.273**), which in Sandmeyer reaction conditions could easily be transformed to corresponding bromides (**1.2.274**), then to nitriles (**1.2.275**) and other derivatives [399–404]. The same 2-aminothiazole derivatives (**1.2.273**) are possible to synthesize from 3-bromo-piperidin-4-ones (**1.2.249**) on its reaction with thiourea [346].

Another type of reaction employed to various piperidine-4-ones (**1.2.177**) is their dehydrogenation to 2,3-dihydropyridin-4-ones – α,β-cyclic enones (**1.2.276**). The reaction was carried out with different oxidants such as *m*-chloroperbenzoic acid [405], with Hg(II)-EDTA (ethylenediaminetetraacetic acid) complex [406], in the presence of palladium(II) bis(trifluoroacetate), and DMSO in acetic acid [407], *O*-iodoxybenzoic acid (IBX) [408,409], with gold nanoparticles supported on a manganese-oxide-based octahedral molecular sieves OMS-2 (Au/OMS-2) [410] (Fig. 1.16).

Some interesting reactions at the first position of 1-methyl-piperidin-4-one (**1.2.177**) take place when it is converted to 1,1-dimethyl-4-oxopiperidinium iodide (**1.2.277**) using methyl iodide in ether or acetone. This quaternary salt on treating with different nuclophiles give a plethora of tetrahydrochalcogeno-1-pyranones-4. With sodium hydrogen sulfide and sodium sulfide tetrahydro-4H-thiopyran-4-one (**1.2.278**) was prepared [411,412]. Tetrahydro-4H-selenopyran-4-one (**1.2.279**) was synthesized by treating 1,1-dimethyl-4-oxopiperidinium iodide (**1.2.277**) with sodium hydrogenselenide in EtOH [413,414]. Unstable tetrahydro-4H-telluropyran-4-one (**1.2.280**) was prepared the same manner using with sodium hydrogentelluride [414]. Transamination of 1,1-dimethyl-4-oxopiperidinium iodide with tert-butylamine affords 1-tert-butylpiperidin-4-one (**1.2.281**) in high yield [415]. Efficient trapping of dimethylamine with sodium acrylate was found as an elegant way for the transamination reaction (Scheme 1.51).

1.2.278 1.2.279 1.2.280 1.2.277 1.2.281

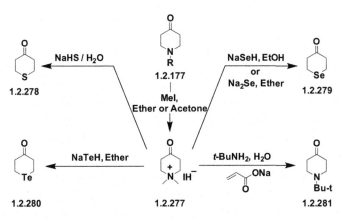

SCHEME 1.51 Reactions at the first position piperidin-4-ones.

Piperidin-3-Ones

The efficient synthetic methods of piperidin-3-ones are limited. Probably the first proposed method is based on the Dieckmann condensation of (**1.2.282**) to β-keto ester (**1.2.283**) with further hydrolysis and decarboxylation to give 1-methyl-piperidin-3-one (**1.2.284**). This remains the main general approach for the synthesis of piperidin-3-ones [416–418] (Scheme 1.52). The structure of obtained β-keto ester (**1.2.283**) was proved and justified.

1.2.282 1.2.283 1.2.284

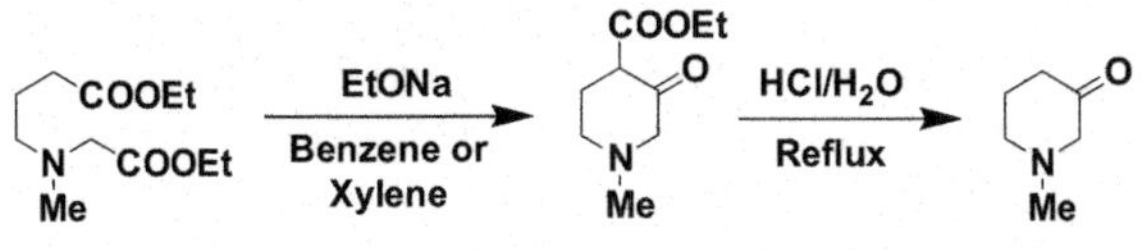

SCHEME 1.52 Synthesis of piperidin-3-ones implementing Dieckmann condensation reaction.

It is interesting to point another paper that mentions the absence of regioselectivity in Dieckmann condensation in the case of the synthesis of piperidin-3-ones from the diester (**1.2.285**), which on Dieckmann cyclization with potassium tert-butoxide in toluene gave a 64:36 mixture of 3-oxopiperidine-4-carboxylate (**1.2.286**) and 3-oxopiperidine-2-carboxylate (**1.2.287**) [419] (Scheme 1.53).

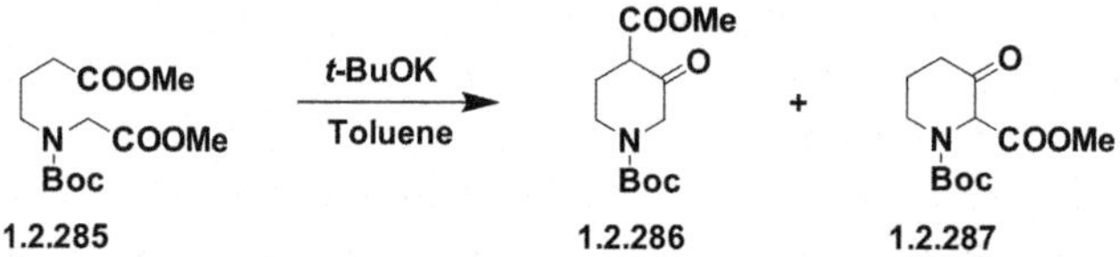

SCHEME 1.53   Synthesis of piperidin-3-ones implementing Dieckmann condensation reaction.

In a closely related method 3-(methylamino)butan-2-one (**1.2.288**) was N-alkylated with propargyl chloride, or bromoacetone, to give propargyl amine (**1.2.289**) or diketone (**1.2.290**), respectively. In attempts to hydrate a triple bond in (**1.2.289**) in Kucherov reaction conditions (HgSO$_4$, 50% H$_2$SO$_4$) 1,2,5-trimethyl-1,6-dihydropyridin-3(2H)-one (**1.2.291**) was obtained. In the same reaction conditions diketone (**1.2.290**) was transformed to unsaturated ketone (**1.2.291**), which was successfully hydrogenated for the desired piperidin-3-one (**1.2.292**) [420] (Scheme 1.54).

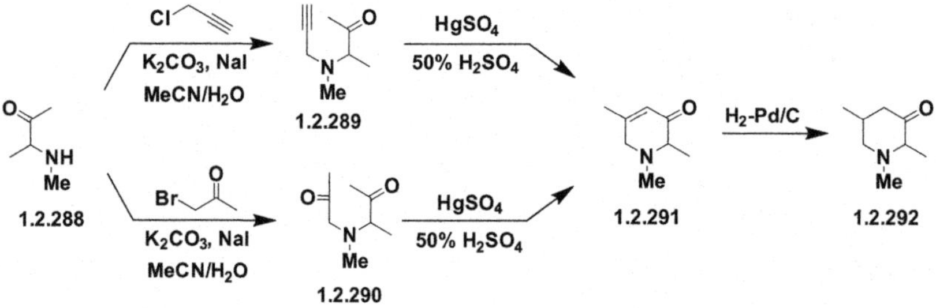

SCHEME 1.54   Synthesis of piperidin-3-ones implementing intramolecular cyclization of 3-(methyl(2-oxopropyl)amino)butan-2-one to 1,6-dihydropyridin-3(2H)-one derivative followed by hydrogenation of double bond.

The selective reduction of pyridinium salts prepared from commercially available 3-hydroxypyridine (**1.2.293**) to 1-alkyl-piperidin-3-ones seems to be a shorter route. For example, after methylation of 3-hydroxypyridine (**1.2.293**) with methyl iodide in methanol in presence of sodium methoxide followed by reduction with NaBH$_4$ 1-methyl-1,2,5,6-tetrahydro-3-pyridyl methyl ether (**1.2.294**) was obtained, which was hydrolyzed with 48% hydrobromic acid under reflux to give 40% of 1-metyl-piperidin-3-one (**1.2.284**) [421] (Scheme 1.55).

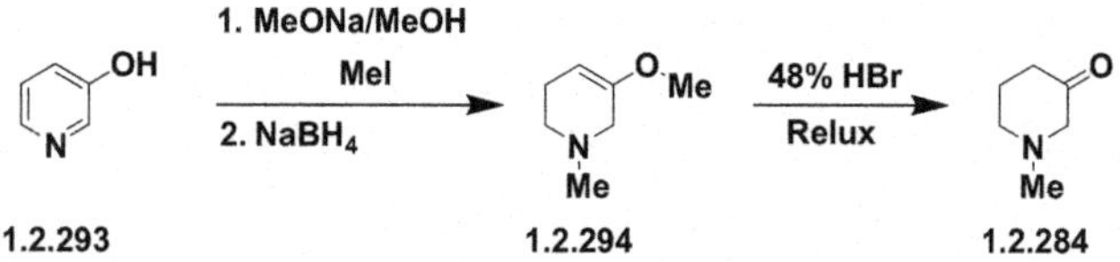

SCHEME 1.55   Reduction of pyridinium salts of 3-hydroxypyridine.

A more sophisticated approach for selective hydrogenation of 3-hydroxypyridinium salts has been demonstrated recently using a homogeneous iridium catalyst — (Bis(1,5-cyclooctadiene) diiridium(I) dichloride), providing a direct access to 2- or 4-substituted piperidin-3-one (**1.2.287**) derivatives with high yields.

For that purpose substituted pyridin-3-ols (**1.2.295**) were *N*-alkylated with benzyl bromide in acetone and obtained 3-hydroxypyridinium salts (**1.2.296**) were hydrogenated under 600 psi of H$_2$ using a mixture of [Ir(COD)Cl]$_2$ and PPh$_3$ stirred in 1,2-dichloroethane [422] as a catalyst to give piperidin-3-ones (**1.2.297**) (Scheme 1.56).

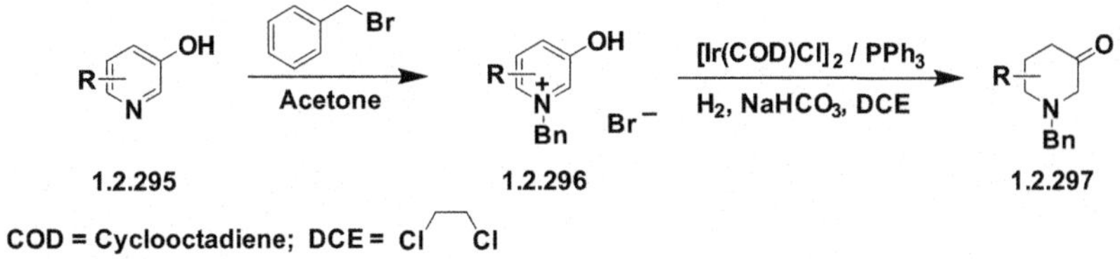

COD = Cyclooctadiene; DCE = Cl⌒Cl

SCHEME 1.56 Reduction of pyridinium salts of 3-hydroxypyridine.

Another method for the synthesis of substituted piperidin-3-ones involves rhodium(2)-catalyzed decomposition of diazo carbonyl compounds bearing dialkylamino substituent six centers away from the carbenoid center.

The desired substrates (**1.2.300**) were obtained by direct alkylation of secondary amines (**1.2.298**) with 5-bromo- 1-diazo- 2-pentanone (**1.2.299**) in ethyl acetate in the presence of trethylamine. Diazoketone (**1.2.300**) was added to Rh$_2$(OAc)$_4$ in CH$_2$Cl$_2$ at room temperature and formed intermediately quaternary salt (**1.2.301**), which underwent a Stevens [1,2]-shift reaction of aryl or carbalkoxy exocyclic group resulting in new piperidin-3-ones (**1.2.302**) in good-to-excellent yield [423] (Scheme 1.57).

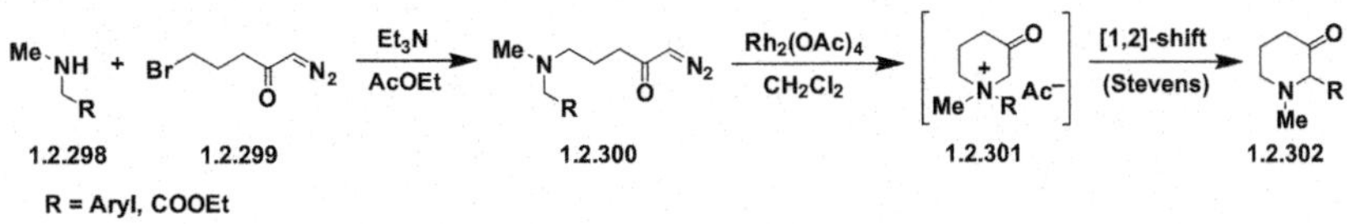

R = Aryl, COOEt

SCHEME 1.57 Synthesis of piperidin-3-ones implementing Stevens [1,2]-shift reaction of aryl or carbalkoxy exocyclic group.

An unprecedented ring expansion-oxidation reaction occurred when 4,4-dialkyl-2-(bromomethyl) pyrrolidines (**1.2.303**) were heated in dimethylsulfoxide in the presence of potassium carbonate giving 5,5-dialkyl-3-piperidones (**1.2.304**).

As the isolation of piperidin-3-ones (**1.2.304**) in analytically pure form was unsuccessful, their immediate reduction was performed by means of sodium borohydride, affording the corresponding 3-hydroxy-piperidines (**1.2.305**) [424] (Scheme 1.58).

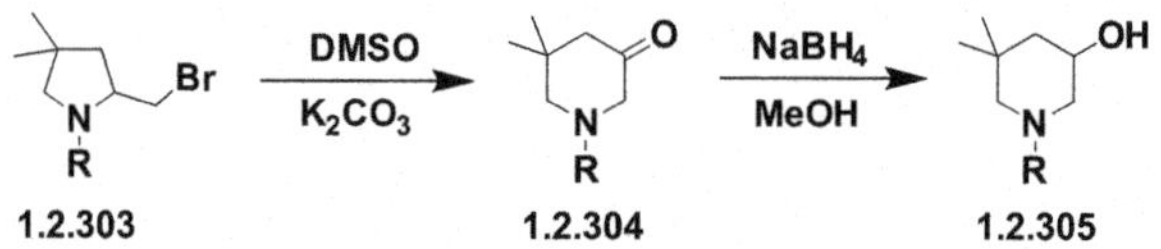

SCHEME 1.58 Synthesis of piperidin-3-ones implementing ring expansion-oxidation reaction.

Another approach to the synthesis of piperidin-3-ones begins with the reaction of phthalimide- protected aldehyde (**1.2.306**) with the anion generated by LDA from a 2-morpholino-2-phenylacetonitrile derivative (**1.2.307**), which resulted in a masked α-hydroxycarbonyl compound (**1.2.308**). Removal of the phthalimide group was followed by carbamate protection of the amine group gave (**1.2.309**).Treatment of the last with 70% acetic acid at reflux released the α-hydroxyketone (**1.2.310**), the precursor for the intramolecular Amadori-type reaction. The cyclization to the N-methoxycarbonylpiperidin-3-one (**1.2.311**) was catalyzed by p-toluenesulfonic acid in toluene at reflux [425] (Scheme 1.59).

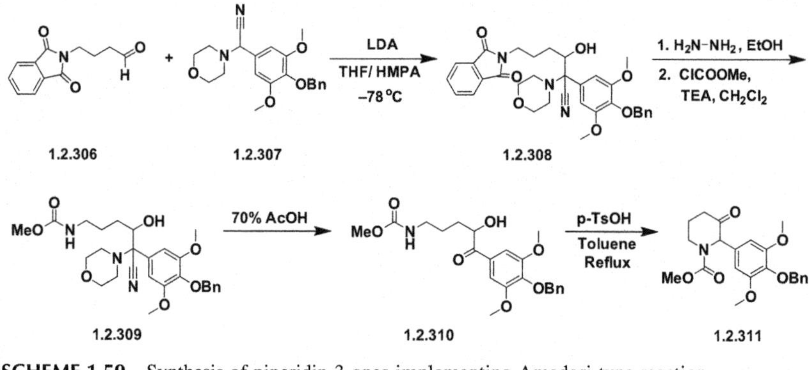

SCHEME 1.59 Synthesis of piperidin-3-ones implementing Amadori-type reaction.

An interesting example for the synthesis of piperidin-3-ones was demonstrated recently. The N-Boc-γ-amino-α-diazoketones (**1.2.312**) on treatment with TFA in dichloromethane at −78°C gave piperidin-3-ones (**1.2.313**) isolated with good-to-excellent yields [426] (Scheme 1.60).

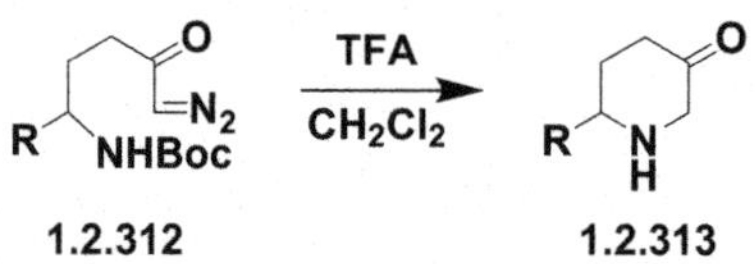

SCHEME 1.60 Synthesis of piperidin-3-ones from γ-amino-α-diazoketones.

The reactions in which have been involved piperidin-3-ones mainly match those described for piperidin-4-ones.

Piperidin-2-Ones

The lactam system to which piperidin-2-ones belong (2-piperidone, δ-valero-lactam) (**1.2.314**) ranks among the most ubiquitous skeletons found in phar-maceuticals and naturally occurring biologically active molecules. The synthesis of piperidin-2-ones have been the focus of many researchers. Piperidin-2-ones have been proposed as a way to prepare the cyclocondensa-tion (lactamization) of 5-aminovaleric acid (**1.2.315**) in different reaction conditions such as autoclaving at high temperatures (100−400°C) [427,428] or to be used for cyclodehydration alumina or silica gel in refluxing PhMe [429,430]. It is possible for derivatives of 5-aminovaleric acid to be cyclized to the lactams with a high yield by refluxing a mixture of the δ-amino acid or its HCl salt with (Me₃Si)₂NH in a solvent like MeOH or EtOH [431]. For the synthesis of piperidin-2-ones from 5-aminovaleric acid different catalysts like tris(2,2,2-trifluoroethyl) borate [432], N-alkyl-2-benzothiazolyl-sulfena-mides [433], triethylgallium [434], 3,4,5-tTrifluorobenzeneboronic acid [435], titanium tetraisopropoxide [436], different organostannyl oxides [437] have been proposed.

Derivatives of 5-aminovaleric acid esters (**1.2.316**) were also cyclized to the lactams with high yield just by stirring for 24 hours in appropriate alco-hol in the presence of a catalytic amount of K_2CO_3 [438,439], or t-BuOK in THF [440], or by refluxing in an appropriate solvent [441,442].

Piperidin-2-one is formed in excellent yields by allowing δ-chlorovaleric acid (**1.2.317**) to react with an excess of concentrated aqueous NH_3 at 275°C, which gives a corresponding δ-valerolactam − piperidin-2-one [443].

1-Cyclohexyl-2-piperidone was prepared from δ-chlorovaleric acid chlo-ride, first transforming it to cyclohexylamide in dichloromethane at 0°C and then cyclizing it to a desired compound using sodium hydride in THF at 70°C [444].

Derivatives of piperidin-2-ones is proposed to prepare by direct lactami-zation of 1,4-azido amides (**1.2.318**). It is supposed that the cyclization occurs via the nucleophilic attack of the amine group to the amide group in the amino amide, which is generated from the azido group by the Staudinger reaction [445]. Another reductant used for the same goal cyclization of azido amides is tributyltin hydride proposed by the same group of authors [446].

δ-Oxocapronitrile (**1.2.319**) in Leuckart reaction conditions (refluxing with HCO_2Na in 100% HCO_2H), then treated with an excess of 40% NaOH, gave 2-methyl-6-piperidone [447].

Hydrogenation of γ-acetobutyric acid (**1.2.320**) in a MeOH solution of $MeNH_2$ at 130−40°C in the presence of Ni catalyst gave 1,2-dimethyl-6piperidone [448].

Piperidin-2-ones were also proposed as a way to synthesize by way of an aluminum triflate-catalyzed cascade cyclization and ionic hydrogenation reaction of substituted ketoamides (**1.2.321**) [449].

Diethyl 2-(cyanomethyl)malonate (**1.2.322**) hydrogenated with Raney nickel at 80°C and 1000 pounds of pressure yielded 3-carboxy-2-piperidone [450].

N-methyl-2-piperidone was prepared by catalytically reducing *N*-methyl-α-pyridone (**1.2.323**) in H_2O using Pt black and H_2 during 2−3 days [451].

An alternative route includes the Beckmann rearrangement. A large number of processes are known for the preparation of 2-piperidone, according to the Beckmann reaction, in the presence of acid. It is known that cyclopentanone oxime (**1.2.324**) can be rearranged into the corresponding lactam by the action of acid reagents such as oleum, concentrated sulfuric acid or acetic anhydride, phosphoric acid, heteropoly-acids, NH_4HSO_4, alkali metal bisulfates, or boric acid. It has been proposed to use aluminum oxide, silicic acid gel, diatomaceous earth, active charcoal, titanium dioxide, and tin dioxide as carrier materials [452−457].

An efficient intramolecular redox lactamization reaction of 1-sustituted pyrrolidine-2-carbaldehydes (**1.2.325**) to piperidin-2-ones catalyzed by a triazolium derived carbene was described recently. Both N-Ts and N-Bn substrates efficiently furnish the desired lactams in high yield [458].

A method for the preparation of piperidin-2-one derivatives is described by the addition of the dianion of 4-(phenylsulfonyl)butanoic acid ($PhSO_2(CH_2)_3CO_2H$) to readily available imines (**1.2.326**), which is followed by cyclization in situ with the aid of $(CF_3CO)_2O$ to the title heterocycles [459].

In another implementation of imines for the synthesis of piperidin-2-ones sulfone-substituted glutaric anhydrides were used, which readily reacted with imines to form γ-lactams [460] (Fig. 1.17).

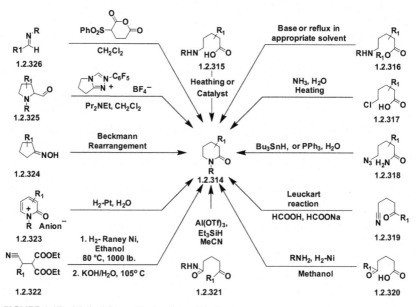

FIGURE 1.17 Method for synthesis of piperidin-2-ones.

Functionalized derivatives 2-oxopiperidine-carboxylic acids such as alkyl 2-oxopiperidine-3-, 4-, 5, and 6-carboxylates have been prepared using different specific methods.

For example, ethyl 2-oxopiperidine-3-carboxylates (**1.2.327**) were synthesized by hydrogenation diethyl 2-(cyanomethyl)malonates (**1.2.322**) on a Raney Ni catalyst [450,461,462] (Scheme 1.61).

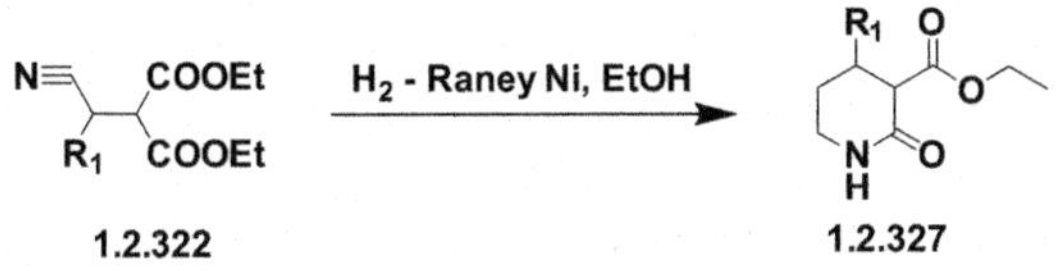

SCHEME 1.61   Synthesis of 2-oxopiperidine-3-carboxylates.

Ethyl 2-oxopiperidine-4-carboxylate was prepared starting from diethyl itaconate (**1.2.328**) to ethanol solution of which was added an aqueous solution of KCN to give diethyl cyanomethylsuccinate (**1.2.329**), which was then hydrogenated using Raney Ni catalyst to give 4-ethoxycarbonyl-2-piperidone (**1.2.330**) [463] (Scheme 1.62).

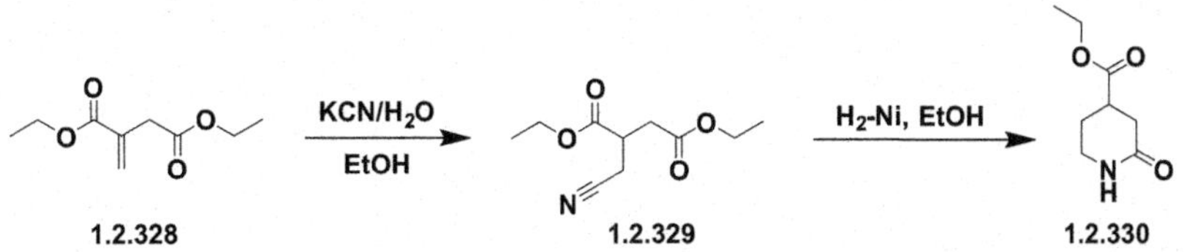

SCHEME 1.62   Synthesis of 2-oxopiperidine-4-carboxylates.

Ethyl 2-oxopiperidine-5-carboxylate (**1.2.333**) was synthesized starting from 6-hydroxynicotinic acid (**1.2.331**), which was dissolved in sodium bicarbonate water solution that was subjected to 500 psi hydrogen at 100°C for 12 hours in the presence of ruthenium on alumina to give 5-carboxy-2-piperidone (**1.2.332**). The addition of thionyl chloride to the solution of the last in ethanol gives the title compound (**1.2.333**) [464] (Scheme 1.63).

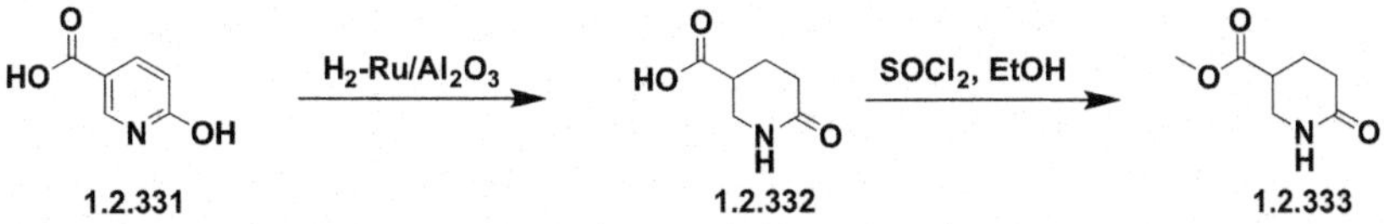

SCHEME 1.63   Synthesis of 2-oxopiperidine-5-carboxylates.

Different approaches for the synthesis of ethyl 2-oxopiperidine-6-carboxylate (**1.2.335**) were proposed.

One of them consists of holding 2-amino-hexanedioic acid diethyl ester (**1.2.334**) over a high vacuum at 110°C to yield desired ethyl 2-oxopiperidine-5-carboxylate (**1.2.335**) [465] (Scheme 1.64).

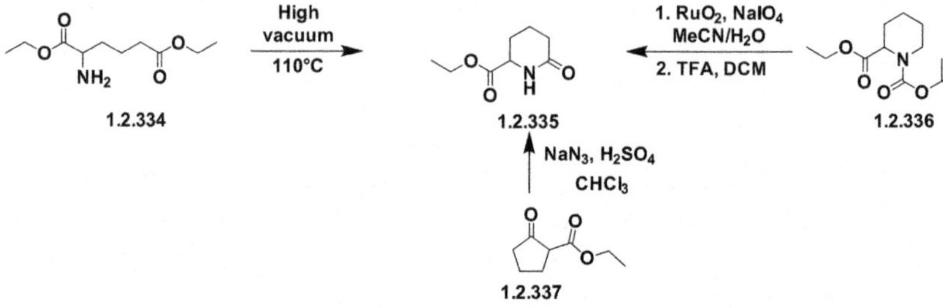

SCHEME 1.64 Synthesis of 2-oxopiperidine-6-carboxylates.

In another publication a solution of 1-(*t*-butyl) 2-ethyl piperidine-1,2-dicarboxylate (**1.2.336**) in a mixture of MeCN and H$_2$O containing sodium periodate at room temperature was added an oxidizing agent such as ruthenium (IV) oxide ruthenium hydrate and obtained intermediate 6-oxo-compound was deprotected with TFA in DCM to give the title compound (**1.2.335**) [466] (Scheme 1.64).

Formation of the same 2-oxopiperidine-6-carboxylate (**1.2.335**) was proposed to realize an implemented Schmidt rearrangement to ethyl 2-oxocyclopentane-1-carboxylate (**1.2.337**) which was dissolved in a chloroform-sulfuric acid mixture to which sodium azide was gradually added to give the desired 2-oxopiperidine-6-carboxylate (**1.2.335**) [467] (Scheme 1.64).

It was shown that the addition of arylsulfonyl isocyanates and alkoxysulfonyl isocyanates to 3,4-dihydro-2-methoxy-2H-pyrans (**1.2.338**) generates the corresponding 3-formyl- or 3-acetyl-6-methoxy-3-methyl-1-(arylsulfonyl)-2-piperidones (**1.2.339 trans 1.2**) and (**1.2.339 cis 1.2**) as a separable mixture in high yield [468] (Scheme 1.65).

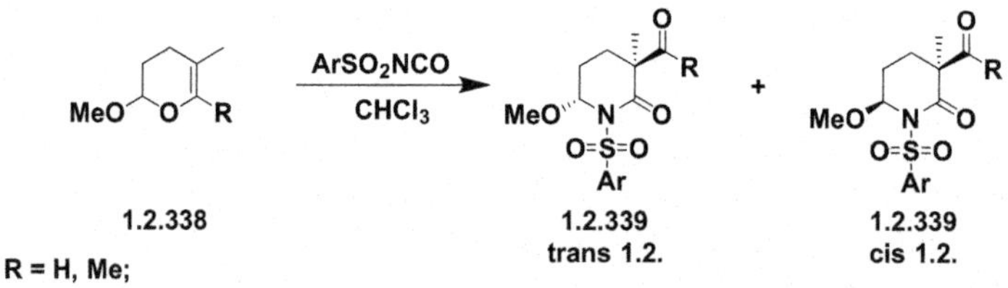

SCHEME 1.65 Synthesis of 1-(arylsulfonyl)-6-methoxy-2-oxopiperidine-3-carboxylates.

Oxazolopiperidone lactams (**1.2.442a, 1.2.442b**) are readily available in both enantiomeric series by cyclocondensation of the chiral amino alcohols (**1.2.440**) with glutaraldehyde [469] or δ-oxo acid derivatives (**1.2.441**) [470,471]. These lactams are exceptionally versatile building blocks for the enantioselective construction of structurally diverse piperidine-containing compounds.

In particular, simple phenylglycinol-derived lactams *trans*- (**1.2.442a**) and *cis*- (**1.2.442b**) form separable (**1.2.443a, 1.2.443b**), which allow the substituents to be introduced at the different ring positions in a regio- and stereocontrolled manner. Phenylethanol moiety is easily removed by catalytic hydrogenation. Further hydrogenation with alumohydrides allow piperidines (**1.2.444a, 1.2.444b**) to be obtained. The use of 3,4-dimethoxyphenylalaninol or tryptophanol in the above cyclocondensation reactions widely expands the potential and scope of the methodology (Scheme 1.66).

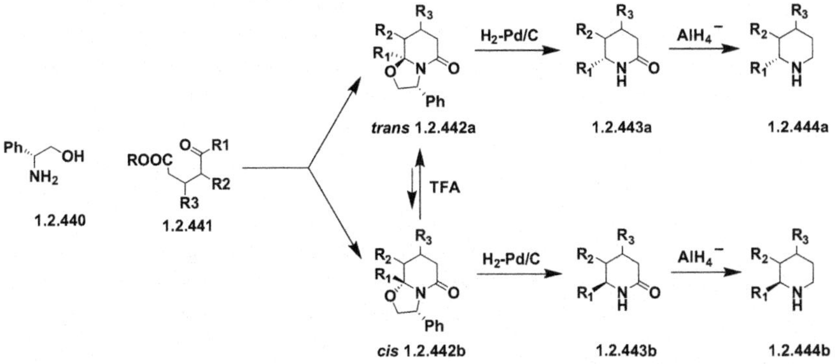

SCHEME 1.66 Synthesis of 2-oxopiperidines implementing cyclocondensation reaction of β-amino alcohols with δ-oxo acid derivatives.

The Pd-catalyzed intramolecular allylic alkylation of unsaturated amides results in 4-vinyl-substituted 2-oxopiperidine-3-carboxylates (**1.2.446**). A 3-Amino-3-oxopropanoate derivative (**1.2.445**) underwent intramolecular allylic alkylation in various conditions using [{Pd(η-allyl)Cl}$_2$] palladium catalyst in the presence of the achiral, bidentate ligand dppe (1,2-bis(diphenylphosphino) ethane), N,O-bis(trimethylsilyl)-acetamide (BSA), and potassium acetate in dichloromethane, or by using the same allylpalladium catalyst and bidentate ligand under biphasic conditions (dichloromethane/50% KOH aqueous solution, tetrabutylammonium bromide) [472] (Scheme 1.67).

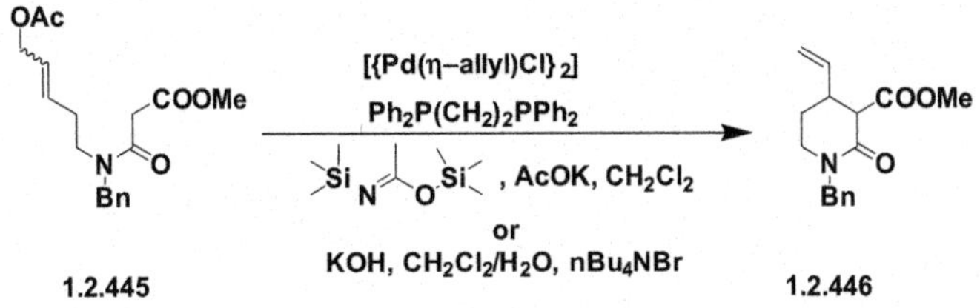

SCHEME 1.67 Synthesis of 2-oxopiperidines implementing intramolecular allylic alkylation reaction.

N-Alkenyl β-ketoamide (**1.2.447**) undergo cyclization to piperidin-2-ones (**1.2.448**) in the presence of Au[P(t-Bu)$_2$(O-biphenyl)]Cl and AgOTf in toluene under mild conditions with an excellent yield [473] (Scheme 1.68).

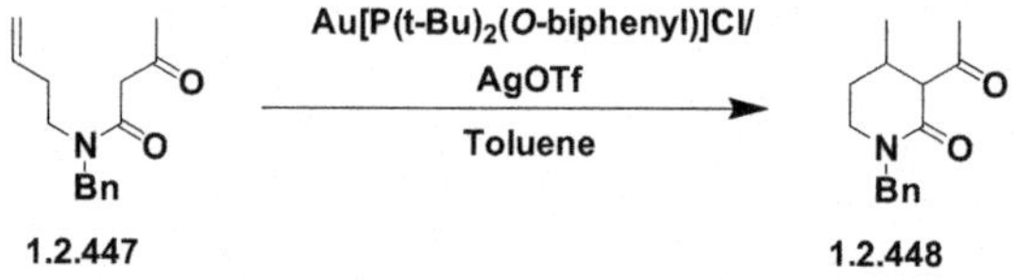

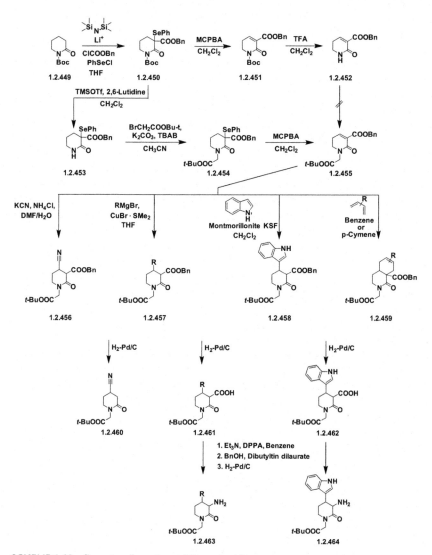

SCHEME 1.68 Synthesis of 2-oxopiperidines implementing intramolecular allylic alkylation reaction.

Piperidin-2-one (**1.2.449**) can be transformed to dihydropyridones such as (**1.2.451**) and (**1.2.454**), which, in turn, can serve as excellent starting materials for the synthesis of compounds with divers biologically activity. Examples of transformation of piperidin-2-ones are presented on Scheme 1.69 [474,475].

SCHEME 1.69 Some transformations of 2-oxopiperidines.

The synthesis of 3-benzyloxycarbonyl-5,6-dihydropyridin-2-one (**1.2.455**) was performed starting from the N-tertbutoxycarbonyl-2-piperidone (**1.2.450**). Insertion of the benzyloxycarbonyl and phenylselanyl groups at position 3 was performed in a one-pot procedure, using lithium bis(trimethyl-silyl)amide as a base and benzyl chloroformate and benzeneselenenyl chloride (PhSeCl) as electrophiles.

N-tert-butyloxycarbonyl group of piperidone (**1.2.450**) have been removed by treatment with trimethylsilyl triflate/2,6-lutidine in dichloro-methane and intermediate N-deprotected piperidone (**1.2.453**) and was N-alkylated with tertbutyl bromoacetate to give piperidone (**1.2.454**) in good yield. Finally, elimination of the phenylselanyl group with meta-chloroperoxybenzoic acid (m-CPBA) as oxidizing agent furnished dihydro-pyridone (**1.2.455**). When the phenylselanyl group elimination (**1.2.450**) → (**1.2.451**) was performed prior to the N-deprotection with TFA, which gave benzyl 2-oxo-1,2,5,6-tetrahydropyridine-3-carboxylate (**1.2.455**) and alkyl-ation steps, intermediate dihydropyridone (**1.2.451**) was obtained that was subsequently treated with TFA to yield N-deprotected dihydropyridone (**1.2. 452**), but attempts to alkylate it to a furnished compound (**1.2.455**) were unsuccessful.

To investigate the potential of 5,6- dihydro-2(1H)-pyridones (**1.2.455**) in conjugate addition reactions, nucleophiles, such as cyanide ion, Grignard reagents, indole and diens were studied to give corresponding compounds (**1.2.456**), (**1.2.457**), (**1.2.458**), and (**1.2.459**) [474]. The obtained compounds (**1.2.456**), (**1.2.457**), and (**1.2.458**) were hydrogenated and the resulting car-boxylic acids (**1.2.461**) and (**1.2.462**) underwent a modified Curtius rear-rangement by treatment with Et_3N and DPPA followed by BnOH and dibutyltin dilaurate to give compounds (**1.2.463**) and (**1.2.464**). It is neces-sary to mention that when 4-cyanopiperidone (**1.2.456**) on hydrogenation occurred in the same conditions, the result was 3-decarboxylated piperidone (**1.2.460**).

The Diels−Alder reaction of 3-benzyloxycarbonyl-5,6-dihydropyridin-2-one (**1.2.455**) → (**1.2.459**) was performed with a variety of diversely substi-tuted buta-1,3-dienes under thermal or catalytic conditions, giving partially reduced isoquinolones (1.2) [475] (Scheme 1.69).

Another interesting derivative of piperidin-2-ones − 5-hydroxypiperidin-2-one (**1.2.470**) − was synthesized stereospecifically from L-glutamic acid (**1.2.465**), which was converted to lactone (S)-5-oxotetrahydrofuran-2-car-boxylic acid (**1.2.466**). The last was reduced by borane dimethylsulfide to give (S)-5-(hydroxymethyl)dihydrofuran-2(3H)-one (**1.2.467**), which was O-mesylated to (**1.2.468**) and then treated with NaN_3 to give azide (**1.2.469**). The latter was hydrogenated over Pd/C to give desired (S) 5-hydroxypiperidine-2-one (**1.2.470**) [476,477] (Scheme 1.70).

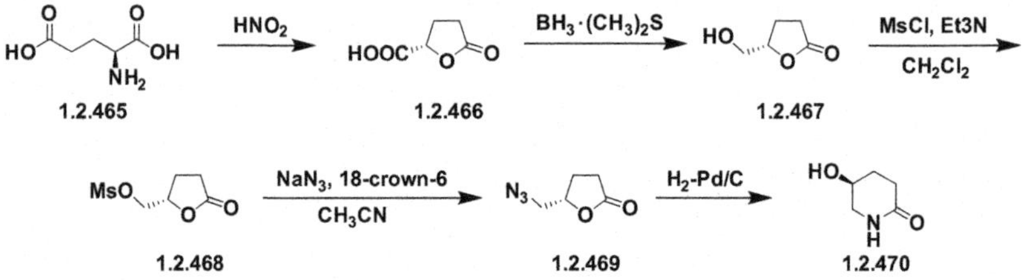

SCHEME 1.70   Synthesis of 5-hydroxypiperidin-2-one.

4-Cyano-4-Phenylpiperidines

Another series of compounds of a great medical importance are 4-cyano-4-phenylpiperidines (**1.2.473**).

One of the first general examples of a synthesis of these type of compounds was published in 1941 and consisted of arylacetonitriles such as (**1.2.471**) that was added to a suspension of sodium amide in benzene (toluene) and then treated with *N*-alkyl-bis(2-chloroethyl)amines (**1.2.472**). After several hours reflux good yield of desired products (**1.2.473**) were obtained [478].

Several modifications and improvements of this reaction were proposed [479−481].

Alternative methods for the synthesis of 4-cyano-4-phenylpiperidines (**1.2.473**) were proposed via condensation of arylacetonitriles of series of (**1.2.471**) with 2-vinyloxyethylchloride (**1.2.474**), which gave α,α-bis(2-vinyloxyethyl)arylacetonitriles (**1.2.476**), or with alkyl 2-chloroethyl formals (**1.2.475**), which gave α,α-bis(2-alkoxymethoxyethyl)arylacetonitriles (**1.2.477**). Hydrolysis of both of them − (**1.2.476**) and (**1.2.477**) to the corresponding dialcohols (**1.2.478**), their chlorination to (**1.2.479**), which on treatment with primary aqueous solution of amines in ethanol in sealed tube at 145°C, gave desired nitriles (**1.2.473**) [479] (Scheme 1.71).

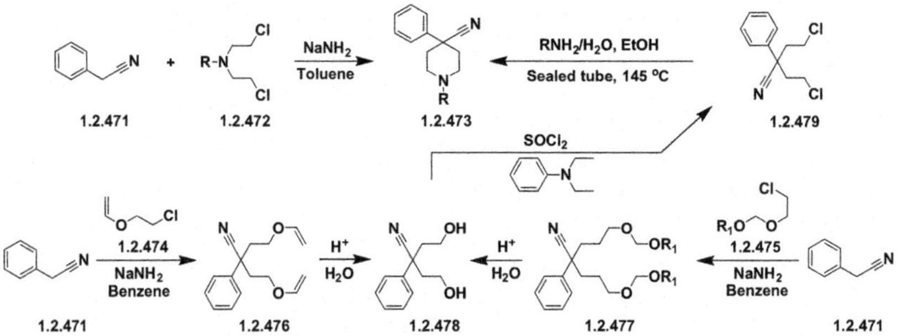

SCHEME 1.71   Synthesis of 4-cyano-4-phenylpiperidines.

Synthesized 4-cyano-4-phenylpiperidines (**1.2.473**) underwent different transformations, some of which are represented below. They easily have been hydrolyzed with mineral acids such as HCl [482,483], HBr [484], H_2SO_4 [485] to 4-phenyl-4-piperidine carboxylic acids (**1.2.480**). Esterification of obtained acids took place on treatment with thionyl chloride in corresponding alcohol. A reduction of prepared esters (**1.2.481**) gave compounds (**1.2.482**), which underwent further transformations to esters (**1.2.483**), etc. [482]. Hydrogenation of 4-cyano-4-phenylpiperidines (**1.2.473**) with different agents gave amines (**1.2.484**) that were also transformed to variety of derivatives such as (**1.2.485**) [482] as well as others mentioned in many other publications.

The same 4-cyano-4-phenylpiperidines (**1.2.473**) were converted to the ester (**1.2.488**) by reduction of the nitrile to an aldehyde (**1.2.486**) with diisobutylaluminium hydride followed by Horner−Emmons elongation using methyl 2-(dimethylphosphono)acetate to give the α,β − unsaturated esters (**1.2.487**). Hydrogenation of the unsaturated ester resulted in compound (**1.2.488**) [486] (Scheme 1.72). Numerous publications describe transformations of 4-cyano-4-phenylpiperidines (**1.2.473**) as building blocks for the synthesis of biologically active molecules with diverse implementations.

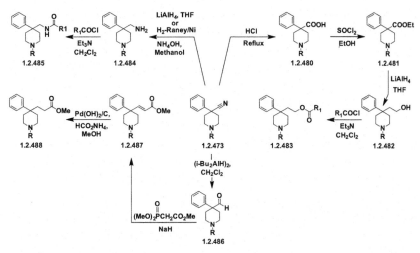

SCHEME 1.72 Transformations of 4-cyano-4-phenylpiperidines.

REFERENCES

[1] Vitaku E, Smith DT, Njardarson JT. Analysis of the structural diversity, substitution patterns, and frequency of nitrogen heterocycles among U.S. FDA approved pharmaceuticals. J Med Chem 2014;57(24):10257−74.

[2] Lipkus AH, Yuan Q, Lucas KA, Funk SA, Bartelt III WF, Schenck RJ, et al. Structural diversity of organic chemistry. A scaffold analysis of the CAS registry. J Org Chem 2008;73(12):4443−51.

[3]   Galemmo Jr RA, Janssens FE, Lewi PJ, Maryanoff BE. In memoriam: Dr. Paul A. J. Janssen (1926-2003). J Med Chem 2005;48(6):1686.

[4]   Black J. A personal perspective on Dr. Paul Janssen. J Med Chem 2005;6(48):1687−8.

[5]   Van Gestel S, Schuermans V. Thirty-three years of drug discovery and research with Dr. Paul Janssen. Drug Dev Res 1986;8(1−4):1−13.

[6]   Vardanyan RS, Hruby VJ. Synthesis of essential drugs. Amsterdam: Elsevier; 2006.

[7]   Vardanyan RS, Hruby VJ. Synthesis of best-seller drugs. Amsterdam: Elsevier; 2016.

[8]   Ullmann's encyclopedia of industrial chemistry. 5th ed. Executive editor, Wolfgang Gerhartz W; Senior editor, Yamamoto YS; Editors Campbell FT, Pfefferkorn R, Rounsaville J.F. Book A2, Wiley-VCH; 1985. p. 13−14.

[9]   Smeykal K, Moll KK. Catalytic hydrogenation of pyridines, GB 1062900; 1967.

[10] Baltz H, Smeykal K, Moll KK, Bruesehaber L. Piperidine production, GB 1109640; 1968.

[11] Ponomarev AA, Chegolya AS, Dyukareva VN. Hydrogenation of some nitrogen-containing heterocyclic compounds on ruthenium catalysts in liquid phase. Khim Geterotsikl Soed 1966;(2):239−42.

[12] Marvel CS, Lazier WA. Benzoyl piperidine. Org Synth Coll, Vol. 1. Organic Syntheses, Inc.; 1941. p. 99.

[13] Scriven EFV, Toomey Jr JE, Murugan R. Pyridine and pyridine derivatives. In: 4th ed. Kroschwitz JI, Howe-Grant M, editors. Kirk-Othmer encyclopedia of chemical technology, Vol. 20. New York: John Wiley; 1996. p. 641−79.

[14] Chichibabin AE, Oparina MP. Synthesis of pyridine from aldehydes and ammonia. J Prakt Chemie (Leipzig) 1924;107:154−8.

[15] Shimizu S, Watanabe N, Kataoka T, Shoji T, Abe N, Morishita S, et al. Pyridine and pyridine derivatives. In: Elvers B, Hawkins S, Russey W, Schulz G, editors. Ullmann's encyclopedia of chemical technology. 5th rev. ed. New York: VCH Publishers; 1993. p. 399−430. Book A22

[16] Veitch J. Pyridine production. GB 1135854; 1968.

[17] Frank RL, Seven RP, Pyridines IV. A study of the chichibabin synthesis. J Am Chem Soc 1949;71(8):2629−35.

[18] Manly DG, O'Halloran JP, Rice FJ, Jr. Piperidines from tetrahydrofur-furyl alcohol, US 3163652; 1964.

[19] Manly DG, O'Halloran JP, Rice FJ Jr. Production of pyridine from tetra-hydrofurfuryl alcohol, US 3238214; 1966.

[20] Chichery G, Perras P, Preparation of piperidine, FR 1475961; 1967.

[21] Gardner C, Silverstone GA. Piperidine and derivatives, GB 971187; 1964.

[22] Mensch F. Hydrodealkylation of pyridine bases at atmospheric pressure. Erdoel & Kohle, Erdgas, Petrochemie 1969;22(2):67−71.

[23] Watanabe Y, Takenaka S. Pyridine and α-picoline, JP 45039545; 1970.

[24] Swift G. Catalytic gas-phase production of pyridine and 3-methylpyridine from acrolein and ammonia, DE 1917037; 1969.

[25] Henry GD. De novo synthesis of substituted pyridines. Tetrahedron 2004;60:6043−61.

[26] Haider S, Saify ZS, Begum N, Ashraf S, Zarreen T, Saeed SMG. Emerging pharmaceutical applications of piperidine, pyrrolidine and its derivaties. World J Pharm Res 2014;3 (Suppl. 7):987−1024.

[27] Ajay KK, Pavithra G, Renuka N, Vasanth KG. Piperidone analogs: synthesis and their diverse biological applications. Int Res J Pharm Appl Sci 2012;2(6):145−54.

[28] Kaellstroem S, Leino R. Synthesis of pharmaceutically active compounds containing a disubstituted piperidine framework. Bioorg Med Chem 2008;16(2):601−35.

[29]   Watson PS, Jiang B, Scott B. A diastereoselective synthesis of 2,4-disubstituted piperidines: scaffolds for drug discovery. Org Lett 2000;2(23):3679−81.

[30]   Laschat S, Dickner T. Stereoselective synthesis of piperidines. Synthesis 2000;13:1781−813.

[31]   O'Hagan D. Pyrrole, pyrrolidine, pyridine, piperidine and tropane alkaloids (1998 to 1999). Nat Prod Rep 2000;17(5):435−46.

[32]   Weintraub PM, Sabol JS, Kane JM, Borcherding DR. Recent advances in the synthesis of piperidones and piperidines. Tetrahedron 2003;59(17):2953−89.

[33]   Felpin F-X, Lebreton J. Recent advances in the total synthesis of piperidine and pyrrolidine natural alkaloids with ring-closing metathesis as a key step. Eur J Org Chem 2003;19:3693−712.

[34]   Buffat MGP. Synthesis of piperidines. Tetrahedron 2004;60(8):1701−29.

[35]   Pearson MSM, Mathe-Allainmat M, Fargeas V, Lebreton J. Recent advances in the total synthesis of piperidine aza-sugars. Eur J Org Chem 2005;11:2159−91.

[36]   Tao H-Y, Wang C-J. A facile access to piperidine derivatives via copper(I)-catalyzed 1,3-Dipolar [6 + 3] cycloadditions of azomethine ylides with fulvenes. Synlett 2014;25(4):461−5.

[37]   Mesganaw T, Ellman JA. Convergent synthesis of diverse tetrahydropyridines via Rh(I)-Catalyzed C-H functionalization sequences. Org Proc Res Dev 2014;18(9):1097−104.

[38]   Girling PR, Kiyoi T, Whiting A. Mannich-Michael versus formal aza-Diels-Alder approaches to piperidine derivatives. Org Biomolecular Chem 2011;9(9):3105−21.

[39]   Kuznetsov VV. γ-Substituted piperidines as pharmaceuticals. Khim-Farm Zh 1991;25 (7):61−75.

[40]   De Risi C, Fanton G, Pollini GP, Trapella C, Valente F, Zanirato V. Recent advances in the stereoselective synthesis of *trans*-3,4-disubstituted-piperidines: applications to (-)-paroxetine. Tetrahedron: Asymmetry 2008;19(2):131−55.

[41]   Barco A, Benetti S, Casolari A, Pollini GP, Spalluto G. Tandem Michael reactions for the construction of pyrrolidine and piperidine ring systems. Tetrahedron Lett 1990;31 (21):3039−42.

[42]   Baumann M, Baxendale IR. An overview of the synthetic routes to the best selling drugs containing 6-membered heterocycles. Beilstein J Org Chem 2013;9:2265−319.

[43]   Bailey PD, Millwood PA, Smith PD. Asymmetric routes to substituted piperidines. Chem Commun 1998;(6):633−40.

[44]   Mitchinson A, Nadin A. Saturated nitrogen heterocycles. Perkin 2000;1(17):2862−92.

[45]   Cossy J. Selective methodologies for the synthesis of biologically active piperidinic compounds. Chem Record 2005;5(2):70−80.

[46]   Pearson MSM, Mathe-Allainmat M, Fargeas V, Lebreton J. Recent advances in the total synthesis of piperidine aza-sugars. Eur J Org Chem 2005;11:2159−91.

[47]   Rubiralta M, Giralt E, Diez A. Piperidine: structure, preparation, reactivity, and synthetic applications of piperidine and its derivatives. Series: studies in organic chemistry. Amsterdam: Elsevier Science; 1991.

[48]   Carroll FI, Dolle RE. The discovery and development of the N-substituted trans-3,4-dimethyl-4-(3'-hydroxyphenyl)piperidine class of pure opioid receptor antagonists. ChemMedChem 2014;9(8):1638−54.

[49]   Companyo X, Alba A-N, Rios R. Most relevant recent enantioselective synthesis of pyrrolidines and piperidines. Targets Heterocycl Syst 2009;13:147−74.

[50]   Troin Y, Sinibaldi M-Eve. Access to spirocyclic piperidines, important building blocks in medicinal chemistry. Targets Heterocycl Syst 2009;13:120−46.

[51]   Kharkar PS, Dutta AK, Reith MEA. Structure-activity relationship study of piperidines derivatives for dopamine transporters. In: Trudell ML, Izenwasser S., editors. Dopamine Transporters; 2008. p. 233−264.

[52]   Vargas-Mendez LY, Kouznetsov VV. 4-aminopiperidines and spiro-4-piperidines: pharmacological importance and synthetic efforts, 12. Universitas Scientiarum (Pontificia Universidad Javeriana, Facultad de Ciencias); 2007. p. 23−45 (2)

[53]   Kurbat NM, Praliev KD, Salita TA, Yu VK, Verina EL. Neuropharmacological activity of piperidine derivatives (a review). Khim-Farm Zh 1991;25(7):20−9.

[54]   Vartanyan RS. Synthesis and biological activity of uncondensed cyclic derivatives of piperidine. Rev Khim-Farm Zh 1984;18(11):1294−309.

[55]   Vartanyan RS. 4,N-Substituted piperidines. Rev Khim-Farm Zh 1983;17(5):540−50.

[56]   Vardanyan RS, Hruby, Victor J. Fentanyl-related compounds and derivatives: current status and future prospects for pharmaceutical applications. Future Med Chem 2014;6(4):385−412.

[57]   Mellor JM, Pathirana RN. Synthesis of bridged benzodiazepines by reaction of amines and hydrazine derivatives with 4,6-dibromomethyl-5,2,8-ethanylylidene-5H-1,9-benzodiazacycloundecine. J Chem Soc Perkin Transactions 1984;(4):753−9 1,(1972-1999)

[58]   Suginome H, Yamada S, Wang JB. Photoinduced molecular transformations. Part 107. A versatile substitution of a carbonyl group of steroidal ketones by a heteroatom. The synthesis of aza-, oxa-, thia-, selena-, and tellurasteroids. J Org Chem 1990;55(7):2170−6.

[59]   Ju Y, Varma RS. Aqueous N-heterocyclization of primary amines and hydrazines with dihalides: microwave-assisted syntheses of N-azacycloalkanes, isoindole, pyrazole, pyrazolidine, and phthalazine derivatives. J Org Chem 2006;71(1):135−41.

[60]   Hanessian S, Griffin AM, Cantin L-D. Asymmetric synthesis of functionalized carbocycles and heterocycles. Chirality 2000;12(5/6):342−5.

[61]   Mao H, Joly GJ, Peeters K, Hoornaert GJ, Compernolle F. Synthesis of 1-deoxymannojirimycin analogues using N-tosyl and N-nosyl activated aziridines derived from 1-amino-1-deoxyglucitol. Tetrahedron 2001;57(32):6955−67.

[62]   Xu F, Simmons B, Reamer RA, Corley E, Murry J, Tschaen D. Chlorination/cyclodehydration of amino alcohols with SOCl$_2$: an old reaction revisited. J Org Chem 2008;73 (1):312−15.

[63]   Kazmaier U, Grandel R. A short synthesis of polyhydroxylated piperidines by aldol reaction of chelated amino acid ester enolates. Eur J Org Chem 1998;(9):1833−40.

[64]   Schneider C, Kazmaier U. Synthesis of 5-epi-isofagomine via asymmetric chelate-enolate Claisen rearrangement. Eur J Org Chem 1998;(6):1155−9.

[65]   Najdi S, Kurth MJ. Synthesis of (2R,6R)-(-)-2,6-lupetidine: 2,6-disubstituted piperidines as potentially useful C2-symmetric chiral reagents. Tetrahedron Lett 1990;31 (23):3279−82.

[66]   Sato M, Gunji Y, Ikeno T, Yamada T. Efficient preparation of optically pure C2-symmetrical cyclic amines for chiral auxiliary. Synthesis 2004;(9):1434−8.

[67]   Monterde MI, Brieva R, Gotor V. Stereocontrolled chemoenzymatic synthesis of 2,3-disubstituted piperidines. Tetrahedron: Asymmetry 2001;12(3):525−8.

[68]   Reding MT, Buchwald SL. Short enantioselective total syntheses of the piperidine alkaloids (S)-Coniine and (2R,6R)-trans-Solenopsin A via catalytic asymmetric imine hydrosilylation. J Org Chem 1998;63(18):6344−7.

[69]   Takahata H, Kubota M, Ikota N. A new synthesis of all four stereoisomers of 2-(2,3-dihydroxypropyl)piperidine via iterative asymmetric dihydroxylation to cause enantiomeric enhancement. application to asymmetric synthesis of naturally occurring piperidine-related alkaloids. J Org Chem 1999;64(23):8594−601.

[70] Takahata H, Kubota M, Takahashi S, Momose T. A new asymmetric entry to 2-substituted piperidines. A concise synthesis of (+)-coniine, (-)-pelletierine, (+)-δ-coniceine, and (+)-epidihydropinidine. Tetrahedron: Asymmetry 1996;7(10):3047−54.

[71] Hirai Y, Terada T, Yamazaki T. Asymmetric intramolecular Michael reaction. Construction of chiral building blocks for the synthesis of several alkaloids. J Am Chem Soc 1988;110(3):958−60.

[72] Golubev A, Sewald N, Burger K. Hexafluoroacetone as activating and protecting reagent in amino acid and peptide chemistry. 19. An efficient approach to the family of 4-substituted pipecolic acids. Syntheses of 4-oxo-, cis-4-hydroxy-, and trans-4-hydroxy-L-pipecolic acids from L-aspartic acid. Tetrahedron Lett 1995;36(12):2037−340.

[73] Carretero JC, Arrayas RG, de Gracia IS. Stereoselective synthesis of hydroxypyrrolidines and hydroxypiperidines by cyclization of γ-oxygenated-α,β-unsaturated sulfones. Tetrahedron Lett 1996;37(19):3379−82.

[74] Deziel R, Malenfant. Asymmetric ring closure reactions mediated by a chiral C2 symmetrical organoselenium reagent. J Org Chem 1995;60(14):4660−2.

[75] Molander GA, Dowdy ED, Pack SKA. Diastereoselective intramolecular hydroamination approach to the syntheses of (+)-, (±)-, and (-)-Pinidinol. J Org Chem 2001;66(12):4344−7.

[76] Ryu J-S, Marks TJ, McDonald FE. Organolanthanide-catalyzed intramolecular hydroamination/cyclization/bicyclization of sterically encumbered substrates. Scope, selectivity, and catalyst thermal stability for amine-tethered unactivated 1,2-disubstituted alkenes. J Org Chem 2004;69(4):1038−52.

[77] Molander GA, Romero JAC. Lanthanocene catalysts in selective organic synthesis. Chem Rev 2002;102(6):2161−85.

[78] Hirai Y, Shibuya K, Fukuda Y, Yokoyama H, Yamaguchi S. 1,4-Asymmetric induction in palladium(II)-catalyzed intramolecular N-alkylation reaction. Construction of 2-functionalized 5-hydroxypiperidine. Chem Lett 1997;(3):221−2.

[79] Boger DL, Weinreb SM, editors. Hetero-Diels−Alder methodology in organic synthesis. Orlando: Academic Press; 1987.

[80] Zunnebeld WA, Speckamp WN. Total synthesis of 13- and 14-azaequilenines by hetero-cycloaddition. Tetrahedron 1975;31(15):1717−21.

[81] Holmes AB, Thompson J, Baxter AJG, Dixon J. Total synthesis of (±)-isoprosopinine B and (±)-desoxoprosopinine. J Chem Soc Chem Commun 1985;1985(1):37−9.

[82] Birkinshaw TN, Tabor AB, Holmes AB, Raithby PR. Imino Diels-Alder reaction of 2-(tert-butyldimethylsilyloxy)cyclohexadiene: isolation of an azabicyclo[2.2.2]octene silyl enol ether adduct. J Chem Soc Chem Commun 1988;(24):1601−2.

[83] Hamley P, Helmchen G, Holmes AB, Marshall DR, MacKinnon JWM, Smith DF, et al. Diastereoselective imino ester cycloadditions. Enantioselective synthesis of azabicyclo [2.2.1]heptenes. J Chem Soc Chem Commun 1992;(10):786−8.

[84] Badorrey R, Cativiela C, Diaz-De-Villegas, Maria D, Galvez, Jose A. Study of the reaction of imines derived from (R)-glyceraldehyde with Danishefsky's diene. Tetrahedron 1999;55(24):7601−12.

[85] Bailey PD, Londesbrough DJ, Hancox TC, Heffernan JD, Holmes AB. Highly enantioselective synthesis of pipecolic acid derivatives via an asymmetric aza-Diels-Alder reaction. J Chem Soc Chem Commun 1994;22:2543−4.

[86] Heintzelman GR, Weinreb SM, Parvez M. Imino Diels-Alder-based construction of a piperidine A-ring unit for total synthesis of the marine hepatotoxin cylindrospermopsin. J Org Chem 1996;61(14):4594−9.

[87]   Kerwin Jr JF, Danishefsky S. On the Lewis acid catalyzed cyclocondensation of imines with a siloxydiene. Tetrahedron Lett 1982;23(37):3739–42.

[88]   Danishefsky S, Langer ME, Vogel C. On the use of the imine-diene cyclocondensation reaction in the synthesis of yohimbine congeners. Tetrahedron Lett 1985;26 (48):5983–6.

[89]   Danishefsky SJ, Vogel C. A concise total synthesis of (±)-ipalbidine by application of the aldimine-diene cyclocondensation reaction. J Org Chem 1986;51(20):3915–16.

[90]   Hattori K, Yamamoto H. Asymmetric aza-Diels-Alder reaction mediated by chiral boron reagent. J Org Chem 1992;57(12):3264–5.

[91]   Hattori K, Yamamoto H. Asymmetric aza-Diels-Alder reaction catalyzed by boron reagent: effect of biphenol and binaphthol ligand. Synlett 1993;2:129–30.

[92]   Shimizu M, Arai A, Fujisawa T. Stereospecific synthesis of piperidine skeleton by [4 + 2] cycloaddition, leading to the synthesis of piperidines of biological interests. Heterocycles 2000;52(1):137–40.

[93]   Wakabayashi R, Kurahashi T, Matsubara S. Cobalt(III) porphyrin catalyzed Aza-Diels-Alder reaction. Org Lett 2012;14(18):4794–7.

[94]   Sankar MG, Mantilli L, Bull J, Giordanetto F, Bauer JO, Strohmann C, et al. Stereoselective synthesis of a natural product inspired tetrahydroindolo[2,3-a]-quinolizine compound library. Bioorg Med Chem 2015;23(11):2614–20.

[95]   Sax M, Ebert K, Schepmann D, Wibbeling B, Wunsch B. Synthesis and NMDA-receptor affinity of 4-oxo-dexoxadrol derivatives. Bioorg Med Chem 2006;14(17):5955–62.

[96]   Barluenga J, Aznar F, Valdes C, Martin A, Garcia-Granda S, Martin E. 2-Amino-1,3-butadienes as chiral building blocks: enantioselective synthesis of 4-piperidones, 4-nitrocyclohexanones, and 1,3-cycloheptadione derivatives. J Am Chem Soc 1993;115 (10):4403–4.

[97]   Barluenga J, Aznar F, Valdes C, Ribas C. Enantioselective synthesis of substituted pipecolic acid derivatives. J Org Chem 1998;63(12):3918–24.

[98]   Barluenga J, Aznar F, Ribas C, Valdes C, Novel A. Approach to the Enantioselective synthesis of nuphar alkaloids: first total synthesis of (-)-(5S,8R,9S)-5-(3-Furyl)-8-methy-loctahydroindolizidine and total synthesis of (-)-nupharamine. J Org Chem 1999;64 (10):3736–40.

[99]   Barluenga J, Mateos C, Aznar F, Valdes C. A concise and convergent route to 5,8-disub-stituted indolizidine and 1,4-disubstituted quinolizidine ring cores by diastereoselective Aza-Diels-Alder reaction. Org Lett 2002;4(11):1971–4.

[100]   Badorrey R, Cativiela C, Diaz-De-Villegas MD, Galvez JA. Asymmetric hetero Diels-Alder reaction of N-benzylimines derived from R-glyceraldehyde: a new approach to homochiral piperidine building blocks and its application to the synthesis of (2R)-4-oxo-pipecolic acid. Tetrahedron Lett 1997;38(14):2547–50.

[101]   Waldmann H, Braun M. Asymmetric tandem Mannich-Michael reactions of amino acid ester imines with Danishefsky's diene. J Org Chem 1992;57(16):4444–51.

[102]   Ratni H, Kuendig EP. Synthesis of (-)-Lasubine(I) via a Planar Chiral [(η6-arene)Cr(CO)3] Complex From. Organic Lett 1999;1(12):1997–9.

[103]   Ratni H, Crousse B, Kundig EP. From planar chiral 2-chloro- and 2-iodobenzaldehyde tricarbonyl chromium complexes to enantiopure fused hydroisoquinolines and hydroqui-nolines. Synlett 1999;1999(5):626–8.

[104]   Bailey PD, Wilson RD, Brown GR. Enantio- and diastereoselective synthesis of pipeco-lic acid derivatives using the aza-Diels-Alder reaction of imines with diens. J Chem Soc Perkin Trans 1991;1(5):1337–40.

[105] Bailey PD, Brown GR, Korber F, Reed A, Wilson RD. Asymmetric synthesis of pipecolic acid derivatives using the aza-Diels-Alder reaction. Tetrahedron: Asymmetry 1991;2 (12):1263–82.

[106] Weymann M, Pfrengle W, Schollmeyer D, Kunz H. Enantioselective syntheses of 2-alkyl-, 2,6-dialkylpiperidines, and indolizidine alkaloids through diastereoselective Mannich-Michael Reactions. Synthesis 1997;1997(10):1151–60.

[107] Stella L, Abraham H, Feneau-Dupont J, Tinant B, Declercq JP. Asymmetric aza-Diels-Alder reaction using the chiral 1-phenylethylimine of methyl glyoxylate. Tetrahedron Lett 1990;31(18):2603–6.

[108] Abraham H, Stella L. Diastereoselective aza-Diels-Alder reaction: use of alkyl glyoxylate (1-phenylethyl)imine in the synthesis of cyclic α-amino acid derivatives. Tetrahedron 1992;48(44):9707–18.

[109] Teng M, Fowler FW. The N-acyl-α-cyano-1-azadienes. Remarkably reactive heterodienes in the Diels-Alder reaction. J Org Chem 1990;55(21):5646–53.

[110] Cheng Y-S, Fowler FW, Lupo Jr. AT. Diels-Alder reaction of 1-azadienes. J Am Chem Soc 1981;103(8):2090–1.

[111] Sisti NJ, Fowler FW, Grierson DS. N-Phenyl-2-cyano-1-azadienes: new versatile heterodienes in the Diels-Alder reaction. Synlett 1991;1991(11):816–18.

[112] Comins DL, Joseph SP, Goehring RR. Asymmetric synthesis of 2-alkyl(aryl)-2,3-dihydro-4-pyridones by addition of grignard reagents to chiral 1-acyl-4-methoxypyridinium salts. J Am Chem Soc 1994;116(11):4719–28.

[113] Trione C, Toledo LM, Kuduk SD, Fowler FW, Grierson DS. Diels-Alder reaction of 2-cyano-1-azadienes. The effect of nitrogen substituents. J Org Chem 1993;58 (8):2075–80.

[114] Serckx-Poncin B, Hesbain-Frisque AM, Ghosez L. 1-Aza-1,3-diones. Diels-Alder reactions with α,β-unsaturated hydrazones. Tetrahedron Lett 1982;23(32):3261–4.

[115] Beaudegnies R, Ghosez L. Asymmetric Diels-Alder reactions with chiral 1-azadienes. Tetrahedron: Asymmetry 1994;5(4):557–60.

[116] Boger DL, Corbett WL, Wiggins JM. Room-temperature, endo-specific 1-aza-1,3-butadiene Diels-Alder reactions: acceleration of the LUMOdiene-controlled [4 + 2] cycloaddition reactions through noncomplementary aza diene substitution. J Org Chem 1990;55 (10):2999–3000.

[117] Tietze LF, Schuffenhauer A. Synthesis of tetrahydro- and dihydropyridines by hetero Diels-Alder reactions of enantiopure α,β-unsaturated sulfinimines. Eur J Org Chem 1998;(8):1629–37.

[118] Sainte F, Serckx-Poncin B, Hesbain-Frisque AM, Ghosez LA. Diels-Alder route to pyridone and piperidone derivatives. J Am Chem Soc 1982;104(5):1428–30.

[119] Terada M. Ene Reactions. In: De Vries JG, Molander GA, Evans PA, editors. Science of synthesis, stereoselective synthesis, vol. 3. Thieme; 2011. p. 309–46.

[120] Brummond KM, Loyer-Drew JA. In: Mingos DMP, Crabtree RH, editors. C-C bond formation reactions: alder-ene (Part 1) by addition reaction in comprehensive organometallic chemistry III, 10. Amsterdam: Elsevier; 2007. p. 557–601.

[121] Mikami K, Aikawa K. Asymmetric ene reactions and cycloadditions. In: Ojima I, editor. Catalytic asymmetric synthesis. 3rd ed. New Jersey: Wiley; 2010. p. 683–737.

[122] Mueller TJJ. In: 2nd ed. Knochel P, Molander GA, editors. Ene reactions with carbon enophiles – metallo-ene reactions in Comprehensive Organic Synthesis, vol. 5. Amsterdam: Elsevier; 2014. p. 1–65.

[123] Mikami K, Shimizu M. Asymmetric ene reactions in organic synthesis. Chem Rev 1992;92(5):1021−50.

[124] Snider BB. Lewis-acid catalyzed ene reactions. Accounts Chem Res 1980;13 (11):426−32.

[125] Oppolzer W. Angew Chem Int Ed 1984;23(11):876−89.

[126] Snider BB. Ene reactions with alkenes as enophiles. In: Trost BM, Fleming I, editors. Comprehensive organic synthesis: selectivity, strategy and efficiency in modern organic chemistry, vol. 5. Amsterdam: Elsevier; 1991.

[127] Yamaura Y, Hyakutake M, Mori M. Synthesis of heterocycles using zirconium-catalyzed asymmetric diene cyclization. J Am Chem Soc 1997;119(32):7615−16.

[128] Takacs JM, Zhu J, Chandramouli S. Catalytic palladium-mediated tetraene carbocyclizations: the cycloisomerizations of acyclic tetraenes to cyclized trienes. J Am Chem Soc 1992;114(2):773−4.

[129] Takacs BE, Takacs JM. Catalytic iron-mediated ene carbocyclizations of trienes: the stereoselective preparation of N-acylpiperidines. Tetrahedron Lett 1990;31(20):2865−8.

[130] Borzilleri RM, Weinreb SM. Imino ene reactions in organic synthesis. Synthesis 1995; (4):347−60.

[131] Wolfling J, Frank E, Schneider G, Tietze LF. Synthesis of unusual bridged steroid alkaloids by an iminium ion induced 1,5-shift of a benzylic hydride. Angew Chem, Int Ed 1999;38(1/2):200−1.

[132] Ruggeri RB, Hansen MM, Heathcock CH. Total synthesis of (±)-methyl homosecodaphniphyl-late. A remarkable new tetracyclization reaction. J Am Chem Soc 1988;110(26):8734−6.

[133] Laschat S, Grehl M. Diastereoselective synthesis of amino-substituted indolizidines and quinolizidines from prolinalimine or 2-piperidinecarbaldimine by the intramolecular hetero ene reaction Angew. Chem., (1994), 106(4), 475-8 Angew Chem, Int Ed Engl 1994;33(4):458−61.

[134] Monsees A, Laschat S, Hotfilder M, Wolff J, Bergander K, Terfloth L, et al. Synthesis and in vitro cytotoxic activity of novel hexahydro-2H-pyrido[1,2-b]isoquinolines against human brain tumor cell lines. Bioorg Med Chem Lett 1997;7(23):2945−50.

[135] Laschat S, Froehlich R, Wibbeling B. Preparation of 2,3,4-trisubstituted piperidines by a formal hetero-ene reaction of amino acid derivatives. J Org Chem 1996;61 (8):2829−38.

[136] Laschat S, Fox T. Enantioselective approach towards potential substance P antagonists via hetero-ene reaction of phenylglycine derivatives. Synthesis 1997;1997(4):475−9.

[137] Rutjes FPJT, Veerman JJN, Meester WJN, Hiemstra H, Schoemaker HE. Synthesis of enantiopure, functionalized pipecolic acids via amino acid-derived N-acyliminium ions. Eur J Org Chem 1999;1999(5):1127−35.

[138] Teng TF, Lin JH, Yang TK. Application of titanium tetrachloride-induced iminium ion cyclizations to the preparations of piperidine alkaloids: total syntheses of (±)-coniine. Heterocycles 1990;31(7):1201−4.

[139] Murata Y, Overman LE. Aprotic chloride and bromide-promoted alkyne-iminium ion cyclizations. Heterocycles 1996;42(2):549−53.

[140] Hoffmann RW, Hense A. Stereoselective synthesis of alcohols. Part XLIX. Tetrahydro-4-pyranols and 4-piperidinols by intramolecular allylboration reaction. Lieb Ann 1996; (8):1283−8.

[141] Mori M, Kubo Y, Ban Y. Palladium catalyzed ene-halocyclization of α-haloester having internal double bond with the low-valent metal complex. Tetrahedron 1988;44(14), 4321−4230.

[142]   Grubbs RH, Miller SJ, Fu GC. Ring-closing metathesis and related processes in organic synthesis. Accounts Chem Res 1995;28(11):446−52.

[143]   Schmalz H-G. Catalytic ring-closing metathesis: a new, powerful technique for carbon-carbon coupling in organic synthesis. Angew Chem, Int Ed Eng 1995;34(17):1833−6.

[144]   Rutjes FPJT, Schoemaker HE. Ruthenium-catalyzed ring closing olefin metathesis of non-natural α-amino acids. Tetrahedron Lett 1997;38(4):677−80.

[145]   Stragies R, Blechert S. Total synthesis of (−)-halosaline by a ruthenium-catalyzed ring rearrangement. Tetrahedron 1999;55(27):8179−88.

[146]   Heintz W. Ammonia derivatives of acetone. Chemisches Zentralblatt 1874;372.

[147]   Francis F. Preparation of triacetoneamine hydrate. J Chem Soc 1927;2897−8.

[148]   Matter E. A new reaction product from acetone and ammonia (acetonine). Helv Chim Acta 1947;30:1114−23.

[149]   Bradbury RB, Hancox NC, Hatt HH. Reaction between acetone and ammonia: the formation of pyrimidine compounds analogous to the aldoxanes of Sp.ovrddot.ath. J Chem Soc 1947;1394−9.

[150]   Hall Jr. HK. Steric effects on the base strengths of cyclic amines. J Am Chem 1957;79:5444−7.

[151]   Sosnovsky G, Konieczny M. Preparation of triacetoneamine (4-oxo-2,2,6,6-tetramethyl-piperidine), an improved method. Synthesis 1976;11:735−6.

[152]   Wu A, Yang W, Pan X. Preparation of triacetoneamine, an improved method. Synth Comm 1996;26(19):3565−9.

[153]   Fisher HL. Rubber vulcanization accelerators, US 1473285; 1923.

[154]   No Inventor data available, (Assignee: Sankyo Co., Ltd.), 4-Oxopiperidines, FR 1535012, (1968).

[155]   Orban I., Rody J. Triacetonamine, DE 2352127; 1974.

[156]   Orban I, Lind H, Brunetti H, Rody J. 2,2,6,6-Tetramethyl-4-piperidone, DE 2429936; 1975.

[157]   Murayama K, Morimura S, Yoshioka T, Kurumada T. Triacetoneamine, DE 2429745; 1975.

[158]   Haruna T, Nishimura A, Sugibuchi K. 2,2,6,6-Tetramethyl-4-oxopiperidine by reacting acetone with ammonia, EP 152934; 1985.

[159]   Malz RE, Son Y-C, Suib SL. Process for the synthesis of 2,2,6,6-tetramethyl-4-oxopiperidine from acetone and ammonia-donor compounds in the presence of CaY zeolite catalysts, US 20020128482; 2002.

[160]   Rozantsev EG, Ivanov VP. New aspects of the chemistry of triacetonamine and its synthesis. Khimiko-Farmatsevticheskii Zhurnal 1971;5(1):47−51.

[161]   Wilcox CS, Pearlman A. Chemistry and antihypertensive effects of tempol and other nitroxides. Pharmacol Rev 2008;60(4):418−69.

[162]   Dagonneau M, Kagan ES, Mikhailov VI, Rozantsev EG, Sholle VD. Chemistry of hindered amines from the piperidine series. Synthesis 1984;11:895−916.

[163]   Petrenko-Kritschenko P. Über die Kondensation des Acetondicarbonsäureesters mit Aldehyden, Ammoniak und Aminen. J für Praktische Chemie 1912;85(1):1−37.

[164]   Petrenko-Kritschenko P. Piperidone Synthesis. In: Jie-Jack L, Corey Ej, editors. Name reactions in heterocyclic chemistry. PART 3, Six-Membered Heterocycles Chapter 8. Pyridines, 8.1.1.4.5. John Wiley & Sons; 2005. p. 313.

[165]   Mannich C. "Open" ecgonine and tropine. Archiv der Pharmazie und Berichte der Deutschen Pharmazeutischen Gesellschaft 1934;272:323−59.

[166]   Mannich C, Schumann P. 3,5-Alkylated 4-oxopiperidines. Berichte der Deutschen Chemischen Gesellschaft [Abteilung] B: Abhandlungen 1936;69B:2299−305.

[167] Noller CR, Baliah V. Preparation of some piperidine derivatives by the Mannich reaction. J Am Chem Soc 1948;70:3853—5.

[168] Prostakov NS, Fedorov VO, Soldatenkov AT. 2,4-Diphenyl-3-azafluorene. Khimiya Geterotsiklicheskikh Soedinenii 1979;(8):1098—100.

[169] Thiel M, Deissner I. The common effect of elementary sulfur and gaseous ammonia on ketones. XXI. Synthesis and behavior of some 2,6-diphenyl-4-piperidones. Justus Liebigs Annalen der Chemie 1959;622:98—106.

[170] Baliah V, Jeyaraman R, Chandrasekaran L. Synthesis of 2,6-disubstituted piperidines, oxanes, and thianes. Chem Rev 1983;83(4):379—423.

[171] Ruzicka L, Fornasir V. Synthesis of γ-piperidone. Helv Chim Acta 1920;3:806—18.

[172] Ruzicka L, Seidel CF. γ-Piperidone ring. II. Helv Chim Acta 1922;5:715—20.

[173] McElvain SM. Piperidine derivatives. A cyclic and an open-chain compound related in structure to cocaine. J Am Chem Soc 1924;46:1721—7.

[174] McElvain SM. Piperidine derivatives. II. 1-Alkyl-3-carbethoxy-4-piperidyl benzoates. J Am Chem Soc 1926;48:2179—85.

[175] Bolyard NW, McElvain SM. Piperidine derivatives. VII. 1-Alkyl-4-piperidyl benzoates and p-aminobenzoates. J Am Chem Soc 1929;51:922—8.

[176] McElvain SM, McMahon RE. Piperidine derivatives. XXI. 4-Piperidone, 4-piperidinol, and certain of their derivatives. J Am Chem Soc 1949;71:901—6.

[177] Kuettel GM, McElvain SM. Piperidine derivatives. XI. 3-Carbethoxy-4-piperidone and 4-piperidone hydrochloride. J Am Chem Soc 1931;53:2692—6.

[178] Schaefer JP, Bloomfield JJ. The dieckmann condensation (including the Thorpe-Ziegler condensation). Organic reactions, vol. 15. New York: John Wiley; 1967. p. 1—203.

[179] Baty JD, Jones G, Moore C. Synthesis of some N-substituted 4-piperidones. J Chem Soc, [Section] C], Organic 1967;(24):2645—7.

[180] Fakhraian H, Riseh MBP. Improved procedure for the preparation of 1-(2-phenethyl)-4-piperidone. Org Prep Proc Int 2008;40(3):307—10.

[181] Ma D, Sun H. General route to 2,4,5-trisubstituted piperidines from enantiopure β-amino esters. Total synthesis of pseudodistomin B triacetate and pseudodistomin F. J Org Chem 2000;65(19):6009—16.

[182] Krapcho AP. Synthetic applications of dealkoxycarbonylations of malonate esters, β-keto esters, α-cyano esters and related compounds in dipolar aprotic media - part I. Synthesis 1982;10:805—22.

[183] Kodimuthali A, Prasunamba PL, Pal M. Synthesis of a novel analogue of DPP-4 inhibitor Alogliptin: introduction of a spirocyclic moiety on the piperidine ring. Beilstein J Org Chem 2010;6(71). Available from: http://dx.doi.org/10.3762/bjoc.6.71.

[184] Wang B. Synthesis of 3-alkylidene-piperidin-4-ones via one-pot cascade transylidation-olefination. Tetrahedron Lett 2009;50(21):2487—9.

[185] Nazarov IN, Rudenko VA. Acetylene derivatives. LXXXIV. Synthesis and investigation of heterocyclic compounds. 5. Action of ammonia and methylamine on vinyl allyl ketones. New synthesis of γ-piperidones, Izvestiya Akademii Nauk SSSR, Seriya Khimicheskaya 1948;610—30.

[186] Makin SM, Nazarova ON, Kundryutskova LA. A novel mode of synthesizing 1,2,5-trimethyl-4-piperidone. Khimiko-Farmatsevticheskii Zhurnal 1989;23(12):1493—5.

[187] Cui L, Peng Y, Zhang L. A two-step, formal [4 + 2] approach toward piperidin-4-ones via Au catalysis. J Am Chem Soc 2009;131(24):8394—5.

[188] Bull JA, Mousseau JJ, Pelletier G, Charette AB. Synthesis of pyridine and dihydropyridine derivatives by regio- and stereoselective addition to N-activated pyridines. Chem Rev 2012;112(5):2642—713.

[189]   Peng Z, Wong JW, Hansen EC, Puchlopek-Dermenci ALA, Clarke HJ. Development of a concise, asymmetric synthesis of a smoothened receptor (SMO) inhibitor: enzymatic transamination of a 4-piperidinone with dynamic kinetic resolution. Org Lett 2014;16 (3):860–3.

[190]   Machetti F, Cordero FM, De Sarlo F, Brandi A. Practical synthesis of both enantiomers of protected 4-oxopipecolic acid. Tetrahedron 2001;57(23):4995–8.

[191]   Davis FA, Chao B, Rao A. Intramolecular Mannich reaction in the asymmetric synthesis of polysubstituted piperidines: concise synthesis of the dendrobate alkaloid (+)-241D and its C-4 epimer. Org Lett 2001;3(20):3169–71.

[192]   Carbonnel S, Troin Y. Stereoselective synthesis of C-6 substituted pipecolic acid derivatives. Formal synthesis of (+)-indolizidine 167B and (+)-indolizidine 209D. Heterocycles 2002;57(10):1807–30.

[193]   Ciblat S, Besse P, Canet J-L, Troin Y, Veschambre H, Gelas J. A practical asymmetric synthesis of 2,6-cis-disubstituted piperidines. Tetrahedron: Asymmetry 1999;10 (11):2225–35.

[194]   Rougnon GS, Canet J-L, Troin YA. rapid stereoselective access to highly substituted piperidines. Tetrahedron Lett 2000;41(50):9797–802.

[195]   Prostakov NS, Gaivoronskaya LA. γ-Piperidinones in organic synthesis. Uspekhi Khimii 1978;47(5):859–99.

[196]   Sahu SK, Dubey BK, Tripathi AC, Koshy M, Saraf SK. Piperidin-4-one: the potential pharmacophore. Mini-Rev Med Chem 2013;13(4):565–83.

[197]   Kuznetsov VV, Prostakov NS. γ-Piperidinone imines and enamines in organic synthesis. Khimiya Geterotsiklicheskikh Soedinenii 1994;1:3–17.

[198]   McElvain SM. Rorig, Kurt Piperidine derivatives. XIX. Esters of substituted 4-piperidinols. J Am Chem Soc 1948;70:1826–8.

[199]   Bowden K, Green PN. Syntheses in the piperidine series. I. A facile synthesis of 4-piperidinol and the preparation of related compounds. J Chem Soc 1952;1164–7.

[200]   Levy J, Bernotsky GA. Hydrogenation of piperidone compounds in hydrocarbon solvents, US 2776293; 1957.

[201]   Coan SB, Jaffe B, Papa D. Parasympathetic blocking agents. III. Phenylglycolic acid esters of N-alkyl-4-piperidinol. J Am Chem Soc 1956;78:3701–3.

[202]   Zhang G, Scott BL, Hanson SK. Mild and homogeneous cobalt-catalyzed hydrogenation of C:C, C:O, and C:N bonds. Angew Chem, Int Ed 2012;51(48):12102–6.

[203]   Fankhauser R, Grob CA, Krasnobajew V. Synthesis of 4-chloropiperidines. Helv Chim Acta 1966;49(1):690–5.

[204]   Morishima S, Takada M, Okada K, Takano Y. N-methyl-4-piperidinol, JP 38008326; 1963.

[205]   Toomey RF, Riegel ER. The synthesis of 1-methyl-4-piperidyl tropylate. J Org Chem 1952;17(11):1492–3.

[206]   Waters JA. Aromatic esters of nonquaternary carbon-4 piperidinols as analgesics. J Med Chem 1978;21(7):628–33.

[207]   Mikhlina EE, Vorob'eva VYa, Rubtsov MV. Synthesis of 3- and 4-hydroxypiperi-dine derivatives. Zhurnal Obshchei Khimii 1960;30:1885–93.

[208]   Dvorak CA, Apodaca R, Barbier AJ, Berridge CW, Wilson SJ, Boggs JD, et al. 4-Phenoxypiperidines: potent, conformationally restricted, non-imidazole histamine H3 antagonists. J Med Chem 2005;48(6):2229–38.

[209]   Nettekoven M, Plancher J-M, Richter H, Roche O, Runtz-Schmitt V, Taylor S. Preparation of indole-2-carboxamides as histamine H3 receptor modulators for the treatment of diabetes, US 20070123515; 2007.

[210] Kawamorita S, Hamasaka G, Ohmiya H, Hara K, Fukuoka A, Sawamura M. Hydrogenation of hindered ketones catalyzed by a silica-supported compact phosphine-Rh system. Org Lett 2008;10(20):4697–700.

[211] Van Luppen J, Lepoivre J, Lemiere G, Alderweireldt F. Enzymic "in vitro" reductions of ketones. Part 12. Reduction of 1-alkyl-4-piperidones in an ethanol-NAD + -HLAD-system. Heterocycles 1984;22(4):749–61.

[212] Harper NJ, Chignell CF. The chemistry and pharmacology of some 4-aminopiperidines and their derivatives. J Med Chem 1964;7(6):729–32.

[213] Diez A, Voldoire A, Lopez I, Rubiralta M, Segarra V, Pages L, et al. Synthetic applications of 2-aryl-4-piperidones. X. Synthesis of 3-aminopiperidines, potential substances P antagonists. Tetrahedron 1995;51(17):5143–56.

[214] Luo Z, Sheng J, Sun Y, Lu C, Yan J, Liu A, et al. Synthesis and evaluation of multi-target-directed ligands against Alzheimer's disease based on the fusion of donepezil and ebselen. J Med Chem 2013;56(22):9089–99.

[215] Altomare C, Summo L, Cellamare S, Varlamov AV, Voskressensky LG, Borisova TN, et al. Pyrrolo[3,2-c]pyridine derivatives as inhibitors of platelet aggregation. Bioorg Med Chem Lett 2000;10(6):581–4.

[216] Moormann AE, Metz S, Toth MV, Moore WM, Jerome G, Kornmeier C, et al. Selective heterocyclic amidine inhibitors of human inducible nitric oxide synthase. Bioorg Med Chem Lett 2001;11(19):2651–3.

[217] Han S-Y, Choi JW, Yang J, Chae CH, Lee J, Jung H, et al. Design and synthesis of 3-(4,5,6,7-tetrahydro-3H-imidazo[4,5-c]pyridin-2-yl)-1H-quinolin-2-ones as VEGFR-2 kinase inhibitors. Bioorg Med Chem Lett 2012;22(8):2837–42.

[218] Praliev KD, Yu VK, Akhmetova GS. Synthesis of some oximes of the piperidine series, Izv. Ministerstva Obrazovaniya i Nauki Respubliki Kazakhstan. Ser Khim 2000;(1):96–101.

[219] Trofimov BA, Mikhaleva AI. Further development of the ketoxime-based pyrrole synthesis. Heterocycles 1994;37(2):1193–232.

[220] Dickerman SC, Lindwall HG. Piperidone chemistry. I. Synthesis of 5-homopiperazinones. J Org Chem 1949;14:530–6.

[221] Sen AB, Shanker K. Search for potential filaricides. II. Synthesis of substituted 1-alkyl homopiperazines. J Prakt Chem 1965;29(3-6):312–14.

[222] Ebnother A, Jucker E, Lindenmann A, Rissi E, Steiner R, Suess R, et al. New basic substituted hydrazine derivatives. Helv Chim Acta 1959;42:533–63.

[223] Jucker E. New basic substituted hydrazines and their application in the synthesis of pharmaceuticals. Angew Chem 1959;71:321–33.

[224] Adachi M, Sasakura K, Sugasawa T. Aminohaloborane in organic synthesis. IX. Exclusive ortho acylation reaction of N-monoaminoalkylanilines. Chem Pharm Bull 1985;33(5):1826–35.

[225] Vardanyan R, Vijay G, Nichol GS, Liu L, Kumarasinghe I, Davis P, et al. Synthesis and investigations of double-pharmacophore ligands for treatment of chronic and neuropathic pain. Bioorg Med Chem 2009;17(14):5044–53.

[226] Doerwald FZ. Preparation of substituted piperidines with selective binding to histamine h3-receptor for treatment and/or prevention of histamine receptor related diseases, WO 2003024929; 2003.

[227] Chan CKL, Pereira OZ, Nguyen-Ba N, Zhang M, Das SK, Poisson C, et al. Thumkunta Jagadeeswar Preparation of thiophene derivatives for the treatment of flavivirus infections, WO 2004052879; 2004.

[228] Maddaford S, Ramnauth J, Rakhit S, Patman J, Annedi SC, Andrews J, et al. Preparation of tetrahydroquinolines and related compounds having NOS inhibitory activity, US 20080234237; 2008.

[229] Liang TJ, Ferrer M, He S, Hu X, Hu Z, Marugan JJ, et al. Preparation of piperidine and piperazine derivatives and their use in treatment of viral infections and cancer, WO 2015080949; 2015.

[230] Carson JR, Susan CRJ, Fitzpatrick LJ, Reitz AB, Jetter MC. Preparation of 4-(diarylamino)piperidines as δ-opioid receptor agonists/antagonists, US 6436959; 2002.

[231] Eiden F, Winkler W, Wanner KT, Markhauser A. Pyran derivatives. 109. Centrally acting 4-phenylpyrans: hydrated phenylchromenes, phenylaza- and phenyloxachromenes, phenyloxa- and phenyldioxaxanthenes obtained by [4 + 2]-cycloaddition. Arch Pharm 1985;318(7):648−55.

[232] Schleimer R, Wuerthwein E-U. 2,6-Disubstituted 4-aminopyridines from 1,3-dialkoxy-2-azapropenylium salts and N-methyl-4-piperidone enamines. Chem Ber 1994;127(8):1437−40.

[233] Nitti P, Pitacco G, Rinaldi V, Valentin E. 1,2-Oxazine N-oxide derivatives from 1-hetera-4-cyclohexanone enamines and nitroolefins. Ring-chain tautomerism. Croatica Chem Acta 1986;59(1):165−70.

[234] Vartanyan SA, Abgaryan EA. Preparation and reduction of enamines of six-membered hydrogenated heterocyclic 4-ketones containing oxygen, nitrogen and sulfur. Arm KhimZh 1984;37(5):316−23.

[235] Cappiello JR, Chow K, Heidelbaugh TM, Takeuchi JA, Garst ME, Gil DW, et al. Preparation of 3-amino-1-(thi)oxo-1,2,5,6,7,8-hexahydro-2, 7-naphthyridine-4-carbonitriles as selective α2B adrenergic antagonists, WO 2010033393; 2010.

[236] Berdini V, Cesta MC, Curti R, D'Anniballe G, Di Bello N, Nano G, et al. A modified palladium-catalyzed reductive amination procedure. Tetrahedron 2002;58(28):5669−74.

[237] Margaretha P. Synthesis of alkyl- and cycloalkylamines by reductive amination of carbonyl compounds. Science of Synthesis 2009;40a:65−89, Volume Date 2008.

[238] Moliner M, Gonzalez J, Portilla MT, Willhammar T, Rey F, Llopis FJ, et al. A new aluminosilicate molecular sieve with a system of pores between those of ZSM-5 and beta zeolite. J Am Chem Soc 2011;133(24):9497−505.

[239] Abbas R, Willette RE, Edwards JM. Piperidine derivatives: synthesis of potential analgesics in 3-substituted 4-phenylpiperidine series. J Pharml Sci 1977;66(11):1583−5.

[240] McElvain SM, McMahon RE. Piperidine derivatives. XXI. 4-Piperidone, 4-piperidin-ol, and certain of their derivatives. J Am Chem Soc 1949;71:901−6.

[241] Brookes P, Walker J. Ready formation of ketals by 4-piperidones. J Chem Soc 1957;3173−5.

[242] Branch RF, Casy AF. 1H-Nuclear magnetic resonance spectra of acetals and thioacetals of 4-piperidinones. J Chem Soc [Section] B: Physical Organic 1968;10:1087−9.

[243] Stetter H, Reinartz W. Compounds with urotropine structure. LI. 1-Azaadamantane. Chem Ber 1972;105(9):2773−9.

[244] Burns BL, Wang H, Lin N, Blasko A. Preparation of heterocycles as anti-inflammatory and analgesic agents, WO 2010051497; 2010.

[245] Gammill RB. A new amine catalyzed synthesis of 2-substituted 2,3-dihydro-4H-1,3-benzoxazin-4-ones. J Org Chem 1981;46(16):3340−2.

[246] Wohl RA. Convenient one-step procedure for the synthesis of cyclic enol ethers. Preparation of 1-methoxy-1-cycloalkenes. Synthesis 1974;(1):38−40.

[247] Engler TA, Wanner J. Lewis acid-directed cyclocondensation of piperidone enol ethers with 2-methoxy-4-(N-phenylsulfonyl)-1,4-benzoquinoneimine: a new regioselective synthesis of oxygenated carbolines. J Org Chem 2000;65(8):2444−57.

[248] Abd Rabo Moustafa MM, Pagenkopf BL. Synthesis of 5-azaindoles via a cyclo-addition reaction between nitriles and Donor-Acceptor cyclopropanes. Org Lett 2010;12 (14):3168–71.

[249] Wanner KT, Eiden F. Benzothiazoles by carbon-carbon cleavage of α-[(2-nitrophenyl) thio] ketones. Lieb Ann Chem 1984;(6):1100–8.

[250] Sun A, Lankin DC, Hardcastle K, Snyder JP. 3-fluoropiperidines and N-methyl-3-fluoropiperidinium salts: the persistence of axial fluorine. Chem-Eur J 2005;11 (5):1579–91.

[251] Suennemann HW, Hofmeister A, Magull J, de Meijere A. An efficient access to novel enantiomerically pure steroidal δ-amino acids. Chem-Eur J 2006;12(32):8336–44.

[252] Wenzel B, Sorger D, Heinitz K, Scheunemann M, Schliebs R, Steinbach J, et al. Structural changes of benzylether derivatives of vesamicol and their influence on the binding selectivity to the vesicular acetylcholine transporter. Eur J Med Chem 2005;40 (12):1197–205.

[253] Zakharevskii AS, Zvonok AM, Lugovskii AP, Melentovich LA, Stanishevskii LS, Yushkevich EV. Synthesis and pharmacological activity of acetylenic piperidinediols, Khim.-Farm. Zh 1989;23(2):178–82.

[254] Ziering A, Berger L, Heineman SD, Lee J. Piperidine derivatives. III. 4-Arylpiperidines. J Org Chem 1947;12:894–903.

[255] Harper NJ, Simmonds AB. Halogen compounds related to the reversed esters of pethidine. J Pharm Pharmacol 1964;12(2):72–8.

[256] Harper NJ, Simmonds AB, Wakama WT, Hall GH, Vallance DK. Some basic ketones with central nervous system depressant activity. J Pharm Pharmacol 1966;18(3):150–60.

[257] Harper NJ, Simmonds AB. Fluoro compounds related to the reversed esters of pethidine. J Med Pharm Chem 1959;1:181–5.

[258] Sui Z, De Voss JJ, DeCamp DL, Li J, Craik CS, Ortiz de Montellano PR. Synthesis of haloperidol ethanedithioketal HIV-1 protease inhibitors: magnesium chloride facilitated addition of Grignard reagents. Synthesis 1993;(8):803–8.

[259] McElvain SM, Dickinson WB, Athey RJ. Piperidine derivatives. XXV. The reaction of certain 3-substituted 1-methyl-4-piperidones with organometallic compounds. J Am Chem Soc 1954;76:5625–33.

[260] McElvain SM, Berger RS. Piperidine derivatives. XXVII. The condensation of 4-piperidones and piperidinols with phenols. J Am Chem Soc 1955;77:2848–50.

[261] Nazarov IN, Prostakov NS, Krasnaya ZhA, Mikheeva NN. Heterocyclic compounds. XLI. Synthetic analgesic substances. 6. Esters of 1,2,5-trimethyl-4-aryl-4-piperidols. Zh Obshch Khim 1956;26:2820–34.

[262] Bell KH, Portoghese PS. Stereochemical studies on medicinal agents. 14. Relative stereochemistries and analgetic potencies of diastereomeric 3-allyl and 3-propyl derivatives of 1-methyl-4-phenyl-4-propionoxypiperidine. J Med Chem 1973;16(3):203–5.

[263] Nitta A, Fujii H, Sakami S, Nishimura Y, Ohyama T, Satoh M, et al. (3R)-3-Amino-4-(2,4,5-trifluorophenyl)-N-{4-[6-(2-methoxyethoxy)benzothiazol-2-yl]tetrahydropyran-4-yl}butanamide as a potent dipeptidyl peptidase IV inhibitor for the treatment of type 2 diabetes. Bioorg Med Chem Lett 2008;18(20):5435–8.

[264] Mayrargue J, Vayssiere M, Miocque M. A new synthesis of quinuclidinium derivatives. Heterocycles 1985;23(9):2173–5.

[265] Rusting N, Frielink JG, van der Beek GF. Piperidines, NL 6408223; 1965.

[266] Frielink JG. Piperidine derivatives, NL 6600523; 1966.

[267] Prost M, Urbain M, Charlier R. Propargyl derivatives. IV. Preparation of propargylpiperidinols and their esters. Helv Chim Acta 1966;49(7):2370–94.

[268]   Kalgutkar AS, Castagnoli Jr. N. Synthesis of novel MPTP analogs as potential mono-amine oxidase B (MAO-B) inhibitors. J Med Chem 1992;35(22):4165–74.

[269]   Carcanague DR, Gravestock MB, Hales NJ, Hauck SI, Weber TP. Preparation of oxazo-lidinone/isoxazoline derivatives as antibacterial agents, WO 2004048392; 2004.

[270]   Nazarov IN, Raigorodskaya VY. Izv. Akad. Nauk SSSR, Ser. Khimicheskaya; 1948. p. 631–641.

[271]   Korablev MV, Praliev KD, Salita TA, Zhilkibaev OT, Kurbat NM, Sydykov AO, et al. Synthesis of piperidine and decahydroquinoline derivatives, their analgesic and psycho-tropic properties. XVIII. 1-Methyl-4-acetylpiperidin-4-ol and its esters. Khim-Farm Zh 1985;19(4):419–22.

[272]   Harper NJ, Fullerton SE. Ethynyl and styryl compounds of the prodine type. J Med Pharm, Chem 1961;4:297–316.

[273]   Grob CA, Brenneisen P. Quinuclidine series. VI. Synthesis of 4-bromo- and 4-hydroxyquinuclidine. Helv Chim Acta 1958;41:1184–90.

[274]   Otto N, Opatz T. Heterocycles from α-Aminonitriles. Chem-Eur J 2014;20(41):13064–77.

[275]   Opatz T. The chemistry of deprotonated α-aminonitriles. Synthesis 2009;2009 (12):1941–59.

[276]   Kim IH, Combrink KD, Ma Z, Chapo K, Yan D, Renick P, et al. Synthesis and antibac-terial evaluation of a novel series of rifabutin-like spirorifamycins. Bioorg Med Chem Lett 2007;17(5):1181–4.

[277]   Bekkali Y, Thomson DS, Betageri R, Emmanuel MJ, Hao M, Hickey E, et al. Identification of a novel class of succinyl-nitrile-based Cathepsin S inhibitors. Bioorg Med Chem Lett 2007;17(9):2465–9.

[278]   Samnick S, Brandau W, Sciuk J, Steinstraesser A, Schober O. Synthesis, characterization and biodistribution of neutral and lipid-soluble 99mTc-bisaminoethanethiol spiperone derivatives: possible ligands for receptor imaging with SPECT. Nucl Med Biol 1995;22 (5):573–83.

[279]   Harnden MR, Rasmussen RR. Synthesis of compounds with potential central nervous system stimulant activity. II. 5-Spiro-substituted 2-amino-2-oxazolines. J Med Chem 1970;13(2):305–8.

[280]   Metwally KA, Dukat M, Egan CT, Smith C, DuPre A, Gauthier CB, et al. Spiperone: influence of spiro ring substituents on 5-HT2A serotonin receptor binding. J Med Chem 1998;41(25):5084–93.

[281]   Marco JL, Ingate ST, Chinchon PM. The CSIC [carbanion mediated sulfonate (sulfona-mido) intramolecular cyclization] reaction: scope and limitations. Tetrahedron 1999;55 (24):7625–44.

[282]   van de Westeringh C, van Daele P, Hermans B, van der Eycken C, Boey J, Janssen PAJ. 4-Substituted piperidines. I. Derivatives of 4-tertiaryamino-4-piperidinecarboxamides. J Med Chem 1964;7(5):619–23.

[283]   O'Reilly MC, Oguin III TH, Scott SA, Thomas PG, Locuson CW, et al. Discovery of a highly selective PLD2 inhibitor (ML395): a new probe with improved physiochemical properties and broad-spectrum antiviral activity against influenza strains. ChemMedChem 2014;9(12):2633–7.

[284]   Shipps GW Jr, Cheng CC, Huang X, Fischmann TO, Duca JS, Richards M, et al. Anilinopiperazine derivatives and methods of use thereof and their preparation, WO 2008054702; 2008.

[285]   Albert JS, Aharony D, Andisik D, Barthlow H, Bernstein PR, Bialecki RA, et al. Design, synthesis, and SAR of tachykinin antagonists: modulation of balance in NK1/NK2 recep-tor antagonist activity. J Med Chem 2002;45(18):3972–83.

[286] Fisher A, Karton Y, Marciano D, Barak D, Meshulam H. Preparation of azaspiro compounds as muscarinic agonists, US 5852029; 1998.

[287] Harnden MR, Rasmussen RR. Synthesis of compounds with potential central nervous system stimulant activity. I. 2-Amino-2-oxazolin-4-one-5-spirocycloalcanes and 2-amino-2-oxazolin-4-one-5-spiro(4'-piperidines). J Med Chem 1969;12:919–21.

[288] Verniest G, Piron K, Van Hende E, Thuring JW, Macdonald G, Deroose F, et al. Synthesis of aminomethylated 4-fluoropiperidines and 3-fluoropyrrolidines. Org Biomol Chem 2010;8(11):2509–12.

[289] Harnden MR. 2-Amino-5-spiro substituted oxazolin-4-one compounds, US 3931198; 1976.

[290] Saunders J, Showell GA, Snow RJ, Baker R, Harley EA, Freedman SB. 2-Methyl-1,3-dioxaazaspiro[4.5]decanes as novel muscarinic cholinergic agonists. J Med Chem 1988;31(2):486–91.

[291] Somanathan R, Rivero IA, Nunez GI, Hellberg LH. Convenient synthesis of 1-oxa-3,8-diazaspiro[4,5]decan-2-ones. Synth Comm 1994;24(10):1483–7.

[292] Hermans B, Daele P, van; Westeringh C, van de; Eycken C, van der; Boey J, Janssen PAJ. 4-Substituted piperidines. II. Reaction of 1-benzyl-4-cyano-4-tertiaryaminopiperidines with organometallic compounds. J Med Chem 1965;8(6):851–5.

[293] Iorio MA, Molinari M, Scotti de Carolis A, Niglio T. Nitrogen analogs of phencyclidine: 1-alkyl-4-phenyl-4-(1-piperidinyl)piperidines. Farmaco, (Ed Sci) 1984;39(7):599–611.

[294] Howard H. R. 4-Arylpiperidine derivatives and uses thereof for preparing a medicament for the treatment of CNS disorders. Exp Opin Ther Pat 2007;17(2):251–3.

[295] Hermans B, Van Daele P, van de Westeringh C, van der Eycken C, Boey J, Janssen PAJ. 4-Substituted piperidines. III. Reduction of 1-benzyl-4-cyano-4-tert-aminopiperidines with lithium aluminum hydride. J Med Chem 1966;9(1):49–52.

[296] Inventor data is not available, 1-Aralkyl-4-aminomethylpiperidin-4-ols, BE 818471; 1974.

[297] Favre H, Hamlet Z, Lanthier R, Menard M. Inclination to ring expansion in piperidines as a function of substitution on the nitrogen atom. Reaction of diazomethane with several 4-piperidinones and nitrous acid deamination of the corresponding aminomethylalcohols. Can J Chem 1971;49(19):3075–85.

[298] Brameld KA, Verner E. Quinolone derivatives, especially oxopyridopirimidines, as fibroblast growth factor inhibitors, WO 2014182829; 2014.

[299] Xu Y, Brenning BG, Kultgen SG, Liu X, Saunders M, Ho K. Preparation of imidazo[1,2-b]pyridazine, pyrazolo[1,5-a]pyrimidine, and [1,2,4]triazolo[4,3-a]pyridine derivatives as protein kinase inhibitors, WO 2013013188; 2013.

[300] Kuehnis HH, Ryf H, Denss R. Substituted (4-hydroxy-4-piperidyl) propanones, CH 447163; 1968.

[301] Cope AC, Hofmann CM, Wyckoff C. Hardenbergh, esther condensation reactions. II. Alkylidene cyanoacetic and malonic esters. J Am Chem Soc 1941;63:3452–6.

[302] Stork G, McElvain SM. Piperidine derivatives. XVI. C-Alkylation of 1-benzoyl-3-carbethoxy-4-piperidone. Synthesis of ethyl 3-ethyl-4-piperidylacetate (di-ethyl cincholoiponate). J Am Chem Soc 1946;68:1053–7.

[303] McElvain SM, Lyle, Robert Jr. E. Piperidine derivatives. XXII. The condensation of 1-methyl-4-piperidone with active methylene compounds. J Am Chem Soc 1950;72:384–9.

[304] Doernyei G, Incze M, Moldvai I, Szantay C. Synthesis of some new 4-substituted N-benzyl-piperidines. Synth Comm 2003;33(13):2329–38.

[305]   Shao D, Zou C, Luo C, Tang X, Li Y. Synthesis and evaluation of tacrine-E2020 hybrids as acetylcholinesterase inhibitors for the treatment of Alzheimer's disease. Bioorg Med Chem Lett 2004;14(18):4639−42.

[306]   Zhou Z-Lin, Keana JFW. A practical synthesis of 4-(substituted-benzyl)piperidines and (±)-3-(substituted-benzyl)pyrrolidines via a wittig reaction. J Org Chem 1999;64 (10):3763−6.

[307]   Eenkhoorn JA, De Silva SO, Snieckus V. Wittig reaction of indol-2-ylmethyltriphenyl-phosphonium iodide with 4-piperidone derivatives and aromatic aldehydes, Can. J Chem 1973;51(5):792−810.

[308]   Kuroyan RA, Panosyan AG, Kuroyan NA, Vartanyan SA. Synthesis of six-membered heterocyclic 4-aldehydes with oxygen, sulfur, and nitrogen hetero atoms. Arm Khim Zh 1974;27(11):945−9.

[309]   Kuroyan RA, Akopyan LA, Vartanyan SA, Durgaryan LK, Vlasenko EV. Synthesis of acetic acid derivatives of the piperidine series. Arm Khim Zh 1981;34(2):142−7.

[310]   Kuroyan RH, Arutyunyan NS, Vartanyan SA. Some reactions of 4-formylpiperidines. Arm Khim Zh 1979;32(7):555−63.

[311]   Bernat V, Brox R, Heinrich MR, Auberson YP, Tschammer N. Ligand-biased and probe-dependent modulation of chemokine receptor CXCR3 signaling by negative allosteric modulators. ChemMedChem 2015;10(3):566−74.

[312]   Sugasawa S, Matsuo H. Synthesis of α,β-unsaturated esters by application of the Wittig reaction. Chem Pharm Bull 1960;8:819−26.

[313]   Etayo P, Badorrey R, Diaz-de-Villegas MD, Galvez JA. Base-controlled diastereodivergent synthesis of (R)- and (S)-2-substituted-4-alkylidenepiperidines by the Wadsworth-Emmons reaction. J Org Chem 2007;72(3):1005−8.

[314]   Anker RM, Cook AH. Piperidine series. IV. J Chem Soc 1948;806−10.

[315]   Jucker E, Suess R. Synthetic drugs. XVI. Substituted 2,8-diazaspiro[4.5]decane-1,3-diones. Helv Chim Acta 1966;49(3):1135−45.

[316]   Van Parys M, Vandewalle M. The synthesis of 1-phenyl-3,8-diazaspiro[4,5]decanes. Bull Soc Chim Belg 1981;90(7):757−65.

[317]   Vartanyan RS, Israelyan RG, Vartanyan SA. A new method of preparation of β- and β,β-substituted α-phenylacrylonitriles. Arm Khim Zh 1978;31(4):245−9.

[318]   Vartanyan SA, Vartanyan RS. Israelyan, R. G. β- and β,β-substituted α-phenylacrylo-nitriles, SU 732250; 1980.

[319]   Schaefer H, Hackmack G, Eistetter K, Krueger U, Menge HG, Klosa J. Synthesis, physico-chemical properties and pharmacological studies of budipine and related 4,4-diphenyl-piperidines. Arzneim-Forsch 1984;34(3):233−40.

[320]   Hackmack G, Klosa J. 4,4-Diphenylpiperidines, DE 2166997; 1977.

[321]   Klumpp DA, Garza M, Jones A, Mendoza S. Synthesis of aryl-substituted piperidines by superacid activation of piperidones. J Org Chem 1999;64(18):6702−5.

[322]   Omodt GW, Gisvold O. Synthesis of some glycidic esters as potential antispasmodics. J Am Pharmaceut Assoc, (1912-1977) 1960;49:153−8.

[323]   Vartanyan SA, Kuroyan RA, Minasyan SA, Dangyan FV, Arutyunyan NS. 2-Carbethoxy-4,6,7-trimethyl-1-oxa-6-azaspiro[2.5]octane, Sintezy Geterotsikl. Soedinenii, Erevan 1975;10:36−7.

[324]   Deshpande PK, Sindkhedkar MD, Phansalkar MS, Yeole RD, Gupte SV, Chugh Y, et al. Preparation of 3-(4-piperidinophenyl)oxazolidinones having antibacterial activity with improved in vivo efficacy, WO 2005054234; 2005.

[325] Pathi SL, Acharya V, Rao DR, Kankan RN. Process for preparation of donepezil and intermediates, WO 2007077443; 2007.

[326] Fishman M, Cruickshank PA, Alkylation I. Synthesis and reactions of spiro[oxirane-2,4′-piperidines. J Het Chem 1968;5(4):467–9.

[327] Satyamurthy N, Berlin KD, Hossain MB, Van der Helm D. Synthesis and stereochemistry of 1-oxa-6-heteraspiro[2.5]octanes. Single-crystal analysis of 6-phenyl-1-oxa-6-phosphaspiro[2.5]octane 6-sulfide. Phosph Sulf Related Elements 1984;19(1):113–29.

[328] Popp FrankD, Watts, Raymond F. The synthesis and ring opening of N-substituted-2-oxa-6-azaspiro[2.5]octanes. J Het Chem 1978;15(4):675–6.

[329] Ng JS. Epoxide formation from aldehydes and ketones - a modified method for preparing the Corey-Chaykovsky reagents. Synth Comm 1990;20(8):1193–202.

[330] Harrak Y, Guillaumet G, Pujol MD. The first synthesis of spiro[1,4-benzodioxin-2,4′-piperidines] and spiro[1,4-benzodioxin-2,3′-pyrrolidines]. Synlett 2003;(6):813–16.

[331] Sheng R, Hu Y. Novel and efficient method to synthesize N-benzyl-4-formylpiperi-dine, Synth. Comm 2004;34(19):3529–33.

[332] Barrera H, Lyle RE. Piperidine derivatives with a sulfur-containing function in the 4-position. J Org Chem 1962;27:641–3.

[333] MacDonald IR, Jollymore CT, Reid GA, Pottie IR, Martin E, Darvesh S. Thioesters for the in vitro evaluation of agents to image brain cholinesterases. J Enzyme Inhibition Med Chem 2013;28(3):447–55.

[334] Meanwell NA, Hewawasam P, Thomas JA, Wright JJK, Russell JW, Gamberdella M, et al. Inhibitors of blood platelet cAMP phosphodiesterase. 4. Structural variation of the side-chain terminus of water-soluble 1,3-dihydro-2H-imidazo[4,5-b]quinolin-2-one derivatives. J Med Chem 1993;36(22):3251–64.

[335] Levy J, Bernotsky GA. Hydrogenation of piperidone compounds in hydrocarbon solvents, US 2776293; 1957.

[336] Okano T, Matsuoka M, Konishi H, Kiji J. Meerwein-Ponndorf-Verley reduction of ketones and aldehydes catalyzed by lanthanide tri-2-propoxides. Chem Lett 1987;(1):181–4.

[337] Adlerova E, Seidlova V, Protiva M. Synthetic ataractics. IX. Analogs of prothiadene with heterocyclic residues in the side chain. Cesko-Slovenska Farmacie 1963;12:122–6.

[338] Jia P, Sheng R, Zhang J, Fang L, He Q, Yang B, et al. Design, synthesis and evaluation of galanthamine derivatives as acetylcholinesterase inhibitors. Eur J Med Chem 2009;44(2):772–84.

[339] Yu VK, Kabdraissova AZh, Praliyev KD, Shin SN, Berlin K. Darrell Synthesis and properties of novel alkoxy- and phenoxyalkyl ethers of secondary and tertiary ethynylpiperidin-4-ols possessing unusual analgesic, anti-bacterial, anti-spasmotic, and anti-allergic properties as well as low toxicity. J Saudi Chem Soc 2009;13(2):209–17.

[340] Jucker E, Schenker E. 4-Chloropiperidine derivatives, CH 463504; 1968.

[341] Amato JS, Chung JYL, Cvetovich RJ, Gong X, McLaughlin M, Reamer R. A. Synthesis of 1-tert-Butyl-4-chloropiperidine: generation of an N-tert-butyl group by the reaction of dimethyliminium salt with methylmagnesium chloride. J Org Chem 2005;70(5):1930–3.

[342] No inventor data available, Synthesis of substituted piperidine derivatives, NL 6413199; 1965.

[343] Liu KG, Lo JR, Comery TA, Zhang GM, Zhang JY, Kowal DM, et al. Identification of a novel series of 3-piperidinyl-5-sulfonylindazoles as potent 5-HT6 ligands. Bioorg Med Chem Lett 2009;19(12):3214–16.

[344]   Luebbert H, Ullmer C, Bellott E, Froimowitz M, Gordon D. Derivatives of 4-(thio- or selenoxanthene-9-ylidene)-piperidine or -acridine and their use as selective 5-HT2B receptor antagonists, WO 2003035646; 2003.

[345]   Van Daele GHP, De Bruyn MFL, Sommen FM, Janssen M, Van Nueten JM, Schuurkes JAJ, et al. Synthesis of cisapride, a gastrointestinal stimulant derived from cis-4-amino-3-methoxypiperidine. Drug Dev Res 1986;8(1-4):225−32.

[346]   Walter M, von Coburg Y, Isensee K, Sander K, Ligneau X, Camelin J-C, et al. Azole derivatives as histamine H3 receptor antagonists, Part I: Thiazol-2-yl ethers. Bioorg Med Chem Lett 2010;20(19):5879−82.

[347]   Serra MJ, Lombardero FJ, Moinserrat VC. Ethyl 3-bromo-4-oxo-1-piperidinecarboxylate, process for its preparation, and new intermediates used in its preparation, ES 2053394; 1994.

[348]   Zhou Y, Malosh C, Conde-Ceide S, Martinez-Viturro CM, Alcazar J, Lavreysen H, et al. Further optimization of the mGlu5 PAM clinical candidate VU0409551/JNJ-46778212: progress and challenges towards a back-up compound. Bioorg Med Chem Lett 2015;25 (17):3515−19.

[349]   Conn PJ, Lindsley CW, Stauffer SR, Jones CK, Bartolome-Nebreda JM, Conde-Ceide S, et al. Bicyclic oxazole and thiazole compounds and their use as allosteric modulators of mglur5 receptors, WO 2012031024; 2012.

[350]   Marigo M, Bachmann S, Halland N, Braunton A, Jorgensen KA. Highly enantioselective direct organocatalytic α-chlorination of ketones. Angew Chem Int Ed 2004;43 (41):5507−10.

[351]   Halland N, Jorgensen KA, Marigo M, Braunton A, Bachmann S, Fielenbach D. Catalytic asymmetric synthesis of optically active α-halo-carbonyl compounds, WO 2005080298; 2005.

[352]   Bryan MC, Chan B, Hanan E, Heffron T, Purkey H, Elliott RL, et al. Aminopyrimidine compounds as inhibitors of T790M containing EGFR mutants and their preparation, WO 2014081718; 2014.

[353]   Abbas R, Willette RE, Edwards JM. Piperidine derivatives: synthesis of potential analgesics in 3-substituted 4-phenylpiperidine series. J Pharm Sci 1977;66(11):1583−5.

[354]   Grishina GV, Potapov VM, Abdulganeeva SA, Ivanova IA. Synthesis of 3-substituted 4-piperidinones. Khimiya Geterotsiklicheskikh Soedinenii 1983;11:1510−14.

[355]   Alam M, Baty JD, Jones Gurnos, Moore C. Alkylation of 4-piperidones; intermediates in the synthesis of reduced 2-pyrindin-6-ones. J Chem Soc C 1969;1969(11):1520−8.

[356]   Nagai Yasutaka, Uno Hitoshi, Umemoto Susumu. Studies on psychotropic agents. II. Synthesis of 1-substituted-3-(p-fluorophenacyl)piperidines and the related compounds. Chem Pharm Bull 1977;25(8):1911−22.

[357]   Schenker E, Salzmann R. Antihypertensive action of bicyclic 3-hydrazinopyridazines. Arzneimit-Forsch 1979;29(12):1835−43.

[358]   Demont EH, Bailey JM, Bit RA, Brown JA, Campbell CA, Deeks N, et al. Discovery of tetrahydropyrazolopyridine as sphingosine 1-phosphate receptor 3 (S1P3)-sparing S1P1 agonists active at low oral doses. J Med Chem 2016;59(3):1003−20.

[359]   Jamieson C, Campbell RA, Cumming IA, Gillen KJ, Gillespie J, Kazemier B, et al. A novel series of positive modulators of the AMPA receptor: structure-based lead optimization. Bioorg Med Chem Lett 2010;20(20):6072−5.

[360]   Chappie, T. Allen; Helal, C.J. Kormos, B.L.; Tuttle, J.B.; Verhoest, P.R., Imidazotriazine derivatives as PDE10 inhibitors and their preparation, WO 2014177977 (2014).

[361] Bialy L. Preparation of indanyl-4,5,6,7-tetrahydro-1H-pyrazolo[4,3-c]pyridines acting on TASK-1 channel, IN 2011CH03129; 2013.

[362] McCabe PH, Milne NJ, Sim G. A. Conformational study of bridgehead lactams. Preparation and x-ray structural analysis of 1-azabicyclo[3.3.1]nonane-2,6-dione. J Chem Soc, Perkin Trans 1989;2(10):1459—62.

[363] Freter K, Fuchs V, Pitner TP. Fischer indole synthesis from cis- and trans-hexahydro-7-methyl-6-isoquinolones. Proton NMR determination of the configuration and conformation of products. J Org Chem 1983;48(24):4593—7.

[364] Danishefsky S. Cavanaugh, Robert Reaction of piperidone enamines with methyl β-vinylacrylate. A route to quinolines and isoquinolines. J Org Chem 1968;33(7):2959—62.

[365] Baruchello R, Simoni D, Marchetti P, Rondanin R, Mangiola S, Costantini C, et al. 4,5,6,7-Tetrahydro-isoxazolo-[4,5-c]-pyridines as a new class of cytotoxic Hsp90 inhibitors. Eur J Med Chem 2014;76:53—60.

[366] Takayasu T, Nitta M. An enamine method for the synthesis of 1-azaazulene derivatives. Reactions of troponimines with enamines. J Chem Soc Perkin Trans 1999;1(6):687—92.

[367] Macor JE, Kuipers W, Lachicotte RJ. An unusual byproduct from a non-synchronous reaction between ethyl 1,2,4-triazine-3-carboxylate and an enamine. Chem Comm 1998;(9):983—4.

[368] Haider N. Inverse electron demand Diels-Alder reactions of condensed pyridazines. Part 1. Synthesis of phthalazine derivatives from pyridazino[4,5-d]pyridazines. Tetrahedron 1991;47(24):3959—68.

[369] Haider N, Mereiter K, Wanko R. Inverse-electron-demand Diels-Alder reactions of condensed pyridazines. 4. Synthesis and cycloaddition reactions of 1,4-bis(trifluoromethyl) pyrido[3,4-d]pyridazine. Heterocycles 1994;38(8):1845—58.

[370] Marcelis ATM, Van der Plas HC. Inverse electron demand Diels-Alder reactions of 5-nitropyrimidine with enamines. Synthesis of 3-nitropyridine derivatives. Tetrahedron 1989;45(0):2693—702.

[371] Branowska D. Synthesis of unsymmetrical annulated 2,2-bipyridine analogues with attached cycloalkene and piperidine rings via sequential Diels-Alder reaction of 5,5-bi-1,2,4-triazines. Mol;ecules 2005;10(1):265—73.

[372] Lipinska T. Experimental and theoretical FMO interaction studies of the Diels-Alder reaction of 5-acetyl-3-methythio-1,2,4-triazine with cyclic enamines. Tetrahedron 2005;61(34):8148—58.

[373] McElvain SM, Rorig K. Piperidine derivatives. XVIII. The condensation of aromatic aldehydes with 1-methyl-4-piperidone. J Am Chem Soc 1948;70:1820—5.

[374] Vatsadze SZ, Manaenkova MA, Sviridenkova NV, Zyk NV, Krut'ko DP, Churakov AV, et al. Synthesis and spectroscopic and structural studies of cross-conjugated dienones derived from cyclic ketones and aromatic aldehydes. Russ Chem Bull 2005;55(7):1184—94.

[375] Sumesh RV, Malathi A, Ranjith KR. A facile tandem Michael addition/O-cyclization/elimination route to novel chromeno[3,2-c]pyridines. Mol Divers 2015;19(2):233—49.

[376] Pati HN, Das U, Das S, Bandy B, De Clercq E, Balzarini J, et al. The cytotoxic properties and preferential toxicity to tumour cells displayed by some 2,4-bis(benzylidene)-8-methyl-8-azabicyclo[3.2.1] octan-3-ones and 3,5-bis(benzylidene)-1-methyl-4-piperidones, Eur. J Med Chem 2009;44(1):54—62.

[377] Yuan L, Sumpter BG, Abboud KA, Castellano RK. Links between through-bond interactions and assembly structure in simple piperidones. New J Chem 2008;32(11):1924—34.

[378] Krapcho J, Turk CF. Bicyclic pyrazolines, potential central nervous system depressants and antiinflammatory agents. J Med Chem 1979;22(2):207−10.

[379] Shadjou N, Hasanzadeh M. Amino functionalized mesoporous silica decorated with iron oxide nanoparticles as a magnetically recoverable nanoreactor for the synthesis of a new series of 2,4-diphenylpyrido[4,3-d]pyrimidines. RSC Advances 2014;4(35):18117−26.

[380] Han Z, Tu S, Jiang B, Yan S, Zhang X, Wu S, et al. An efficient and chemoselective synthesis of 1,6-naphthyridines and pyrano[3,2-c]pyridines under microwave irradiation. Synthesis 2009;10:1639−46.

[381] Moore TW, Zhu S, Randolph R, Shoji M, Snyder JP. Liver S9 fraction-derived metabolites of curcumin analogue UBS109. ACS Med Chem Lett 2014;5(4):288−92.

[382] Mojtahedi MM, Abaee MS, Khakbaz M, Alishiri T, Samianifard M, Mesbah AW, et al. An efficient procedure for the synthesis of α,β-unsaturated ketones and its application to heterocyclic systems. Synthesis 2011;(23):3821−6.

[383] Majewski M, Gleave DM. Diastereoselective aldol reactions of cyclohexanone lithium enolate. Tetrahedron Lett 1989;30(42):5681−4.

[384] Soni V, Sharma M, Agarwal A, Kishore D. Exploration of newer possibilities to the synthesis of diazepine and quinoline carboxylic acid derivatives. J Chem 2013;149270(8).

[385] Tyagi R, Kaur N, Singh B, Kishore D. A novel synthetic protocol for the heteroannulation of oxocarbazole and oxoazacarbazole derivatives through corresponding oxoketene dithioacetals. J Het Chem 2014;51(1):18−23.

[386] Ivachtchenko AV, Golovina ES, Kadieva MG, Kysil VM, Mitkin OD, Vorobiev AA. Antagonists of 5-HT6 receptors. Substituted 3-(phenylsulfonyl)pyrazolo[1,5-a]pyrido[3,4-e]pyrimidines and 3-(phenylsulfonyl)pyrazolo[1,5-a]pyrido[4,3-d]pyrimidines- Synthesis and 'structure-activity' relationship. Bioorg Med Chem Lett 2012;22(13):4273−80.

[387] Roychowdhury A, Sharma R, Gupte A, Kandre S, Gadekar PK, Chavan S, et al. Heterocyclic compounds as EZH2 inhibitors and their preparation, WO 2015110999; 2015.

[388] Winters G, Sala A, Barone D. Pharmacologically active pyrazolo[4,3-c]pyridines EP 86422; 1983.

[389] Duncan KW, Chesworth R, Munchhof MJ, Shapiro G. Heterocyclic compounds as PRMT5 inhibitors and their preparation, WO 2015200677; 2015.

[390] Nagase H, Imaide S, Yamada T, Hirayama S, Nemoto T, Yamaotsu N, et al. Essential structure of opioid κ receptor agonist nalfurafine for binding to κ receptor 1: synthesis of decahydroisoquinoline derivatives and their pharmacology. Chem Pharm Bull 2012;60(8):945−8.

[391] Bai R, Liu P, Yang J, Liu C, Gu Y. Facile Synthesis of 2-Aminothiophenes Using NaAlO2 as an Eco-Effective and Recyclable Catalyst. ACS Sustainable Chem Eng 2015;3(7):1292−7.

[392] Wang T, Huang X, Liu J, Li B, Wu J, Chen K, et al. An efficient one-pot synthesis of substituted 2-aminothiophenes via three-component Gewald reaction catalyzed by L-proline. Synlett 2010;(9):1351−4.

[393] Mekheimer RA, Ameen MA, Sadek KU. Solar thermochemical reactions. Part II. Synthesis of 2-aminothiophenes via Gewald reaction induced by solar thermal energy. Chin Chem Lett 2008;19(7):788−90.

[394] Aurelio L, Valant C, Figler H, Flynn BL, Linden J, Sexton PM, et al. J. 3- and 6-Substituted 2-amino-4,5,6,7-tetrahydrothieno[2,3-c]pyridines as A1 adenosine receptor allosteric modulators and antagonists. Bioorg Med Chem 2009;17(20):7353−61.

[395] El-Sherbeny MA, Youssef KM, Al-Shafeih FS, Al-Obaid AMA. Novel pyridothienopyrimidine and pyridothienothiazine derivatives as potential antiviral and antitumor agents. Med Chem Res 2000;10(2):122–35.

[396] Hassan AY, Mohamed HA. Synthesis and biological evaluation of thieno[2,3-c]pyridines and related heterocyclic systems. Asian J Chem 2009;21(5):3947–61.

[397] Sinyakov YuV, Sidorin DN, Boikova NV, Kryukov LN. Reaction of 1-methyl-4-piperidone with malonodinitrile. Zhurnal Obshchei Khimii 1990;60(8):1931–2.

[398] Lakshmi NV, Suganya JGA, Perumal PT. Novel route to spiropiperidines using N-methyl-4-piperidone, malononitrile and electrophiles. Tetrahedron Lett 2012;53 (10):1282–6.

[399] Crawford JJ, Ortwine DF, Wei B, Young WB. Process for the preparation of aromatic compound by nitroxide-catalyzed Sandmeyer-like reaction, WO 2013067274; 2013.

[400] Ueda T. Preparation of activated blood coagulation factor X (FXa) inhibitor compounds under mild conditions, WO 2015125710; 2015.

[401] Crawford JJ, Ortwine DF, Wei B, Young WB. 8-Fluorophthalazin-1(2H)-one compounds as inhibitors of BTK activity and their preparation, WO 2013067264; 2013.

[402] Barbosa AJM, Blomgren PA, Currie KS, Krishnamoorthy R, Kropf JE, Lee SH, et al. Preparation of pyridone and azapyridone compounds as Btk inhibitors for treating immune disorders, inflammation, cancer, and other Btk-mediated diseases, WO 2011140488; 2011.

[403] Berry A, Chen Z, De Lombaert S, Emmanuel MJ, Loke PL, Man CC, et al. Biarylamide as inhibitors of leukotriene production and their preparation, WO 2012082817; 2012.

[404] Kawanami K. Process for the preparation of aromatic compound by nitroxide-catalyzed Sandmeyer-like reaction, WO 2012017932; 2012.

[405] Stuetz P, Stadler PA. Novel approach to cyclic β-carbonyl enamines. Δ7,8-Lysergic acid derivatives via the Polonovski reaction. Tetrahedron Letters 1973;14(51):5095–8.

[406] Moehrle H, Claas M. Mercury(II) dehydrogenations of N-tertiary piperidine derivatives with different types of substitution. Pharmazie 1988;43(11):749–53.

[407] Diao T, Stahl SS. Synthesis of cyclic enones via direct palladium-catalyzed aerobic dehydrogenation of ketones. J Am Chem Soc 2011;133(37):14566–9.

[408] Mather P, Whittall J. (D)-codeinone from (D)-dihydrocodeinone via the use of modified o-iodoxybenzoic acid (IBX). A convenient oxidation of ketones to enones. In: Roberts SM, Whittall J, editors. Catalysts for Fine Chemical Synthesis, vol. 5. Wiley; 2007. p. 263–6.

[409] Nicolaou KC, Montagnon Tamsyn, Baran, Phil S. Modulation of the reactivity profile of IBX by ligand complexation: ambient temperature dehydrogenation of aldehydes and ketones. Angew Chem, Int Ed 2002;41(6):993–6.

[410] Jin X, Yamaguchi K, Mizuno N. Gold-catalyzed heterogeneous aerobic dehydrogenative amination of α,β-unsaturated aldehydes to enaminals. Angew Chem, Int Ed 2014;53 (2):455–8.

[411] Johnson PY, Berchtold GA. Photochemical reactions of γ-keto sulfides. J Org Chem 1970;35(3):584–92.

[412] Garst ME, McBride BJ, Johnson AT. Epoxyannulation. 4. Reactions of 1,5-, 1,6-, and 1,7-oxosulfonium salts. J Org Chem 1983;48(1):8–16.

[413] Thompson MD, Holt EM, Berlin KD, Scherlag BJ. Preparation and single-crystal x-ray characterization of 4-selenanone. J Org Chem 1985;50(14):2580–1.

[414] Evers M, Christiaens L, Renson M. Synthesis of tetrahydro-1-seleno- and -1-telluro-4-pyranones. Tetrahedron Lett 1985;26(44):5441–544.

[415] Amato JS, Chung JYL, Cvetovich RJ, Reamer RA, Zhao D, Zhou G, et al. Acrylate as an efficient dimethylamine trap for the practical synthesis of 1-tert-Butyl-4-piperidone via transamination. Org Proc Res Dev 2004;8(6):939–41.

[416] Prill EA, McElvain SM. Cyclization of a series of ω,ω'-dicarbethoxydialkylmethyl-amines through the acetoacetic ester condensation. J Am Chem Soc 1933;55:1233–41.

[417] McElvain SM, Vozza, John F. Piperidine derivatives. XX. The preparation and reactions of 1-methyl-3-piperidone. J Am Chem Soc 1949;71:896–900.

[418] Leonard NnJ, Barthel Jr. E. Rearrangement of α-aminoketones during Clemmensen reduction. V. Influence of alkyl substitution on the α-carbon. J Am Chem Soc 1950;72:3632–5.

[419] Knight DW, Lewis N, Share AC, Haigh D. β-Hydroxypiperidinecarboxylates: additions to the chiral pool from bakers' yeast reductions of β-ketopiperidinecarboxylates. J Chem Soc, Perkin Trans 1998;1(22):3673–84.

[420] Katvalyan GT, Shashkov AS, Mistryukov EA. Conformational analysis of 3-ketopiperidines. Vicinal N1C(2)-alkyl interactions. J Het Chem 1985;22(1):53–5.

[421] Lyle RE, Adel RE, Lyle GG. Hydrates of 1-methyl-3-and-4-piperidone hydrochlorides. J Org Chem 1959;24:342–5.

[422] Huang W-X, Wu B, Gao X, Chen M-W, Wang B, Zhou Y-G. Iridium-catalyzed selective hydrogenation of 3-hydroxypyridinium salts: a facile synthesis of piperidin-3-ones. Org Lett 2015;17(7):1640–3.

[423] West FG, Naidu BN. New route to substituted piperidines via the Stevens [1,2]-shift of ammonium ylides. J Am Chem Soc 1993;115(3):1177–8.

[424] D'hooghe M, Baele J, Contreras J, Boelens M, De Kimpe N. Reduction of 5-(bromo-methyl)-1-pyrrolinium bromides to 2-(bromomethyl)pyrrolidines and their transformation into piperidin-3-ones through an unprecedented ring expansion-oxidation protocol. Tetrahedron Lett 2008;49(42):6039–42.

[425] Guzi TJ, Macdonald TL. A novel synthesis of piperidin-3-ones via an intramolecular Amadori-type reaction. Tetrahedron Lett 1996;37(17):2939–42.

[426] Yang H, Jurkauskas V, Mackintosh N, Mogren T, Stephenson CRJ, Foster K, et al. Trifluoroacetic acid-mediated intramolecular formal N-H insertion reactions with amino-α-diazoketones: a facile and efficient synthesis of optically pure pyrrolidinones and piperidinones. Can J Chem 2000;78(6):800–8.

[427] Schotten C. Die unwandlung des piperidins in amidovalerianasaure und in oxypiperidins. Ber Deutsch Chem Gesellschaft 1888;21:2235–54.

[428] Uosaki Y, Moriyoshi T, Zota H, Hata K, Asuka S. Catalyst-free cyclization of amino acids, JP 2002121183; 2002.

[429] Blade-Font A. Facile synthesis of γ-, δ-, and ε-lactams by cyclodehydration of ω-amino acids on alumina or silica gel. Chem Lett 1989;5:797–800.

[430] Blade-Font A. Facile synthesis of γ-, δ-, and ε-lactams by cyclodehydration of ω-amino acids on alumina or silica gel. Tetrahedron Lett 1980;21(25):2443–6.

[431] Pellegata R, Pinza M, Pifferi G. An improved synthesis of γ-, δ-, and ε-lactams. Synthesis 1978;(8):614–16.

[432] Lanigan RM, Starkov P, Sheppard TD. Direct synthesis of amides from carboxylic acids and amines using B(OCH2CF3)3. J Org Chem 2013;78(9):4512–23.

[433] Murayama T, Kobayashi T, Miura T. A convenient preparative method for β-lactams from β-amino acids using sulfenamide/triphenylphosphine. Tetrahedron Lett 1995;36(21):3703–6.

[434] Yamamoto Y, Furuta T. Triethylgallium-mediated lactamization of α,ω-amino carboxylic acids. Chem Lett 1989;5:797–800.

[435] Ishihara K, Ohara S, Yamamoto H. 3,4,5-Trifluorobenzeneboronic Acid as an Extremely Active Amidation Catalyst. J Org Chem 1996;61(13):4196−7.

[436] Mader M, Helquist P. Titanium tetraisopropoxide-mediated lactamizations. Tetrahedron Lett 1988;29(25):3049−52.

[437] Steliou K, Szczygielska-Nowosielska A, Favre A, Poupart MA, Hanessian S. Reagents for organic synthesis: use of organostannyl oxides as catalytic neutral esterification agents in the preparation of macrolides. J Am Chem Soc 1980;102(25):7578−9.

[438] Taylor PJM, Bull SD, Andrews PC. An aza-enolate alkylation strategy for the synthesis of α-alkyl-δ-amino esters and α-alkyl valerolactams. Synlett 2006;9:1347−50.

[439] Huang SB, Nelson JS, Weller DD. Preparation of optically pure ω-hydroxymethyl lactams, Synth. Comm 1989;19(20):3485−96.

[440] Covey DF, Reddy PA, Ferrendelli JA. Anticonvulsant and anxiolytic lactam and thiolactam derivatives, US 5776959; 1998.

[441] Allen JR, Hitchcock SA, Turner W, Wilson Jr., Liu B. Preparation of indanylamide derivatives as muscarinic M1 agonists for the treatment of cognitive disorders, WO 2004094382; 2004.

[442] Mimura M, Hayashida M, Nomiyama K, Ikegami S, Iida Y, Tamura M, et al. Synthesis and evaluation of (piperidinomethylene)bis(phosphonic acid) derivatives as antiosteoporosis agents. Chem Pharm Bull 1993;41(11):1971−86.

[443] Fischer K, Oberrauch H, Lactams DE. 935485; 1955.

[444] Aicher TD, Chicarelli MJ, Gauthier CA, Hinklin RJ, Tian H, Wallace OB, et al. Cycloalkyl lactam derivatives as inhibitors of 11-beta-hydroxysteroid dehydrogenase 1, WO 2006068992; 2006.

[445] Heo I-J, Lee S-J, Cho C-W. Direct Lactamization of Azido Amides via Staudinger-Type Reductive Cyclization. Bull Korean Chem Soc 2012;33(2):333−6.

[446] Lee S-J, Heo I-J, Cho C-W. One-pot synthesis of five-, six-, and seven-membered lactams via Bu3SnH-mediated reductive cyclization of azido amides. Bull Korean Chem Soc 2012;33(2):739−41.

[447] Kost AN, Shchegoleva TA, Yudin LG. Reduction with formic acid and its derivatives. III. Synthesis of substituted α-piperidones. Zh Obshch Khim 1955;25:2464−9.

[448] No Inventor data available, Lactams, DE 609244; 1935.

[449] Qi J, Sun C, Tian Y, Wang X, Li G, Xiao Q, et al. Highly Efficient and Versatile Synthesis of Lactams and N-Heterocycles via Al(OTf)3-Catalyzed Cascade Cyclization and Ionic Hydrogenation Reactions. Org Lett 2014;16(1):190−2.

[450] Albertson NF, Fillman JL. A synthesis of DL proline. J Am Chem Soc 1949;71:2818−20.

[451] Ruzicka L. Derivatives of δ- and ε- amino acids. Helv Chim Acta 1921;4:472−82.

[452] Stephen T, Stephen H. The Beckmann rearrangement of cyclopentanone oxime. J Chem, Soc 1956;4694−5.

[453] Horning EC, Stromberg VL. Beckmann rearrangements. A new method. J Am Chem Soc 1952;74:2680−1.

[454] Matache S. Lactams by Beckmann rearrangement of oximes, GB 1125246; 1968.

[455] Sato H, Yoshioka H, Izumi Y. Homogeneous liquid-phase Beckmann rearrangement of oximes catalyzed by phosphorous pentoxide and accelerated by a fluorine-containing strong acid. J Mol Catalysis A: Chemical 1999;149(1-2):25−32.

[456] Takahashi T. Preparation of d-valerolactam from cyclopentanone oxime, JP 071145145; 1995.

[457]   Takahashi T, Ueno K, Kai T. Vapor phase Beckmann rearrangement of cyclopentanone oxime over high silica HZSM-5 zeolites. Microporous Mater 1993;1(5):323−7.

[458]   Thai K, Wang L, Dudding T, Bilodeau F, Gravel M. NHC-Catalyzed Intramolecular Redox Amidation for the Synthesis of Functionalized Lactams. Org Lett 2010;12 (24):5708−11.

[459]   Thompson CM, Green DLC, Kubas R. Remote dianions. 3. Novel synthesis of substituted 2-piperidones from imines. J Org Chem 1988;53(22):5389−90.

[460]   Sorto NA, Di Maso MJ, Munoz MA, Dougherty RJ, Fettinger JC, Shaw JT. Diastereoselective Synthesis of gamma- and delta-Lactams From Imines and Sulfone-Substituted Anhydrides. J Org Chem 2014;79(6):2601−4.

[461]   Koelsch CF. Synthesis of 3-alkylpiperidones. J Am Chem Soc 1943;65:2458−9.

[462]   Koelsch CF. Synthesis of 4-phenylpiperidines. J Am Chem Soc 1943;65:2459−60.

[463]   Schmitt GJ, Klein KP, Reimschuessel HK. Preparation and polymerization of 4-carboxy-2-piperidone and its esters, DE 2006063; 1970.

[464]   Pearce HL, Winter MA. Preparation of anticancer 5-deaza-10-oxo- and 5-deaza-10-thio-5,6,7,8-tetrahydrofolic acids and intermediates, US 5159079; 1992.

[465]   Allen JR, Hitchcock SA, Turner WW Jr, Liu B. Preparation of indanylamide derivatives as muscarinic M1 agonists for the treatment of cognitive disorders, WO 2004094382; 2004.

[466]   Lawson EC, Maryanoff BE. Preparation of bicyclic triazole amino acid derivatives as α4 integrin inhibitors, WO 2006052962; 2006.

[467]   Biaggio FC, Barreiro EJ. The synthesis of a new 8-azaprostanoid. J Het Chem 1989;26 (3):725−8.

[468]   Jao E, Slifer PB, Lalancette R, Hall SS. Chemistry of 3,4-dihydro-2-alkoxy-2H-pyrans. 10. Addition-Rearrangement of Aryl- and Alkoxysulfonyl Isocyanates with 5-Methyl-Substituted 3,4-Dihydro-2-methoxy-2H-pyrans. Selective Synthesis of Functionalized 2-Piperidones. J Org Chem 1996;61(8):2865−70.

[469]   Husson H-P, Royer J. Chiral non-racemic N-cyanomethyloxazolidines: the pivotal system of the CN(R,S) method. Chem Soc Rev 1999;28(6):383−94.

[470]   Escolano C, Amat M, Bosch J. Chiral oxazolopiperidone lactams: versatile intermediates for the enantioselective synthesis of piperidine-containing natural products. Chem Eur J 2006;12(32):8198−207.

[471]   Amat M, Escolano C, Gomez-Esque A, Lozano O, Llor N, Griera R, et al. Stereoselective α-amidoalkylation of phenylglycinol-derived lactams. Synthesis of enantiopure 5,6-disubstituted 2-piperidones, Tetrahedron: Asymmetry 2006;17(10):1581−8.

[472]   Bantreil X, Prestat G, Moreno A, Madec D, Fristrup P, Norrby P-O, et al. γ- and δ-Lactams through palladium-catalyzed intramolecular allylic alkylation: enantioselective synthesis, NMR investigation, and DFT rationalization. Chem Eur J 2011;17 (10):2885−96 S2885/1-S2885/122

[473]   Zhou C-Y, Che C-M. Highly efficient Au(I)-catalyzed intramolecular addition of β-ketoamide to unactivated alkenes. J Am Chem Soc 2007;129(18):5828−9.

[474]   Ecija M, Diez A, Rubiralta M, Casamitjana N, Kogan MJ, Giralt E. Synthesis of 3-Aminolactams as X-Gly Constrained Pseudodipeptides and Conformational Study of a Trp-Gly Surrogate. J Org Chem 2003;68(25):9541−53.

[475]   Casamitjana N, Lopez V, Jorge A, Bosch J, Molins E, Roig A. Diels-Alder reactions of 5,6-dihydro-2(1H)-pyridones. Tetrahedron 2000;56(24):4027−42.

[476]   Herdeis C. Chirospecific synthesis of (S)-(+)- and (R)-(-)-5-amino-4-hydroxypentanoic acid from L- and D-glutamic acid via (S)-(+)- and (R)-(-)-5-hydroxy-2-oxopiperidine. Synthesis 1986;(3):232−3.

[477] Olsen RK, Bhat KL, Wardle RB, Hennen WJ, Kini GD. Syntheses of (S)-(-)-3-piperidi-nol from L-glutamic acid and (S)-malic acid. J Org Chem 1985;50(6):896−9.

[478] Eisleb O. New syntheses with sodium amide. Chem Ber 1941;74B:1433−50.

[479] Bergel F, Morrison AL, Rinderknecht H. Synthetic analgesics. II. New synthesis of pethidine and similar compounds. J Chem Soc 1944;265−7.

[480] Blicke FF, Faust JA, Krapcho John, Tsao EuPhang. The preparation of 4-cyano-1-methyl-4-phenylpiperidine. J Am Chem Soc 1952;74:1844−5.

[481] Thompson D, Reeves PC. Facile synthesis of N-substituted-4-cyano-4-phenylpiperidines via phase-transfer catalysis. J Het Chem 1983;20(3):771−2.

[482] Lee N, Zhang X, Darna M, Dwoskin LP, Zheng G. Muscarinic acetylcholine receptor binding affinities of pethidine analogs. Bioorg Med Chem Lett 2015;25(22):5032−5.

[483] No Inventor data available, Piperidine derivatives, CH 262428; 1949.

[484] Kagi H, Miescher K. New synthesis of 4-phenyl-4-piperidyl alkyl ketones and related compounds with morphinelike action. Helv Chim Acta 1949;32:2489−507.

[485] Kaegi, H.; Miescher, K., Preparation of 4-aryl-4-carbalkoxy-piperidines, US 2486795; 1949.

[486] Wu W, Burnett DA, Spring R, Qiang L, Sasikumar TK, Domalski MS, et al. Synthesis and structure-activity relationships of piperidine-based melanin-concentrating hormone receptor 1 antagonists. Bioorg Med Chem Lett 2006;16(14):3668−73.

Chapter 2

1-Substituted Piperidines

This book was conceived as an attempt to demonstrate the capabilities of functionally substituted piperidines for the creation of different classes of medicines. Therefore, this chapter does not formally refer to the concept of the book. Nevertheless, for completeness of the panorama of piperidine ring containing drugs, we have included here medicines — representatives of different classes of compounds — where piperidine moiety exists as just one of the possible side chain amino substituents.

One example of these type of drugs is the family 3-dialkylaminopropanols, which are effectively used for controlling various central nervous system disorders such as anxiety, depression, and sleep disorders. These compounds also possess an analgesic effect. Their central anticholinergic properties are also used in treating Parkinsonism. The therapeutic value of these drugs is relatively small, and they are used either in combination with levodopa, or in cases of minor Parkinsonism.

2.1 DERIVATIVES OF 1-PHENYL-3-(PIPERIDIN-1-YL)PROPAN-1-OL

The drugs which belong to the compounds of the series 1-phenyl-3-(piperidin-1-yl)propan-1-ols and represented by trihexyphenidyl (**2.1.1**), biperiden (**2.1.2**), pridinol (**2.1.3**), and cycrimine (**2.1.4**) (Fig. 2.1).

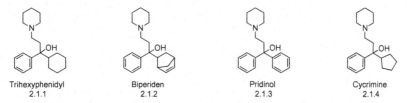

Trihexyphenidyl
2.1.1

Biperiden
2.1.2

Pridinol
2.1.3

Cycrimine
2.1.4

FIGURE 2.1 Derivatives of 1-phenyl-3-(piperidin-1-yl)propan-1-ol.

Trihexyphenidyl (2226)

Trihexyphenidyl (**2.1.1**) is classified as an antiparkinsonian agent. It is available under its generic name and under the brand names Artane and

Piperidine-Based Drug Discovery. DOI: http://dx.doi.org/10.1016/B978-0-12-805157-3.00002-8

Apo-Trihex, Parkin, Pacitane. Trihexyphenidyl is used to treat a group of parkinsonian side effects that include tremors, difficulty walking, and slack muscle tone. The exact mechanism of action of trihexyphenidyl is not precisely understood, but it is known that it blocks efferent impulses in para-sympathetically innervated structures like smooth muscles, salivary glands, and eyes. Recent observations suggest the existence of *trihexyphenidyl* abuse linked to its hallucinogenic and euphoric effects [1].

Trihexyphenidyl (**2.1.1**), was synthesized by the Grignard reaction of 2-(1-piperidino)-propiophenone (**2.1.6**) with cyclohexylmagnesiumbromide (chloride). This initial ketone (**2.1.6**) in turn was prepared by the amino-methylation of benzophenone (**2.1.5**) in Mannich reaction conditions using paraformaldehyde and piperidine [2–4] (Scheme 2.1).

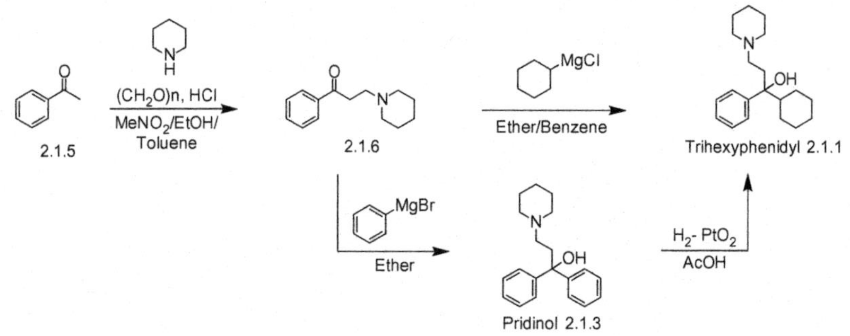

SCHEME 2.1 Synthesis of trihexyphenidyl.

An alternate way by catalytic hydrogenation of only one phenyl group in 3-piperidino-1,1-diphenyl-1-propanol—pridinol (**2.1.7**) in acetic acid with H_2 (1 atm.) in the presence of Adams's catalyst (PtO_2) has been reported [5].

The absolute (*R*) configuration has been established for the more active (−)-enantiomer of the anticholinergic trihexyphenidyl hydrochloride [6].

Biperiden (1329)

Biperiden (**2.1.2**) (Akineton) is another centrally active anticholinergic drug. It is a potent, selective M1 receptor antagonist used as an adjunct in all forms of parkinsonism [7,8]. Biperiden is prepared by method analogs and is implemented for the synthesis for trihexyphenidyl, where exo-1-(bicyclo [2.2.1]hept-5-en-2-yl)ethanone (**2.1.7**) with paraformaldehyde and piperidine hydrochloride is converted in Manich reaction conditions into exo/endo-1-(bicyclo[2.2.1]hept-5-en-2-yl)-3-piperidino-1-propanone (**2.1.8**) followed by a Grignard reaction with phenylmagnesium bromide (Scheme 2.2).

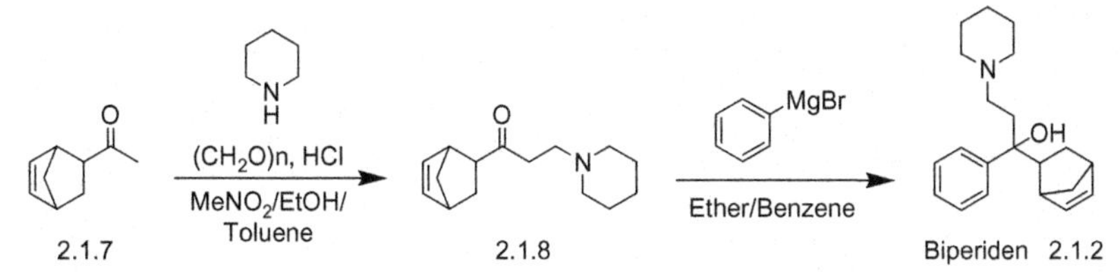

SCHEME 2.2 Synthesis of biperiden.

Pridinol (157)

Pridinol (Nonplesin) (**2.1.3**) is another antiparkinsonian drug used also for the treatment of a variety of other clinical conditions that have in common only the presence of skeletal muscle hyperactivity.

It was synthesized in three different methods. The first includes analogs that are implemented for the synthesis of *trihexyphenidyl*, starting from ketone (**2.1.6**), which reacts with phenyl magnesium bromide to give pridinol (**2.1.3**) [9] (Scheme 2.3).

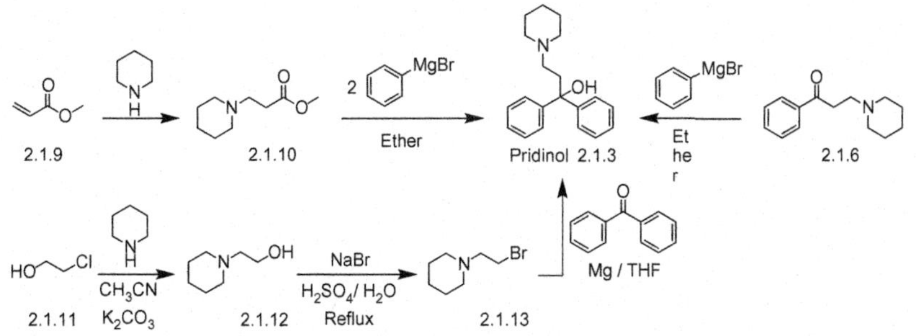

SCHEME 2.3 Synthesis of pridinol.

The second method comprises the steps of Michael addition of piperidine to methyl acrylate (**2.1.9**) to obtain 3-(1-piperidino) propionic acid methyl ester (**2.1.10**) followed by a reaction with phenyl magnesium bromide, which results in the preparation of the desired pridinol (**2.1.3**) [9−11].

The third, a patented method that includes the alkylation of piperidine with chlorohydrin (**2.1.11**) to obtain 2-piperidinyl ethanol (**2.1.12**) and the performing of its bromination in sodium bromide sulfuric acid media, which results in bromide (**2.1.13**). Prepared bromide (**2.1.13**) is described as reacting with magnesium to obtain a Grignard reagent, which reacts with benzophenone gaing pridinol (**2.1.3**) [12] (Scheme 2.3).

Cycrimine (82)

Cycrimine (**2.1.4**) (Pagitane) is another central anticholinergic used to reduce levels of acetylcholine and to return a balance with dopamine, which is rarely employed in the management and treatment of Parkinson's disease. It is prepared by the addition of 1-phenyl-3-(piperidin-1-yl)propan-1-one (**2.1.6**) to a solution of cyclopentylmagnesium bromide by way of an analog with the synthesis of trihexyphenidyl (**2.1.1**) ((**2.1.6**) → (**2.1.1**)), as shown on the Scheme 2.1 [2,13].

2.2 DERIVATIVES OF 1-PHENYL-4-(PIPERIDIN-1-YL)BUTAN-1-OL

The closest homologs of 1-phenyl-3-(piperidin-1-yl)propan-1-ols are derivatives of 1-phenyl-4-(piperidin-1-yl)butan-1-ols represented by two drugs: diphenidol (**2.2.1**) and pirmenol (**2.2.2**) (Fig. 2.2).

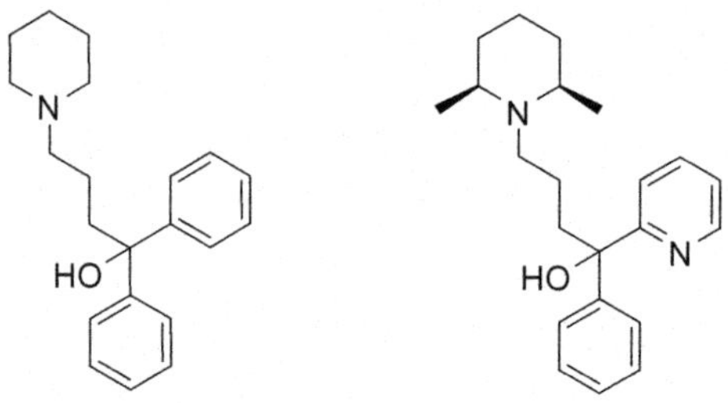

Diphenidol 2.2.1   Pirmenol 2.2.2

FIGURE 2.2   Derivatives of 1-phenyl-4-(piperidin-1-yl)butan-1-ol.

Diphenidol (323)

Diphenidol (Cephadol, Vontrol) (**2.2.1**), an antiemetic agent which exerts anticholinergic effect due to interactions with M1, M2, M3, and M4 and used in the treatment of vomiting and vertigo [14].

It has been reported to cause various adverse effects including drowsiness, hypotension, confusion, hallucination, restlessness, and other antimuscarinic effects [15]. The antiarrhythmic activity of diphenidol has been also demonstrated [16].

Diphenidol has been synthesized by Grignard reaction of (3-(piperidin-1-yl)propyl)magnesium chloride with benzophenone (**2.2.5**) in ether or tetrahydrofuran (THF). Starting N-[L-chloropropyl-(3)]piperidine (**2.2.4**) was prepared by alkylation of piperidine with 1-bromo-3-chloropropane (**2.2.3**) in benzene in the presence of triethylamine [17,18] (Scheme 2.4).

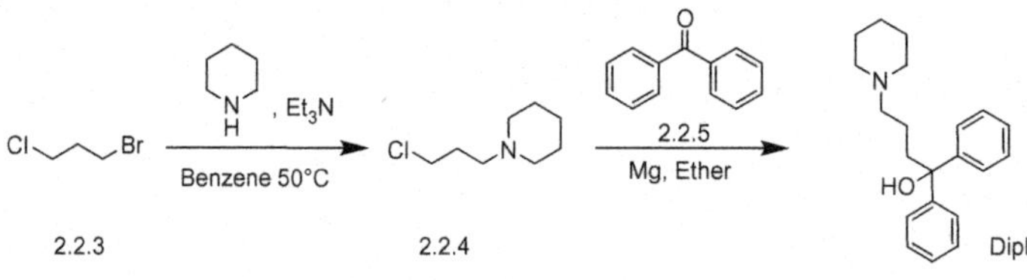

SCHEME 2.4   Synthesis of diphenidol.

Pirmenol (234)

Pirmenol (**2.2.2**) is a new and long-acting class Ia antiarrhythmic drug that has been shown to be effective in both intravenous and oral form for suppression of ventricular ectopic depolarizations, ventricular tachycardia, and supraventricular arrhythmia [19−23].

   Pirmenol (**2.2.2**) was first synthesized by direct addition of 2-lithiopyridine (**2.2.11**) to 4-(2,6-dimethyl-piperidinyl)-1-phenyl-1-butanon (**2.2.10**), which, in turn, was prepared by alkylation of cis-2,6-dimethylpiper-idine (**2.2.8**) with γ-chlorobutyrophenone ethylene ketal (**2.2.7**) followed by acidic hydrolysis of obtained product (**2.2.9**) to (**2.2.10**), which, in turn, was prepared in traditional way from γ-chlorobutyrophenone (**2.2.6**) and ethylene glycol by refluxing their mixture in benzene in the presence of p-toluenesul-fonic acid [24] (Scheme 2.5).

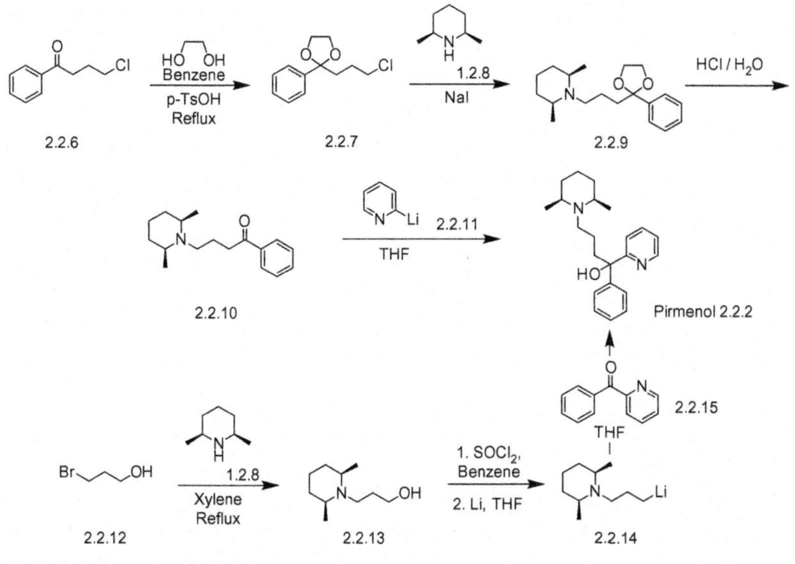

SCHEME 2.5 Synthesis of pirmenol.

   Another approach for the synthesis of pirmenol (**2.2.2**) was demonstrated in the same patent where cis-1-(3-chloropropyl)-2,6-dimethylpiperidine was prepared by alkylation of cis-2,6-dimethylpiperidine (**2.2.8**) with 3-bromopropanol (**2.2.12**) in xylene was chlorinated with thionyl chloride in benzene to give an intermediate cis-1-(3-chloropropyl)-2,6-dimethylpiperidine, (**2.2.13**), which was then involved in a reaction with lithium in THF to prepare a lithium derivative (**2.2.14**). Adding to this (3-((2S,6R)-2,6-dimethylpiperidin-1-yl)propyl)lithium (**2.2.14**) 2-benzoylpyridine (**2.2.15**) allowed the preparation of the desired pirmenol (**2.2.2**) (Scheme 2.5).

Some modifications of the two proposed general methods were later published [25,26].

2.3 DERIVATIVES OF 2-(PIPERIDIN-1-YL)ETHAN-1-OL AND 3-(PIPERIDIN-1-YL)PROPAN-1-OL

Generalizing derivatives of 2-(piperidin-1-yl)ethan-1-ol or 3-(piperidin-1-yl)propan-1-ol is possible as a way to classify them as ethers and esters.

Ether derivatives of 2-(piperidin-1-yl)ethan-1-ol are presented in the pharmaceutical market as cloperastine (**2.3.1**), benproperine (**2.3.2**), pipazetate (**2.3.3**), and raloxifene (**2.3.4**) (Fig. 2.3).

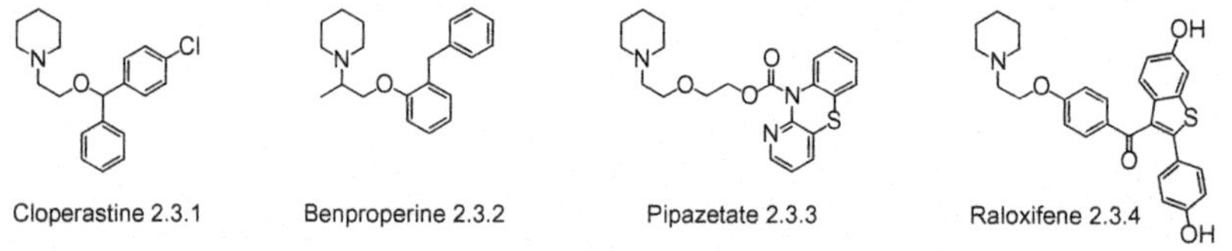

Cloperastine 2.3.1   Benproperine 2.3.2   Pipazetate 2.3.3   Raloxifene 2.3.4

FIGURE 2.3   Derivatives of 2-(piperidin-1-yl)ethan-1-ol and 3-(piperidin-1-yl)propan-1-ol.

The first three are cough suppressants, and raloxifene (**1.3.4**), an antiresorptive agent, is a new representative of a class of drugs that prevent the loss of bone mass, as it is used to treat osteoporosis and similar diseases.

Cloperastine (199)

Cloperastine (**2.3.1**) is a cough suppressant that has been shown to possess dual activity. It is a drug used in the treatment of acute and chronic cough having a dual mechanism of action at the central bulbar cough center and at peripheral receptors in the tracheobronchial tree showing also mild bronchorelaxant and antihistaminic properties [27]. It was synthesized by reaction of 4-chlorobenzhydrol (**2.3.5**) with N-(2-chlororthyl)piperidine (**2.3.6**) in benzene in the presence of sodium amide [28] (Scheme 2.6). Later the optical isomers of cloperastine, which exists in dextrorotatory and levorotatory forms, were prepared [29]. The levorotatory enantiomer L(−)cloperastine has been shown to be a less toxic product having fewer side effects compared to the D(+) isomer and the racemic mixture [30].

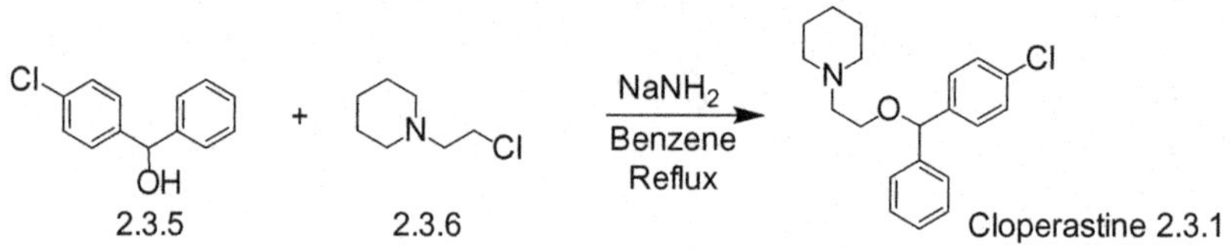

2.3.5   2.3.6   Cloperastine 2.3.1

SCHEME 2.6   Synthesis of cloperastine.

Benproperine (140)

Benproperine (Cofrel) (**2.3.2**) is a racemate comprising equimolar amounts of (R)- and (S)-benproperine. It is used as a cough suppressant in the treatment of acute and chronic bronchitis, a cough that is due to varieties of causes [31].

Synthesis of benproperine (**2.3.2**) consists of a reaction of O-benzylphenol (**12.3.7**) and propylene oxide (**2.3.8**) in the presence of sodium at 140°C, which gave 1-O-benzylphenoxy-2-propanol (**2.3.9**). Obtained product was tosylated with toluenesulfonylchloride in pyridine to give O-benzylphenoxy-2-propyl p-tosylate (**2.3.10**), or chlorinated with thionyl chloride to give O-benzylphenoxy-2-propyl chloride (**2.3.11**). Prepared tosylate (**2.3.10**) or chloride (**2.3.11**) were used for N-alkylation of piperidine methanol to yield the desired benproperine (**2.3.2**) [32,33].

The synthesis of R-(+) and S-(−)-benproperine using chiral synthons is described [34,35] (Scheme 2.7). The antitussive effect of the R-(+)- and S-(−)-enantiomers of benproperine did not show any advantage over the racemate with regard to their antitussive effect [36].

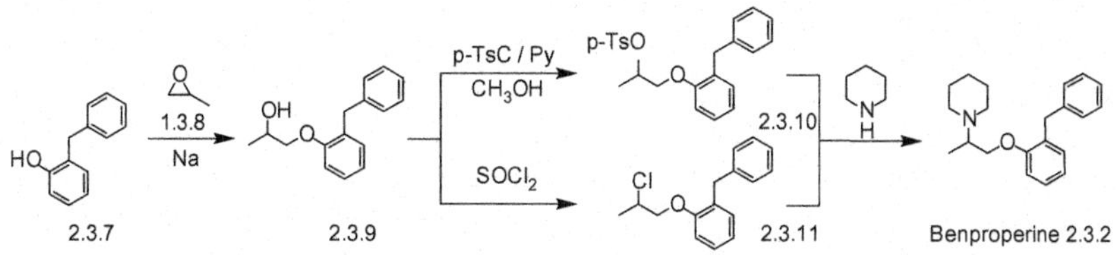

SCHEME 2.7   Synthesis of benproperine.

Pipazethate (98)

Pipazethate (Pipazetate, Theratuss, Lenopect) (**2.3.3**) is a centrally acting antitussive drug which also has bronchodilatory and local anesthetic activities and has been introduced as a cough remedy [37,38]. Synthesis of pipazethate had been described in details [37,39,40].

Pipazethate (**2.3.3**) was prepared by reaction of 1-azaphenothiazine (**2.3.12**) with phosgene (COCl$_2$) in the presence of pyridine to give 1-azaphenothiazine-10-carbonyl chloride (**2.3.13**). Obtained (**2.3.13**) was condensed with (2-piperidinoethoxy)ethanol (**2.3.14**) giving desired pipazethate (**2.3.3**) (Scheme 2.8).

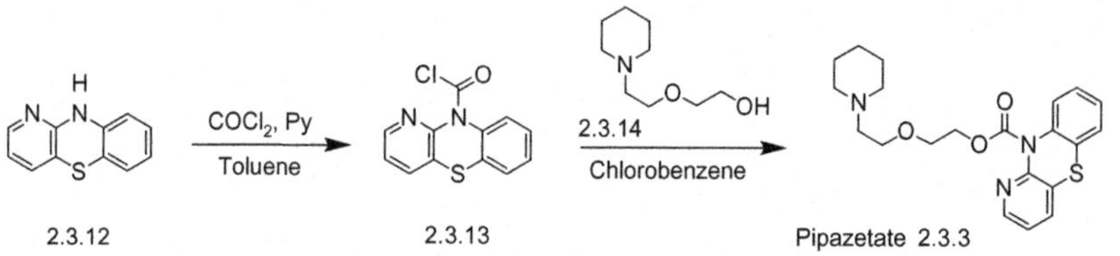

SCHEME 2.8 Synthesis of pipazethate.

Raloxifene (7685)

Raloxifene (Evista) (1.3.4) is a second-generation selective estrogen receptor modulator that functions as an estrogen antagonist on breast and uterine tissues, and an estrogen agonist on bone. Raloxifene is an antiresorptive agent, a new representative of a class of drugs that prevent the loss of bone mass, i.e., used to treat osteoporosis and similar diseases in postmenopausal women and those postmenopausal women at increased risk of invasive breast cancer [41−53].

It was shown that raloxifene can have some affect on cognition, mental health, sleep, and sexual function in menopausal women [54]. Raloxifene was used also as an adjuvant treatment in postmenopausal women with schizophrenia [55].

The first reported synthesis of the raloxifene scaffold consists in Friedel-Crafts aroylation in 1,2-dichloroethane and using AlCl₃ as a catalyst by coupling of 4-(2-(piperidin-1-yl)ethoxy)benzoyl chloride (**2.3.15**) with benzothiophene derivative (**2.3.16**) followed by alkaline hydrolysis of mesyl groups, which give the desired raloxifene (**2.3.4**) [56−58] (Scheme 2.9).

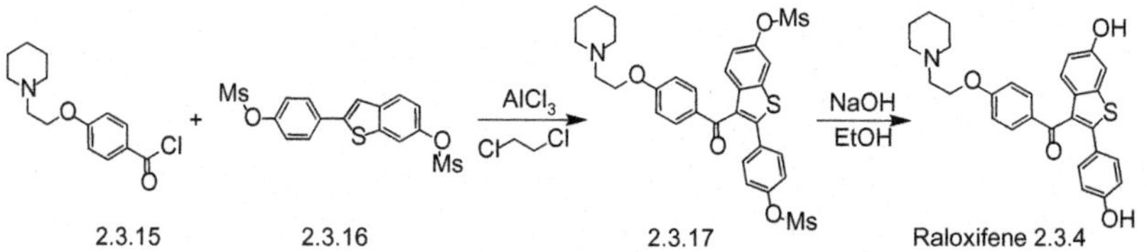

SCHEME 2.9 Synthesis of raloxifene.

The key intermediate − 6-methoxy-2-(4-methoxyphenyl)benzo[*b*]thiophene (**2.3.16**) − was prepared by the cyclization-rearrangement of 1-(4-methoxyphenyl)-2-((3-methoxyphenyl)thio)ethan-1-one (**2.3.20**) induced by polyphosphoric acid (PPA). This rearrangement (Kost rearrangement [59]) is general for 3-(*R*-substituted)indoles, -benzofurans, and -benzothiophenes, which are converted to the corresponding 2-isomers by heating with PPA.

The synthesis started from thiophenol (**2.3.18**) and bromoketone (**2.3.19**), which were coupled in presence of KOH in ethanol/water solution. Obtained (**2.3.20**) was heated with PPA to give a mixture that is easily separable by crystallization isomeric 2-phenylbenzo[*b*]thiophenes (**2.3.21**) and (**2.3.22**), where preferable, isomer (**2.3.22**) predominates. Cleavage of the methoxy groups in (**2.3.22**) was done conveniently with pyridine hydrochloride to give (**2.3.23**), which was easily converted to mesylate (**2.3.16**) with methanesulfonyl chloride in pyridine and 4-dimethylaminopyridine as a catalyst (Scheme 2.10).

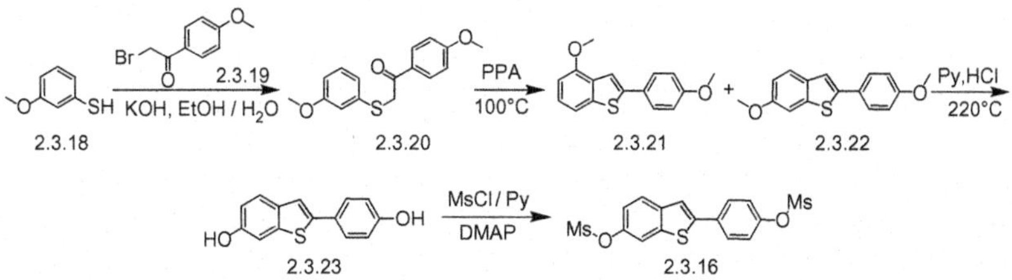

SCHEME 2.10 Synthesis of key intermediate – 6-methoxy-2-(4-methoxyphenyl)benzo[*b*] thiophene.

The   second   reagent—4-(2-(piperidin-1-yl)ethoxy)benzoyl   chloride (**2.3.15**)—was prepared starting with 4-hydroxybenzoate (**2.3.24**), which with 1-(2-chloroethyl)piperidine (**2.3.25**) in anhydrous DMF, and K$_2$CO$_3$ or sodium hydride, gave methyl 4-(2-(piperidin-1-yl)ethoxy)benzoate (**2.3.26**) hydrolyzed in MeOH/water NaOH solution. The acid (**2.3.26**) was converted to its chloride (**2.3.15**) with SOCl$_2$ in 1,2-dichloroethane and a catalytic amount of DMF (Scheme 2.11).

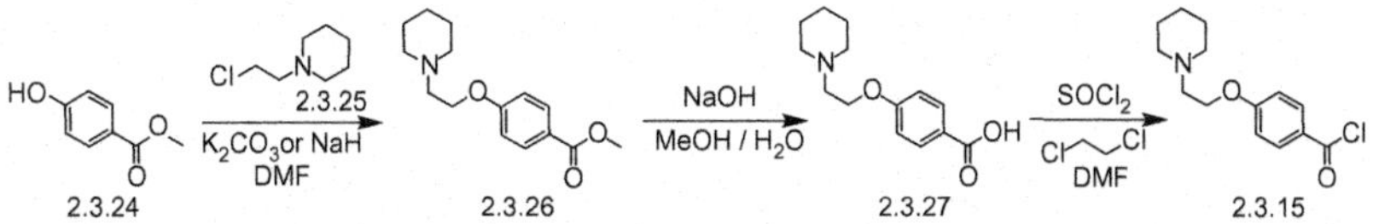

SCHEME 2.11   Synthesis of the second key intermediate—4-(2-(piperidin-1-yl)ethoxy)benzoyl chloride.

Another novel convenient synthesis of raloxifene (**2.3.4**) have been proposed [60]. According to this method anisaldehyde (**2.3.28**) was transformed to corresponding cyanohydrin (**2.3.29**) using a mixture of sodium cyanide ethanol containing triethylamine through which HCl gas was passed over 30 minutes at 5−10°C.

Gaseous HCl was added to the solution of prepared cyanohydrin (**2.3.29**) in ethanol at room temperature over 30 minutes in order to give *p*-methoxy-benzaldehyde cyanohydrin iminoether hydrochloride (**2.3.30**). Then, hydrogen sulfide was bubbled into a solution of the methyl imidate (**2.3.30**) and triethylamine in methanol at 0°C to give α-(4-methoxy phenyl)-α-hydroxy-*N,N* dimethylthioacetamide (**2.3.31**).

To the obtained α-hydroxythioamide (**2.3.31**) dissolved-in-methylene chloride methanesulfonic acid was slowly added, which transformed the starting material to 2-*N,N*-dimethylamino-6-methoxy benzo[β]thiophene (**2.3.32**).

The obtained 2-dimethylaminobenzothiophene (**2.3.32**) and known 4-(2-piperidinoethoxy)-benzoyl chloride (**2.3.15**) were partially dissolved in chlorobenzene and the mixture was warmed in a 100—105°C to give 2-(4-methoxyphenyl)-6-methoxy-3-[4-(piperidinoethoxy)benzoyl]-benzo[β]thiophene (**2.3.33**). 4-Methoxyphenylmagnesium bromide (**2.3.34**) in THF was added to chilled to 0°C prepared compound (**2.3.33**) in THF, which gave 2-(4-methoxyphenyl)-6-methoxy-3-[4-(piperidinoethoxy)benzoyl] benzo[β] thiophene (**2.3.35**). To the prepared benzothiophene (**2.3.35**) suspended in chlorobenzene was added AlCl$_3$, followed by the addition of *n*-propanethiol, and the mixture was heated at 35°C. After the workup with aqueous HCl, the desired raloxifene (**2.3.4**) was separated [60] (Scheme 2.12).

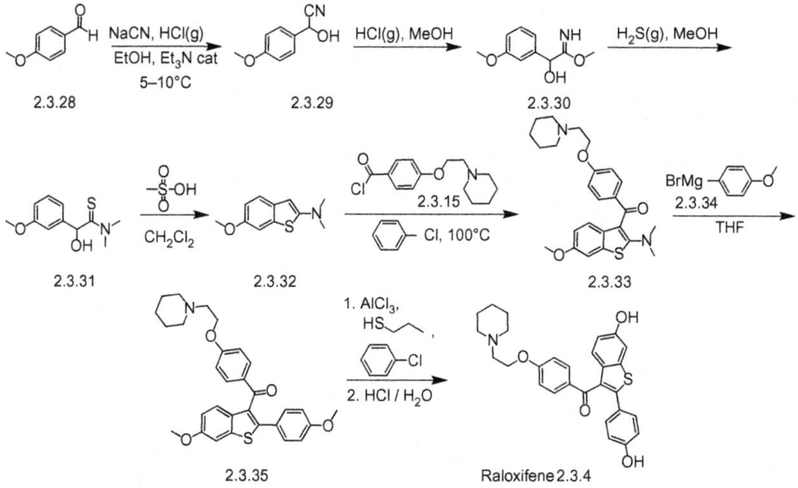

SCHEME 2.12 Synthesis of raloxifene.

There exist plenty of modifications for these two approaches, as reviewed in [61,62].

Flavoxate (452)

Flavoxate (**2.3.30**), a drug with anticholinergic with antimuscarinic, exerts a selective and direct muscle relaxant activity by counteracting smooth muscle spasms of the urinary tract. Due to antispasmodic activity on the bladder, it relieves the pain and discomfort accompanying a variety of urological disorders, and is indicated for symptomatic relief of frequent urination, dysuria, urgency, nocturia, cystitis, prostatitis, urethritis, urethrocystitis/urethrotrigonitis, and incontinence [63−68]. The mechanism of its action is not yet clearly defined. In addition, flavoxate has been shown to have an analgesic and local anesthetic properties.

The synthesis of flavoxate (**2.3.30**) is described starting from salicylic acid (**2.3.32**), which was acylated with propionyl chloride (**2.3.33**) in a standard Schotten−Baumann reaction conditions, in the presence of aqueous sodium or potassium hydroxide to give 2-(propionyloxy)benzoic acid (**2.3.34**). The obtained propionated salicylic acid (**2.3.34**) was transformed to 2-hydroxy-3-carboxypropiophenone (**2.3.35**) via Fries rearrangement of (**2.3.34**). For that purpose anhydrous $AlCl_3$ and CS_2 were added to (**2.3.34**), and after evolving of HCl and removal of CS_2, the mixture was heated at 150−60°C to give 2-hydroxy-3-carboxypropiophenone (**2.3.35**). The reaction of the last with benzoic anhydride in the presence of sodium benzoate at 180−190°C gave 3-methylflavone-8-carboxylic acid (**2.3.36**), which upon treatment with $SOCl_2$ in benzene, gave 3-methylflavone-8-carboxylic acid chloride (**2.3.37**). Obtained acid chloride (**2.3.37**) was reacted with 2-(piperidin-1-yl)ethan-1-ol (**2.3.38**) in benzene to give the desired flavoxate (**2.3.30**) [69−72] (Scheme 2.13).

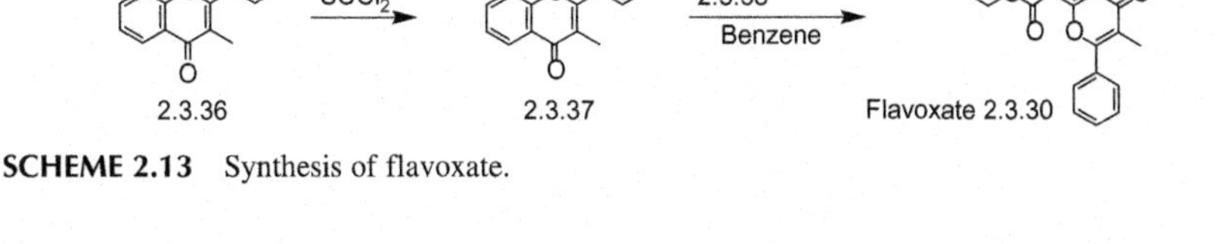

SCHEME 2.13 Synthesis of flavoxate.

Piperocaine (266)

Piperocaine (Metycaine; Neothesin; Isocaine) (**2.3.31**)—3-(piperidin-1-yl) propyl benzoate—is a local anesthetic drug [73,74] and belongs to a big family of "-caine" compounds, which were developed in the 1920s and are used mainly as anesthetics and analgesics (lidocaine, procaine, tetracaine, etc.), decongestants and antiinflammatory agents, circulatory stimulants, and cardio-protective agents as well as fungicides, insecticides, flavorants, repellents, antifoulants, etc.

Piperocaine (**2.3.31**) was prepared starting with benzoyl chloride (**2.3.39**), which, at 0°C, was added to a solution of 3-chloropropan-1-ol (**2.3.40**) and triethylamine in dichloromethane and obtained 3-chloropropyl benzoate (**2.3.41**), which was then heated at 100°C with 3-methyl piperidine to give the desired piperocaine (**2.3.31**) [75] (Scheme 2.14).

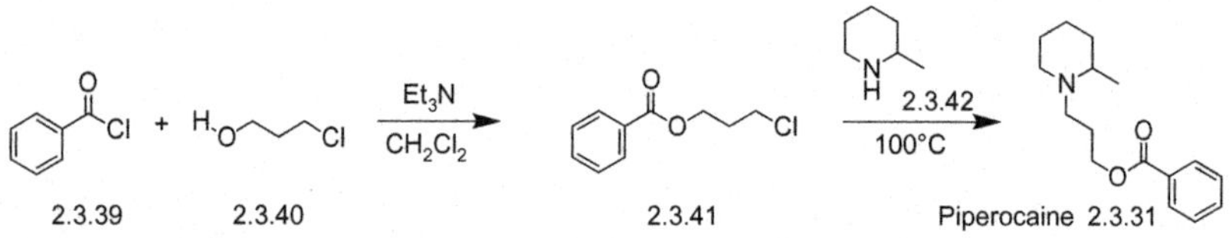

SCHEME 2.14 Synthesis of piperocaine.

2.4 DERIVATIVES OF 3-(PIPERIDIN-1-YL)PROPANE-1,2-DIOL AND 3-(PIPERIDIN-1-YL)PROPANE-1,1-DIOL

There are per-single representatives of 3-(piperidin-1-yl)propane-1,2-diol-diperodon (**2.4.1**) and 3-(piperidin-1-yl)propane-1,1-diol - pipoxolan (**2.4.2**) on the pharmaceutical market (Fig. 2.4).

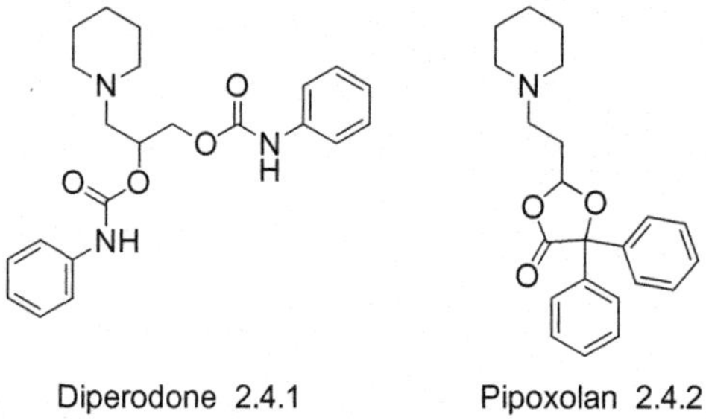

Diperodone 2.4.1   Pipoxolan 2.4.2

FIGURE 2.4 Derivatives of 3-(piperidin-1-yl)propane-1,2-diol and 3-(piperidin-1-yl)propane-1,1-diol.

Diperodon (147)

Diperodon (**2.4.1**) is a local anesthetic used in topical ointment preparations for skin abrasions, irritations, pruritus, and is used intrarectally to bring pain relief from hemorrhoids.

The synthesis of diperodon (**2.4.1**) is based on reaction of piperidine (**2.4.3**) with glycidol (**2.4.4**) [76−78], or α-chlorohydrin (**2.4.5**) [79], both of which give good yields of 3-(piperidin-1-yl)propane-1,2-diol (**12.4.6**) followed by reaction of ethereal solution of phenyl isocyanate (**2.4.7**) [76−79] to give the desired diperodon (**2.4.1**) (Scheme 2.15).

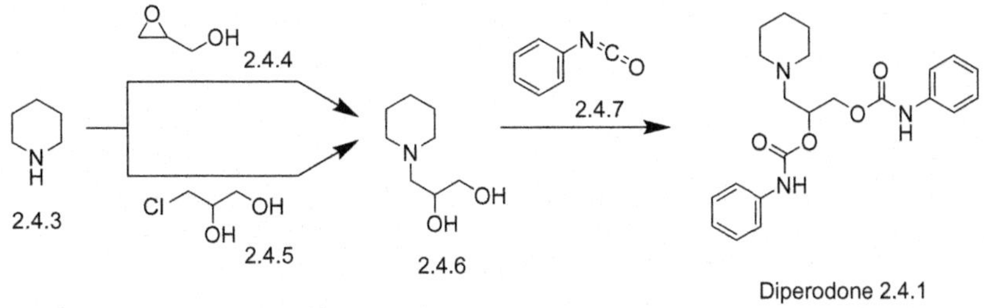

SCHEME 2.15 Synthesis of diperodon.

Pipoxolan (25)

Pipoxolan (**1.4.2**) has been proposed as an antispasmodic medication (animuscarinic) for relief of smooth muscle spasms in the digestive tract, bronchial tree, urinary tract, and gynecological system [80−82].

It was effective in treating dysmenorrhea, renal colic, bilateral urinary lithiasis, cholelithiasis, chronic gastritis, postnatal uterine pain, urolithiasis, and hydronephrosis [83].

Pipoxolan is a useful therapeutic drug for pathological conditions caused by vascular smooth muscle cell proliferation and migration to relieve a body appeared vascular injury, cerebrovascular ischemia, intimal hyperplasia, atherosclerotic stenosis, cerebral ischemia, and stroke [84]. Pipoxolan has been reported to have antitumor activity [85,86].

The synthesis of pipoxolan (**2.4.2**) started from a reaction of 3-chloropropionaldehyde diethylacetal (**2.4.8**) with piperidine (**2.4.3**) to give 1-(3,3-diethoxypropyl)piperidine (**2.4.9**). The mixture of synthesized diethyl acetal (**2.4.9**) and benzylic acid (**2.4.10**) in glacial acetic acid was treated with dry gaseous HCl until saturation to give pipoxolan (**2.4.2**) [87,88] (Scheme 2.16).

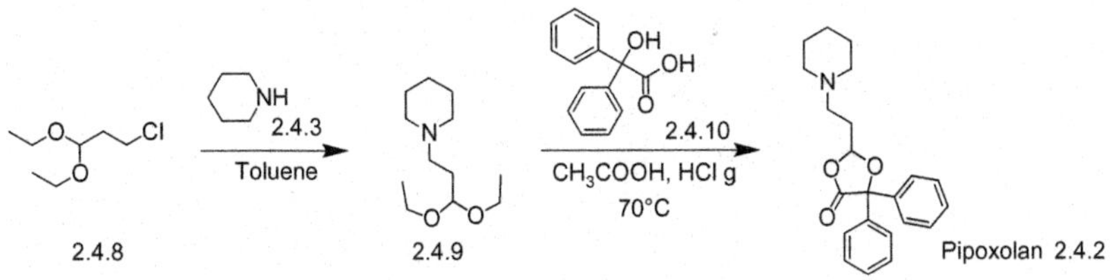

SCHEME 2.16 Synthesis of pipoxolan.

2.5   DERIVATIVES OF 1-PHENYL-3-(PIPERIDIN-1-YL)PROPAN-1-ONE

The single drug represented by this series of piperidine derivatives is dyclonine (**1.5.1**).

Dyclonine (584)

Dyclonine (**2.5.1**) is a local anesthetic with bactericidal and fungicidal properties with strong mucous membrane penetration. It has been used as a local anesthetic agent prior to laryngoscopy, bronchoscopy, esophagoscopy, or endotracheal intubation, and so on. It is used in combination with menthol for temporary relief of occasional minor irritation, pain, or soreness of mouth or throat, and for cough associated with a cold or inhaled irritants. It is used also as ingredient of Sucrets, an over-the-counter throat lozenge [89]. Some publications mention the implementation of dyclonine for the treatment of premature ejaculation, which is recognized to be the most common male sexual disorder [90,91].

Dyclonine was synthesized starting with *p*-butoxyacetophenone (**2.5.2**), which in Mannich reaction with piperidine hydrochloride (**2.5.3**) and paraformaldehyde (**2.5.4**) in a refluxing mixture of nitroethane, toluene, and ethanol, and a catalytic amount of HCl, provided the desired product (**2.5.1**) [92−95] (Scheme 2.17).

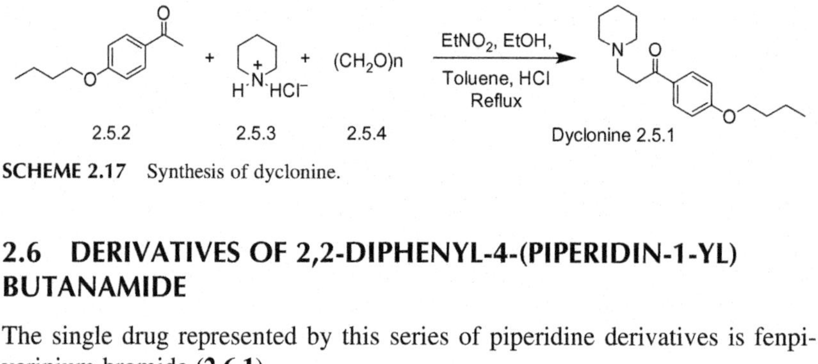

SCHEME 2.17   Synthesis of dyclonine.

2.6   DERIVATIVES OF 2,2-DIPHENYL-4-(PIPERIDIN-1-YL) BUTANAMIDE

The single drug represented by this series of piperidine derivatives is fenpiverinium bromide (**2.6.1**).

Fenpiverinium Bromide (83)

Fenpiverinium bromide (**2.6.1**) is an anticholinergic and antispasmodic compound. Data concerning fenpiverinium is controversial. On the one hand, it is classified as an active pharmaceutical ingredient. On the other, it is intended for research use only, not for use in human, therapeutic, or diagnostic applications without the expressed written authorization of United States Biological.

At the same time it is used in the Bulgarian drug Spasmalgon and Indian Spasmalgin, which together are a combination of fenpiverinium bromide, pitofenone, and metamisol sodium. Spasmalgon releases pain and spasms of smooth muscles showing definite synergism between the antispasmodics pitofenone and fenpiverinium [96−98].

The fenpiverinium bromide (**2.6.1**) is prepared by quaternization of 2,2-diphenyl-4-(piperidin-1-yl)butanamide (**2.6.5**) with methyl bromide (**2.6.6**) in an appropriate solvent. In turn, 2,2-diphenyl-4-(piperidin-1-yl)butanamide (**2.6.5**) was synthesized by alkylation of diphenylacetonitrile (**12.6.2**) with 1-(2-chloroethyl)piperidine (**2.6.3**) in refluxing toluene using as a base lithium amide followed by hydrolysis of an obtained nitrile (**2.6.4**) group to amide on heating with concentrated sulfuric acid [99−102] (Scheme 2.18).

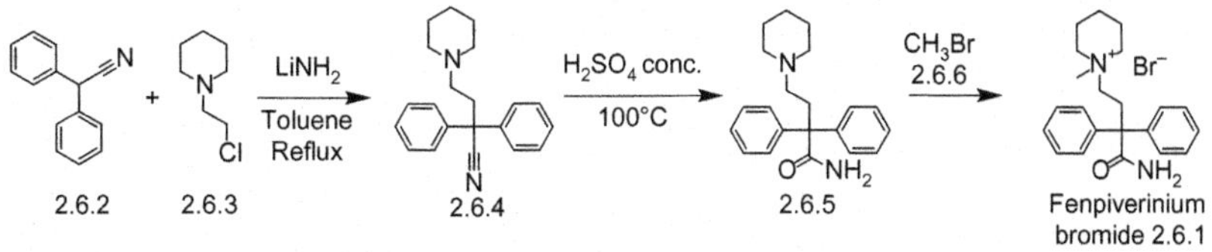

SCHEME 2.18   Synthesis of fenpiverinium bromide.

REFERENCES

[1]   Frauger E, Thirion X, Chanut C, Natali F, Debruyne D, Saillard C, et al. Misuse of trihexyphenidyl (Artane, Parkinane): recent trends. Therapie 2003;58(6):541−7.

[2]   Ruddy AW, Becker TJ. Substituted tertiary aminoalkyl carbinols, US 2680115; 1954.

[3]   Denton JJ. Basic tertiary piperidino alcohols, US 2716121; 1956.

[4]   Baltzly R, Billinghurst JW. Reaction of Grignard reagents with β-tertiary amino ketones. J. Org. Chem. 1965;30(12):4330−2.

[5]   Adamson DW, Wilkinson S. Catalytic reduction of diphenylalkanolamines, US 2682543; 1954.

[6]   Adamson DW, Duffin WM. Resolution of cyclohexylphenylaminopropanols, GB 750156; 1956.

[7]   Hanna C. Toxicologic studies on biperiden (Akineton) (1-bicyclo-heptenyl-1-phenyl-3-piperidino-1-propanol). Toxicol. Appl. Pharmacol. 1960;2:379−91.

[8]   Ammon S. Biperiden. Deutsche Apotheker Zeitung 1989;129(21):1092−3.

[9]   Eisleb O. γ,γ-Diphenylallylamines and their hydrogenation products, DE 875660; 1953.

[10]   Adamson DW. Substituted γ-hydroxypropylamines, GB 624118; 1949.

[11]   Wang W, Zhou B, Sun D, Hu L, Li L, Miao J, et al. Preparation of pridinol mesylate, CN 104262290; 2015.

[12]   Wang L. Process for preparation of pridinol mesylate, CN 103254154; 2013.

[13]   Denton JJ, Schedl HP, Lawson Virginia A, Neier WB. Antispasmodics. VII. Additional morpholinyl and piperidyl tertiary alcohols. J. Am. Chem. Soc. 1950;72:3795−6.

[14]   Leonard CA, Fujita T, Tedeschi DH, Zirkle CL, Fellows EJ. The pharmacology of diphenidol, a potent antiemetic. J. Pharmacol. Exp. Ther. 1966;154(2):339−45.

[15]   Yang CC, Deng JF. Clinical experience in acute overdosage of diphenidol From Journal of toxicology. Clin. Toxicol. 1998;36(1−2):33−9.

[16]   Wasserman AJ, Horgan JH, ul-Hassan Z, Proctor JD. Diphenidol treatment of arrhythmias. Chest 1975;67(4):422−4.

[17]   Miescher K, Marxer A. Amino alcohols, US 2411664; 1946.

[18] Craig PN, Zirkle CL. 1,1-Diphenyl-4-piperidinobutanol, BE 620404; 1961.

[19] Salerno DM, Fifield J, Farmer C, Hodges M. Pirmenol: an antiarrhythmic drug with unique electrocardiographic features--a double-blind placebo-controlled comparison with quinidine. Clin. Cardiol. 1991;14(1):25−32.

[20] Toivonen LK, Nieminen MS, Manninen V, Frick MH. Pirmenol in the long-term treatment of chronic ventricular arrhythmias: a placebo-controlled study. J. Cardiovasc. Pharmacol. 1986;8(1):156−60.

[21] Kaplan HR, Mertz TE, Steffe TJ, Toole JG. Pirmenol. New Drugs Annual: Cardiovascular Drugs, 1. 1983. p. 133−50.

[22] Kaplan HR,, Mertz TE, Steffe TJ. Preclinical pharmacology of pirmenol. Am. J. Cardiol. 1987;59(16):2H−9H.

[23] Reiter MJ. Clinical pharmacology and pharmacokinetics of pirmenol. Angiol. 1988;39(3 Pt 2):293−8.

[24] Fleming RW. (+,-)-cis-α-[3-(2,6-Dimethyl-1-piperidinyl)propyl]-α-phenyl-2-pyridine-methanol, DE 2806654; 1978.

[25] Hoefle ML, Blouin LT, Fleming RW, Hastings S, Hinkley JM, Mertz TE, et al. Synthesis and antiarrhythmic activity of cis-2,6-dimethyl-α,α-diaryl-1-piperidinebutanols. J. Med. Chem. 1991;34(1):12−19.

[26] Hicks JL, Huang CC. Synthesis of carbon-13- and -14-labeled pirmenol hydrochloride. J. Labell. Comp. Radiopharm. 1983;20(6):771−8.

[27] Catania MA, Cuzzocrea S. Pharmacological and clinical overview of cloperastine in treatment of cough. Therapeut. Clin. Risk Manage. 2011;7:83−92.

[28] No Inventor data available, p-Halogenated benzhydryl basic ethers, GB 670622; 1952.

[29] Puricelli L. Preparation of optical isomers of cloperastine, EP 894794; 1999.

[30] Aliprandi P, Cima L, Carrara M. Therapeutic use of levocloperastine as an antitussive agent: an overview of preclinical data and clinical trials in adults and children. Clin. Drug Investigat. 2002;22(4):209−20.

[31] Baehring-Kuhlmey SR. Benproperine embonate. Drugs of Today 1977;13(6):221−4.

[32] Inventor data available, β-Piperidinopropane derivatives, their preparation and antitussive compositions containing them, GB 914008; 1962.

[33] Mortensen Th H. Procedure for preparation of 1-(o-benzylphenoxy)-2-(piperidine)propane or acid-addition salts thereof, DK 96113; 1963.

[34] Li Y, Chen S, Zhong D, Gan C, Min L, Sun Y, et al. Synthesis of R(+)- and S(-)-benproperine phosphate and their antitussive activity. Zhongguo Yaowu Huaxue Zazhi 2004;14 (1):19−22.

[35] Wang X, Guo L, Chen G, Xu Y, Guo X, Xu H, et al. A new method for synthesis of (R)-(+)- and (S)-(-)-benproperine phosphate enantiomers. Youji Huaxue 2011;31(2):212−15.

[36] Chen S, Min L, Li Y, Li W, Zhong D, Kong W. Anti-tussive activity of benproperine enantiomers on citric-acid-induced cough in conscious guinea-pigs. J. Pharm. Pharmacol. 2004;56(2):277−80.

[37] Schuler WA, Klebe H, Schlichtegroll AV. Syntheses of 4-azaphenothiazines. II. Derivatives of 4-azaphenothiazine-10-carboxylic acid. Justus Liebigs Ann. Chem. 1964;673:102−12.

[38] Vakil BJ, Mehta AJ, Prajapat KD. Trial of pipazethate as an antitussive. Clin. Pharmacol. Ther. 1966;7(4):515−19.

[39] Schuler WA. 4-Azaphenothiazines, DE 1055538; 1959.

[40] Lehmann B, Petrat B. Process for the preparation of pipazethate, EP 527298; 1993.

[41] Lewis-Wambi J, Jordan VC. Raloxifene. In: Taylor JB, Triggle DJ, editors. Comprehensive medicinal chemistry II, 8. Oxford: Elsevier; 2006. p. 103−21.

[42] Bryant HU, Glasebrook AL, Yang NN, Sato M. A pharmacological review of raloxifene. J. Bone Mineral Metabo. 1996;14(1):1−9.

[43] Jordan VC, Morrow M. Tamoxifen, raloxifene, and the prevention of breast cancer. Endocrine Rev. 1999;20(3):253−78.

[44] Heringa M. Review on raloxifene: profile of a selective estrogen receptor modulator. Int. J. Clin. Pharmacol. Therapeut. 2003;41(8):331−45.

[45] Khovidhunkit W, Shoback DM. Clinical effects of raloxifene hydrochloride in women. Ann. Internal Med. 1999;130(5):431−9.

[46] Balfour JA, Goa KL. Raloxifene. Drugs & Aging 1998;12(4):335−41.

[47] Snyder KR, Sparano N, Malinowski JM. Raloxifene hydrochloride. Am. J. Health-System Pharmacy 2000;57(18):1669−78.

[48] Barrett-Connor E. Raloxifene: risks and benefits. Ann. New York Acad.Sci. 2001;949:295−303 (Selective Estrogen Receptor Modulators (SERMs))

[49] Kellen JA. Raloxifene. Curr. Drug Targets 2001;2(4):423−5.

[50] Dodge JA, Bryant HU. Raloxifene: a selective estrogen receptor modulator (SERM). In: Huang X, Aslanian RG, editors. Case studies in modern drug discovery and development. New York: Wiley; 2012. p. 392−415.

[51] Moen MD, Keating GM. Raloxifene: a review of its use in the prevention of invasive breast cancer. Drugs 2008;68(14):2059−83.

[52] Gennari L, Merlotti D, De Paola V, Nuti R. Raloxifene in breast cancer prevention. Exp. Opin. Drug Safety 2008;7(3):259−70.

[53] Vestergaard P. Raloxifene for the treatment of postmenopausal osteoporosis. Int. J. Clin. Rheumatol. 2012;7(3):261−9.

[54] Yang ZD, Yu J, Zhang Q. Effects of raloxifene on cognition, mental health, sleep and sexual function in menopausal women: a systematic review of randomized controlled trials. Maturitas 2013;75(4):341−8.

[55] Rodante DE, Usall J. Raloxifene as adjuvant in the treatment of schizophrenia: a review of efficacy and safety issues. J. Sympt. Signs 2014;3(4):229−37.

[56] Jones CD, Jevnikar MG, Pike AJ, Peters MK, Black LJ, Thompson AR, et al. Antiestrogens. 2. Structure-activity studies in a series of 3-aroyl-2-arylbenzo[b]thiophene derivatives leading to [6-hydroxy-2-(4-hydroxyphenyl)benzo[b]thien-3-yl]-[4-[2-(1-piperidinyl)ethoxy]phenyl]methanone hydrochloride (LY 156758), a remarkably effective estrogen antagonist with only minimal intrinsic estrogenicity. J. Med. Chem. 1984;27 (8):1057−66.

[57] Jones CD, Goettel ME. 3-(4-Aminoethoxybenzoyl)benzo[b]thiophenes, EP 62504; 1982.

[58] Vicenzi JT, Zhang TY, Robey RL, Alt CA. Org. Proc. Res. Dev. 1999;3(1):56−9.

[59] Kost AN, Budylin VA, Matveeva ED, Sterligov DO. Isomerization of 3-substituted indoles, benzofurans, and benzothiophenes. Zh. Org. Khim. 1970;6(7):1503−5.

[60] Godfrey AG. 2-amino-3-aroylbenzo[b]thiophenes and methods for preparing and using same to produce 6-hydroxy-2-(4-hydroxyphenyl)-3-[4-(2-aminoethoxy)benzoyl]benzo[b] thiophene, US 5420349; 1995.

[61] Dadiboyena S. Recent advances in the synthesis of raloxifene: a selective estrogen receptor modulator. Eur. J. Med. Chem. 2012;51:17−34.

[62] Pineiro-Nunez M. Raloxifene, Evista: a selective estrogen receptor modulator (SERM). In: Li JJ, Johnson DS, editors. Modern drug synthesis. New York: Wiley; 2010. p. 309−27.

[63] Guarneri L, Robinson E, Testa R. A review of flavoxate: pharmacology and mechanism of action. Drugs of Today 1994;30(2):91−8.

[64] Zor M, Aydur E, Dmochowski RR. Flavoxate in urogynecology: an old drug revisited. Int. Urogynecol. J. 2015;26(7):959−66.

[65] Arcaniolo D, Conquy S, Tarcan T. Flavoxate: present and future. Eur. Rev. Med. Pharmacol. Sc. 2015;19(5):719−31.

[66] Ruffmann RA. Review of flavoxate hydrochloride in the treatment of urge incontinence. J. Inter. Med. Res. 1988;16(5):317−30.

[67] Tang HC, Lam WP, Zhang X, Leung PC, Yew DT, Liang W. Short-term flavoxate treatment alters detrusor contractility characteristics: renewed interest in clinical use? Lower Urinary Tract Symptoms 2015;7(3):149−54.

[68] Cazzulani P, Panzarasa R, Luca C, Oliva D, Graziani G. Pharmacological studies on the mode of action of flavoxate. Arch. Int. Pharmacodynam. Ther. 1984;268(2):301−12.

[69] Da Re P. Basic esters of 3-methylflavone-8-carboxylic acid, US 2921070; 1960.

[70] Da Re P, Verlicchi L, Setnikar I. Basic derivatives of 3-methylflavone-8-carboxylic acidFrom. J. Med. Pharm. Chem. 1960;2:263−9.

[71] Da Re P. Preparation of 3-methylflavone-8-carboxylic acid, US 3350411; 1967.

[72] No Inventor data available, β-Piperidinoethyl 3-methylflavone-8-carboxylate, GB 1343119; 1974.

[73] Beyer KH, Latven AR. Effect of (1-cyclohexylamino-2-propyl) benzoate (cyclaine, hexylcaine) and other local anesthetic agents administered intravenously on the cardiovascular-respiratory systems of the dog. J. Pharmacol. Expe. Ther. 1952;106:37−48.

[74] Koelzer PP, Wehr KH. Relations between chemical constitution and pharmacological activity in various types of new local anesthetics. VII. Various ω-aminoalkyl derivatives. Arzneim. -Forsch. 1958;8:708−16.

[75] McElvain SM. Piperidine derivatives (local anesthetics), US 1784903; 1930.

[76] Rider TH, Hill AJ. Glycidol. II. Reactions with secondary amines. J. Am. Chem. Soc. 1930;52:1528−30.

[77] Rider TH. Local anesthetics derived from dialkylaminopropanediols. I. Phenylurethans. J. Am. Chem. Soc. 1930;52:2115−18.

[78] Rider TH. Phenylurethan anesthetics. J. Am. Chem. Soc. 1930;52:2583.

[79] Dofek R, Vrba C. Local anesthetics. XIX. Substituted phenylcarbamates of piperidylpropanediol. Archiv der Pharmazie 1959;292:44−50.

[80] Felix W, Rimbach G, Wengenroth H. Spasmolytic effect and activity on circulation and respiration of a new dioxolane derivative (BR 18) [5,5-diphenyl-2-(2-piperidinoethyl)-1,3-dioxolan-4-one hydrochloride]. Arzneimit. -Forsch. 1969;19(11):1860−3.

[81] Moersdorf K, Wengenroth H. 5,5-Diphenyl-2-(2'-piperidinoethyl)-1,3-dioxolane-4-one, a new drug with potent antispasmolytic activity. Pharmacology 1970;3(4):193−200.

[82] Chen Y-F, Tsai H-Y, Wu K-J, Siao L-R, Wood WG. Pipoxolane ameliorates cerebral ischemia via inhibition of neuronal apoptosis and intimal hyperplasia through attenuation of VSMC migration and modulation of matrix metalloproteinase-2/9 and Ras/MEK/ERK signaling pathways. PLoS One 2013;8(9):e75654.

[83] Chlud K. Clinical experiences with a new spasmolytic, Rowapraxin. Die Medizinische Welt 1969;33:1801−3.

[84] Chen Y-F, Tsai H-Y. Novel use of pipoxolan for treatment of pathological conditions caused by vascular smooth muscle cell proliferation and migration, US 20150037399; 2015.

[85] Lee M-M, Chen Y-Y, Liu P-Y, Hsu S, Sheu M-J. Pipoxolane inhibits CL1-5 lung cancer cells migration and invasion through inhibition of MMP-9 and MMP-2. Chemico-Biol. Interac. 2015;236:19–30.

[86] Chen Y-F, Yang J-S, Huang W-W, Tsai H-Y. Novel anti-leukemia activities of pipoxolan operate via the mitochondria-related pathway in human leukemia U937 cells and attenuate U937 cell growth in an animal model. Mol. Med. Rrep. 2010;3(5):851–6.

[87] Inventor data available, Dioxolane, oxathiolane, and dithiolane spasmolytics, GB 1109959; 1968.

[88] Pailer M, Streicher W, Takacs F, Moersdorf K. Derivatives of 1,3-dioxolan-4-one and 1,3-oxathiolan-5-one. Monatshefte Chem. 1968;99(3):891–901.

[89] Shelmire B, Gastineaug FM, Shields TL. Evaluation of a new topical anesthetic, dyclonine hydrochloride. AMA Arch. Derm. 1955;71(6):728–30.

[90] Gurkan L, Oommen M, Hellstrom WJG. Premature ejaculation: current and future treatments. Asian J. Androl. 2008;10(1):102–9.

[91] Morales A, Barada J, Wyllie MG. A review of the current status of topical treatments for premature ejaculation. BJU Int. 2007;100(3):493–501.

[92] Profft E. The preparation of more active anesthetics having a broad spectrum. Chemische Technik (Leipzig, Germany) 1952;4:241–6.

[93] Bockstahler ER. p-Alkoxy-β-piperidinopropiophenones, US 2771391; 1956.

[94] Florestano HJ, Jeffries SF, Osborne CE, Bahler ME. Pharmaceutical preparations containing dyclonine hydrochloride and chlorobutanol, US 2868689; 1959.

[95] Babu BR. Preparation of dyclonine hydrochloride, IN 172270; 1993.

[96] Zharov OV, Novikov SV, Ustinova TA, Yudina TI. Spasmolytic tablets baralgin, RU 2203041; 2003.

[97] Jain R, Singh A. Novel anti-spasmodic and antiinflammatory pharmaceutical composition, CA 2202425; 1997.

[98] Golhar KB, Gupta RL. Open labelled evaluation of injection Manyana (a combination of diclofenac + pitofenone + fenpiverinium) in ureteric, biliary and intestinal spasm—a preliminary report. J. Ind. Med. Associat. 1999;97(9):398–400.

[99] Bockmuhl M, Ehrhart G. Basic acid amides, DE 731560; 1943.

[100] Inventor data available, Quaternary compounds, GB 708859; 1954.

[101] Moffett RB, Aspergren BD. Antispasmodics. X. α,α-Diphenyl-γ-amino amides. J. Am. Chem. Soc. 1957;79:4451–7.

[102] Fisnerova LA, Brunova B, Gajewski K. Process for preparing 2-(3-carbamoyl-3,3-diphenylpropyl)-1-methylpiperidinium bromide [fenpiverenium bromide], CZ 278022; 1993.

Chapter 3

2-Substituted and 1,2-Disubstituted Piperidines

3.1 METHYLPHENIDATE (11968)

Methylphenidate (**3.1.6**) (Ritalin) is a commonly prescribed central nervous system (CNS) stimulant. Methylphenidate is used to treat attention deficit disorder, attention deficit hyperactivity disorder, and narcolepsy, a chronic sleep disorder. However, a growing number of young individuals misuse or abuse methylphenidate to sustain attention, enhance intellectual capacity, and increase memory [1−4]. Side effects of methylphenidate include trouble sleeping, loss of appetite, weight loss, dizziness, nausea, vomiting, and headache.

Methylphenidate (**3.1.6**) has been synthesized via condensation of phenylacetonitrile (**3.1.1**) with a 2-chloropyridin (**3.1.2**) at 110−112°C in toluene in the presence of $NaNH_2$, which gave 2-phenyl-2-(pyridin-2-yl)acetonitrile (**3.1.3**). The last was hydrolyzed to corresponding amide (**3.1.4**), which on treatment with HCl in methanol on heating gave methyl 2-phenyl-2-(pyridin-2-yl)acetate (**3.1.5**). Hydrogenation of the pyridine ring to a piperidine ring in the obtained product in acetic acid on the Pt or PtO_2 catalyst gave the desired methylphenidate (**3.1.6**) [5−7] (Scheme 3.1).

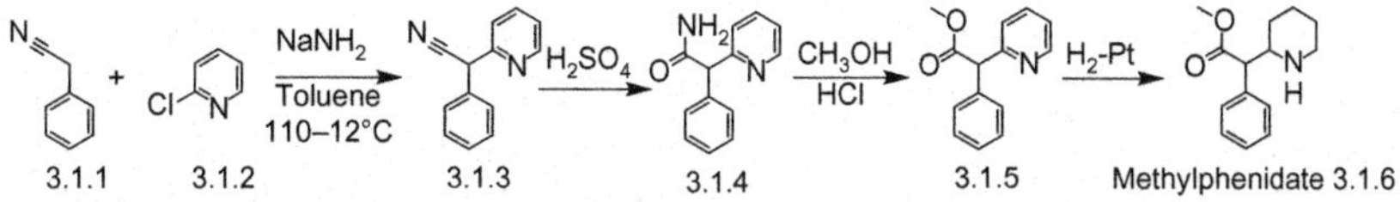

SCHEME 3.1   Synthesis of methylphenidate.

Alternatively, 2-bromopyridine can be used instead of 2-chloropyridine [8]. A huge amount of chemical work is described on the separation and interconversion stereoisomers of methylphenidate [9−14]. The absolute (2R,2′R; threo) stereochemistry of the most active enantiomer, (2R,2′R)-threo-methylphenidate, was proven [15,16].

3.2 PERHEXILINE (1216)

Perhexiline (**3.1.11**) was originally developed as an antianginal drug and was launched on the UK market as a racemate in 1975 under the trade name

Piperidine-Based Drug Discovery. DOI: http://dx.doi.org/10.1016/B978-0-12-805157-3.00003-X

103

Pexid. It rapidly gained a reputation for efficacy in the management of angina pectoris. However, hepatic and neurological adverse effects in a small proportion of patients led to a marked decline in its use in 1985.The drug was originally classified as a coronary vasodilator, and later as a calcium channel antagonist. Recent data suggests that it acts as a cardiac metabolic agent through inhibition of the enzyme, carnitine palmitoyltransferase-1 [17−22].

Perhexiline (**3.1.11**) consists of a piperidine framework with a 2,2-dicyclohexylethyl substituent at the 2-position. The synthesis of racemic perhexiline is based on nucleophilic addition of lithiated 2-picoline (**3.1.7**) to dicyclohexyl ketone (**3.1.8**) to give the corresponding tertiary alcohol (**3.1.9**), which undergoes HCL mediated dehydration forming alkene (**3.1.10**), the subsequent hydrogenation of which catalyzed by PtO_2 gives desired perhexiline (**3.1.11**) [23,24]. An alternative approach was demonstrated, using as starting ketone, bezophenone (**3.1.12**), which on reaction with lithiated 2-picoline gives tertiary alcohol (**3.1.13**), which after dehydration using hydrochloric acid gives alkene (**3.1.14**), the hydrogenation of which catalyzed by PtO_2 [25] or in presence Raney-Ni [26] or $Rh-Al_2O_3$ [27] gives desired perhexiline (**3.1.11**) (Scheme 3.2).

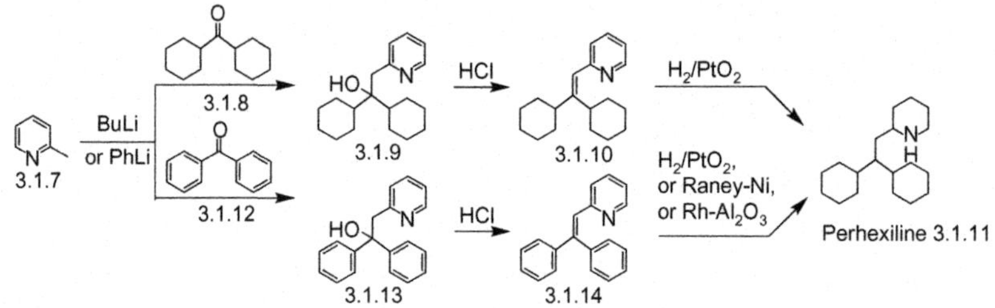

SCHEME 3.2 Synthesis of perhexiline.

Two enantiomers, (+)- and (−)-perhexiline have different pharmacodynamic profiles. It has been suggested that the (−)-enantiomer is primarily responsible for the therapeutic effects, whereas the (+)-enantiomer is primarily responsible for the toxic effects [17]. Optically enriched perhexiline has been obtained by resolution of the 1,1′-binaphthyl-2,2′-diyl(hydrogen)phosphate diastereomeric salts of perhexiline [28].

Synthesis of both enantiomers of perhexiline in high enantiomeric excess through a stereoselective catalytic hydrogenation of the 2-(oxazolidin-2-one)-substituted-pyridine and the elucidation of the absolute configurations of the two enantiomers of perhexiline which was unknown, was recently reported [29].

3.3 PIPRADROL (259)

Pipradrol (**3.1.17**) is a dopamine reuptake inhibitor and norepinephrine reuptake inhibitor, a mild amphetamine type psychostimulant with action similar to methylphenidate. Pipradrol was developed in the 1950s as an antidepressant and was used for treatment of obesity and dementia, but the adverse effects associated with its use and its abuse potential led to its withdrawal and international control [30].

Pipradrol (**3.1.17**) was synthesized from pyridyl Grignard reagent prepared from 2-pyridyl bromide (**3.1.15**) and bezophenone (**3.1.12**), which gave diphenylpydinemethanol (**3.1.16**) reduced catalytically to desired pipradrol (3.1.17) [31,32]. Enantiomers of pipradrol were synthesized from (*R*)- and (*S*)- pipecolic acid ethers (**3.1.18**) and the probable conformation of the base was deduced. All of the central stimulant activity resided in (*R*)-pipradrol, but both the (*R*) and (*S*) isomers possessed anticonvulsant properties [33] (Scheme 3.3).

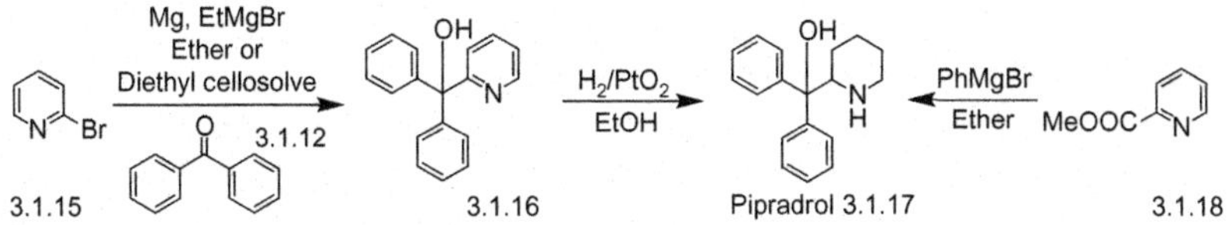

SCHEME 3.3 Synthesis of pipradrol.

3.4 MEFLOQUINE (5370)

Mefloquine (**3.1.27**), sold under the brand name Lariam, is an orally administered very potent blood schizontocide that has been marketed since 1990 for both malaria prophylaxis and for acute treatment of falciparum malaria. It is a long-acting antimalarial drug known for its efficacy against chloroquine- and SP-resistant Plasmodium falciparum [34−37]. Mefloquine can cause serious side effects that include nervous system changes.

The synthesis of mefloquine (**3.1.27**) began with the synthesis quinolin-4-ol (**3.1.21**) obtained by polyphosphoric acid condensation of the ethyl 4,4,4-trifluoroacetoacetate (**3.1.19**) with *O*-trifluoromethylaniline (**3.1.20**). A further conversion of prepared (**3.1.21**) by POBr$_3$ into the 4-bromoquinoline (**3.1.22**) led to the transformation of the last 4-Li derivative (**3.1.23**) followed by CO$_2$ carboxylation gave cinclioninic acid (**3.1.24**). Addition of 2-pyridyllithium (**3.1.25**) gave the pyridyl ketone (**3.1.26**). Hydrogenation with H$_2$-PtO$_2$ gave a good yield of desired mefloquine (**3.1.27**) [38,39] (Scheme 3.4).

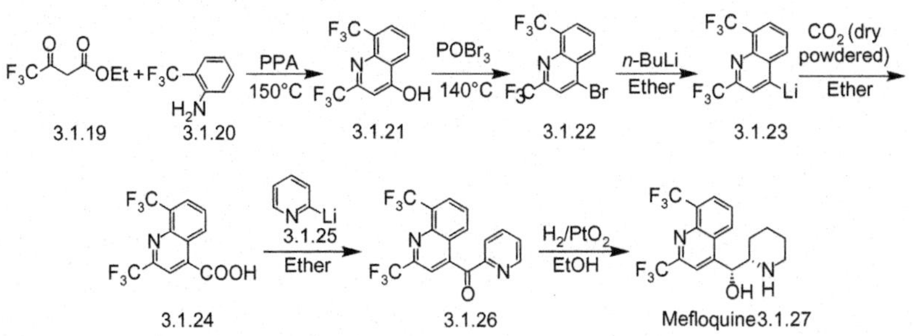

SCHEME 3.4 Synthesis of mefloquine.

None of the optically active forms of mefloquine (**3.1.27**) resolved via its hydrochloride salt with (+)- and (−)-3-bromo-8-camphorsulfonic acid ammonium salts showed any significant differences in antimalarial activity [40].

3.5 MEPIVACAINE (4176)

Many local anesthetics are presently available for clinical use, and among them many derivatives of 2-substituted piperidines. The choice of a particular agent for a particular case is based mainly on its clinical and pharmacological features.

Mepivacaine (**3.1.31**), launched on the market as Carbocaine and Polocaine, is a local anesthetic with a reasonably rapid onset and medium duration of action that became available in the 1960s. Mepivacaine exerts its local anesthetic effect by blocking voltage-gated sodium channels in peripheral neurons, which creates temporary anesthesia (lack of feeling or numbness). Mepivacaine is used for causing numbness during surgical procedures, labor, or delivery [41,42]. It may cause dizziness, drowsiness, or blurred vision.

Two basic methods for the synthesis of mepivacaine are proposed. The first comprises the transformation of ethyl 1-methylpipecolate (**3.1.30**) to 1-methylpiperidine-2-carboxylic acid amide with magnesium (2,6-dimethylphenyl) amide bromide (**3.1.29**) under reflux in ether. A magnesium derivative (**3.1.29**), in turn, was prepared via interaction of 2,6-xylidine (**3.1.28**) with ethylmagnesium bromide [43−45].

In another method, picolinic acid was converted to its amide (**3.1.32**), hydrogenated over platinum on carbon catalyst, and alkylated at the piperidine ring nitrogen with formalin using palladium on carbon [43,45,46] (Scheme 3.5).

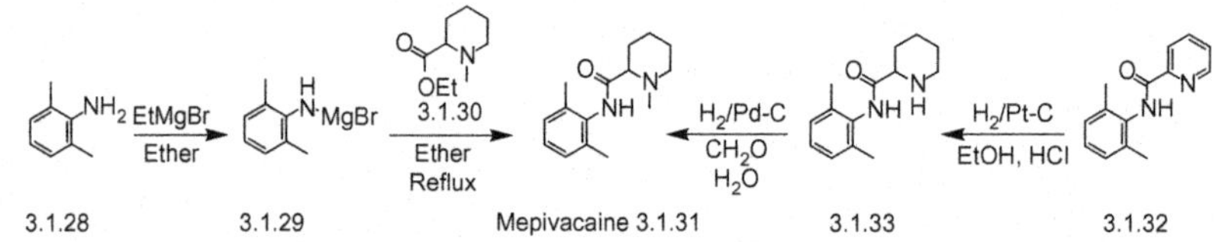

SCHEME 3.5 Synthesis of mepivacaine.

(+)-Mepivacaine − (*S*)-configuration is a longer-acting local anesthetic than the mixture enantiomers obtained during synthesis [47].

3.6 ROPIVACAINE (6847)

Ropivacaine (**3.1.37**) (Naropin) is the pure *S*(−)-enantiomer of propivacaine released for clinical use in 1996. It is a long-acting, well tolerated local anesthetic agent and first produced as a pure enantiomer. Its effects and mechanism of action are similar to other local anesthetics working via reversible inhibition of sodium ion influx in nerve fibers. It may be a preferred option among other drugs among this class of compounds because of its reduced CNS and cardiotoxic potential and its lower propensity for motor block in the management of postoperative pain and labor pain [48−58].

The synthesis of ropivacaine (**3.1.37**) was carried out starting with L-pipecolic acid (**3.1.34**), prepared by a resolution of (±)-pipecolic acid with (+)-tartaric acid, which was dissolved in acetyl chloride and converted to acid chloride (**3.1.35**) with phosphorus pentachloride. The obtained compound (**3.1.35**) dissolved in toluene a solution of 2,6-xylidine (**3.1.28**) dissolved in the mixture of equal volumes of acetone, and N-methyl-2-pyrrolidone was added at 70°C to give (+)-L-pipecolic acid-2,6-xylidide (**3.1.36**). Reaction of this compound with propyl bromide in presence of potassium carbonate in i-PrOH/H$_2$O gave the desired ropivacaine (**3.1.37**) [59] (Scheme 3.6).

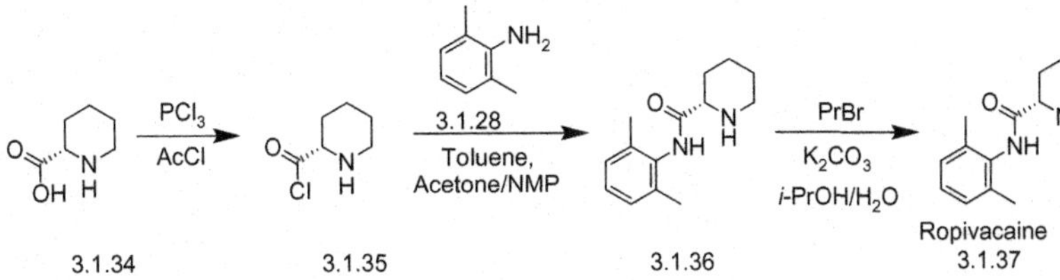

SCHEME 3.6   Synthesis of ropivacaine.

Another approach for the synthesis of ropivacaine (**3.1.37**) was proposed via a resolution of enantiomers of chiral pipecolic acid-2,6-xylide [60].

3.7   BUPIVACAINE (21293) AND LEVOBUPIVACAINE (1976)

Bupivacaine (**3.1.41**) (Marcaine) is a local anesthetic of great potency and long duration that has been widely used for years, but it has cardio and CNS toxic sideeffects. For many years it was nearly the only local anesthetic applicable to almost all kinds of loco-regional anesthetic techniques, and nowadays, in many occasions, it is still the only alternative available [61−64].

Bupivacaine is currently used in racemic form. At high doses, however, the racemate is potentially hazardous due to toxicity problems.

Currently, racemic bupivacaine (**3.1.41**) is produced from picolinic acid (**3.1.38**) either by reduction to pipecolic acid (**3.1.39**) and then, after conversion to corresponding acid chloride (3.1.40) coupling with 2,6-xylidine to give pipecolic acid-2,6-xylide (**3.1.33**), or by reducing the pyridyl amide (**3.1.43**) prepared from picolinic acid chloride (**3.1.42**) over platinum oxide. The amide intermediate (**3.1.33**), which can also be used to prepare the anesthetics ropivacaine (**3.1.37**) and mepivacaine (**3.1.31**), was transformed to desired bupivacaine (**3.1.41**) either by direct alkylation using butyl bromide and potassium carbonate or by reductive amination using butyraldehyde [45,59,65−69] (Scheme 3.7).

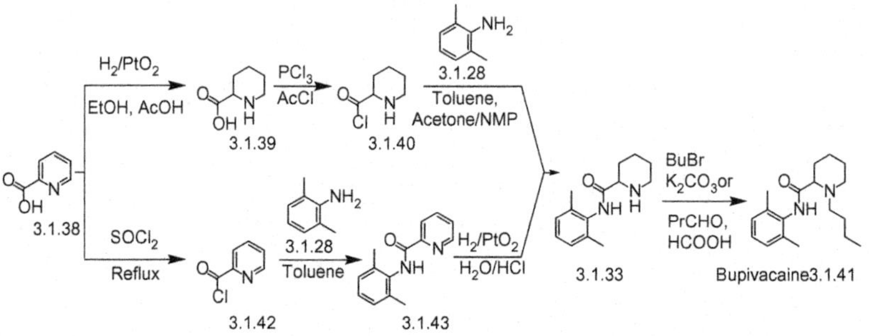

SCHEME 3.7   Synthesis of bupivacaine.

Enantiomers of bupivacaine can be prepared via diastereomeric salt resolution with tartaric acid or by resolution of the amide (**3.1.33**) with O, O-dibenzoyl tartaric acid followed by alkylation [47,70].

One of enantiomers, S(−) isomer of the racemic bupivacaine (levobupivacaine), has equal potency but less cardiotoxic and CNS effects in comparison with both R(+) bupivacaine and bupivacaine racemate. The reduced toxicity of levobupivacaine (**3.1.48**) gives a wider safety margin in clinical practice [71,72].

Stereospecific synthesis of levobupivacaine from (S)-lysine have been proposed (Scheme 3.8).

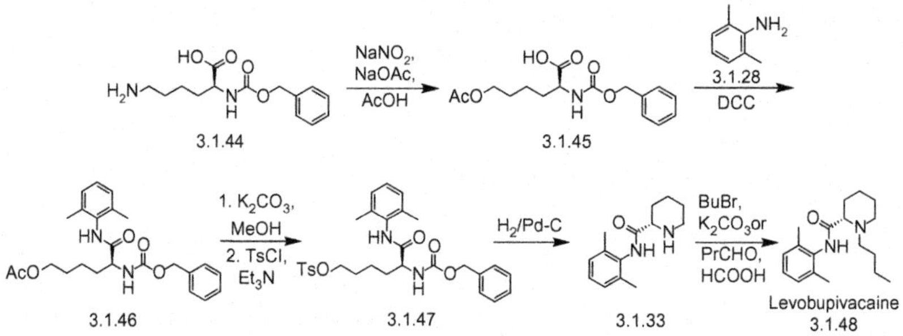

SCHEME 3.8   Synthesis of levobupivacaine.

Treatment of N-CBZ (S)-lysine (**3.1.44**) with sodium nitrite in acetic acid yields the acetate (**3.1.45**). The prepared acetate (**3.1.45**) was then coupled with dimethyl aniline using N,N′-dicyclohexylcarbodiimide to give the amide (**3.1.46**) in good yield. The acetate group was then converted into the tosylate (**3.1.47**), which was deprotected and cyclized stereospecifically in one-pot reaction to give the amide (**3.1.33**) in high yield. Alkylation is easily achieved using an alkyl bromide and K_2CO_3 without any racemization. Alkylation can also be carried out using butyraldehyde/formic acid although the former is a much simpler process [73] (Scheme 3.8).

3.8 FLECAINIDE (3638)

Flecainide (**3.1.54**), sold under the trade name (Tambocor), is an antiarrhythmic drug used to prevent and treat tachyarrhythmias, a wide variety of cardiac arrhythmias including paroxysmal atrial fibrillation, paroxysmal supraventricular tachycardia and ventricular tachycardia, and has been used extensively worldwide over the last 25 years. It is a sodium channel blocker that is effective medicine for tachyarrhythmia, for which the other antiarrhythmic medication is not effective. Flecainide is also effective in the treatment of catecholaminergic polymorphic ventricular tachycardia but, in this condition, its mechanism of action is contentious. It can be given either intravenously or orally and its pharmacokinetic properties allow for a relatively long (12 hours) effect. Flecainide is an antiarrhythmic agent that has the potential to be considered an narrow therapeutic index drug that has a narrow window between its effective dose and a dose at which it can produce adverse toxic effects (dizziness, vision problems, shortness of breath, headache, nausea, vomiting, stomach pain, diarrhea, constipation, tremor or shaking, tiredness, weakness, anxiety, depression, numbness, or tingling) [74−79].

The original synthesis of flecainide is described on the Scheme 3.9, where a solution of 2,5-dihydroxybenzoic acid (**3.1.49**) in acetone was added to a suspension of $KHCO_3$ in acetone followed by a solution of 2,2,2-trifluoroethyl trifluoromethanesulfonate (**3.1.50**) and stirred under reflux for 72 hours to give 2,2,2-trifluoroethyl 2,5-bis(2,2,2-trifluoroethoxy)benzoate (**3.1.51**). The key step in this route is aminolysis. For that purpose obtained benzoate (**3.1.51**) was added to a stirred solution of 2-aminomethylpyridine (**3.1.52**) in glyme to give (**3.1.53**). Catalytic hydrogenation of the resultant N-(2-pyridylmethyl)-2,5-bis(2,2,2-trifluoroethoxy)benzamide (**3.1.54**) was carried out in AcOH over PtO_2 to give desired flecainide (**3.1.54**) [80−83].

A significant simplification in the synthesis was achieved by the use of 2-aminmethylpiperidine (**3.1.55**) [84] (Scheme 3.9).

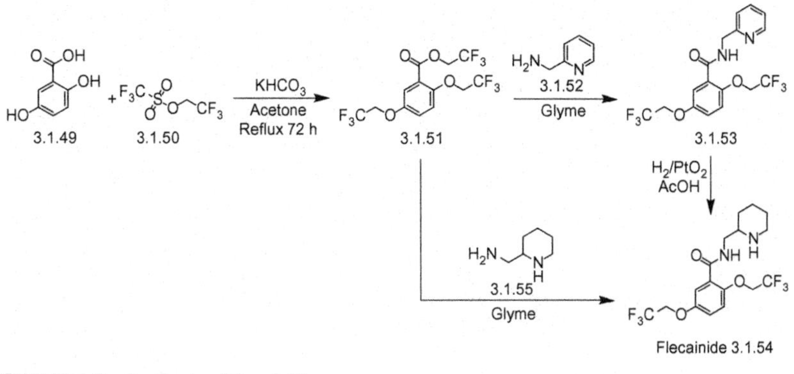

SCHEME 3.9 Synthesis of flecainide.

3.9 ENCAINIDE (682)

Encainide (**3.1.62**), formerly marketed as Enkaid, is an antiarrhythmic drug with class IC activity and has been used in the treatment of life-threatening ventricular arrhythmias, symptomatic ventricular arrhythmias, and supraventricular arrhythmias. The most common noncardiac side effects were dizziness and blurred vision and proarrhythmic effects. Encainide was associated with increased death rates in patients who had asymptomatic heart rhythm abnormalities after a recent heart attack and was withdrawn from the US market in 1991 [85−89].

The first step in practically all proposed methods for encainide synthesis is based on condensation of picoline (**3.1.56**) or picolinium salts (**3.1.63**) (methiodide, methsulfate) with 2-nitrobenzaldehyde (**3.1.57**). In some patents and papers a mixture of picoline as well as the aforementioned aldehyde was refluxed in acetic anhydride to give 2-[2-(2-Nitrophenyl)ethenyl-2]pyridine (**3.1.57**) [90−93]. In others a mixture of previously prepared picolinium salts (**3.1.63**) and 2-nitrobenzaldehyde (**3.1.57**) was refluxed in methanol in the presence of a catalytic amount of piperidine to give product (**3.1.64**), which was dehydrated on reflux in the mixture of acetic acid, acetic anhydride and potassium acetate to give (**3.1.65**) [94,95].

A solution of 2-[2-(2-nitrophenyl)ethenyl-2]pyridine (**3.1.57**) in ethanol was hydrogenated over Pd-C catalyst forming 2-[2-(2-aminophenyl)ethyl-2] pyridine (**3.1.58**). The obtained product was dissolved in pyridine and acylated with 4-anisoyl chloride (**3.1.59**) at 70°C to give amide (**3.1.60**). The last was dissolved in warm acetonitrile, treated with dimethyl sulfate (or methyl iodide), and heated to 70°C. The resulted crystalline salt (**3.1.61**) was hydrogenated in ethanol using platinum oxide to give desired encainide (**3.1.62**).

In the method started from picolinium salts (**3.1.63**) (methiodide, methsulfate), the salt (**3.1.65**) obtained after dehydration of (**3.1.64**) was hydrogenated over a platinum oxide catalyst to give 2-(2-(1-methylpiperidin-2-yl) ethyl)aniline (**3.1.66**), which was acylated with 4-anisoyl chloride 4- (**3.1.59**) in acetone to give desired encainide (**3.1.62**) (Scheme 3.10).

Another interesting method for the synthesis of encainide was proposed, according to which methyl anthranilate (**3.1.67**) was acylated with 4-anisoyl chloride (**3.1.59**) in dichloromethane solution to give amide (**3.1.68**). The obtained product on reaction with 2-pyridyllithium (**3.1.69**) formed product (**3.1.70**) underwent one-pot hydrogenation with Pt-C AcOH, followed by addition of Pd-C catalyst, and reductive methylation with formalin in the presence of Pd-C to give encainide (**3.1.62**) with good yield [96] (Scheme 3.11).

3.10 THIORIDAZINE (6558)

Thioridazine (**3.1.73**) (Mellaril), is one of the older, first-generation typical antipsychotic oral medications used not only for management of

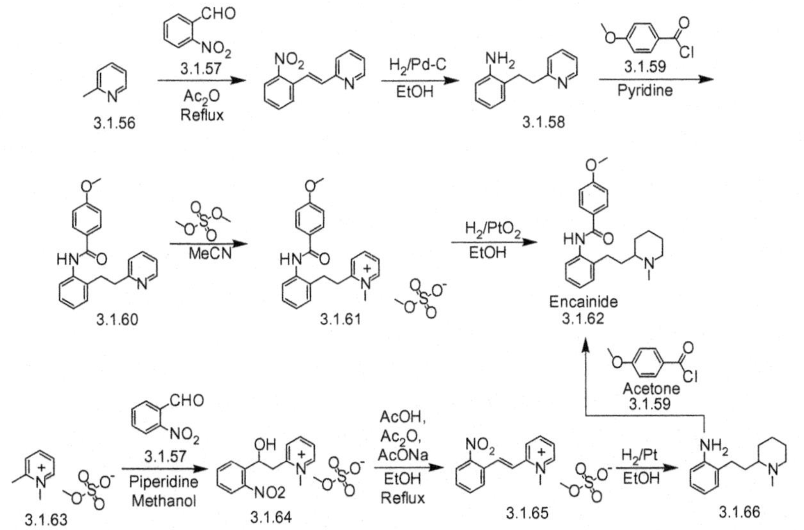

SCHEME 3.10   Synthesis of encainide.

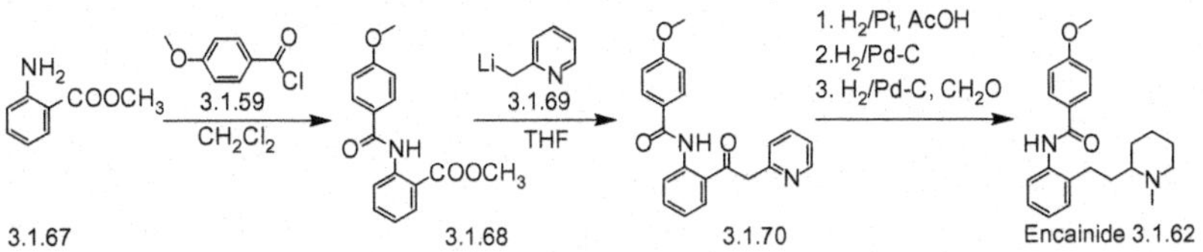

SCHEME 3.11   Synthesis of encainide.

schizophrenia but is also widely used for the relief of anxiety, agitation, mania, manic depressive psychosis, and behavioral problems. However, evidence of cardiac complications led to the restriction of its use from 2000 and withdrawal worldwide in 2005 because it caused severe cardiac arrhythmias and excessive death rates associated with its use [97−103].

It has been shown recently that thioridazine has in vitro activity against multidrug-resistant (MDR) and extensively drug-resistant (XDR) strains of Mycobacterium tuberculosis, and is able to cure antibiotic-susceptible and -resistant pulmonary tuberculosis infections. Under proper cardiac evaluation procedures, it is safe and does not produce any known cardiopathy [104−107].

The synthesis of thioridazine was achieved through reaction of 2-(methylthio)-10*H*-phenothiazine (**3.1.71**) and 2-(2-chloroethyl)-*N*-methylpiperidine (**3.1.72**) in refluxing xylene in the presence of sodium amide to give desired thioridazine (**3.1.73**) [108−110] (Scheme 3.12).

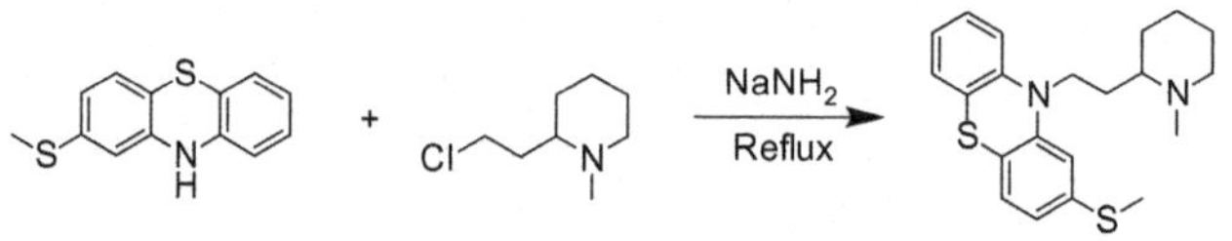

SCHEME 3.12   Synthesis of thioridazine.

3.11   RIMITEROL (219)

Rimiterol (**3.1.82**) is a third-generation, short-acting selective β2-adrenoreceptor agonist used for the treatment of bronchospasm.

It is not effective by the oral route of administration, but may be of value in the intravenous therapy of severe asthma. Rimiterol is available in pressurized aerosols. There have been no reports of significant subjective side effects following the acute administration of rimiterol by aerosol. There are no specific contraindications to rimiterol, but it should be given with care to patients with thyrotoxicosis, cardiovascular disease, diabetes mellitus, renal, or hepatic dysfunction [111–118].

Synthetic routes to rimiterol (**3.1.82**) are described. For that purpose 3,4-dimethoxyphenyl-2-pyridylcarbinol (**3.1.77**) was prepared by reaction between veratraldehyde (**3.1.74**) and picolinic acid (**3.1.75**) on reflux in *p*-cumene, or by treatment of the same aldehyde (**3.1.74**) with 2-pyridyllithium (**3.1.76**) prepared from 2-bromopyridine and *n*-butyllithium in ether. Obtained carbinol (**3.1.77**) was oxidized to the corresponding ketone (**3.1.80**) by potassium permanganate in water at 70°C, or by dimetliylsulphoxide-acetic anhydride mixtures, or by air in boiling nitrobenzene. An alternative approach for the synthesis of ketone (**3.1.81**) was proposed via acylation of veratrole (**3.1.78**) with picolinic acid chloride (**3.1.79**) in nitrobenzene in the presence of aluminum chloride.

The protecting methoxy groups in (**3.1.80**) were changed to hydroxyl groups in boiling hydrobromic acid and the product (**3.1.81**) hydrogenated in methanol over platinum oxide to give desired rimiterol (**3.1.82**) [119–122] (Scheme 3.13).

3.12   LOBELINE (1914)

Lobeline (**3.1.90**) is a plant alkaloid, the main constituent of the 20 known of Lobelia inflata, known as Indian tobacco because native Americans smoked the dried leaves as a substitute for tobacco.

Lobeline has a long history of therapeutic use and, and during the 19th century it was prescribed as an emetic or a respiratory stimulant used to treat asthma, collapse, and anesthetic accidents. Lobeline has multiple mechanisms of action.

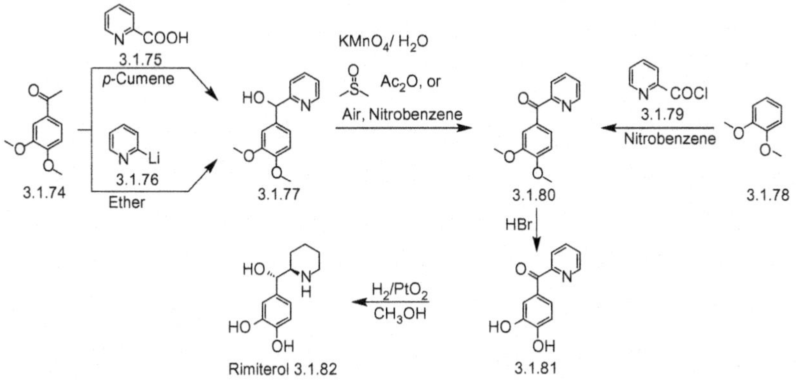

SCHEME 3.13   Synthesis of rimiterol.

Lobeline is a high affinity compound for nicotinic acetylcholine receptors, and it is considered a promising candidate for pharmacotherapy of addiction and abuse (smoking, cocaine, amphetamines) and is used in tablets as a smoking cessation remedy.

It has been classified as a compound having many nicotine-like effects working as both an agonist and an antagonist at nicotinic receptors, having many nicotine-like effects including hypertension, bradycardia and hypotension, anxiolytic effects, enhancement of cognitive performance. Lobeline inhibits the function of vesicular monoamine and dopamine transporters and diminishes the behavioral effects of nicotine and amphetamines. Lobeline binds to μ-opiate receptors, blocking the effects of opiate receptor agonists. Lobeline esters was shown be useful for treating neurodegenerative diseases of the CNS, which include Alzheimer's disease, Parkinson's disease, Huntington's disease, etc. [123−129].

Isolation from plants is uneconomical procedure, and many different routes of synthesis have been considered.

The first synthesis of lobeline (**3.1.90**) started with a Claisen condensation between ethyl glutarate (**3.1.83**) and acetophenone (**3.1.84**) to give tetraketone (**3.1.85**). The last on treatment with ammonia was cyclized to piperidine-2,6-diylidene derivative (**3.1.86**). The carbonyl groups in (**3.1.86**) were reduced by hydrogenation over platinum oxide giving a mixture of two diastereoisomers that were separated by crystallization to give β-norlobelanidiene (**3.1.87**). The double bonds in obtained compound (**3.1.87**) were reduced with aluminum amalgam to give (**3.1.88**), which was converted into its methylated derivative (**3.1.89**) by treatment with methyl tosylate. The produced compound (**3.1.89**) was converted into (+/−)-lobeline by treatment with an oxidizing agent such as permanganate, or chromium trioxide, and then was dissolved by D-tartaric acid giving (−)-lobeline (**3.1.90**) [130] (Scheme 3.14).

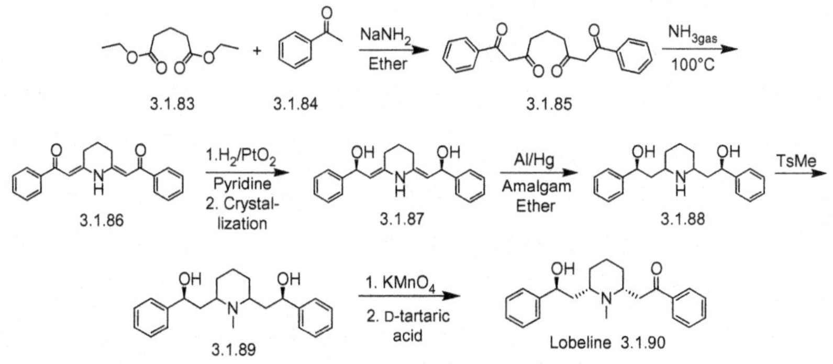

SCHEME 3.14 Synthesis of lobeline.

Another strategy for the synthesis of lobeline (**3.1.90**) was proposed the same year. For that purpose 2,6-lutidine (**3.1.91**) was condensed with benzaldehyde (**3.1.92**) at high temperature in presence of zinc chloride to give 2,6-distyrylpyridine (**3.1.93**). Bromination of the last with an excess of bromine rand further dehydrobromination with potassium hydroxide gave diphenylethynylpyridine (**3.1.95**). Hydration of (**3.1.95**) with concentrated sulfuric acid furnished the diphenacylpyridine (**3.1.96**). Alkylation of the pyridine ring with methyl tosylate gave a quaternary salt (**3.1.97**) that was reduced to lobelanidine (**3.1.98**). A mild oxidation of (+/−)-lobelanidine (**3.1.98**) by potassium permanganate gave a mixture of the (+/−)-lobeline (**3.1.90a**) [131] Scheme 3.15.

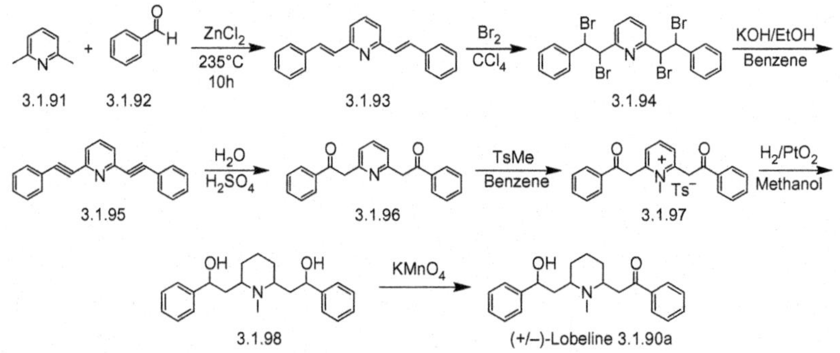

SCHEME 3.15 Synthesis of (+/-)-lobeline.

Described synthetic routes might not be implemented for large scale preparations as they are of low efficiency.

An efficient process for the preparation of (−)-lobeline (**3.1.90**) from lobelanine (**3.1.99**) on industrial scale have been developed using asymmetric hydrogenation with a catalyst system consisting of cyclooctadiene rhodium chloride dimer ([RhCl(COD)]$_2$) and (2R,4R)-4-(dicyclohexyl-phosphino)-2-(diphenylphosphinomethyl)-N-methylaminocarbonyl pyrrolidine [132,133] (Scheme 3.16).

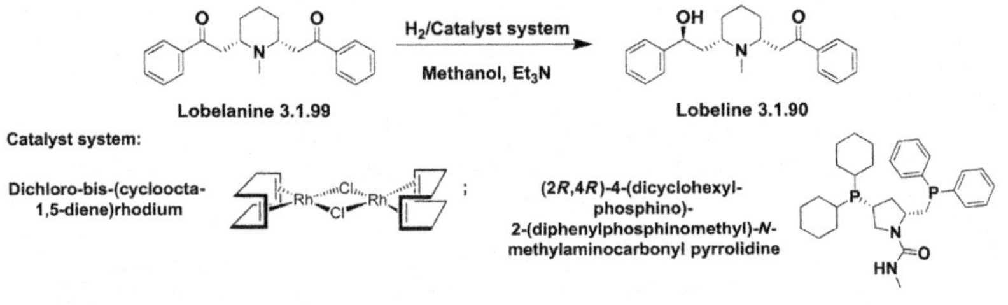

Catalyst system:

Dichloro-bis-(cycloocta-1,5-diene)rhodium

(2R,4R)-4-(dicyclohexyl-phosphino)-2-(diphenylphosphinomethyl)-N-methylaminocarbonyl pyrrolidine

SCHEME 3.16 Synthesis of lobeline.

The chemical precursor for the biosynthesis of lobeline, 2,6-*cis*-lobelanine (**3.1.99**), was easily obtained by an elegant one-pot synthesis which involves Mannich condensation and a Robinson type biomimetic reaction of glutaric dialdehyde (**3.1.100**), benzoylacetic acid (**3.1.101**), and methylamine hydrochloride (**3.1.102**) in acetone and citrate buffer [134] (Scheme 3.17).

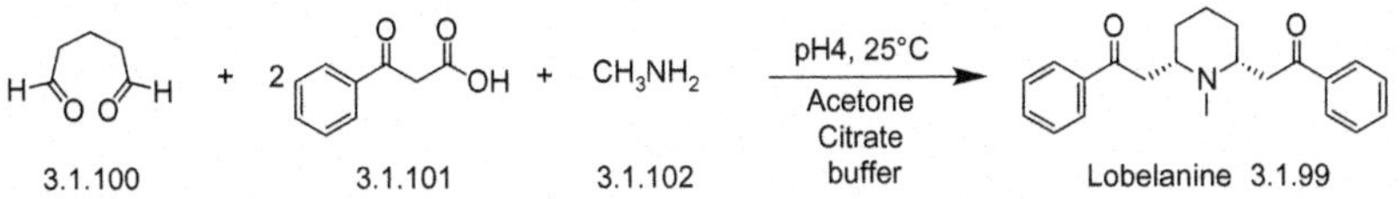

SCHEME 3.17 Synthesis of lobelanine.

Perfect reviews on synthesis and chemistry of lobeline are published [135−138].

3.13 ARGATROBAN (2627)

Argatroban (**3.1.111**) (Arganova) is a highly selective direct thrombin inhibitor indicated for use as an anticoagulant for the treatment and prophylaxis of thrombosis in patients with heparin-induced thrombocytopenia (HIT), a devastating, life-threatening, immune-mediated complication of therapy with heparin and in patients undergoing percutaneous coronary intervention who have, or are at risk for HIT.

Argatroban does not generate antibodies, is not susceptible to degradation by proteases and is cleared hepatically.

It is a reversible antithrombin agent and therefore exhibits a considerably different pharmacological profile. Its mechanisms of action include several other processes that have not been explored fully to date. These include the inhibition of nonthrombin serine proteases, a direct effect on endothelial cells and the vasculature (generation of nitric oxide), and downregulation of various inflammatory and thrombotic cytokines. Argatroban is more effective than heparins and hirudins in the antithrombotic management of microvascular disorders (Heparin has historically been used as the anticoagulant of choice in the management of a number of thrombotic diseases.) [139–153].

The improved synthesis of argatroban (**3.1.111**) started with preparation of racemic (±)-*trans*-benzyl 4-methyl pipecolic acid ester (**3.1.104**), which was synthesized via α-lithiation and further benzyloxycarbonylation sequence of *N*-Boc-4-methylpiperidine (**3.1.103**) using for that purpose s-BuLi and tetramethylethylene-diamine in ether followed by addition of benzyl chloroformate, which yielded the afforded compound (**3.1.104**). After *N*-Boc deprotection (HCl, AcOEt), the desired benzyl 4-methyl pipecolic acid ester (**3.1.105**) was obtained. The condensation of compound (**3.1.105**) with N^{α}-Boc-N^{ω}-nitro-L-arginine (**3.1.106**), two diastereomers (**3.1.107a**) and (**3.1.107b**) were obtained and separated by flash chromatography on silica gel to afford the desired (**3.1.107a**). After removal of the Boc group in (**3.1.107a**) by (HCl, AcOEt), the obtained compound (**3.1.108**) was transformed to (**3.1.110**) by treatment with 3-methyl-8-quinoline sulfonyl chloride (**3.1.109**) in dichloromethane in the presence of trimethylamine. The hydrogenation of the last over Pd/C catalyst in ethanol/acetic acid mixture effected the debenzylation of the ester group, the cleavage of the nitro group, and the hydrogenation of the pyridine ring affording desired argatroban (**3.1.111**) [154] (Scheme 3.18). The first patents on the synthesis of argatroban are based on ethyl 4-methyl pipecolic acid ester [155–157].

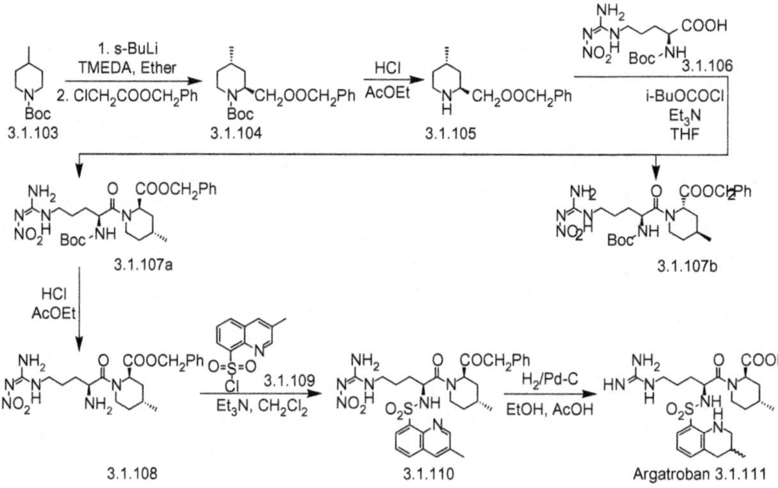

SCHEME 3.18 Synthesis of argatroban.

3.14 ASCOMYCIN (752), PIMECROLIMUS (1645), TACROLIMUS (32436), SIROLIMUS (22468), EVEROLIMUS (8972), AND TEMSIROLIMUS (2859)

Series of Ascomycin (**3.1.112**) derivatives – Pimecrolimus (**3.1.113**), Tacrolimus (**3.1.114**) as well as Sirolimus (**3.1.115**) and its derivatives Everolimus (**3.1.116**) and Temsirolimus (**3.1.117**) – can formally be considered 1,2-disubstituted piperidines (Fig. 3.1).

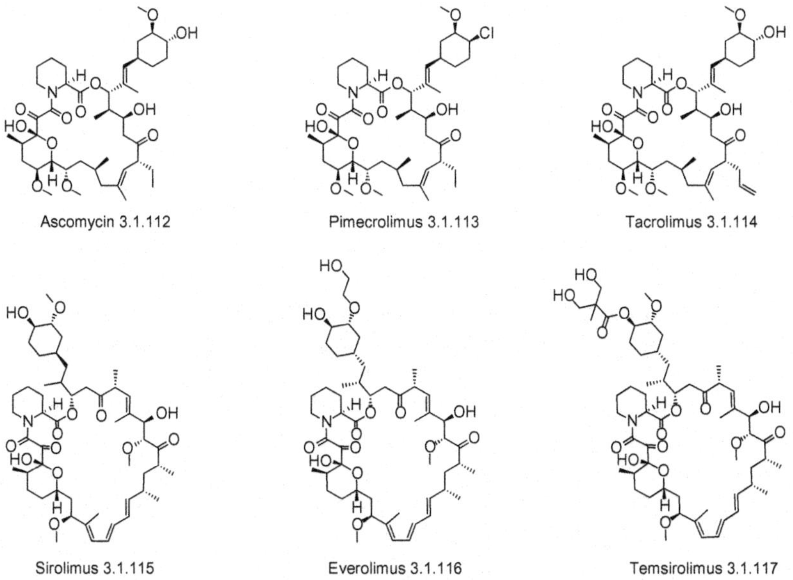

Ascomycin 3.1.112   Pimecrolimus 3.1.113   Tacrolimus 3.1.114

Sirolimus 3.1.115   Everolimus 3.1.116   Temsirolimus 3.1.117

FIGURE 3.1   Ascomycin, pimecrolimus, tacrolimus, sirolimus, everolimus, and temsirolimus.

The 23-membered macrolactam Ascomycin (**3.1.112**) and 31-membered macrocyclic polyketide Sirolimus (**3.1.115**) are fermentation products originally isolated from the cultured broth of Streptomyces hygroscopicus.

Ascomycin and its derivatives are powerful calcium-dependent serine/threonine protein phosphatase (calcineurin (CaN), protein phosphatase 2B inhibitors and have been used therapeutically mainly as immunosuppressants in inflammatory skin diseases. Calcineurin inhibitors (CNIs) have been also proposed for the treatment of inflammatory and degenerative brain diseases. Ascomycin and its derivatives may be useful in preventing ischemic brain damage and neuronal death in the treatment of CNS and exhibit anticonvulsant activity. Nonimmunosuppressant activity of its derivatives as CNS drugs probably should be further explored.

Pimecrolimus (**3.1.113**) prepared by the substitution of 32-hydroxy group in ascomycin with a chlorine with an inversion of configuration and tacrolimus (**3.1.114**), which was obtained by using the mutant Streptomyces species. These compounds have been successfully introduced in the treatment of atopic dermatitis. They inhibit T cell proliferation, mast cell degranulation, production, and the release of IL-2, IL-4, IF-γ, and TNF-α. They do not effect endothelial cells and fibroblasts, so they do not induce skin atrophy and consequently are well tolerated and safe.

They have been used also for treatment of other inflammatory skin diseases including psoriasis, lichen planus, seborrheic dermatitis, allergic contact dermatitis, vitiligo, pyoderma gangrenosum, alopecia areata, graft-versus-host disease, akne rosacea, etc. The pharmacology, use, and modifications of Ascomycin and its derivatives are reviewed [158−167].

Sirolimus (**3.1.115**) also known as Rapamycin, and its derivatives, Everolimus (**3.1.116**) and Temsirolimus (**3.1.117**), are a class of immunosuppressive drugs approved for solid organ transplantation.

Sirolimus, a mammalian target of rapamycin (mTOR) inhibitors, are a kind of macrolide antibiotics, producing by Streptomyces hygroscopicus in appropriate fermenting culture had been found to have also potent antiinflammatory, antineoplasms, antiatherosclerosis, antiaging, neuroprotection properties.

Sirolimus and its derivates, everolimus and temsirolimus, have a similar structure that inhibits the proliferation of T cells by interfering with a serine-threonine kinase, called mTOR. By inhibiting the ubiquitous mTOR pathway, they present a peculiar safety profile. Apart from their immunosuppressive effects, these agents may also inhibit endothelial intimal proliferation, the replication of cytomegalovirus, and the development of certain cancers. They are used not only as immunosuppressants after organ transplantation in combination with CNIs but also as proliferation signal inhibitors coated on drug-eluting stents [168−173].

REFERENCES

[1] Busardo FP, Kyriakou C, Cipolloni L, Zaami S, Frati P. From clinical application to cognitive enhancement: the example of methylphenidate. Curr Neuropharmacol 2016;14 (1):17−27.

[2] Gadoth N. Methylphenidate (ritalin): what makes it so widely prescribed during the last 60 years? Curr Drug Therapy 2013;8(3):171−80.

[3] Devos D, Moreau C, Delval A, Dujardin K, Defebvre L, Bordet R. Methylphenidate. CNS Drugs 2013;27(1):1−14.

[4] Patrick KS, Markowitz JS. Pharmacology of methylphenidat, amphetamine enantiomers and pemoline in attention-deficit hyperactivity disorder. Human Psychopharmacol 1997;12 (6):527−46.

[5] Hartmann M, Panizzon L. Pyridine and piperidine derivatives, US 2507631; 1950.

[6] Huntley CFM, Kataisto EW, La Lumiere KD, Reisch HA. A process for the preparation of methylphenidate hydrochloride, WO 2012080834; 2012.

[7] Stefanick SM, Smith BJ, Barr C, Dobish MC. A process for the preparation of methylphenidate, WO 2015069505; 2015.

[8] Deutsch HM, Shi Q, Gruszecka-Kowalik E, Schweri MM. Synthesis and pharmacology of potential cocaine antagonists. 2. Structure-activity relationship studies of aromatic ring-substituted methylphenidate analogs. J Med Chem 1996;39(6):1201−9.

[9] Prashad M, Kim H-Y, Lu Y, Liu Y, Har D, Repic O, et al. The first enantioselective synthesis of (2R,2′R)-threo-(+)-methylphenidate hydrochloride. J Org Chem 1999;64(5):1750−3.

[10] Rometsch R. Stereoisomers of α-phenyl-α-(2-piperidyl)acetic acid, US 2838519; 1958.

[11] Rometsch R. Conversion of stereoisomers, US 2957880; 1960.

[12] Khetani V, Luo Y, Ramaswamy S. Resolution of piperidylacetamide stereoisomers, WO 9852921; 1998.

[13] Prashad M, Har D, Repic O, Blacklock TJ, Giannousis P. Enzymic resolution of (±)-threo-methylphenidate. Tetrahedron: Asymmetry 1998;9(12):2133−6.

[14] Renalson KS, Kalambe AB, Panhekar DY. Efficient method for enantioselective synthesis of dexmethylphenidate hydrochloride (Focalin). Int J Res Develop Pharm Life Sci 2014;3 (4):1066−9.

[15] Weisz I, Dudas A. Stereoisomeric 2-piperidylphenylacetic acid esters. The spatial structure of a 7-phenylazabicyclo[4.2.0]octane. Monatsh Chem 1960;91:840−9.

[16] Shafi'ee A, Hite G. Absolute configurations of the pheniramines, methyl phenidates, and pipradrols. J Med Chem 1969;12(2):266−70.

[17] Ashrafian H, Horowitz JD, Frenneaux MP. Perhexiline. Cardiovasc Drug Rev 2007;25 (1):76−97.

[18] Horowitz JD, Chirkov YY. Perhexiline and hypertrophic cardiomyopathy: a new horizon for metabolic modulation. Circulation 2010;122(16):1547−9.

[19] Killalea SM, Krum H. Systematic review of the efficacy and safety perhexiline in the treatment of ischemic heart disease. Am J Cardiovasc Drugs 2001;1(3):193−204.

[20] Hudak WJ, Lewis RE, Lucas RW, Kuhn WL. Review of the cardiovascular pharmacology of perhexiline. Postgraduate Medical J Suppl 1973;49(3):16−25.

[21] Winsor T. Clinical evaluation of perhexiline maleate. Clin Pharmacol Therapeut 1970;11 (1):85−9.

[22] Hudak WJ, Lewis RE, Kuhn WL. Cardiovascular pharmacology of perhexiline. J Pharmacol Exp Therapeut 1970;173(2):371−82.

[23] Palopoli FP, Kuhn WL. 1,1-Dicyclohexyl-2-(2-piperidyl)ethylene, US 3038905; 1962.

[24] No Inventor data available, 1,1-Dicyclohexyl-2-(2-piperidyl)ethane, GB 1025578; 1966.

[25] Bianchetti G, Viscardi R. 1,1-Dicyclohexyl-2-(2′-piperidyl)ethane, DE 2643473; 1977.

[26] Horgan SW, Palopoli FP, Schwoegler EJ. 2-(2,2-Dicyclohexylethyl)piperidines, DE 2713500; 1977.

[27] Horgan SW, Palopoli FP, Wenstrup DL. DE 2714081; 1977.

[28] Davies BJ, Herbert MK, Culbert JA, Pyke SM, Coller JK, Somogyi AA, et al. J Chromatogr B Analyt Technol Biomed Life Sci 2006;832(1):114−20.

[29] Tseng C, Greig IR, Harrison WTA, Zanda M. Asymmetric synthesis and absolute configuration of (+)- and (−)-perhexiline. Synthesis 2016;48(1):73−8.

[30] White MW, Archer JRH. Pipradrol and pipradrol derivatives. In: Dargan PI, Wood DM, editors. Novel psychoactive substances. Amsterdam: Elsevier Inc.; 2013. p. 233−59.

[31] Tilford CH, Shelton RS, Van Campen Jr. MG. Histamine antagonists. Basically substituted pyridine derivatives. J Am Chem Soc 1948;70:4001−9.

[32] Werner HW, Tilford CH. α,α-Diarylpiperidinemethanols, US 2624739; 1953.

[33] Portoghese PS, Pazdernik TL, Kuhn WL, Hite G, Shafi'ee A. Stereochemical studies on medicinal agents. V. Synthesis, configuration, and pharmacological activity of pipradol enantiomers. J Med Chem 1968;11(1):12−15.

[34] Singh SK, Singh S. A brief history of quinoline as antimalarial agents. Int J Pharm Sci Rev Res 2014;25(1):295−302.

[35] Wongsrichanalai C, Prajakwong S, Meshnick SR, Shanks GD, Thimasarn K. Mefloquine - its 20 years in the Thai Malaria Control Program. Southeast Asian J Tropical Med Publ Health 2004;35(2):300−8.

[36] Winstanley P. Mefloquine: the benefits outweigh the risks. Brit J Clin Pharmacol 1996;42 (4):411−13.

[37] Palmer KJ, Holliday SM, Brogden RN. Mefloquine: a review of its antimalarial activity, pharmacokinetic properties and therapeutic efficacy. Drugs 1993;45(3):430−75.

[38] Lutz RE, Ohnmacht CJ, Patel AR. Antimalarials. 7. Bis(trifluoromethyl)-α-(2-piperidyl)-4-quinolinemethanols. J Med Chem 1971;14(10):926−8.

[39] Bomches H, Hardegger B. High purity preparation of mefloquine hydrochloride, EP 92185; 1983.

[40] Carroll FI, Blackwell JT. Optical isomers of aryl-2-piperidylmethanol antimalarial agents. Preparation, optical purity, and absolute stereochemistry. J Med Chem 1974;17(2):210−19.

[41] Tagariello V, Caporuscio A, De Tommaso O. Mepivacaine: update on an evergreen local anaesthetic. Minerva Anestesiol 2001;67(9 Suppl. 1):5−8.

[42] Helmy R, Youssef MH, Rasheed MH, El-Shirbini A, Mandour A. Pharmacological properties of a new local anesthetic drug (mepivacaine hydrochloride). J Egypt Med Assoc 1967;50(11−12):688−701.

[43] Ekenstam BT, Egner BPH. Amides of N-alkylpiperidinemonocarboxylic acid and N-alkyl-pyrrolidine-α-monocarboxylic acids, US 2799679; 1957.

[44] Rinderknecht H. New local anesthetics. Helv Chim Acta 1959;42:1324−7.

[45] Ekenstam BT, Egner B, Pettersson G. Local anesthetics. I. N-Alkylpyrrolidine and N-alkylpiperidinecarboxylic acid amides. Acta Chem Scand 1957;11:1183−90.

[46] Pettersson BG. N-Methylpiperidine-2-carboxylic acid 2,6-xylidide hydrochloride, DE 2726200; 1978.

[47] Tullar BF. Optical isomers of mepivacaine and bupivacaine. J Med Chem 1971;14(9):891−2.

[48] McClure JH. Ropivacaine. Brit J Anaesth 1996;76(2):300−7.

[49] McClellan KJ, Faulds D. Ropivacaine: an update of its use in regional anesthesia. Drugs 2000;60(5):1065−93.

[50] Simpson D, Curran MP, Oldfield V, Keating GM. Ropivacaine: a review of its use in regional anaesthesia and acute pain management. Drugs 2005;65(18):2675−717.

[51] Hansen TG. Ropivacaine: a pharmacological review. Exp Rev Neurotherapeut 2004;4 (5):781−91.

[52] Wang RD, Dangler LA, Greengrass RA. Update on ropivacaine. Exp Opin Pharmacother 2001;2(12):2051−63.

[53] Bansal T, Hooda S. Ropivacaine - a novel and promising local anaesthetic drug. Asian J Pharmaceut Clin Res 2012;5(Suppl. 1):13−15.

[54] Kuthiala G, Chaudhary G. Ropivacaine: a review of its pharmacology and clinical use. Ind J Anaesth 2011;55(2):104−10.

[55] Owen MD, Dean LS. Ropivacaine. Exp Opin Pharmacother 2000;1(2):325−36.

[56] Markham A, Faulds D. Ropivacaine: a review of its pharmacology and therapeutic use in regional anesthesia. Drugs 1996;52(3):429−49.

[57] Casati A, Putzu M. Best Pract Res Clin Anaesthesiol 2005;19(2):247−68.

[58] De Jong RH. Ropivacaine: white knight or dark horse? Reg Anesth 1995;20(6):474−81.

[59] Ekenstam BT, Bovin C. L-N-n-Propylpipecolic acid 2,6-xylidide, WO 8500599; 1985.

[60] Federsel HJ, Jaksch P, Sandberg R. An efficient synthesis of a new, chiral 2′,6′-pipecoloxylidide local anesthetic agent. Acta Chem Scand 1987;B41(10):757–61.

[61] Tuominen M. Bupivacaine spinal anaesthesia. Acta anaesthesiol Scand 1991;35(1):1–10.

[62] Chapman PJ. Review: bupivacaine -a long-acting local anaesthetic. Australian Dental J 1987;32(4):288–91.

[63] Babst CR, Gilling BN. Bupivacaine: a review. Anesthesia Progress 1978;25(3):87–91.

[64] Gazzotti F, Bertellini E, Tassi A. Best indications for local anaesthetics: bupivacaine. Minerva anestesiologica 2001;67(9 Suppl. 1):9–14.

[65] Tullar BF, Bolen CH. 1-Butyl-2′,6′-pipecoloxylidide, GB 1166802; 1969.

[66] No Inventor data available. Amides of N-alkylpiperidinemonocarboxylic acid and N-alkylpyrrolidine-α-monocarboxylic acids, GB 775749; 1957.

[67] No Inventor data available. Amides of N-alkylpiperidinemonocarboxylic acid and N- alkylpyrrolidine-α-monocarboxylic acids, GB 775750; 1957.

[68] Bofors A. N-Alkyl- and cycloalkylpiperidinecarboxylic acid amides, GB 869978; 1958.

[69] Ekesntam BT, Pettersson BG. Synthesis of N-alkylpiperidine and N-alkylpyrrolidine-α-carboxylic acid amides, US 2955111; 1960.

[70] Luduena FP, Tullar BF. Optical isomers of an aminoacyl xylidide, ZA 6802611; 1968. From S. African (1968), ZA 6802611 19681212.

[71] Ivani G, Borghi B, van Oven H. Levobupivacaine. Minerva Anestesiologica 2001;67(9 Suppl. 1):20–3.

[72] McLeod GA, Burke D. Levobupivacaine. Anaesthesia 2001;56(4):331–41.

[73] Adger B, Dyer U, Hutton G, Woods M. Stereospecific synthesis of the anesthetic levobupivacaine. Tetrahedron Lett 1996;37(35):6399–402.

[74] Hudak JM, Banitt EH, Schmid JR. Discovery and development of flecainide. Am J Cardiol 1984;53(5):17–20.

[75] Aliot E, Capucci A, Crijns HJ, Goette A, Tamargo J. Twenty-five years in the making: flecainide is safe and effective for the management of atrial fibrillation. Europace 2011;13 (2):161–73.

[76] Roden DM, Woosley RL. Flecainide: a new agent for the treatment of ventricular arrhythmias. N Engl J Med 1986;315(1):36–41.

[77] Tamargo J, Le Heuzey J, Mabo P. Narrow therapeutic index drugs: a clinical pharmacological consideration to flecainide. Eur J Clin Pharmacol 2015;71(5):549–67.

[78] Apostolakis S, Oeff M, Tebbe U, Fabritz L, Breithardt G, Kirchhof P. Flecainide acetate for the treatment of atrial and ventricular arrhythmias. Exp Opin Pharmacother 2013;14 (3):347–57.

[79] Holmes B, Heel RC. Flecainide: a preliminary review of its pharmacodynamic properties and therapeutic efficacy. Drugs 1985;29(1):1–33.

[80] Banitt EH, Coyne WE, Schmid JR, Mendel A. Antiarrhythmics. N-(Aminoalkylene)trifluoroethoxybenzamides and N-(aminoalkylene)trifluoroethoxynaphthamides. J Med Chem 1975;18(11):1130–4.

[81] Banitt EH, Bronn WR, Coyne WE, Schmid JR. Antiarrhythmics. 2. Synthesis and antiarrhythmic activity of N-(piperidylalkyl)trifluoroethoxybenzamides. J Med Chem 1977;20 (6):821–6.

[82] Banitt EH, Bronn WR. Derivatives of pyrrolidine and piperidine, US 3900481; 1975.

[83] Banitt EH, Bronn WR. Derivatives of pyrrolidine and piperidine, US 4013670; 1977.

[84] Banitt EH, Bronn WR. Antiarrhythmic method utilizing fluoroalkoxy-N-piperidyl and pyridyl benzamides, US 4005209; 1977.

[85] Woosley RL, Wood AJJ, Roden DM. Encainide. N Engl J Med 1988;318(17):1107–15.

[86] Mitchell LB, Winkle RA. Encainide. New Drugs Annual: Cardiovasc Drugs 1983;1:93–107.

[87] Antonaccio MJ, Gomoll AW, Byrne JE. Encainide. Cardiovasc Drugs Ther 1989;3 (5):691–710.

[88] Brogden RN, Todd PA. Encainide. A review of its pharmacological properties and therapeutic efficacy. Drugs 1987;34(5):519–38.

[89] Somberg JC, Zanger D, Levine E, Tepper D. Encainide: a new and potent antiarrhythmic. Am Heart J 1987;114(4 Pt 1):826–35.

[90] Dykstra SJ, Minielli JL. Pharmacologically active substituted piperidin, DE 2210154; 1972.

[91] Dykstra SJ, Minielli JL. Substituted piperidines, US 4000143; 1976.

[92] Dykstra SJ, Minielli JL, Lawson JE, Ferguson HC, Dungan KW. Lysergic acid and quinidine analogs. 2-(o-Acylaminophenethyl)piperidines. J Med Chem 1973;16(9):1015–20.

[93] Johnson PC, Covington RR. Synthesis of encainide-13C hydrochloride from 2-nitrobenzaldehyde-formyl-13C. J Labelled Comp Radiopharm 1982;19(8):953–8.

[94] No Inventor data available, Preparation of the antiarrhythmic encainide, BE 1000112; 1988.

[95] Dillon JL, Spector RH. Process for the preparation of encainide hydrochloride, US 4675409; 1987.

[96] No Inventor data available, Encainide, NL 8204775; 1983.

[97] Gold N. Experience with *thioridazine* ("Melleril") therapy. J Mental Sci 1961;107:523–8.

[98] Jones AL, Lewis DJ, Miller A. Use of *thioridazine* (mellaril) in a variety of clinical settings. CMAJ 1960;83:948–9.

[99] Barsa JA, Saunders JC. *Thioridazine* (mellaril) in the treatment of chronic schizophrenics. Am J Psychiatry 1960;116:1028–9.

[100] Kirchner V, Kelly CA, Harvey RJ. *Thioridazine* for dementia. Cochrane Database Syst Rev 2001;(3):CD000464.

[101] Sultana A, Reilly J, Fenton M. *Thioridazine* or schizophrenia. Cochrane Database Syst Rev 2000;(3):CD001944.

[102] Fenton M, Rathbone J, Reilly J, Sultana A. *Thioridazine* for schizophrenia. Cochrane Database Syst Rev 2007;(3):CD001944.

[103] Abdel-Moety EM, Al-Rashood, Khalid A. Analytical profile of *thioridazine and thioridazine* hydrochloride. Anal Profiles Drug Subst 1989;18:459–525.

[104] Dutta NK, Karakousis PC. Thioridazine for treatment of tuberculosis: promises and pitfalls. Tuberculosis 2014;94(6):708–11.

[105] Amaral L, Molnar J. Why and how the old neuroleptic thioridazine cures the XDR-TB. Pharmaceuticals 2012;5:1021–31.

[106] Thanacoody RHK. Thioridazine: the good and the bad. Rec Pat Anti-Infect Drug Disc 2011;6(2):92–8.

[107] Thanacoody HKR. Thioridazine: resurrection as an antimicrobial agent? Brit J Clin Pharmacol 2007;64(5):566–74.

[108] Bourquin JP, Schwarb G, Gamboni G, Fischer R, Ruesch L, Guldimann S, et al. Syntheses in the phenothiazine family. II. N-Substituted phenothiazinethiol derivatives. Helv Chim Acta 1958;41:1072–108.

[109] No Inventor data available. 2-Mercaptophenothiazine derivatives, GB 863550; 1961.

[110] No Inventor data available. 2-Mercaptophenothiazine derivatives, GB 863551; 1961.

[111] Pinder RM, Brogden RN, Speight TM, Avery GS. Rimiterol: a review of its pharmaco-
       logical properties and therapeutic efficacy in asthma. Drugs 1977;14(2):81−104.
[112] Muittari A. A brief review of sympathomimetic bronchodilators and a description of a
       new selective agent, rimiterol hydrobromide. Respiration 1978;35(3):173−80.
[113] Carney I, Daly MJ, Lightowler JE, Pickering RW. Comparative pharmacology of WG
       253 (rimiterol hydrobromide), a new bronchodilator. Arch Int Pharmacodyn, Therapie
       1971;194(2):334−45.
[114] Griffin JP, Turner P. New bronchodilator (WG 253) in man. J Clin Pharmacol New
       Drugs 1971;11(4):280−7.
[115] Bowman WC, Rodger IW. Actions of the sympathomimetic bronchodilator, rimiterol
       (R798), on the cardiovascular, respiratory, and skeletal muscle systems of the anesthe-
       tized cat. Brit J Pharmacol 1972;45(4):574−83.
[116] Palma-Carlos AG, Palma-Carlos GS. Beta-2-agonists of third generation. Allergie et
       Immunologie 1986;18(4):31−32, 35.
[117] Lai CK, Twentyman OP, Holgate ST. The effect of an increase in inhaled allergen dose
       after rimiterol hydrobromide on the occurrence and magnitude of the late asthmatic
       response and the associated change in nonspecific bronchial responsiveness. Am Rev
       Resp Dis 1989;140(4):917−23.
[118] Ydreborg SO, Svedmyr N, Thiringer G. Comparison of rimiterol and terbutaline, given
       by aerosol, in a long-term study. Scand J Resp Diseases 1977;58(2):117−24.
[119] Sankey GH, Whiting KDE. Cyclic adrenaline derivatives. Aryl-2-piperidylcarbinols. J
       Het Chem 1972;9(5):1049−55.
[120] Sankey GH, Whiting KDE. α-(Hydroxy- and alkoxy-substituted)phenyl-α-(2-piperidinyl)
       methanols, useful as bronchodilators, DE 2024049; 1970.
[121] Sankey GH, Whiting KDE. α-(Hydroxy and alkoxy substituted)phenyl-α-(2-piperidinyl)-
       methanols, US 3910934; 1975.
[122] Kaiser C, Ross ST. Bronchodilating hydroxyphenyl-2-piperidinylcarbinols, DE 2047937;
       1971.
[123] Aldaheff M. Lobeline sulfate. Drugs of Today 1977;13(6):236−9.
[124] Damaj MI, Patrick GS, Creasy KR, Martin BR. Pharmacology of lobeline, a nicotinic
       receptor ligand. J Pharmacol Exp Ther 1997;282(1):410−19.
[125] Dwoskin LP, Crooks PA. A novel mechanism of action and potential use for lobeline as
       a treatment for psychostimulant abuse. Biochem Pharmacol 2002;63(2):89−98.
[126] Miller DK, Lever JR, Rodvelt KR, Baskett JA, Will MJ, Kracke GR. Lobeline, a poten-
       tial pharmacotherapy for drug addiction, binds to µ opioid receptors and diminishes the
       effects of opioid receptor agonists. Drug Alcohol Depend 2007;89(2−3):282−91.
[127] Willyard C. Pharmacotherapy: quest for the quitting pill. Nature 2015;522(7557):
       S53−5.
[128] McCurdy CR, Miller RL, Beach JW. Lobeline: a natural product with high affinity for
       neuronal nicotinic receptors and a vast potential for use in neurological disorders.
       In: Cutler SJ, Cutler HG, editors. Biologically active natural products. Boca Raton, FL:
       CRC Press; 2000. p. 151−62.
[129] Stead LF, Hughes JR. Lobeline for smoking cessation. Cochrane Database System Rev
       2012;(2):CD000124.
[130] Wieland H, Drishaus I. Lobelia alkaloids. IV. Synthesis of lobelia alkaloids. Just Lieb
       Ann 1929;473:102−18.
[131] Scheuing G, Winterhalder L. Synthesis of lobelia alkaloids. Just Lieb Ann
       1929;473:126−36.

[132] Klingler F, Sobotta R. Process for manufacturing of chiral lobeline, US 200600114791; 2006.

[133] Klingler FD. Asymmetric hydrogenation of prochiral amino ketones to amino alcohols for pharmaceutical use. Acc Chem Res 2007;40(12):1367–76.

[134] Schopf C, Lehmann G. Syntheses and transformations of natural products under physiological conditions (model experiments on the question of the biogenesis of natural products). IV. Syntheses of tropinone, pseudopelletierine, lobelanine and related alkaloids under physiological conditions. Just Lieb Ann 1935;518:1–37.

[135] Felpin F, Lebreton J. History, chemistry and biology of alkaloids from Lobelia inflate. Tetrahedron 2004;60(45):10127–53.

[136] Felpin F, Lebreton J. A highly stereoselective asymmetric synthesis of (-)-Lobeline and (-)-Sedamine. J Org Chem 2002;67(26):9192–9.

[137] Klingler FD. Development of efficient technical processes for the production of enantiopure amino alcohols in the pharmaceutical industry, Synthesis of Lobeline, Lobelane and their Analogues. In: Blaser H, Federsel H, editors. A review, asymmetric catalysis on industrial scale. 2nd Edition Weinheim: Wiley; 2010. p. 171–85.

[138] Zheng G, Crooks PA. Synthesis of lobeline, lobelane and their analogues. Rev Org Prep Procedures Int 2015;47(5):317–37.

[139] Jeske WP, Fareed J, Hoppensteadt DA, Lewis B, Walenga JM. Pharmacology of argatroban. Exp Rev Hematol 2010;3(5):527–39.

[140] McKeage K, Plosker GL. Argatroban. From Drugs 2001;61(4):515–22.

[141] Kondo LM, Wittkowsky AK, Wiggins BS. Argatroban for prevention and treatment of thromboembolism in heparin-induced thrombocytopenia. Ann Pharmacother 2001;35 (4):440–51.

[142] Jeske W, Walenga JM, Lewis BE, Fareed J. Pharmacology of argatroban. Exp Opin Invest Drugs 1999;8(5):625–54.

[143] Dhillon S. Argatroban. A review of its use in the management of heparin-induced thrombocytopenia. Am J Cardiovasc Drugs 2009;9(4):261–82.

[144] Boggio LN, Oza VLM. Argatroban use in heparin-induced thrombocytopenia. Exp Opin Pharmacother 2008;9(11):1963–7.

[145] Rice L, Hursting MJ. Argatroban therapy in heparin-induced thrombocytopenia. Exp Rev Clin Pharmacol 2008;1(3):357–67.

[146] Escolar G, Bozzo J, Maragall S. Argatroban: a direct thrombin inhibitor with reliable and predictable anticoagulant actions. Drugs of Today 2006;42(4):223–36.

[147] Yeh RW, Jang I. Argatroban: update. Am Heart J 2006;151(6):1131–8.

[148] Lewis BE, Hursting MJ. Argatroban therapy in heparin-induced thrombocytopenia. Fund Clin Cardiol 2004;47:437–74.

[149] Fareed J, Jeske WP. Small-molecule direct antithrombins: argatroban. Best Practice Res, Clinical Haematol 2004;17(1):127–38.

[150] Warkentin TE. Management of heparin-induced thrombocytopenia: a critical comparison of lepirudin and argatroban. Thromb Res 2003;110(2–3):73–82.

[151] Bambrah RK, Pham DC, Rana F. Argatroban in heparin-induced thrombocytopenia: rationale for use and place in therapy. Therapeut Adv Chronic Disease 2013;4(6):302–4.

[152] Kathiresan S, Shiomura J, Jang I, Argatroban J. Thrombos. Thrombolys 2000;13(1):41–7.

[153] Saugel B, Schmid RM, Huber W. Safety and efficacy of argatroban in the management of heparin-induced thrombocytopenia. Clin Med Insights: Blood Disorders 2011;4:11–19.

[154] Cossy J, Belotti DA. Short synthesis of argatroban a potent selective thrombin inhibitor. Bioorg Med Chem, Lett 2001;11(15):1989−92.

[155] Okamoto S, Hijikata A, Kikumoto R, Tamao Y, Ohkubo K, Tezuka T, et al. N2-Arylsulfonyl-1-argininamides and pharmaceutically acceptable salts, US 4258192; 1981.

[156] Kikumoto R, Tamao Y, Ohkubo K, Tezuka T, Tonomura S, Hijikata A, et al. α-(N-Arylsulfonyl-L-argininamides and pharmaceutical compositions containing these substances, EP 8746; 1980.

[157] Cossy J. Selective methodologies for the synthesis of biologically active piperidinic compounds. Chem Record 2005;5(2):70−80.

[158] Ermertcan AT, Ozturkcan S. Topical calcineurin inhibitors, pimecrolimus and tacrolimus. Antiinflamm Antiallergy Agents Med Chem 2007;6(3):237−43.

[159] Bulusu MARC, Baumann K, Stuetz A. Chemistry of the immunomodulatory macrolide ascomycin and related analogues. Prog Chem Org Nat Prod 2011;94:59−126.

[160] Mollison KW, Fey TA, Gauvin DM, Sheets MP, Smith ML, Pong M, et al. Discovery of ascomycin analogs with potent topical but weak systemic activity for treatment of inflammatory skin diseases. Curr Pharm Des 1998;4(5):367−79.

[161] Paul C, Graeber M, Stuetz A. Ascomycins: promising agents for the treatment of inflammatory skin diseases. Exp Opin Invest Drugs 2000;9(1):69−77.

[162] Griffiths CEM. Ascomycin: an advance in the management of atopic dermatitis. Br J Dermatol 2001;144(4):679−81.

[163] Hersperger R, Keller TH. Ascomycin derivatives and their use as immunosuppressive agents. Drugs Fut 2000;25(3):269−77.

[164] Simpson D, Noble S. Tacrolimus ointment: a review of its use in atopic dermatitis and its clinical potential in other inflammatory skin conditions. Drugs 2005;65(6):827−58.

[165] Zimmer R, Baumann K, Sperner H, Schulz G, Haidl E, Grassberger MA. Synthetic modifications of ascomycin. Part V. Access to novel ascomycin derivatives by replacement of the cyclohexylvinylidene subunit. Croat Chem Acta 2005;78(1):17−27.

[166] Zimmer R, Grassberger MA, Baumann K, Horvath A, Schulz G, Haidl E. Synthetic modifications of ascomycin. II. A simple and efficient way to modified iso-ascomycin derivatives. Tetrahedron Lett 1995;36(42):7635−8.

[167] Ferraboschi P, Colombo D, De Mieri M, Grisenti P. First chemoenzymatic synthesis of immunomodulating macrolactam pimecrolimus. Tetrahedron Lett 2009;50 (30):4384−8.

[168] Vezina C, Kudelski A, Sehgal SN. Rapamycin (AY-22,989), a new antifungal antibiotic. I. Taxonomy of the producing streptomycete and isolation of the active principle. J Antibiot 1975;28(10):721−6.

[169] Klawitter J, Nashan B, Christians U. Everolimus and sirolimus in transplantation-related but different. Exp Opin Drug Safety 2015;14(7):1055−70.

[170] Moes DJAR, Guchelaar H, de Fijter JW. Everolimus and sirolimus in kidney transplantation. Drug Discov Today 2015;20(10):1243−9.

[171] Augustine JJ, Hricik DE. Experience with everolimus, Transplant. Proc., 2004, 36 2S0, 500S-503S.

[172] Webster AC, Lee VW, Chapman JR, Craig JC. Target of rapamycin inhibitors (TOR-I; sirolimus and everolimus) for primary immunosuppression in kidney transplant recipients. Cochrane Database Syst Rev 2006;(2):CD004290.

[173] Schulze M, Stock C, Zaccagnini M, Teber D, Rassweiler JJ. Temsirolimus. Recent Results Cancer Res 2014;201:393−403.

Chapter 4

3-Substituted and 1,3-Disubstituted Piperidines

4.1 METHIXENE (92)

Methixene (**4.1.3**) is an anticholinergic agent and has antihistamine properties. It has a pronounced inhibitory action on intestinal hypermotility and may be useful in the symptomatic treatment of functional bowel disorders. It can be used also for the symptomatic treatment of parkinsonism, for the alleviation of the extrapyramidal syndrome, but it is of no value against tardive dyskinesias.

Methixene is a medicine available in a number of countries worldwide. It was approved for use in the United States in 1982, but now it seems to be discontinued [1−5].

Methixene (**4.1.3**) was synthesized via alkylation of 9-thioxanthylsodium formed by interaction of phenylsodium with thioxanthene (**4.1.1**) and with N-methyl-3-chloromethylpiperidine (**4.1.2**) [6,7] (Scheme 4.1).

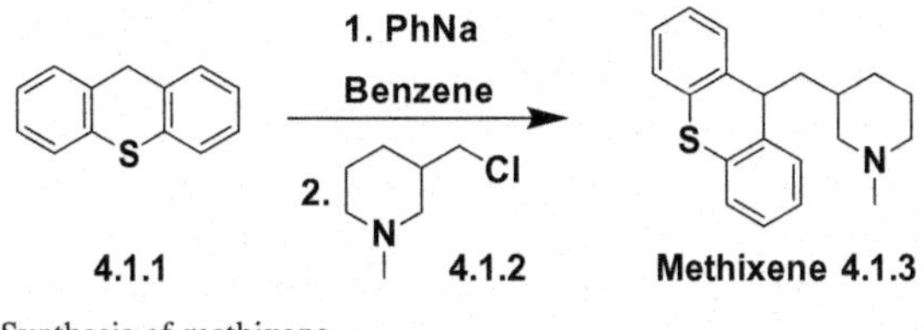

SCHEME 4.1   Synthesis of methixene.

4.2 TIPEPIDINE (132)

Tipepidine (**4.1.7**) has been widely used solely as an antitussive and expectorant in Japan since 1959 in the management of cough. It is approximately equal to codeine in its efficacy, and the drug never caused tolerance and seemed to establish no addiction liabilities. Tipepidine does not cause any adverse side effects such as euphoria, craving, or tolerance [8].

Piperidine-Based Drug Discovery. DOI: http://dx.doi.org/10.1016/B978-0-12-805157-3.00004-1

It was reported recently that tipepidine inhibits G-protein—coupled inwardly rectifying potassium (GIRK)-channel currents, which is expected to modulate the catecholaminergic system, the level of catecholamines in the brain. Consequently, it was hypothesized and proved that tipepidine may have an antidepressant-like effect and could be used in the management of obsessive—compulsive disorder and could improve attention deficit/hyperactivity disorder symptoms [9—14].

The synthesis of tipepidine was started from the ethyl ester of *N*-methyl-nipecotic acid (**4.1.4**) which, in turn, was prepared by hydrogenation of ethyl nicotinate on Raney nickel. To the solution of (**4.1.4**) in ether was added an ethereal solution of a Grignard reagent (**4.1.5**) prepared from 2-bromothiophene. The synthesized carbinol (**4.1.6**) was dehydrated on heating at 100°C in the mixture of concentrated hydrochloric and acetic acids to give desired tipepidine (**4.1.7**) [15] (Scheme 4.2).

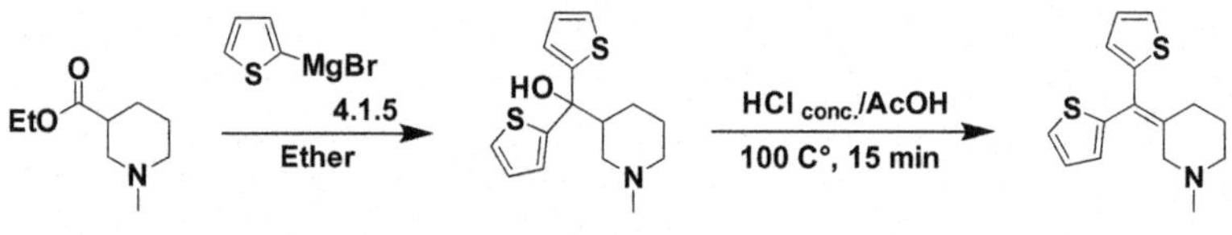

SCHEME 4.2   Synthesis of tipepidine.

4.3   TIMEPIDIUM (83)

Timepidium (**4.1.15**) is an anticholinergic agent with antispasmodic effects used for the relief of muscular and visceral spasms available in a number of countries worldwide. It reduces muscle spasms of the intestine, biliary system, uterus, and urinary bladder, and is indicated for the management of pain in gastrointestinal, biliary, urological, and gynecological problems such as gastroenteritis, diarrhea, dysentery, biliary colic, enterocolitis, cholecystitis, colonopathies, mild cystitis, and spasmodic dysmenorrhea [16,17].

Synthesis of timepidium (**4.1.15**) was carried from 5-bromonicotinic acid (**4.1.8**) yield by refluxing I in NaOH solution in the presence of $CuSO_4$ and Cu powder yielded 5-hydroxy-nicotinic acid (**4.1.9**). Esterification of this acid with methanol and HCl gave (**4.1.10**). Treatment of methyl 5-hydroxynicotinate (**4.1.10**) with dimethylsulfate in the presence of anhydrous K_2CO_3 in benzene/ methanol mixture afforded crude (**4.1.11**), which was directly hydrogenated without purification to give (**4.1.12**). The carbinol (**4.1.13**) was synthesized by a procedure similar to that employed for tipepidine (**4.1.7**). Dehydration with dilute hydrochloric acid gave the product (**4.1.14**) that on reaction with MeBr gave desired timepidium (**4.1.15**) [18,19] (Scheme 4.3).

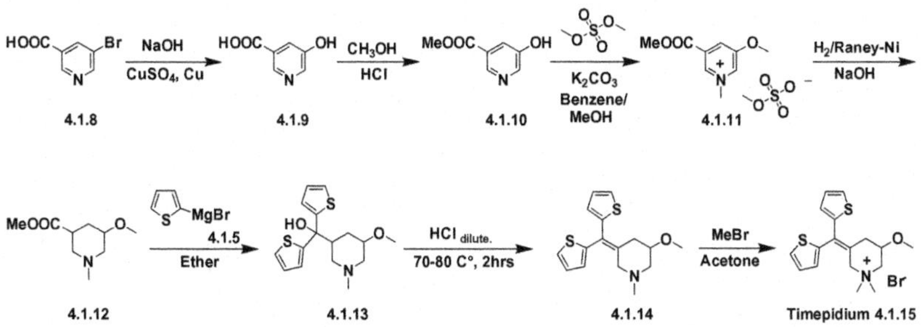

SCHEME 4.3 Synthesis of timepidium.

4.4 TIAGABINE (2008)

Tiagabine (**4.1.23**) (Gabitril) is an anticonvulsant medication, an approved antiepileptic drug, to treat partial seizures, which inhibits γ-aminobutyric acid (GABA) reuptake into neurons and glia. The regulation of neuronal signal transduction in the nervous system is mainly controlled by the interplay of excitatory neurotransmitter glutamate and the major inhibitory neurotransmitter GABA. A failure of either one of these neurotransmitter systems leads to neurological disorders. It is thought that tiagabine blocks GABA uptake into presynaptic neurons, permitting more GABA to be available for receptor binding on the surfaces of postsynaptic cells. Particularly low GABA levels are connected not only with epilepsy but Alzheimer's disease and depression. Enhancing the amount of GABA in the synaptic cleft is believed to be beneficial in the treatment of these diseases. It has been reported that tiagabine possesses anxiolytic properties and effective in prophylactic treatment of bipolar disorder.

The most common side effects are dizziness, asthenia, somnolence, accidental injury, infection, headache, and nausea [20−35].

The strategy used for the synthesis of tiagabine (**4.1.23**) is based on N-alkylation of ethyl nipecotate (**4.1.20**) with 4-bromo-l,l-bis(3-methyl-2-thienyl)-l-butene (**4.1.19**) prepared by different methods. In particular bis(3-methyl-2-thienyl)methanone (**4.1.16**) was reacted with cyclopropylmagnesium bromide (**4.1.17**) to give cyclopropylbis-(3-methyl-2-thienyl)methanol (**4.1.18**) followed by the acid-catalyzed opening of the cyclopropyl ring to give bromide (**4.1.19**). The obtained bromide was used to alkylate ethyl nipecotate (**4.1.20**) in standard conditions (acetone, KI, K_2CO_3) to give ester (**4.1.21**). The last was hydrolyzed with base giving racemic 1-(4,4-bis(3-methylthiophen-2-yl)but-3-en-1-yl)piperidine-3-carboxylic acid (**4.1.22**) from which crystals of the desired tiagabine (**4.1.23**)- (−)-(R)-1-[4,4-bis(3-methyl-2-thienyl)-3-butenyl]nipecotic acid hydrochloride were separated just via dissolving in dilute hydrochloric acid [36−38] (Scheme 4.4).

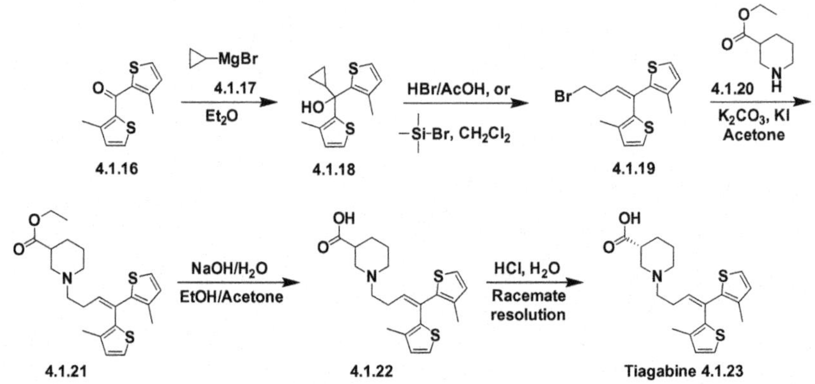

SCHEME 4.4 Synthesis of tiagabine.

Other methods for the synthesis of tiagabine were proposed [39−41].

4.5 PIPERIDOLATE (72)

Piperidolate (**4.1.28**) (Dactiran) is a muscarinic receptor antagonist (anticholinergic) which inhibits smooth muscle contraction in digestive organs to relieve pain due to convulsion of the gastrointestinal tract. It also suppresses the abnormal contraction of the uterus. It is usually used to treat cramp-like pain in patients with gastric or duodenal ulcer, gastritis, enteritis, cholelithiasis, cholecystitis or biliary dyskinesia, and improves various symptoms of threatened abortion or threatened premature labor. Piperidolate is a medicine available in a number of countries worldwide and was initially marketed in Japan in 1967. Administration of this product may induce mydriasis, dizziness, and others [42−44].

Piperidolate (**4.1.28**) was prepared by interaction of diphenylacetyl chloride (**4.1.27**) with N-ethyl-3-hydroxypiperidine (**4.1.2**), which, in turn, was synthesized by the reductive aminolysis of furfural (**4.1.24**) with ethylamine on hydrogenation in the presence Raney nickel catalyst at 100°C, which gave N-ethyltetrahydrofurfurylamine (**4.1.25**) converted to N-ethyl-3-hydroxypiperidine (**4.1.26**) on passing hydrogen bromide in its solution in glacial acetic acid at 100−105°C [45,46] (Scheme 4.5).

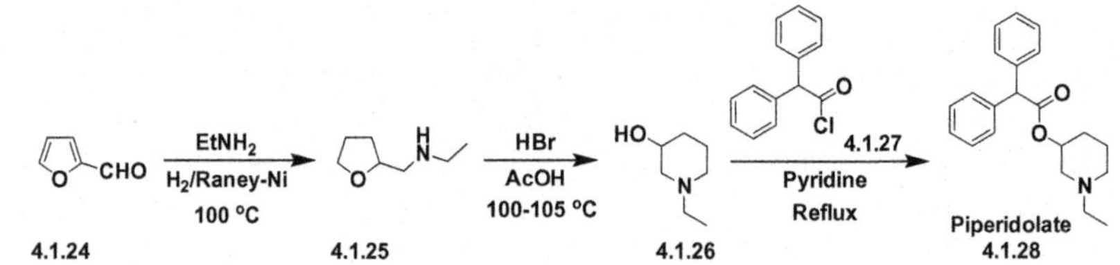

SCHEME 4.5 Synthesis of piperidolate.

4.6 MEPENZOLATE (137)

Mepenzolate (**4.1.32**), brand name Cantil, is a specific muscarinic receptor antagonist originally used for gastrointestinal disorders in a clinical setting such as peptic ulcers and irritable bowel syndrome [47−53].

Recently, it has been reported that mepenzolate displays beneficial effects in chronic obstructive pulmonary disease and pulmonary fibrosis by preventing inflammatory responses and reducing oxidative stress [54−56] and for treating diabetic ulcers and other chronic wounds in clinics [57].

For the synthesis of mepenzolate (**4.1.32**) N-methyl-3-chloropiperidine (**4.1.29**) and benzylic acid (**4.1.30**) in anhydrous iso-propyl alcohol were refluxed for three days to give ester (**4.1.31**), which was converted to piperidinium salt via adding methyl bromide to iso-propyl alcohol solution of (**4.1.31**) at room temperature to give desired mepenzolate (**4.1.32**) [58] (Scheme 4.6).

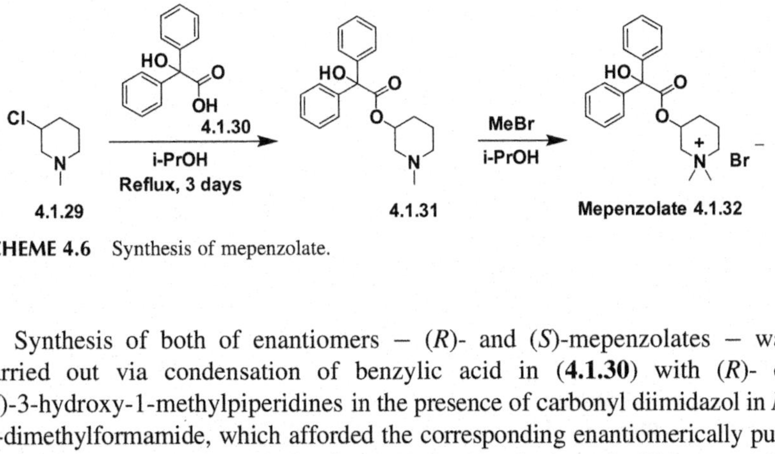

SCHEME 4.6   Synthesis of mepenzolate.

Synthesis of both of enantiomers − (R)- and (S)-mepenzolates − was carried out via condensation of benzylic acid in (**4.1.30**) with (R)- or (S)-3-hydroxy-1-methylpiperidines in the presence of carbonyl diimidazol in N, N-dimethylformamide, which afforded the corresponding enantiomerically pure (R)- and (S)-mepenzolates. In in vivo experiments, the bronchodilatory activity of (R)-mepenzolate was superior to that of (S)-mepenzolate, whereas antiinflammatory activity was indistinguishable between the two enantiomers [59].

4.7 PIPENZOLATE (109)

Pipenzolate (**4.1.33**), a medicine with anticholinergic and antispasmodic action, is available in a number of countries; it is a close analog of mepenzolate (**4.1.32**). It is usually used as an adjunct in the treatment of gastrointestinal disorders characterized by smooth muscle spasm [60−62].

Its synthesis was carried out by the full analogy with the synthesis of mepenzolate (**4.1.32**) starting from N-ethyl-3-chloropiperidine instead of N-methyl-3-chloropiperidine (**4.1.29**) [45,63]. (Fig. 4.1.)

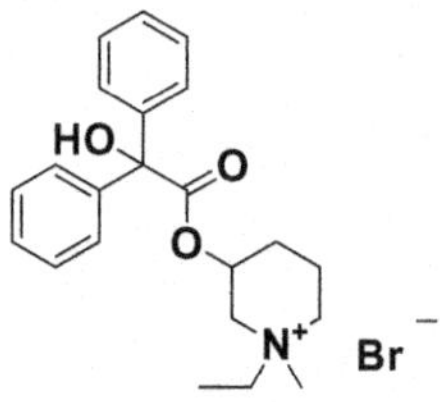

Pipenzolate 4.1.33

FIGURE 4.1 Pipenzolate.

4.8 BENIDIPINE (927)

Benidipine (**4.1.39**) (Coniel) was developed in Japan and is a long-acting orally active antihypertensive agent that displays a wide range of activities [64]. It inhibits three different types, L-, N-, and T-type of Ca^{2+} channels, and inhibits aldosterone-induced mineralocorticoid receptor activation.

Benidipine exhibits cardioprotective and antiartherosclerotic effects.

It is used clinically as a racemate, containing the $-(a)$- and $+(a)$- isomers of benidipine.

There are two chiral atoms in the molecule of benidipine hydrochloride, which locate on site 4 of the dihydropyridine ring and site 3′ of the side chain piperidine ring. Accordingly, benidipine hydrochloride has four optical isomers: (S)-(S)-$(+)$-α, (R)-(R)-$(-)$-α, (R)-(S)-$(+)$-β and (S)-(R)-$(-)$-β, and the active ingredients for drug are a mixture of (S)-(S)-$(+)$-α and (R)-(R)-$(-)$-α. The α-isomer (benidipine) showed a very strong hypotensive effect but little activity was observed in the β-isomer.

Fractional crystallization of the mix of hydrochlorides allowed to separate α- and β-diastereoisomers.

So, benidipine (**4.1.39**) consists of two optical isomers (S)-(S)-$(+)$-α and (R)-(R)-$(-)$-α of the four possible optical isomers because the β-isomers fortunately and easily are removed during the crystallization step on synthesis [65]. The (S)-(S)-$(+)$-α isomer was shown to be 30- to 100-fold more active than the (R)-(R)-$(-)$-α isomer in terms of the antihypertensive effect [66].

Benidipine (**4.1.39**) enlarges the peripheral arteries and the coronary vessels, decreases blood pressure by reducing influx of calcium ion into cells, and consequently prevents or relieves episodes of angina. It is usually used for the treatment of hypertension, angina pectoris, and renal parenchymal hypertension.

Several cardiocerebrovascular protective effects of benidipine, such as maintaining stable heart rate, renal protective effects, high vascular selectivity, vascular endothelial cell protection, and inhibition of aldosterone activities have also been noted [67−71]. Common side effects are: palpitation, flash, headache, rash, itch, photosensitivity, gynecomastia, etc.

Several synthetic routes of benidipine hydrochloride and its analogs have been disclosed and generalized in in two methods.

According to one of them benidipine (**4.1.39**) was synthesized using the Hantzch reaction methodology involving into the reaction acetoacetic acid-l-benzyl-3-piperidyl ester (**4.1.35**) prepared by reaction of 1-benzyl-3-hydroxypiperidine (**4.1.34**) and diketene, with m-nitrobenzaldehyde (**4.1.36**) and methyl β-aminocrotonate (**4.1.37**) on reflux in THF gave expected racemic product of cyclization (**4.1.38**), which on recrystallization from acetone-ethanol (9:1) gave (±)-α-isomeric form of the desired compound — benidipine (**4.1.39**) [72,73].

In the second method the title compound was prepared by condensation of acid chloride (**4.1.41**) prepared from 2,6-dimethyl-4-(3-nitrophenyl)-1,4-dihydropyridine-3,5-dicarboxylic acid monomethyl ester (**4.1.40**) and thionyl chloride and 1-benzyl-3-hydroxypiperidine (**4.1.34**) to give the same racemic (±)-a product (**4.1.38**) [73] (Scheme 4.7). Variations of these methods are proposed [73,74].

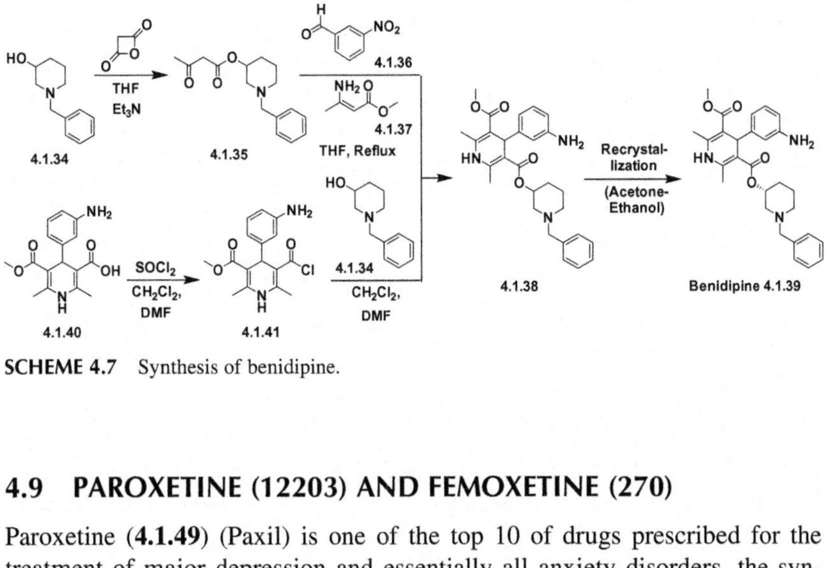

SCHEME 4.7   Synthesis of benidipine.

4.9 PAROXETINE (12203) AND FEMOXETINE (270)

Paroxetine (**4.1.49**) (Paxil) is one of the top 10 of drugs prescribed for the treatment of major depression and essentially all anxiety disorders, the synthesis of which was described in one of our previous books [75] and here we have borrowed and repeated the main points outlined in that book.

Paroxetine is a specific serotonin reuptake inhibitor, which treats depression, obsessive—compulsive disorder, panic disorder, social anxiety disorder, premenstrual dysphoric disorder, generalized anxiety disorder, and posttraumatic stress disorder. Paroxetine is effective for the psychic symptoms of anxiety, which include worry, tension, irritability, and concentration difficulties, and carry tolerable and safe side effect profiles [76−85].

Several strategies for the synthesis of paroxetine have been described in the recent review [86]. One of the relatively large-scale synthesis method is based on the − (+) enantiomer of the *N*-benzyl trans-4-(4-fluorophenyl)-3-piperidinemethanol (**4.1.46**), which was prepared starting from 1-benzyl-4-piperidone (**4.1.42**). Grignard reaction of the last with 4-fluorophenyl magnesium bromide give tertiary alcohol (**4.1.43**) dehydration of which, promoted by *p*-toluenesulfonic acid, gave tetrahydropyridine derivative (**4.1.44**). Transformation of (**4.1.44**) in the Prins reaction conditions afforded racemic mixture of tetrahydropyridine-3-methanol, which was resolved with (−)-L-dibenzoyltartaric acid to give (−) (**4.1.45**). The stereoselective reduction of which on Pd/C catalyst, under acidic conditions in water with retention of *N*-benzyl protective group and led to cis- (3*R*,4*R* isomer of piperidine-3-methanol (**4.1.46**), obtained cis-alcohol (**4.1.46**) was converted into cis-mesylate (**4.1.47**) with methanesulfonyl chloride. The reaction of the last with sodium sesamolate resulted in the formation of trans *N*-benzylparoxetine (**4.1.48**), which was debenzylated on hydrogenation on Pd/C catalyst to give desired paroxetine (**4.1.49**) [86] (Scheme 4.8).

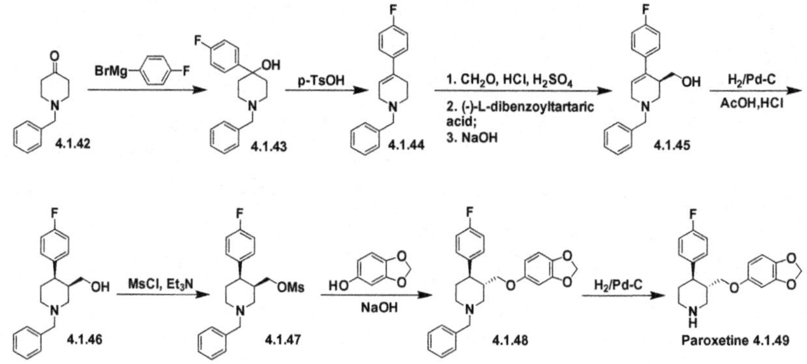

SCHEME 4.8 Synthesis of paroxetine.

Another method for the synthesis of paroxetine is based on conjugate addition reaction of 4-fluorophenylmagnesium bromide to arecoline (**4.1.50**). This additional reaction gave racemic esters (**4.1.51**) (+/ −)-cis and (+/ −)-trans and what's more, it was found that the cis-form could be converted to thermodynamically more stable trans-form on heating with a strong base [36]. The obtained trans-form of 1-methylpiperidine-3-carboxylate was hydrolyzed to acid (**4.1.52**) with concentrated hydrochloric acid and then converted to corresponding acid chloride (**4.1.53**). The acid chloride was esterified with (−)-menthol and pure (−)-trans-(3*S*,4*R*) menthol ester (**4.1.54**) was separated as hydrobromide. The base liberated from hydrobromide was reduced with lithium aluminum hydride to give (−)-(3*S*,4*R*)-piperidin-3-yl-methanol (**4.1.55**), which was transferred to the (3*S*,4*R*)-3-(chloromethyl)-piperidine (**4.1.56**), which reaction with sodium sesamolate resulted in formation of (3S,4R)-3-(1,3-benzdioxolyl-3-oxymethyl)-piperidine. The last was desmethylated with phenylchloroformate to give desired paroxetine (**4.1.49**) [87] (Scheme 4.9).

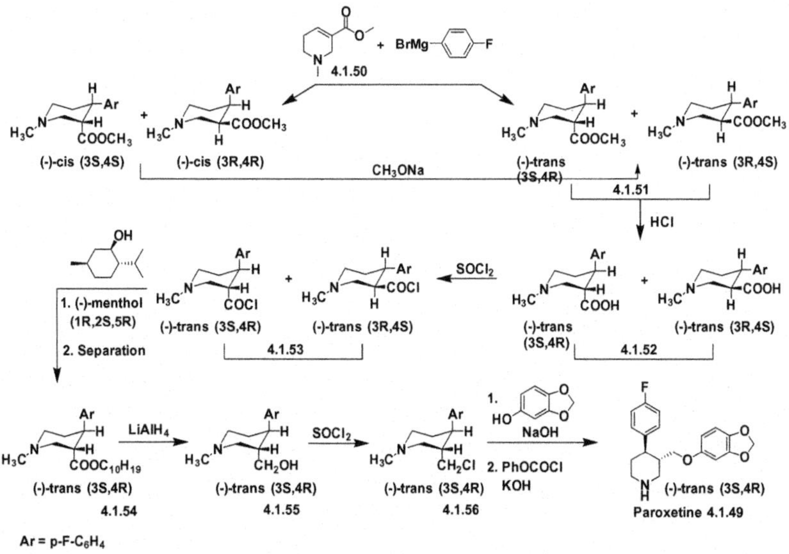

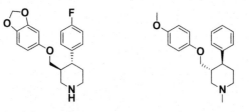

SCHEME 4.9   Synthesis of paroxetine.

Very similar variations of proposed method based on the 4-arylpiperidine-3-carboxylates synthesis from arecoline are published [88−92].

Femoxetine (**4.1.57**) − (3R,4S)-3-((4-methoxyphenoxy)methyl)-1-methyl-4-phenylpiperidine, is a close structural analog of paroxetine (**4.1.49**) − (3S,4R)-3-((benzo[d][1,3]dioxol-5-yloxy)methyl)-4-(4-fluorophenyl)piperidine, but with opposite stereochemistry, and methylated nitrogene in piperidine ring is another selective serotonin uptake inhibitor with antidepressant properties that was invented in the 1970s but for some reasons was gradually abandoned from practice (Fig. 4.2).

Paroxetine 4.1.49   **Femoxetine 4.1.57**

FIGURE 4.2   Paroxetine and femoxetine.

Several other methods for the synthesis of paroxetine have been proposed. They generally concluded in preparation of ((3S,4R)-4-phenylpiperidin-3-yl)methanols from different starting materials and are partly represented on Fig. 4.3 [86,93–96].

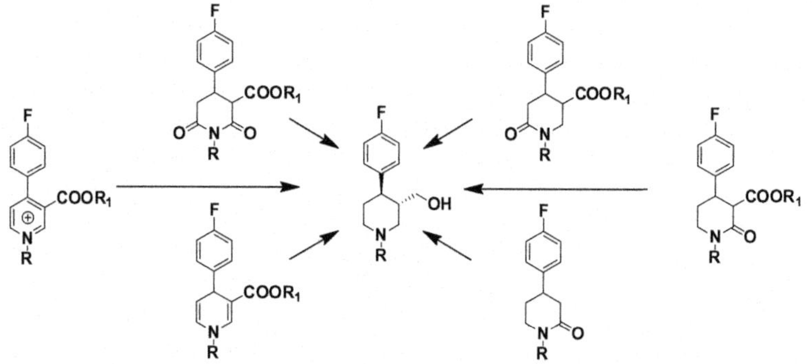

FIGURE 4.3 Preparation of key starting compounds—((3S,4R)-4-phenylpiperidin-3-yl)methanols from different starting materials.

4.10 TROXIPIDE (110)

Troxipide (**4.1.60**) (Aplace) is a gastro protective agent with antiulcer, antiinflammatory, and mucus-secreting properties. Troxipide has cytoprotective properties on the gastric mucosa. It is used in gastric ulcers' amelioration of gastric mucosal lesions (erosion, hemorrhage, redness, and edema) in the following diseases: acute gastritis, acute exacerbation stage of chronic gastritis. The most frequently observed adverse reactions were gastrointestinal symptoms including constipation in 0.19% patients, increased aspartate aminotransferase levels in 0.17% patients, and increased alanine aminotransferase levels in 0.17% patients. The gastro protective action of troxipide includes increase in mucus production, cytoprotective prostaglandin secretion, regeneration of collagen fibers, and gastric mucosal metabolism enhancement [97,98].

Synthesis of troxipide (**4.1.60**) is based on condensation of 3-aminopyridine (**4.1.58**) with 3,4,5-trimethoxybenzoic acid chloride to give 3,4,5-trimethoxy-N-(pyridin-3-yl)benzamide (**4.1.59**) followed by catalytic hydrogenation gave desired product (**4.1.60**) as hydrochloride salt [99,100] (Scheme 4.10).

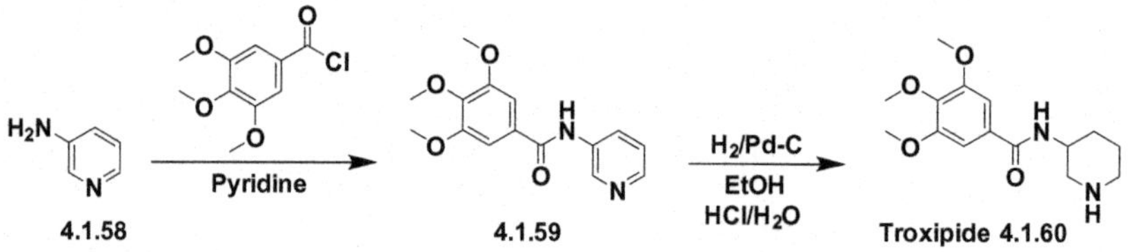

SCHEME 4.10 Synthesis of troxipide.

4.11 LINAGLIPTIN (1133)

Linagliptin (**4.1.70**) (Trajenta) is an oral, highly selective inhibitor of dipeptidyl peptidase-4 (DPP-4). It is the first agent of its class to be eliminated predominantly via a nonrenal route. DPP-4 inhibitors act by increasing endogenous glucagon-like peptide-1 (GLP-1) and gastric inhibitory polypeptide (GIP), also known as the glucose-dependent insulinotropic peptide concentrations, which in turn improves glucose homeostasis with a low risk of hypoglycemia and potential for disease modification. Via this mechanism, insulin secretion is glucose-dependently stimulated and inhibits glucagon secretion. This results in a low risk for hypoglycemia. Linagliptin is indicated for once-daily use for the treatment of adults with type 2 diabetes mellitus. More common side effects are: back pain, difficulty with moving, muscle aches, muscle pain or stiffness, pain in the joints, sore throat, stuffy, or runny nose. Linagliptin may cause other side effects such as rash, itching, flaking, or peeling of the skin, hives, swelling of the face, lips, tongue, or throat, difficulty breathing or swallowing, hoarseness, nausea, vomiting, and loss of appetite [101−113].

Linagliptin (**4.1.70**) has been discovered through a high-throughput screening campaign that involved about 500,000 compounds in the search for an effective DPP-4 inhibitor [114].

The industrial process for the manufacturing of linagliptin (**4.1.70**) started from condensation of 8-bromo-3-methylxantine (**4.1.61**) with 1-bromobut-2-yne (**4.1.62**) in the presence of trimethylamine (TEA) or diisopropylethylamine (DIEA) in dimethylformamide (DMF) to give 8-bromo-7-(but-2-yn-1-yl)-3-methyl-1H-purine-2,6(3H,7H)-dione (**4.1.63**). The second reagent − 2-(chloromethyl)-4-methylquinazoline (**4.1.66**) − was prepared via cyclization of 2′-amino-acetophenon (**4.1.64**) with 2-chloroacetonitrile (**4.1.65**) in dioxane in the presence of hydrogen chloride. Condensation of these two reagents − (**4.1.64**) and (**4.1.66**) − in N-methyl-pyrrolydone (NMP) in the presence of sodium carbonate gave 8-bromo-7-(but-2-yn-1-yl)-3-methyl-1-((4-methylquinazolin-2-yl)methyl)-1H-purine-2,6(3H,7H)-dione (**4.1.67**).

Coupling of prepared product (**4.1.67**) with (R)-3-(Boc-amino)piperidine (**4.1.68**) by means of K_2CO_3 in DMF gave compound (**4.1.69**), which on deprotection with trifluoroacetic acid in dichloromethane gave the desired linagliptin (**4.1.70**)

Another variant for preparation of linagliptin consists in condensation of the same purine derivative (**4.1.67**) with (R)-3-phtalimidopiperidin D-tartrate (**4.1.71**) in NMP using diisopropylethylamine gives compound (**4.1.72**) which was converted to the desired linagliptin (**4.1.70**) via aminolysis using ethanolamine in boiling toluene or tetrahydrofurane [115−119] (Scheme 4.11).

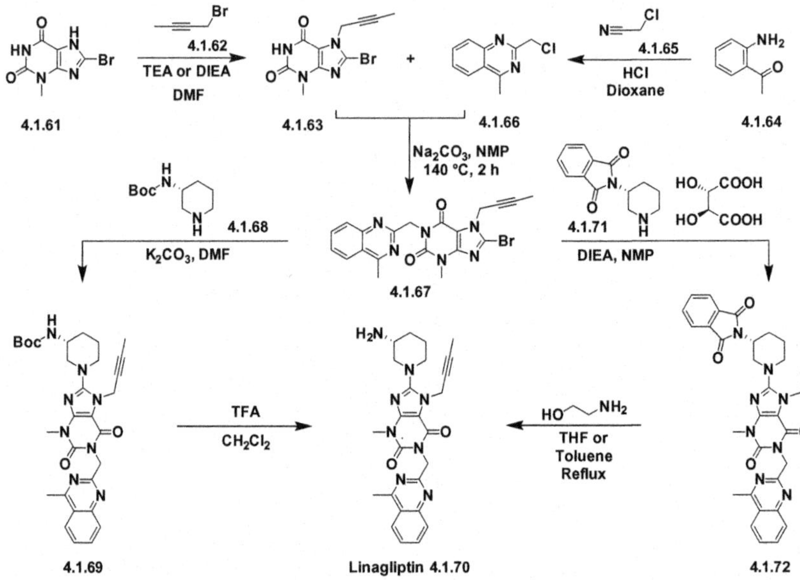

SCHEME 4.11 Synthesis of linagliptin.

An alternative methodology for the synthesis of linagliptin is based on ester (**4.1.74**) prepared via condensation of (**4.1.67**) with ethyl nipecotate (**4.1.73**), which can be transformed to variety of carboxylic acid derivatives able to transformation to amines via Hofmann reaction, Curtius, Lossen, or Schmidt rearrangements. Desired compounds of formulas (**4.1.75, 4.1.76, 4.1.70**), could be prepared either as a single enantiomer, or a mixture thereof, typically as a racemic mixture [120] (Fig. 4.4).

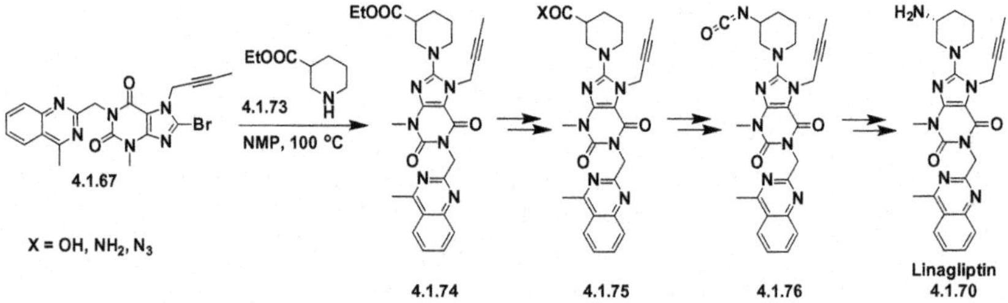

FIGURE 4.4 Synthesis of linagliptin.

4.12 TOFACITINIB (1130)

Tofacitinib (**4.1.83**), an oral Janus kinase (JAK) inhibitor, was approved in 2012 by the US FDA with the trade name Xeljanz for the treatment of rheumatoid arthritis.

Tofacitinib, an inhibitor of the JAK family, blocks intracellular signaling of multiple key cytokines involved in the inflammatory cascade. It primarily inhibits signaling through receptors associated with JAK3, JAK1, or both of them, with functional selectivity over JAK2-paired receptors. Inhibition of JAK1 and JAK3 by tofacitinib blocks signaling for several cytokines, including interleukins 2, 4, 7, 9, 15, and 21. These cytokines are integral to lymphocyte activation, function, and proliferation [121−126].

Tofacitinib is being investigated as a treatment for treatment of psoriasis [127,128], ulcerative colitis [129], Crohn's disease [130], kidney transplantation [131].

Commonly reported side effects of tofacitinib include: headache and diarrhea, nausea, indigestion, upper respiratory tract infection (nasopharyngitis), increase of cholesterol levels. Tofacitinib carries a boxed warning of risk of developing infections that may lead to hospitalization or death.

Synthesis of tofacitinib (**4.1.83**) started with the coupling of (3R,4R)-1-benzyl-3-(methylamino)-4-methylpiperidine (**4.1.78**) obtained by the resolution of racemic (**4.1.77**) by the use of (−)-di-p-toluoyl-L-tartaric acid (DTTA), or (S)-(+)-Phencyphos, or chiral HPLC with 4-chloro-7H-pyrrolo [2,3-d]-pyrimidine (**4.1.79**) on heating in a sealed tube at 100°C for 4 days which gave (3R,4R)-1-benzyl-4-methylpiperidin-3-yl)methyl(7H-pyrrolo[2,3-d] pyrimidin-4-yl-amine (**4.1.80**). The last was hydrogenated on 20% palladium hydroxide on carbon in ethanol to give debenzylated product (**4.1.81**), which was amidated using cyanoacetic acid 2,5-dioxopyrrolidin-1-yl ester (**4.1.82**) to give desired tofacitinib (**4.1.83**) [132] (Scheme 4.12).

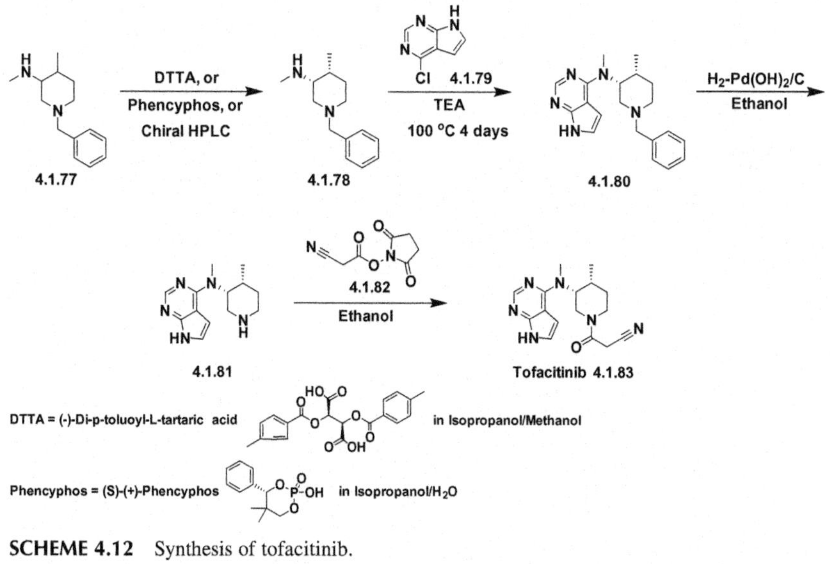

SCHEME 4.12 Synthesis of tofacitinib.

Other routes for the synthesis of tofactinib, which differs mainly in the approaches for the synthesis of (3R,4R)-1-benzyl-3-(methylamino)-4-methyl-piperidine (**4.1.78**) or in details of the process, are available in the literature [133−140].

REFERENCES

[1] Anonymous. Evaluation of an antispasmodic agent. Methixene hydrochloride (Trest). JAMA 1966;195(10):851.

[2] Volles E, Fredrich H. The treatment of Parkinson tremors and forms of extrapyramidal tremors with methixene. A multicenter study. Med Welt 1983;34(24):707−9.

[3] Maller O, Heller S. Neutralization of extrapyramidal side-effects with methixene. Dis Nerv Syst 1971;32(6):409−15.

[4] Clarke S, Hay GA, Vas CJ. Therapeutic action of methixene hydrochloride on Parkinsonian tremor and a description of a new tremor-recording transducer. Br J Pharmacol Chemother 1966;26(2):345−50.

[5] Lehner H, Lauener H, Schmutz J. The metabolism of 9[(N-methyl-3-piperidyl)methyl] thioxanthene hydrochloride. Arzneimittel-Forsch 1964;14:89−92.

[6] Caviezel R, Eichenberger E, Kunzle F, Schmutz J. Pharmacology and chemistry of basic substituted thioxanthene derivatives. Pharm Acta Helv 1958;33:447−57.

[7] Schmutz, J. Thioxanthene derivatives. US 2905590; 1959.

[8] Kase Y, Yuizono T, Yamasaki T, Yamada T, Io S, Tamiya M, et al. A new potent non-narcotic antitussive, 1-methyl-3-bis(2-thienyl)methylenepiperidine. Pharmacol Clin Efficacy Chem Pharm Bull 1959;7:372−7.

[9] Kawaura K, Ogata Y, Inoue M, Honda S, Soeda F, Shirasaki T, et al. The centrally acting non-narcotic antitussive tipepidine produces antidepressant-like effect in the forced swimming test in rats. Behav Brain Res 2009;205(1):315−18.

[10] Sasaki T, Hashimoto K, Tachibana M, Kurata T, Kimura H, Komatsu H, et al. Tipepidine in adolescent patients with depression: a 4 week, open-label, preliminary study. Neuropsychiatr Dis Treat 2014;10:719−22.

[11] Honda S, Kawaura K, Soeda F, Shirasaki T, Takahama K. The potent inhibitory effect of tipepidine on marble-burying behavior in mice. Behav Brain Res 2011;216(1):308−12.

[12] Hamao K, Kawaura K, Soeda F, Hamasaki R, Shirasaki T, Takahama K. Tipepidine increases dopamine level in the nucleus accumbens without methamphetamine-like behavioral sensitization. Behav Brain Res 2015;284:118−24.

[13] Sasaki T, Hashimoto K, Tachibana M, Kurata T, Okawada K, Ishikawa M, et al. Tipepidine in children with attention deficit/hyperactivity disorder: a 4-week, open-label, preliminary study. Neuropsychiatr Dis Treat 2014;10:147−51.

[14] Hashimoto K, Sasaki T. Old drug tipepidine as new hope for children with ADHD. Aust NZ J Psychiatry 2015;49(2):181−2.

[15] No Inventor data available. Piperidine derivative. ES 272195; 1962.

[16] Tamaki H, Tanaka M, Murata S, Harigaya S, Kiyomoto A. Pharmacological properties of a new anticholinergic agent, 1,1-dimethyl-5-methoxy-3-(dithien-2-ylmethylene)piperidinium bromide (SA-504). Jpn J Pharmacol 1972;22(5):685−99.

[17] Sweetman SC. 37th ed. Martindale: the complete drug reference: the extra pharmacopoeia, vol. A. London: The Pharmaceutical Press; 2011. pp. 1932, 1853−1854, 2642.

[18] Kawazu M, Kanno T, Saito S, Tamaki H. 5-Hydroxy-3-piperidylidenemethane derivatives as spasmolytics. J Med Chem 1972;15(9):914−18.

[19] No Inventor data available. Bis(2-thienyl)(N-methyl-5-methoxy-3-piperidylidene)methane quaternary derivatives. FR 2100750; 1972.

[20] Adkins JC, Noble S. Tiagabine: a review of its pharmacodynamic and pharmacokinetic properties and therapeutic potential in the management of epilepsy. Drugs 1998;55 (3):437−60.

[21] Leach JP, Brodie MJ. Tiagabine. Lancet 1998;351(9097):203−7.

[22] Kalviainen R. In: Shorvon S, Perucca E, Engel Jr. J, editors. Tiagabine, from treatment of epilepsy. 3rd ed. Wiley-Blackwell; 2009. p. 663−72.

[23] Bauer J, Cooper-Mahkorn D. Tiagabine: efficacy and safety in partial seizures - current status. Neuropsychiatr Dis Treat 2008;4(4):731−6.

[24] Schwartz TL, Nihalani N. Tiagabine in anxiety disorders. Expert Opin Pharmacother 2006;7(14):1977−87.

[25] Walker MC. The mechanism of action of tiagabine. Rev Contempor Pharmacother 2002;12(5):213−23.

[26] Meldrum BS, Chapman AG. Basic mechanisms of gabitril (tiagabine) and future potential developments. Epilepsia 1999;40(Suppl. 9):S2−6.

[27] Schachter SC. Pharmacology and clinical experience with tiagabine. Expert Opin Pharmacother 2001;2(1):179−87.

[28] Schmidt D, Gram L, Brodie M, Kramer G, Perucca E, Kalviainen R, et al. Tiagabine in the treatment of epilepsy − a clinical review with a guide for the prescribing physician. Epilepsy Res 2000;41(3):245−51.

[29] Schachter SC. A review of the antiepileptic drug Tiagabine. Clin Neuropharmacol 1999;22(6):312−17.

[30] Schachter SC. Tiagabine. Epilepsia 1999;40(Suppl. 5):S17−22.

[31] Luer MS, Rhoney DH. Tiagabine: a novel antiepileptic drug. Ann Pharmacother 1998;32 (11):1173−80.

[32] Schachter SC. Tiagabine. Drugs Today 1998;34(3):283−8.

[33] Schachter SC. Tiagabine: current status and potential clinical applications. Expert Opin Invest Drugs 1996;5(10):1377−87.

[34] Mengel H. Tiagabine. Epilepsia 1994;35(Suppl. 5):581−4.

[35] Gram L. Tiagabine: a novel drug with a GABAergic mechanism of action. Epilepsia 1994;35(Suppl. 5):585−7.

[36] Groenvald FC, Braestrup C. Diheterocyclylbutenylamino acids as GABA uptake inhibitors. WO 8700171; 1987.

[37] Andersen KE, Braestrup C, Groenwald FC, Joergensen AS, Nielsen EB, Sonnewald U, et al. The synthesis of novel GABA uptake inhibitors. 1. Elucidation of the structure-activity studies leading to the choice of (R)-1-[4,4-bis(3-methyl-2-thienyl)-3-butenyl]-3-piperidinecarboxylic acid (Tiagabine) as an anticonvulsant drug candidate. J Med Chem 1993;36(12):1716−25.

[38] Petersen H, Nielsen P, Cain M, Patel RS. Crystalline tiagabine hydrochloride monohydrate, a method for its preparation and use as antiepileptic. WO 9217473; 1992.

[39] Chorghade MS, Deshpande MN, Pariza RJ. Tying a GABA from Copenhagen to Chicago: the chemistry of tiagabine. In: Chorghade MS, editor. Drug discovery and development, vol. 2. USA: John Wiley & Sons, Inc; 2007. p. 279−308.

[40] Chorghade MS, Petersen H, Lee EC, Bain S. Efficient syntheses of regioisomers of tiagabine. Pure Appl, Chem 1996;68(3):761−3.

[41] Quandt G, Hofner G, Wanner KT. Synthesis and evaluation of N-substituted nipecotic acid derivatives with an unsymmetrical bis-aromatic residue attached to a vinyl ether spacer as potential GABA uptake inhibitors. Bioorg Med Chem 2013;21(11):3363−78.

[42] Long JP, Keasling HH. The comparative anticholinergic activity of a series of derivatives of 3-hydroxypiperidine. J Am Pharmaceut Assoc 1954;43:616–19.

[43] Chen JYP, Beckman H. Antispasmodic effects and toxicity of 1-ethyl-3-hydroxypiperidine diphenylacetic ester hydrochloride (JB-305) with special reference to its action on the sphincter of Oddi of the dog. J Pharmacol Exp Therapeut 1952;104:269–76.

[44] Pomeranze J, Beinfield WH, Goldbloom AA, Lucariello RJ, Chessin M. Human and experimental appraisal of a new antispasmodic, (1-ethyl-3-piperidyl) diphenylacetate hydrochloride (Dactil). NY State J Med 1955;55:233–6.

[45] Biel JH, Friedman HL, Leiser HA, Sprengeler EP. Antispasmodics. I. Substituted acetic acid esters of 1-alkyl-3-hydroxypiperidine. J Am Chem Soc 1952;74:1485–8.

[46] Biel JH. Antispasmodics. US 2918407; 1959.

[47] Bennett HD. Drugs affecting digestion: appentents, absorbents, antacids, carminatives, bile derivatives and others. In: Dipalma JR, editor. Drills pharmacology in medicine. New York, NY: McGraw-Hill Book Company; 1971. p. 967–9.

[48] Harvey SC. Antimuscarinic and antispasmodic drugs. In: Oslo A, editor. Remington's pharmaceutical sciences. 16th ed. Easton, Pennsylvania: Marck Printing Company; 1980. p. 850–60.

[49] Buckley JP, DeFeo JJ, Reif EC. The comparative antispasmodic activity of N-methyl-3-piperidyl diphenylglycolate methobromide (JB-340) and atropine sulfate. J Am Pharm Assoc 1957;46:592–4.

[50] Chen JYP. Antispasmodic activity of JB-340 (N-methyl-3-piperidyl diphenylglycolate methobromide) with special reference to a relative selective action on the sphincter of Oddi, colon, and urinary bladder of the dog. Arch Int Pharmacodyn Ther 1959;121:78–84.

[51] Anonymous. Mepenzolate bromide in mucomembranous colitis. A report from the General Practitioner research group. Practitioner 1974;212(1272):890–4.

[52] Kleckner Jr. MS. Clinical evaluation of a 3-hydroxypiperidine (cantil) in the therapy of intestinal disturbances. A double-blind, controlled study. Am J Gastroenterol 1959;32:609–19.

[53] Riese JA. Clinical observations with cantil, a new anticholinergic for colon disorders. Am J Gastroenterol 1957;28(5):541–7.

[54] Tanaka K, Ishihara T, Sugizaki T, Kobayashi D, Yamashita Y, Tahara K, et al. Mepenzolate bromide displays beneficial effects in a mouse model of chronic obstructive pulmonary disease. Nat Commun 2013;4:2686.

[55] Tanaka K, Kurotsu S, Asano T, Yamakawa N, Kobayashi D, Yamashita Y, et al. Superiority of pulmonary administration of mepenzolate bromide over other routes as treatment for chronic obstructive pulmonary. Sci Rep 2014;4:4510/1–4510/11.

[56] Kurotsu S, Tanaka K, Niino T, Asano T, Sugizaki T, Azuma A, et al. Ameliorative effect of mepenzolate bromide against pulmonary fibrosis. J Pharmacol Exp Therapeut 2014;350 (1):79–88.

[57] Zheng Y, Wang X, Ji S, Tian S, Wu H, Luo P, et al. Mepenzolate bromide promotes diabetic wound healing by modulating inflammation and oxidative stress. Am J Translat Res 2016;8(6):2738–47.

[58] Biel JH. Antispasmodics. US 2918408; 1959.

[59] Yamashita Y, Tanaka K, Asano T, Yamakawa N, Kobayashi D, Ishihara T, et al. Synthesis and biological comparison of enantiomers of mepenzolate bromide, a muscarinic receptor antagonist with bronchodilatory and anti-inflammatory activities. Bioorg Med Chem 2014;22(13):3488–97.

[60]   Duggan JM. A controlled trial of an anticholinergic drug, pipenzolate methylbromide ("piptal"), in the management of peptic ulcer. Med J Aust 1965;2(20):826−7.

[61]   Anonymous. Pipenzolate methylbromide. JAMA 1957;164(3):280−1.

[62]   Vincent PC, Fenton BH, Beeston D. Effect of pipenzolate on gastric secretion in man. Med J Aust 1967;1(11):546−8.

[63]   Beil JH. Antispasmodics. US 2918406; 1959.

[64]   Yao K, Nagashima K, Miki H. Pharmacological, pharmacokinetic, and clinical properties of benidipine hydrochloride, a novel, long-acting calcium channel blocker. J Pharmacol Sci 2006;100(4):243−61.

[65]   Kobayashi H, Kobayashi S. Absorption, distribution, metabolism and excretion of 14C-labeled enantiomers of the calcium channel blocker benidipine after oral administration to rat. Xenobiotica 1998;28(2):179−97.

[66]   Muto K, Kuroda T, Kawato H, Karasawa A, Kubo K, Nakamizo N. Synthesis and pharmacological activity of stereoisomers of 1,4-dihydro-2,6-dimethyl-4-(3-nitrophenyl)-3,5-pyridine-dicarboxylic   acid   methyl   1-(phenylmethyl)-3-piperidinyl   ester. ArzneimittelForsch 1988;38(11A):1662−5.

[67]   Suzuki H, Saruta T. Benidipine. Cardiovasc Drug Rev 1989;7:25−38.

[68]   Kosaka H, Hirayama K, Yoda N, Sasaki K, Kitayama T, Kusaka Hi, et al. The L-, N-, and T-type triple calcium channel blocker benidipine acts as an antagonist of mineralocorticoid receptor, a member of nuclear receptor family. Eur J Pharmacol 2010;635 (1−3):49−55.

[69]   Kitakaze M, Karasawa A, Kobayashi H, Tanaka H, Kuzuya T, Hori M. Benidipine: a new Ca^{2+} channel blocker with a cardioprotective effect. Cardiovasc Drug Rev 1999;17 (1):1−15.

[70]   Abe M, Soma M. Multifunctional L-/N- and L-/T-type calcium channel blockers for kidney protection. Hypertens Res 2015;38(12):804−6.

[71]   Tomino Y. Renoprotective effects of the L-/T-type calcium channel blocker benidipine in patients with hypertension. Curr Hypertens Rev 2013;9(2):108−14.

[72]   Muto K, Watanabe M, Hatta T, Sugaya T, Takemoto Y, Nakamizo N. 1,4-Dihydropyridine derivative and pharmaceutical composition containing it. EP 63365; 1982.

[73]   Muto K, Kuroda T, Karasawa A, Yamada K, Nakamizo N. (±)-2,6-Dimethyl-4-(3-nitro-phenyl)-1,4-dihydropyridine-3,5-dicarboxylic acid 1-benzyl-3-piperidyl methyl ester. EP 106275; 1984.

[74]   Kutsuma T, Ikawa H, Sato Y. 1,4-Dihydropyridine derivatives and pharmaceutical compositions comprising them. EP 161877; 1985.

[75]   Vardanyan RS, Hruby VJ. Synthesis of best-seller drugs. Amsterdam: Elsevier; 2016. p. 119−21.

[76]   Dechant KL, Clissold SP. Paroxetine. A review of its pharmacodynamic and pharmacokinetic properties, and therapeutic potential in depressive illness. Drugs 1991;41(2):225−53.

[77]   Bourin M, Chue P, Guillon Y. Paroxetine: a review. CNS Drug Rev 2001;7(1):25−47.

[78]   Wagstaff AJ, Cheer SM, Matheson AJ, Ormrod D, Goa KL. Paroxetine: an update of its use in psychiatric disorders in adults. Drugs 2002;62(4):655−703.

[79]   Heydorn WE. Paroxetine: a review of its pharmacology, pharmacokinetics and utility in the treatment of a variety of psychiatric disorders. Expert Opin Invest Drugs 1999;8 (4):417−41.

[80]   Gunasekara NS, Noble S, Benfield P. Paroxetine: an update of its pharmacology and therapeutic use in depression and a review of its use in other disorders. Drugs 1998;55 (1):85−120.

[81] Tang SW, Helmeste D. Paroxetine. Expert Opin Pharmacother 2008;9(5):787−94.

[82] Pae C, Patkar AA. Paroxetine: current status in psychiatry. Expert Rev Neurotherapeut 2007;7(2):107−20.

[83] Green B. Focus on paroxetine. Curr Med Res Opin 2003;19(1):13−21.

[84] Snyderman SH, Rynn MA, Bellew K, Rickels K. Paroxetine in the treatment of generalised anxiety disorder. Expert Opin Pharmacother 2004;5(8):1799−806.

[85] Caley CF, Weber SS. Paroxetine: a selective serotonin reuptake inhibiting antidepressant. Ann Pharmacother 1993;27(10):1212−22.

[86] De Risi C, Fanton G, Pollini GP, Trapella C, Valente F, Zanirato V. Recent advances in the stereoselective synthesis of trans-3,4-disubstituted-piperidines: applications to (-)-paroxetine. Tetrahedron: Asymmetry 2008;19(2):131−55.

[87] Czibula L, Nemes A, Seboek F, Szantay Jr. C, Mak M. A convenient synthesis of (-)-paroxetine. Eur J Org Chem 2004;(15):3336−9.

[88] Christensen JA, Squires RF. 4-Phenylpiperidine compounds. US 4007196; 1977.

[89] Murthy KSK, Rey AW. Stereoselective preparation of 3-acyl-4-arylpiperidines. WO 9907680; 1999.

[90] Engelstoft M, Hansen JB. Synthesis and 5HT modulating activity of stereoisomers of 3-phenoxymethyl-4-phenylpiperidines. Acta Chem Scand 1996;50(2):164−9.

[91] Murthy KSK, Rey AW, Tjepkema M. Enantioselective synthesis of 3-substituted-4-aryl piperidines useful for the preparation of paroxetine. Tetrahedron Lett 2003;44 (28):5355−8.

[92] Shinji Y, Jahan I. A new route to 3,4-disubstituted piperidines: formal synthesis of (−)-paroxetine and (+)-femoxetine. Tetrahedron Lett 2005;46(50):8673−6.

[93] Cossy J. Selective methodologies for the synthesis of biologically active piperidinic compounds. Chem Record 2005;5(2):70−80.

[94] Kim M, Park Y, Jeong B, Park H, Jew S. Synthesis of (-)-paroxetine via enantioselective phase-transfer catalytic monoalkylation of malonamide ester. Org Lett 2010;12 (12):2826−9.

[95] Bonifacio F, Mancinetti D, Crescenzi C, De Iasi G, Donnarumma M, Mastrangeli C. Asymmetric hydrogenation applied to industrial processes: a convenient synthesis of paroxetine. PharmaChem 2003;2(11−12):13−15.

[96] Lawrie KWM, Rustidge DC. The synthesis of [methylenedioxy-14C] paroxetine BRL 29060A. J Labelled Compd Radiopharm 1993;33(8):777−81.

[97] Kusugami K, Ina K, Hosokawa T, Kobayashi F, Kusajima H, Momo K, et al. Troxipide, a novel antiulcer compound, has inhibitory effects on human neutrophil migration and activation induced by various stimulants. Dig Liver Dis 2000;32(4):305−11.

[98] Dewan B, Shah D. An open-label, multicenter post marketing study to assess the symptomatic efficacy and safety of troxipide [Troxip™] in the management of acid peptic disorders in indian patients. Br J Med Med Res 2013;3(4):1881−92.

[99] Irikura T, Kasuga K. New antiulcer agents. 1. Synthesis and biological activities of 1-acyl-2-, -3-, and -4-substituted benzamidopiperidines. J Med Chem 1971;14(4):357−61.

[100] Irikura T, Kasuga K, Segawa M. Benzoylamino-substituted 1-benzoylpiperidines for treating gastric ulcers. US 3647805; 1972.

[101] Deacon CF, Holst JJ. Linagliptin, a xanthine-based dipeptidyl peptidase-4 inhibitor with an unusual profile for the treatment of type 2 diabetes. Expert Opin Invest Drugs 2010;19(1):133−40.

[102] Deeks ED. Linagliptin: a review of its use in the management of type 2 diabetes mellitus. Drugs 2012;72(13):1793−824.

[103]   Scott LJ. Linagliptin: in type 2 diabetes mellitus. Drugs 2011;71(5):611−24.

[104]   Barnett AH. Linagliptin: a novel dipeptidyl peptidase 4 inhibitor with a unique place in therapy. Adv Ther 2011;28(6):447−59.

[105]   Neumiller JJ, Setter SM. Review of linagliptin for the treatment of type 2 diabetes mellitus. Clin Ther 2012;34(5):993−1005.

[106]   Gallwitz B. Linagliptin a novel dipeptidyl peptidase inhibitor for type 2 diabetes therapy. Clin Med Insights Endocrinol Diabetes 2012;5:1−11.

[107]   Barnett AH. Linagliptin for the treatment of type 2 diabetes mellitus: a drug safety evaluation. Expert Opin Drug Safety 2015;14(1):149−59.

[108]   McKeage K. Linagliptin: an update of its use in patients with type 2 diabetes mellitus. Drugs 2014;74(16):1927−46.

[109]   Tella SH, Akturk HK, Rendell M. Linagliptin for the treatment of type 2 diabetes. Diabetes Manag 2014;4(1):85−101.

[110]   Pal D, De T, Baral A. DPP-4 inhibitor linagliptin: a new anti-diabetic drug in the treatment of type-2 diabetes. Int J Pharm Pharm Sci 2013;5(Suppl. 2):58−62.

[111]   Gallwitz B. Emerging DPP-4 inhibitors: focus on linagliptin for type 2 diabetes. Diabetes Metab Syndr Obes 2013;6:1−9.

[112]   Neumiller JJ. Pharmacology, efficacy, and safety of linagliptin for the treatment of type 2 diabetes mellitus. Ann Pharmacother 2012;46(3):358−67.

[113]   Scheen AJ. Linagliptin for the treatment of type 2 diabetes (pharmacokinetic evaluation). Expert Opin Drug Metabol Toxicol 2011;7(12):1561−76.

[114]   Kanstrup AB, Christiansen LB, Lundbeck JM, Sams CK, Kristiansen M. Preparation of piperazinylpurinediones as inhibitors of dipeptidylpeptidase IV. WO 20020022560; 2002.

[115]   Pfrengle W, Pachur T, Nicola T, Duran A. Method for producing chiral 8-(3-amino-piperidin-1-yl)-xanthines. WO 2006048427; 2006.

[116]   Wang Y. BI-1356: dipeptidyl-peptidase IV inhibitor antidiabetic agent. Drugs Future 2008;33(6):473−7.

[117]   Huang Y, He X, Wu T, Zhang F. Synthesis and characterization of process-related impurities of antidiabetic drug, linagliptin. Molecules 2016;21(8):1041−50.

[118]   Pfrengle W, Pachur T. Preparation of chiral 8-(3-amino-piperidin-1-yl)xanthin. DE 102004054054; 2006.

[119]   Eckhardt M, Langkopf E, Mark M, Tadayyon M, Thomas L, Nar H, et al. 8-(3-(R)-Aminopiperidin-1-yl)-7-but-2-ynyl-3-methyl-1-(4-methyl-quinazolin-2-ylmethyl)-3,7-dihydropurine-2,6-dione (BI 1356), a highly potent, selective, long-acting, and orally bioavailable dpp-4 inhibitor for the treatment of type 2 diabetes. J Med Chem 2007;50(26):6450−3.

[120]   Allegrini P., Attolino E, Artico M. Process for the preparation of linagliptin. EP 2468749; 2012.

[121]   Vyas D, O'Dell KM, Bandy JL, Boyce EG. Tofactinib: the first Janus kinase (JAK) inhibitor for the treatment of rheumatoid arthritis. Ann Pharmacother 2013;47(11):1524−31.

[122]   Scott LJ. Tofactinib: a review of its use in adult patients with rheumatoid arthritis. Drugs 2013;73(8):857−74.

[123]   Kaur K, Kalra S, Kaushal S. Systematic review of tofactinib: a new drug for the management of rheumatoid arthritis. Clin Therapeut 2014;36(7):1074−86.

[124]   Lopez-Olivo MA, Lu H, Tayar JH. Review of tofactinib in rheumatoid arthritis. Clin Invest 2015;5(1):23−38.

[125]   Rakieh C, Conaghan PG. Tofactinib for treatment of rheumatoid arthritis. Adv Ther 2013;30(8):713−26.

[126] Zerbini CAF, Lomonte ABV. Tofactinib for the treatment of rheumatoid arthritis. Expert Rev Clin Immunol 2012;8(4):319−31.

[127] Galluzzo M, D'Adamio S, Servoli S, Bianchi L, Chimenti S, Talamonti M. Tofactinib for the treatment of psoriasis. Expert Opin Pharmacother 2016;17(10):1421−33.

[128] Chiricozzi A, Faleri S, Saraceno R, Bianchi L, Buonomo O, Chimenti S, et al. Tofactinib for the treatment of moderate-to-severe psoriasis. Expert Rev Clin Immunol 2015;11(4):443−55.

[129] Izzo R, Bevivino G, Monteleone G. Tofactinib for the treatment of ulcerative colitis. Expert Opin Invest Drugs 2016;25(8):991−7.

[130] Sandborn WJ, Ghosh S, Panes J, Vranic I, Wang W, Niezychowski W. A phase 2 study of tofactinib, an oral Janus kinase inhibitor, in patients with Crohn's disease. Clin Gastroenterol Hepatol 2014;12(9):1485−93.

[131] Wojciechowski D, Vincenti F. Tofactinib in kidney transplantation. Expert Opin Invest Drugs 2013;22(9):1193−9.

[132] Flanagan ME, Blumenkopf TA, Brissette WH, Brown MF, Casavant JM, Chang S, et al. Discovery of CP-690,550: a potent and selective Janus Kinase (JAK) inhibitor for the treatment of autoimmune diseases and organ transplant rejection. J Med Chem 2010;53 (24):8468−84.

[133] Zhi S, Liu D, Liu Y, Liu B, Wang D, Chen L. An efficient method for synthesis of tofacitinib citrate. J. Heterocyclic Chem. 2016;53(4):1259−63.

[134] Patil YS, Bonde NL, Kekan AS, Sathe DG, Das A. An improved and efficient process for the preparation of tofacitinib citrate. Org Proc Res Dev 2014;18(12):1714−20.

[135] Ripin DHB, Abele S, Cai W, Blumenkopf T, Casavant JM, Doty JL, et al. Development of a scaleable route for the production of cis-N-benzyl-3-methylamino-4-methylpiperidine. Org Proc Res Dev 2003;7(1):115−20.

[136] Ripin DHB. Method for the preparation of 3-aminopiperidine derivatives. US 20040102627; 2004.

[137] Wilcox GE, Koecher C, Vries T, Flanagan ME, Munchhof MJ. Optical resolution of (1-benzyl-4-methylpiperidin-3-yl)-methylamine and the use thereof for the preparation of pyrrolo[2,3-d]pyrimidines as protein kinase inhibitors. WO 2002096909; 2002.

[138] Stavber G, Cluzeau J. Process for the preparation of 3-aminopiperidine compounds useful as intermediates in synthesis of pharmaceutically active agents. WO 2014016338; 2014.

[139] Ruggeri SG, Hawkins JM, Makowski TM, Rutherford JL, Urban FJ. Process for preparation of piperidinylaminopyrrolopyrimidines from activated pyrrolopyrimidines and piperidinylamines. WO 2007012953; 2007.

[140] Rao TS, Zhang C. Deuterated tasocitinib derivatives as Janus kinase 3 inhibitors and their preparation and use for the treatment and prevention of Janus kinase 3-mediated diseases. WO 2010123919; 2010.

Chapter 5

4-Substituted and 1,4-Disubstituted Piperidines

This group of compounds is the most common in piperidine based drugs series.

5.1 4-PIPERIDINOLS AND THEIR DERIVATIVES

Periciazine (112)

Periciazine (**5.1.3**) (Neuleptil) is an old drug used in the treatment of various psychiatric illnesses in many countries but which is not approved for sale in the United States. It is mainly used for the treatment of symptoms or prevention of relapse of psychosis, such as schizophrenia and bipolar mania. It is often used to help anxiety, tension, and agitation. It can also be used to help the symptoms of attention deficit hyperactivity disorder and psychotic depression. The drug induces a noradrenergic rather than a dopaminergic blockade.

It also has adrenolytic, anticholinergic, metabolic, and endocrine effects and an action on the extrapyramidal system [1−4]. Propericiazine has been utilized also for the treatment of cannabis dependence [5].

The desired pericyazine (**5.1.3**) was prepared by direct coupling of 4-piperidinol (**5.1.1**) with 3-cyano-10-(3-*p*-toluenesulfonyloxypropyl)phenothiazine (**5.1.2**) on refluxing in toluene [6,7] (Scheme 5.1).

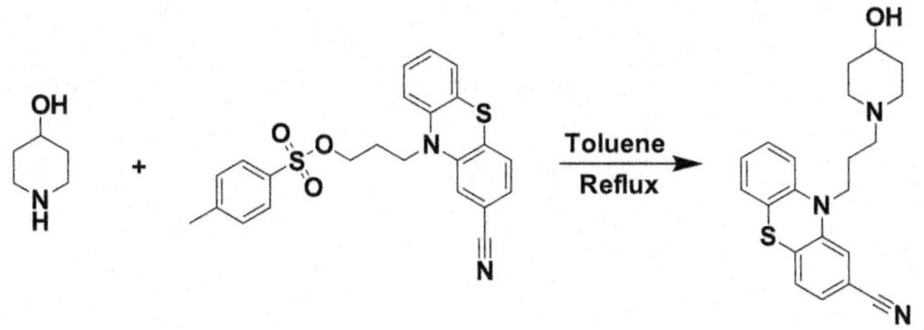

| 5.1.1 | 5.1.2 | Periciazine 5.1.3 |

SCHEME 5.1 Synthesis of periciazine.

Sedation, hypothermia, muscle relaxation, and other signs of central depressant effects were observed on investigations of close analog of

Piperidine-Based Drug Discovery. DOI: http://dx.doi.org/10.1016/B978-0-12-805157-3.00005-3

periciazine (**5.1.3**) − perimetazine (**5.1.3a**), which was prepared by the same method and by the same authors. It did not become a widely used drug by several reasons [8,9] (Fig. 5.1).

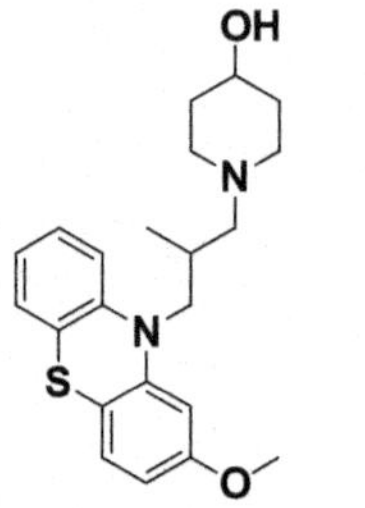

Perimetazine 5.1.3a

FIGURE 5.1 Perimetazine.

Diphenylpyraline (453)

Diphenylpyraline (**5.1.6**) (Allergen) is a first generation antihistaminic agent marketed in Europe that is used to treat rash, hives, watery eyes, runny nose, itching, and sneezing due to allergies.

Diphenylpyraline is an H1 receptor antagonist, which antagonizes most of the pharmacological actions of histamine and temporarily reduces allergic symptoms. It also has anticholinergic properties, reduces secretions, and may thus provide a drying effect on the nasal mucosa [10].

In addition, it possesses psychostimulant [11], antimycobacterial [12] properties, and has been shown to be useful in the treatment of parkinsonism [13], and could be considered a potential candidate for treatment of cocaine addiction [14].

Diphenylpyraline may cause, sedation, inability to concentrate, motor incoordination, fatigue, insomnia, somnolence, urinary hesitancy, and other side effects characteristic for first generation antihystamines. Synthesis of diphenylpyraline (**5.1.6**) was carried out by coupling on reflux in benzene or xylene of 1-methyl-4-hydroxypiperidine (**5.1.4**) obtained by catalytic hydrogenation of 1-methyl-4-piperidone along with diphenylbromo(chloro)methane (**5.1.5**) [15,16]. Another patent describes the same process in the presence of K_2CO_3 [17] (Scheme 5.2).

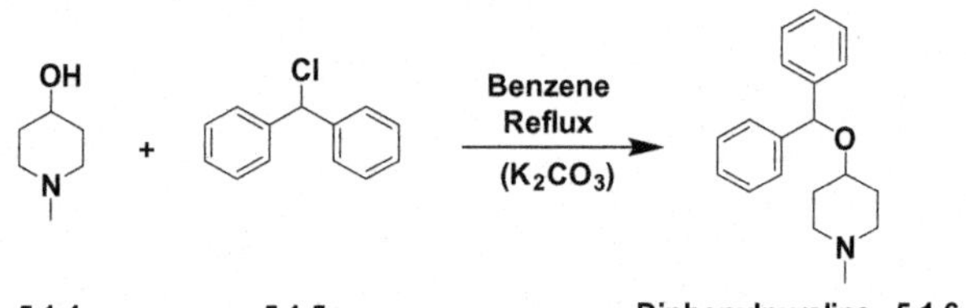

SCHEME 5.2 Synthesis of diphenylpyraline.

Ebastine (990)

Ebastine (**5.1.10**) is structurally related to diphenylpyraline (**5.1.6**) and is a non-sedating, selective, long-acting, second-generation antihistamine. It is used for the treatment of patients suffering from intermittent and persistent allergic rhinitis and chronic idiopathic urticarial [18−21]. Side effects include: headache, dry mouth, drowsiness, inflammation of pharynx/nose/sinus, abdominal pain, indigestion, weakness, nosebleed, nausea, and sleeplessness.

Ebastine (**5.1.10**) is generally synthesized from 1-(4-(tert-butyl)phenyl)-4-(4-hydroxypiperidin-1-yl)butan-1-one (**5.1.9**), and is in turn prepared from 4-hydroxypiperidine (**5.1.7**) and 1-(4-(tert-butyl)phenyl)-4-chlorobutan-1-one (**5.1.8**) via coboiling in methyl isobutyl ketone (MIBK) in the presence of sodium bicarbonate and a crystal of potassium iodide. The obtained product is coupled with benzhydryl bromide (**5.1.5**) on reflux in the same solvent in presence of sodium carbonate to give desired ebastine (**5.1.10**).

In a second method a process for the preparing ebastine (**5.1.10**) was carried out by refluxing a mixture of 4-diphenylmethoxy piperidine (**5.1.12**) with 1-(4-tert-butylphenyl)-4-chlorobutan-1-one (**5.1.8**) in the presence of sodium carbonate and MIBK as a solvent. 4-Diphenylmethoxy piperidine (**5.1.12**) for this process was prepared via hydrolysis of 1-ethoxycarbonyl-4-diphenylmethoxypiperidine (**5.1.11**) with 85% potassium hydroxide in isopropanol [22] (Scheme 5.3).

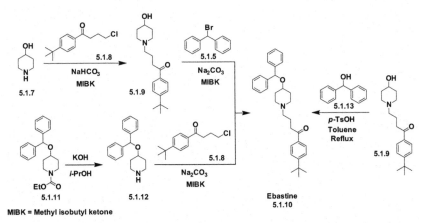

MIBK = Methyl isobutyl ketone

SCHEME 5.3   Synthesis of ebastine.

Several modification of these basic methods for the preparation ebastine are known in the literature. Conceptually, another method consists in the reaction of 1-[4-(1,1-dimethylethyl)phenyl]-4-(4-hydroxypiperidin-1-yl)butan-1-one (**5.1.9**) and benzhydrol (**5.1.13**). Thus, compound (**5.1.9**) and benzhydrol (**5.1.13**) in refluxing toluene in the presence of *p*-toluenesulfonic acid give desired ebastine (**5.1.10**) [23] (Scheme 5.3).

Bepotastine (236)

Bepotastine (**5.1.20**) is a second-generation, highly selective histamine H1-receptor antagonist that also suppresses some allergic inflammatory processes. It is a widely used medication in the treatment of allergic rhinitis and other allergic diseases [24−28].

Bepotastine eye drops is used to treat itching of the eye caused by allergic conjunctivitis.

Common side effects of bepotastine include an unusual or unpleasant taste in the mouth, headache, eye irritation, stuffy nose, and sore throat.

Bepotastine (**5.1.20**) was prepared starting with 4-hydroxypiperidine (**5.1.7**), which was *N*-acylated with acetyl chloride in dichloromethane in the presence of triethylamine to give *N*-acetyl-4-hydroxypiperidine (**5.1.14**). The obtained compound (**5.1.14**) was coupled with 2-[chloro(4-chlorophenyl)methyl]pyridine) (**5.1.15**) on heating for 5 hours at 130°C, followed by *N*-acetyl-group removal via acidic hydrolysis with 10% HCl at 80°C to give product (**5.1.16**). The crude (**5.1.16**) was purified via conversion to bis-(*RS*)-4-[(4-chlorophenyl)(2-pyridyl) methoxy]-piperidine-2,3-dibenzoyl-DL-tartrate followed by separation of free base with NaOH. The pure racemic (**5.1.16**) was alkylated with D-menthyl-4-chlorobutanoate (**5.1.17**) in acetonitrile in presence of potassium carbonate and sodium iodide to give the racemic (*RS*)-bepotastine L-menthyl ester (**5.1.18**). Optical resolution of the last was achieved by treating the racemate (**5.1.18**) with L-tartaric acid in the acetone/*n*-hexane mixture, which gave (*S*)-4-{4-[(4-chlorophenyl) (2-pyridyl)methoxy]piperidino}butyric acid D-menthyl ester L-tartrate (**5.1.19**) isolated by filtration. The desired final compound (+)-(*S*)-4-[4-[(4-chlorophenyl) (2-pyridyl)-methoxy]piperidino]butyric acid − bepotastine (**5.1.20**) − was prepared via basic hydrolysis of L-menthyl ester group with 10% NaOH aqueous solution in ethanol at room temperature [29] (Scheme 5.4).

Variations of the described method are proposed in the literature [30−32].

Lamifiban (257)

Lamifiban (**5.1.31**) is an intravenously administered, selective, reversible, nonpeptide glycoprotein IIb/IIIa receptor antagonist that is used instead of standard heparin and aspirin therapy that inhibits platelet aggregation and thrombus formation by preventing the binding of fibrinogen to platelets in patients with non-Q wave myocardial infarction or unstable angina pectoris preventing forming the blood clots that can cause heart attacks. The new drug is useful as an antithrombotic for the treatment of cerebral stroke, coronary infarction, inflammation, arteriosclerosis, osteoporosis, as neoplasm inhibitors or agents for the promotion of wound healing [33,34].

Synthesis of lamifiban (**5.1.31**) started from 4-hydroxypiperidine (**5.1.7**), which was transformed to tert-butyl(4-piperidinyloxy)acetate (**5.1.24**) via *N*-coupling with benzyl chloroformate (**5.1.21**), which resulted in carbamate (**5.1.22**). Obtained product (**5.1.22**) was alkylated with tert-butyl

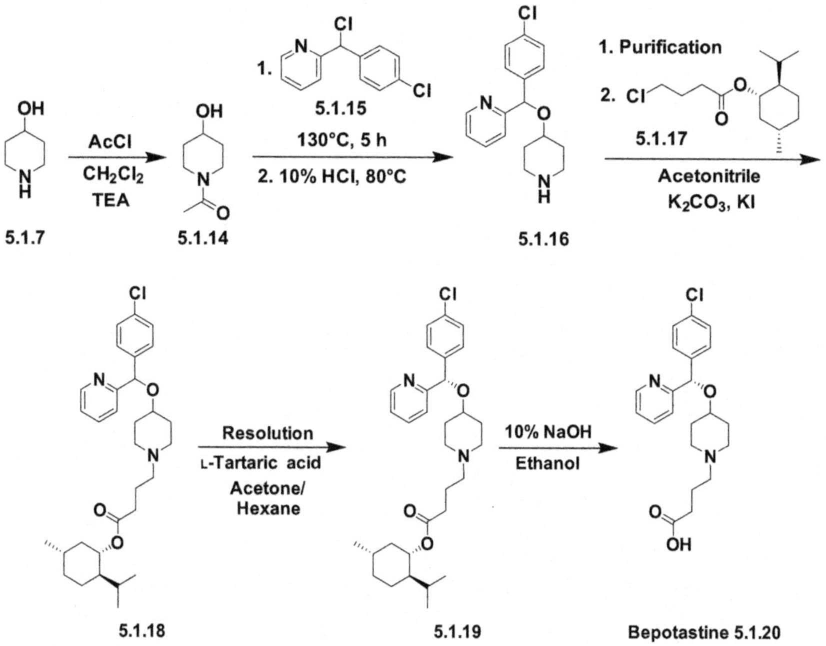

SCHEME 5.4 Synthesis of bepotastine.

bromoacetate (**5.1.23**) to give benzyl 4-(2-(tert-butoxy)-2-oxoethoxy)piperi-dine-1-carboxylate (**5.1.24**). The *N*-protecting group was removed on hydrogenation over a Pd-C catalyst in EtOH to give tert-butyl 2-(piperidin-4-yloxy)acetate (**5.1.25**). *N*-Acylation of the last with *N*-(benzyloxycarbonyl)-*O*-tert-butyltyrosine (**5.1.26**) in standard coupling conditions (CDMT or HBTU, NMM, DMF or CH$_2$Cl$_2$) gives tert-butyl (*S*)-2-((1-(2-(((benzyloxy)carbonyl)amino)-3-(4-(tert-butoxy)phenyl)propanoyl)piperidin-4-yl)oxy)acetate (**5.1.27**), *N*-deprotection of which was carried out over 10% Pd/C in EtOH/AcOH media to give tert-butyl (*S*)-2-((1-(2-amino-3-(4-(tert-butoxy)phenyl)propanoyl)piperidin-4-yl)oxy)acetate (**5.1.28**). The obtained product was acylated with *p*-amidinobenzoyl chloride (**5.1.29**) in CH$_2$Cl$_2$ and saturated aqueous NaHCO$_3$ to give compound (**5.1.30**); two protective tert-butoxy- groups were removed with trifluoroacetis acid in CH$_2$Cl$_2$ to give desired lamifiban (**5.1.31**) [35−37] (Scheme 5.5).

Eucatropine (90)

Eucatropine (**5.1.36**), an anticholinergic and antimuscarinic, is applied topically to the eye and produces prompt mydriasis free from anesthetic action, pain, corneal irritation, and it is also routinely used for inspection of the eye fundus [38,39].

The synthesis of eucatropine is accomplished starting from 2,2,6-tri-methylpiperidin-4-one (**5.1.33**) prepared from diacetonamine (**5.1.32**) on

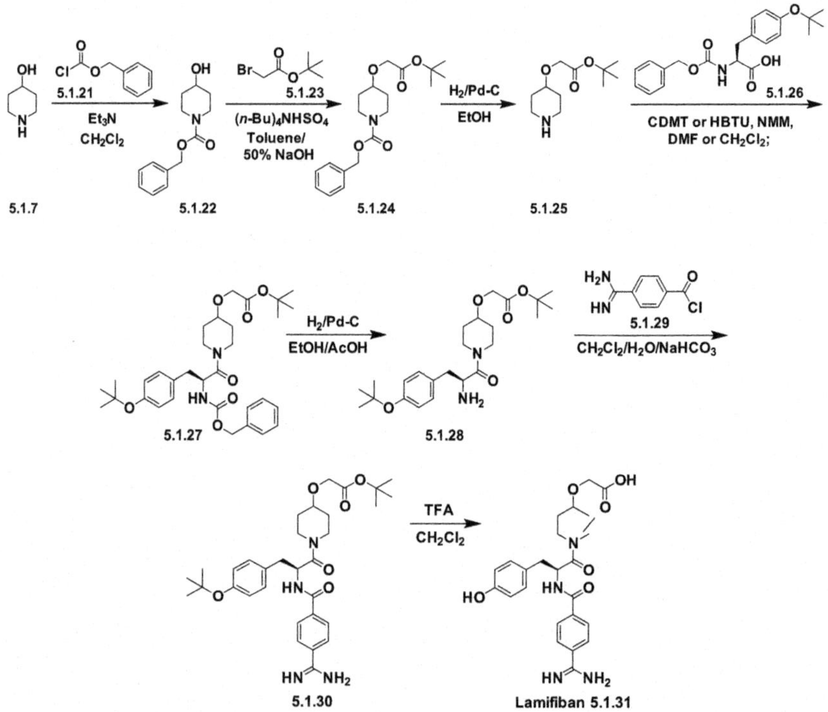

SCHEME 5.5 Synthesis of lamifiban.

CDMT = (2-chloro-4,6-dimethoxy-1,3,5-triazine)
HBTU = (2-(1H-benzotriazol-1-yl)-1,1,3,3-tetramethyluronium hexafuorophosphate)
NMM (N-methylmorpholine)

reaction with paraldehyde or acetaldehyde diethyl acetal. A reduction of the carbonyl group of the obtained product (**5.1.33**) was carried out with sodium amalgam and water or via hydrogenation over Raney-Ni catalyst to give compound (**5.1.34**). For the preparation of the N-methyl derivative (**5.1.35**), the 2,2,6-trimethylpiperidin-4-ol (**5.1.34**) underwent reductive alkylation on heating with concentric solution of formaldehyde which gave 1,2,2,6-tetramethylpiperidin-4-ol (**5.1.35**).The desired tropic ester − eucatropine (**5.1.36**) − was prepared by adding an excess of tropic acid to a solution of (**5.1.35**) in hydrochloric acid on heating [40−43] (Scheme 5.6).

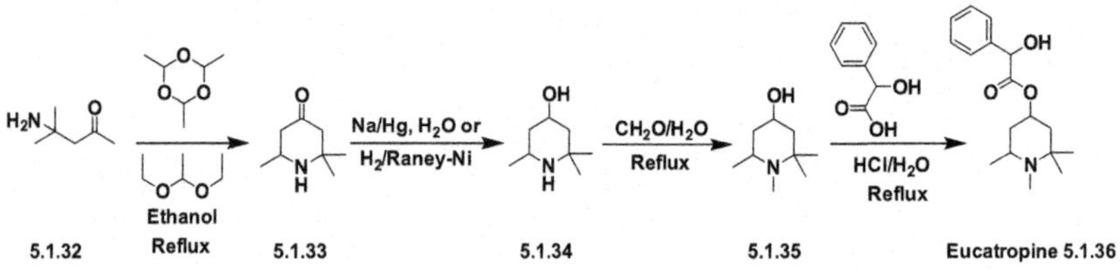

SCHEME 5.6 Synthesis of eucatropine.

Propiverine (664)

Propiverine (**5.1.41**) (Detrunorm) is an antimuscarinic agent with a mixed mode of action (neurotropic and musculotropic) in the treatment of symptoms associated with overactive bladder syndrome or detrusor hyperreflexia (it is one of the few drugs recommended for the treatment of detrusor overactivity). Propiverine is used for the symptomatic treatment of urinary incontinence and of increased urinary frequency and urgency [44,45]. Common side effects of taking propiverine are: dryness of the mouth, constipation, and blurred vision.

Several modifications of the methods for the preparation of propiverine (**5.1.41**) are proposed. One is started from the esterification of benzilic acid (**5.1.38**) with 1-methyl-4-chloropiperidine (**5.1.37**) on reflux in *i*-propoanol, which gave 1-methyl-4-piperidinyl ester (**5.1.39**). The last was transformed to chloride (**5.1.40**), which on reflux with propoanol gave desired propiverine (**5.1.41**). Another method for the synthesis of ester (**5.1.39**) included reaction of 1-methyl-4-piperidinol (**5.1.4**) with methyl benzilate (**5.1.42**). An alternate way for the synthesis of chloride (**5.1.40**) was the acylation of 1-methyl-4-piperidinol (**5.1.4**) with diphenylchloroacetylchloride (**5.1.43**) [46,47] (Scheme 5.7).

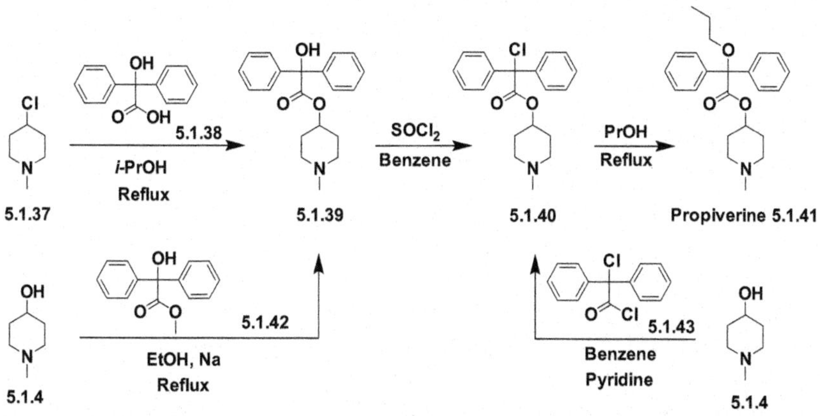

SCHEME 5.7   Synthesis of propiverine.

Modified methods for the synthesis of propiverine are also known [48−50].

Pentapiperide (22)

Pentapiperide (**5.1.45**), (Quilene) is an anticholinergic drug with very few and old references with antisecretory and antimotility activity on the upper gastrointestinal tract, which is used as adjunctive therapy for relieving the symptoms of peptic ulcer [51,52].

For the synthesis of *pentapiperide* (**5.1.45**), 2-phenyl-3-methylpentanoic acid chloride (**5.1.44**) was mixed with 1-methyl-4-piperidinol (**5.1.4**) in pyridine and allowed to stand 24 hours to give desired product (**5.1.45**) [53] (Scheme 5.8).

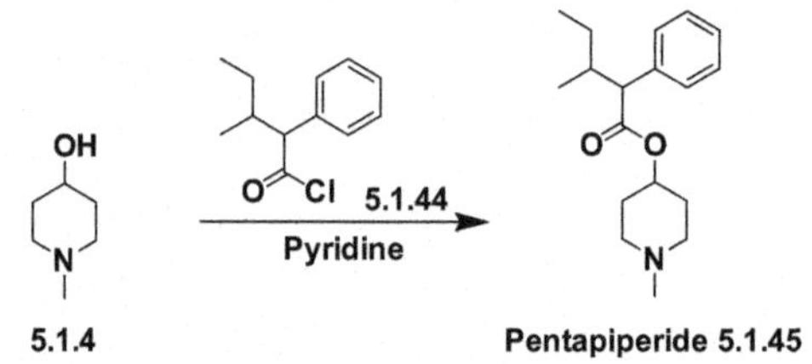

SCHEME 5.8 Synthesis of pentapiperide.

5.2 4-PHENYLPIPERIDIN-4-OLS AND THEIR DERIVATIVES

Haloperidol (41140)

Haloperidol (**5.2.7**), discovered at the end of the 1950s, constitutes one of the greatest advances of 20th century psychiatry. Marketed under the trade name Haldol, haloperidol is the most commonly used and widely prescribed for more than 40 years psychotropic agent indicated for treatment of schizophrenia and other psychoses. It is also used in schizoaffective disorder, delusional disorders, ballism, and Tourette syndrome and, occasionally, as adjunctive therapy in mental retardation and the chorea of Huntington disease. It is a potent antiemetic and is used in the treatment of intractable hiccups. Haloperidol has been included in the World Health Organization's list of essential medicines.

The precise mechanism whereby the therapeutic effects of haloperidol are produced is not known, but it is proved that haloperidol like most antipsychotics blocks the D2 receptors. The drug binds also to α1 and 5-HT2 receptors, and has negligible affinity for histamine H1 receptors and muscarinic M1 acetylcholine receptors.

Common side effects of haloperidol are nausea, vomiting, diarrhea dizziness, fainting, fast or pounding heartbeat, restless muscle movements of eyes, tongue, jaw, or neck, tremor, seizure, pale skin, easy bruising or bleeding, flu symptoms, rigid muscles, high fever, sweating, confusion, fast or uneven heartbeats, tremors, stabbing chest pain, feeling short of breath, sudden mood changes, agitation, hallucinations, unusual thoughts or behavior, or jaundice [54–62].

The synthesis of haloperidol (**5.2.7**) started from one of the varieties of Mannich reaction – aminomethylation of olefins, particularly of a derivative of α-methylstyrene – 1-chloro-4-isopropenylbenzene (**5.2.1**).

A mixture of ammonium chloride and formaldehyde was stirred and heated to 60°C and 1-chloro-4-isopropenylbenzene (**5.2.1**) was added with cooling to maintain this temperature. After the addition was complete, the intermediate 6-(4-chlorophenyl)-6-methyl-1,3-oxazinane (**5.2.2**) was formed. The conversion of 1,3-oxazine (**5.2.2**) to 4-(4-chlorophenyl)-1,2,3,6-tetrahydropyridine (**5.2.3**) has been carried out by heating the obtained mixture with excess of hydrochloric acid.

Anhydrous hydrogen bromide gas was passed through a 4-(4-chlorophenyl)-1,2,3,6-tetrahydropyridine (**5.2.3**) solution in acetic acid at room temperature to give 4-bromo-4-phenylpiperidine derivative (**5.2.4**).

A solution of 4-bromo-4-(4-chlorophenyl)piperidine hydrobromide (**5.2.4**) in water was treated with an excess 20% NaOH to give 4-(4-chlorophenyl)piperidin-4-ol (**5.2.5**). The obtained product (**5.1.5**), 4-chlorobutyrophenone (**5.1.6**), and potassium iodide in toluene were heated in a closed reaction vessel at 100−110°C to give desired haloperidol [63−69] (Scheme 5.9).

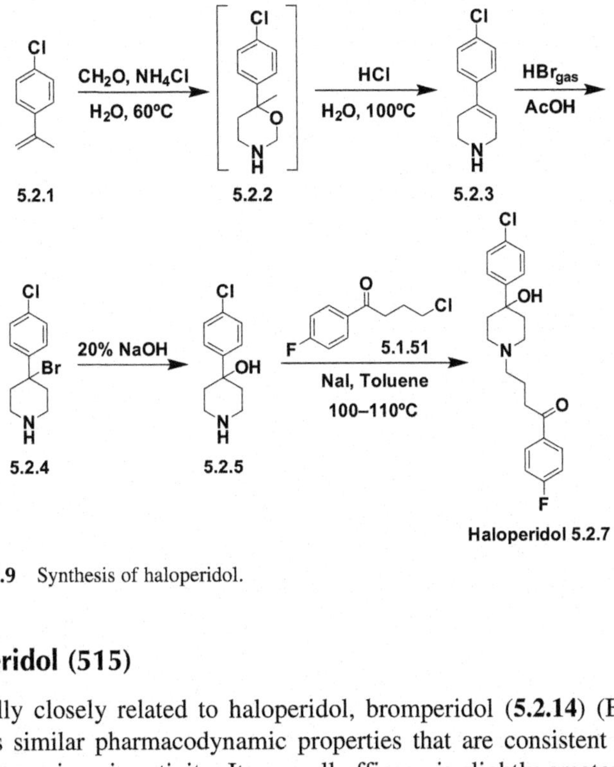

SCHEME 5.9 Synthesis of haloperidol.

Bromperidol (515)

Chemically closely related to haloperidol, bromperidol (**5.2.14**) (Bromodol) possesses similar pharmacodynamic properties that are consistent with central antidopaminergic activity. Its overall efficacy is slightly greater than that of haloperidol. Bromperidol frequently displays a faster onset of action than that of haloperidol as well as other certain differences, particularly because of its inherent long duration of action in which adequate doses of bromperidol, administered once daily, control or relieve psychotic symptoms. It was also hoped that the incidence of extrapyramidal stimulation would be low [70−73].

Bromperidol (**5.2.14**) was synthesized according to the same method, which was emoloyed for the synthesis of haloperidol (Scheme 5.9) and replaced 1-chloro-4-isopropenylbenzene (**5.2.1**) for 1-bromo-4-isopropenylbenzene [65,66].

In attempts for the synthesis Br-labeled bromperidol (**5.2.13**) other methods were published based on substitution of an aromatic amino group in "aminoperidol" (**5.2.12**) for bromine by Sandmeyer-type or Gatterman reactions [75.76]. In one of them the synthetic sequence started from the 4-chloro-1,1-(ethylenedioxy)-1-(4-fluorophenyl) butane (**5.2.8**) was prepared via ketaliztion of 4-chlorobutyrophenone (**5.2.6**) with ethylene glycol in benzene in the presence of p-toluenesulfonic acid. The obtained compound was coupled with 4-hydroxypiperidine (**5.1.7**) at 110−120°C using a common protocol (DMF, K_2CO_3, KI). A reaction product (**5.2.9**) was oxidized to 1-[4-(4-Fluorophenyl)-4,4-(ethylenedioxy)butyl]-4-piperidinone (**5.2.11**) by a modified Oppenauer oxidation, where alcohol (**5.2.9**), 9-fluorenone (**5.2.10**) and potassium tert-butoxide were kept in benzene in vacuum conditions to give product (**5.2.11**). In the next step p-lithium-N,N-bis(trimethylsilyl)aniline solution was prepared from p-bromo-N,N-bis(trimethylsilyl)-aniline (**5.2.12**) and n-butyllithium in hexane to which piperidin-4-one (**5.2.11**) dissolved in tetrahydrofuran (THF) was added to give, after overnight hydrolysis with 4 N HCl at room temperature, "aminoperidol" (**5.2.13**).

The last step of the synthetic route was the incorporation of bromine into the "aminoperidol" (**5.2.13**) molecule accomplished through Brackman and Smit's modification of the Sandmeyer reaction. This involved the formation of a complex between $CuBr_2$ and nitric oxide in acetonitrile, which on reaction with "aminoperidol" (**5.2.13**) gave desired bromperidol (**5.2.14**) [74]. In an analogous method the substitution of aromatic amino group for bromine was carried out in Gattermann reaction conditions. For that purpose "aminoperidol" (**5.2.13**) was diazotated in 2 N sulfuric acid by adding a slightly excess amount of sodium nitrite and sodium bromide, after which the catalyst amount of copper powder was added and the mixture was heated to 80°C for 10 minutes [75] (Scheme 5.10).

Moperone (218)

Moperone (**5.2.15**) (Luvatren) is discontinued in the United States, but the medicine is available in a number of countries worldwide, particularly in Japan for the treatment of schizophrenia. It has a higher antagonist affinity for D2- than 5-HT2A-receptors and has a high binding affinity for sigma receptors. It displays generally low toxicity but can induce extrapyramidal motor side effects such as insomnia and thirst [76−78].

Moperone has been synthesized via methods implemented to haloperidol [63,79] (Fig. 5.2).

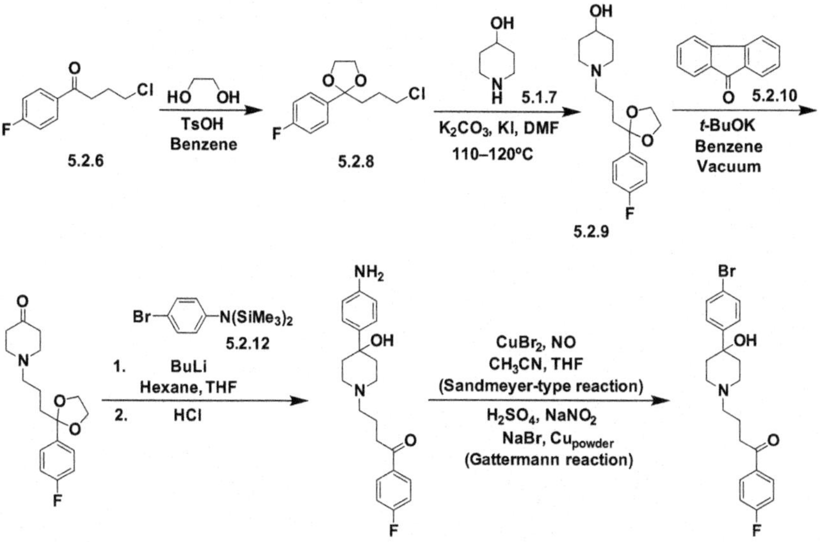

SCHEME 5.10   Synthesis of bromperidol.

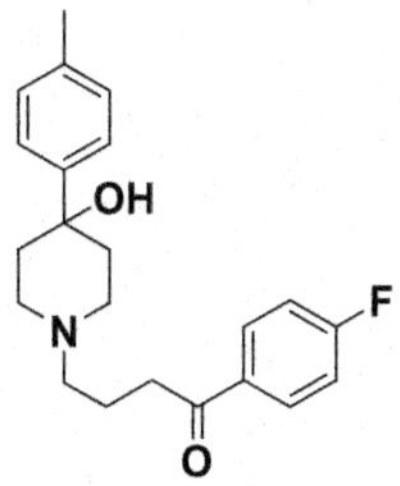

Moperone 5.2.15

FIGURE 5.2   Moperone.

Trifluperidol (613)

Trifluperidol (**5.2.20**) (Triperidol) is another first generation antipsychotics of the same butirophenone series with general properties similar to those of haloperidol (**5.2.7**). It is considerably more potent but causes relatively more severe extrapyramidal and tardive dyskinesiaside side effects. Trifluperidol is used in the treatment of psychoses including mania and schizophrenia [80–82].

It was synthesized via condensing Grignard's reagent (**5.2.17**) prepared from 3-trifluoromethylphenyl bromide with 1-benzyl-4-piperidin-4-one (**5.2.16**) in ether, followed by a removal of the benzyl group in obtained compound (**5.2.18**) using standard hydrogenation technique with Pd-C catalyst and *i*-PrOH as a solvent, which gave compound (**5.2.19**). The last was alkylated with γ-chloro-4-fluorobutyrophenone (**5.2.6**) in regular conditions (Na$_2$CO$_3$, KI, MIBK) to give desired (**5.2.20**) [65,66] (Scheme 5.11).

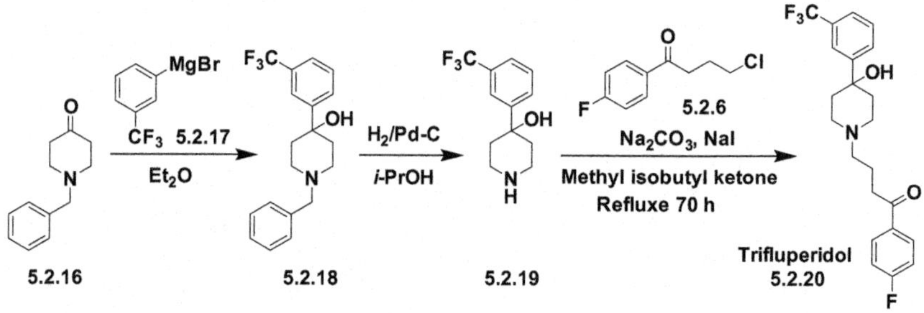

SCHEME 5.11   Synthesis of trifluperidol.

Penfluridol (633)

Penfluridol (**5.2.26**) (Flupidol) (Semap) is a long-acting, oral neuroleptic belonging to the diphenylbutylpiperidines that was discovered at Janssen Pharmaceutical at the same time of the discovery of typical antipsychotic medications of the butyrophenone series (Haloperidol, Bromoperidol, etc.).

Penfluridol is a highly potent, first generation antipsychotic with extremely long elimination half-life (66 hours), and its effects last for many days after single oral dose. Penfluridol is often prescribed to be taken orally only once a week. Its antipsychotic potency is similar to haloperidol and pimozide. Penfluridol is indicated for treatment of chronic schizophrenia and similar psychotic disorders. It is, however, like most typical antipsychotics, being increasingly replaced by the atypical antipsychotics.

It is only slightly sedative but often causes extrapyramidal side effects, such as akathisia, dyskinesiae, and pseudo-Parkinsonism, delirium, agitation, anxiety, depression, euphoria, anorexia, constipation or diarrhea, alopecia, hypersalivation, nausea, vomiting [83−88].

Penfluridol (**5.2.26**) was prepared starting with 3-methyl-4-oxo-1-(alkoxy-carbonyl)piperidine (**5.2.21**), which was condensed with a Grignard reagent − 4-chloro-3-trifluoromethyl-phenyl magnesium bromide (**5.2.22**) in THF to give methyl 4-(4-chloro-3-(trifluoromethyl)phenyl)-4-hydroxypiperidine-1-carboxylate (**5.2.23**). On refluxing in *i*-propanol in the presence of potassium hydroxide the obtained product (**5.2.23**) was deprotected to give 4-(4-chloro-3-(trifluoromethyl)phenyl)piperidin-4-ol (**5.2.24**), which was alkylated with 4,4-bis(*p*-fluorophenyl) butyl chloride (**5.2.25**) on reflux in MIBK in the presence of sodium carbonate and few crystals of potassium iodide to give penfluridol (**5.2.26**) [89].

Other authors have described a method where 4,4-bis(4-fluorophenyl) butyl iodide (**5.2.28**) was treated with 4,4-ethylenedioxypiperidine (**5.2.27**) and potassium carbonate in refluxing diethylketone. The resulting 1-[4,4-bis (4-fluorophpenyl)butyl]-4,4-ethylenedioxypiperidine (**5.2.29**) was hydrolyzed with aqueous HCl to give ketone (**5.2.30**) which on Grignard reaction with 4-chloro-3-trifluoromethyl-phenyl magnesium bromide (**5.2.22**) in THF gave desired penfluridol (**5.2.26**) [90,91] (Scheme 5.12).

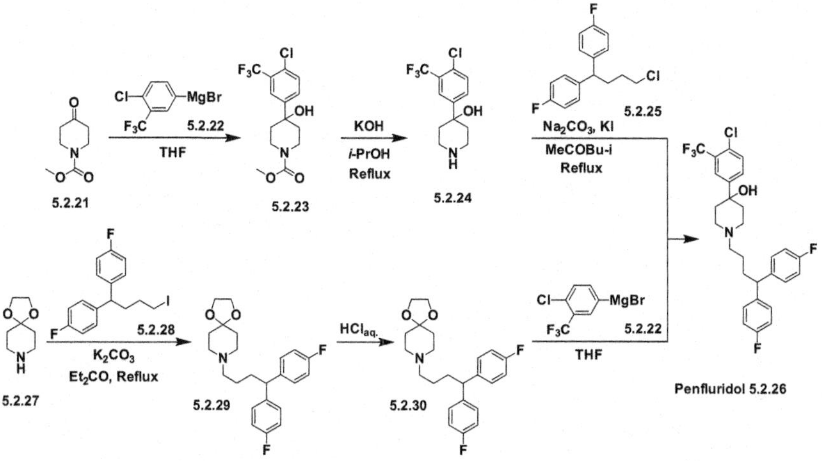

SCHEME 5.12   Synthesis of loperamide.

A Grignard reaction of (**5.2.22**) with *N*-Boc piperidone-4 smoothly gave a corresponding compound − Boc analog of (**5.2.23**). However, removal of the Boc- group with trifluoroacetic acid only led to the dehydrated products. After several trials, (**5.2.24**) was prepared via deprotection of Boc- group in the mixture of aqueous HCl/and ethyl acetate at room temperature [92].

Loperamide (3756)

Loperamide (**5.2.35**) (Imodium) is a peripherally acting μ-opioid receptor agonist and an avid substrate for P-glycoprotein. It is an over-the-counter medication that is used for the relief of diarrheal syndromes, including acute, nonspecific (infectious) diarrhea; traveler's diarrhea; and chemotherapy-related and protease inhibitor-associated diarrhea. Loperamide is effective for the "gut-directed" symptom of diarrhea in patients with painless diarrhea or diarrhea-predominant irritable bowel syndrome. Loperamide acts on receptors along the small intestine to decrease circular and longitudinal muscle activity reducing diarrhea by slowing the forward propulsion of intestinal contents by the intestinal muscles, and perhaps also by directly inhibiting fluid and electrolyte secretion and/or stimulating salt and water absorption [93−105].

Loperamide treats only the symptoms, not the cause of the diarrhea (e.g., infection). Dizziness, drowsiness, tiredness, or constipation may occur as side effects. The US Food and Drug Administration (FDA) has issued a warning that high doses of the antidiarrheal medication loperamide are associated with a risk of heart problems [106].

Loperamide (**5.2.35**) was synthesized via the coupling of 4-(4-chlorophenyl)piperidin-4-ol (**5.2.5**) with dimethyl(tetrahydro-3,3-diphenyl-2-furylidene)ammonium bromide (**5.2.34**) on refluxing reagents in isobutyl methyl ketone in the presence of excess sodium carbonate. Synthesis of 4-(4-chlorophenyl)-piperidin-4-ol (**5.2.5**) was described above (see haloperidol). Compound (**5.2.34**) was prepared via ring opening of 2,2-diphenyl- 4-hydroxybutyric acid γ-lactone (**5.2.31**) with 48% HBr in acetic acid which afforded 4-bromo-2,2-diphenylbutyric acid (**5.2.32**). The reaction of the last with thionyl chloride in chloroform gave crude 4-bromo-2,2-diphenylbutyroyl chloride (**5.2.33**), which was dissolved in toluene and cautiously added to a solution of dimethylamine and sodium carbonate in water while the temperature was kept between 0°C and 5°C to give tetrahydro-3,3-diphenyl-2-furylidene ammonium salt (**5.2.34**) [107,108] (Scheme 5.13).

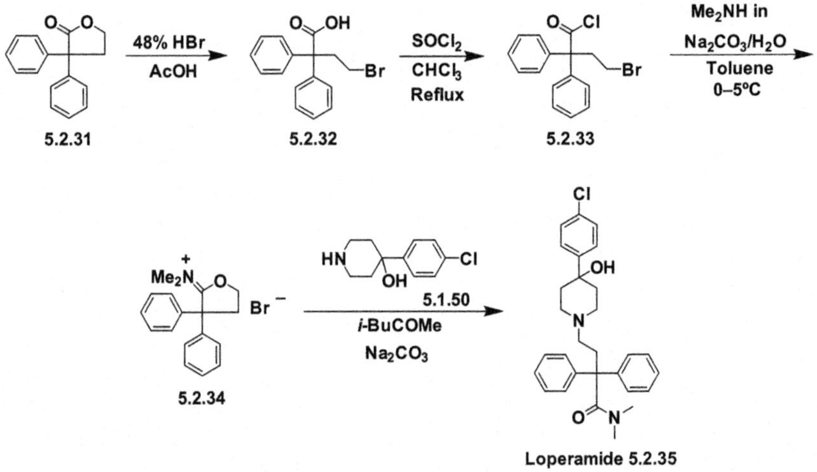

SCHEME 5.13 Synthesis of loperamide.

Alpha-Prodine (502)

Several 1-alkyl-4-phenyl-4-acyloxypiperidines, such as desmethylprodine (**5.2.36**), prodine (**5.2.37**) (Nisentil), allylprodine (**5.2.38**), and trimeperidine (**5.2.39**) (Promedol) have been synthesized and found to possess significant analgesic actions [109−112] (Fig. 5.3). Compounds with the 4-propionoxy substituent appear to be the potent analgesics.

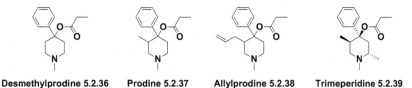

Desmethylprodine 5.2.36   Prodine 5.2.37   Allylprodine 5.2.38   Trimeperidine 5.2.39

FIGURE 5.3   Desmethylprodine, prodine, allylprodine, and trimeperidine.

Desmethylprodine (**5.2.36**), a derivative of pethidine, is an opioid analgesic with the potency of morphine. It has been listed as a Schedule I controlled drug in the United States, and is not used clinically.

There are two isomers of prodine (**5.2.37**), alpha-prodine (α-prodine) (**5.2.37a**) and beta-prodine (β-prodine) (**5.2.37b**) [113]. Beta-prodine is more potent than alpha-prodine, but is metabolized more rapidly, and only alpha-prodine (Nisentil), which has a duration of action of 1−2 hours, was developed for medicinal use for minor surgical procedures for pain relief [114−117] and particularly for obstetric analgesia [118] and in dentistry [119].

Side effects of α-prodine (**5.2.37a**) can include itching, nausea, and potentially serious respiratory depression, which can be life-threatening. Respiratory depression can be a problem with α-prodine even at normal therapeutic doses [120].

Allylprodine (**5.2.38**) produces analgesia and sedation and the same side effects characteristic of the entire 4-phenylpiperidin-4-yl propionate series such as respiratory depression, nausea, vomiting, itching, and it is more potent as an analgesic than α-prodine [121,122].

Trimeperidine (**5.2.39**) is another analog of prodine with additional methyl group at the second position of piperidine ring [123,124].

The stereoselectivity of trimeperidine and that reported for α-prodine is qualitatively the same and the conformational features of the more potent enantiomers are also very similar [125].

An entire row of drugs of l-alkyl-4-phenyl-4-propionoxypiperidines series was prepared in a similar way. Prodine (**5.2.37**), in particular, was synthesized by a reaction of phenyllithium or phenyl magnesium bromide with 1,3-dimethyl-4-piperidone (**5.2.40**), which gave a mixture of diastereomeric alcohols (**5.2.41a, 5.2.41b**), which were acylated with propionic anhydride in the presence of sulfuric acid or in pyridine under reflux [126−128], or with propionic acid chloride in chloroform in presence of pyridine [121] to give desired product (**5.2.37**) (Scheme 5.14).

After the reaction of phenyllithium on 1,3-dimethyl-4-piperidone a mixture of diastereomeric alcohols was obtained.

There are four possible *cis-trans* arrangements of the methyl group and the H atom on C-3, and the propionoxy chain and the phenyl ring in the desired product − prodine (**5.2.37**). Fractional crystallization of the tartaric or dibenzoyltartaric acid salts of the corresponding piperidinol diastereomers allowed for the separation of two distinct fractions named α-isomer dl-α-1,3-dimethyl-4-phenyl-4-propionoxypiperidine and β-isomer dl-β-1,3-dimethyl-4-phenyl-4-propionoxypiperidine. They were esterificated with proprionyl chloride to give two isomeric products: major − α-prodine (**5.2.37a**) and minor − β-prodine (**5.2.37b**).

An X-ray crystallographic study of crystals of dl-α-prodine (**5.2.37a**) hydrochloride and hydrobromide had demonstrated that the propionyloxy chain is in an axial, and the phenyl group is in an equatorial position on the chair form of piperidine ring and that the methyl group on C-3 is trans to the phenyl ring on C-4.

Crystals of the corresponding salts of the dl-β-prodine (**5.2.37b**) demonstrated that the phenyl ring still in an equatorial and the propionyloxy chain in an axial position on C-4 but the methyl group is axial and the H atom equatorial on C-3. Thus the methyl on C-3 is cis to the phenyl on C-4 [121,128−135] (Scheme 5.14)

After the relative stereochemistries of dl-α-prodine (**5.2.37a**) and dl-β-prodine (**5.2.37b**) were established, the absolute configurations and potencies of the optically pure enantiomers were reported.

The complete stereochemistry of (+)-α-prodine (**5.2.37a**) hydrochloride is (3R:4S) and the stereochemistry of (−)-α-prodine hydrochloride is designated as (3S:4R). Configurational assignment of (+)- and (−)-β-prodines (**5.2.37b**) are (3S:4S) and (3R:4R), respectively. It was found that most of the analgesic activity of both α- and β-prodines resides in the (+)-antipodes [136].

analog assignments were carried out for allylprodine (**5.2.38**) [137] and for trimeperidine (**5.2.39**) [124,138].

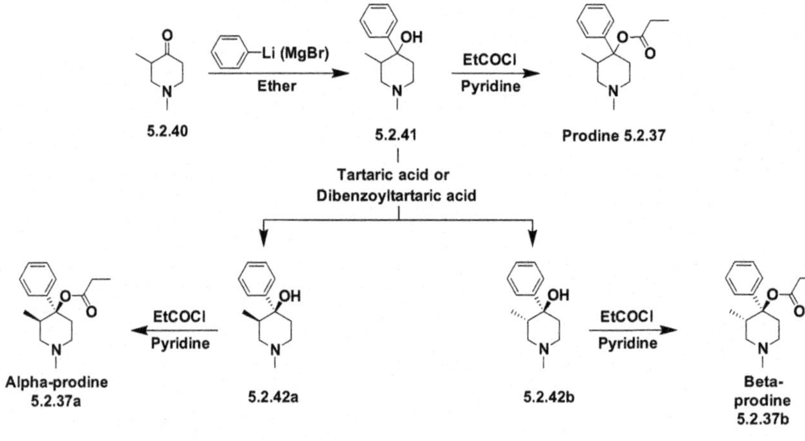

SCHEME 5.14 Synthesis of α- and β-prodines.

5.3 DERIVATIVES OF 4-PHENYLPIPERIDINE-4-CARBOXYLIC ACIDS

Pethidine (11164)

Pethidine (**5.3.5**) (Meperidine, Demerol) is used as an analgesic for the relief of moderate-to-severe pain including obstetric analgesia, and is used as a preoperative medication and analgesia during anaesthesia and postoperative analgesia. Analgesic effects usually last 3−4 hours; parenteral administration pethidine is not recommended as a first choice analgesic by The American Pain Society. The most serious adverse effects of pethidine are respiratory depression and hypotension [139−141].

Pethidine was portrayed in practice and teaching as having unique clinical advantages. But in a recently published paper it has been criticized, with the paper mentioning that "analgesic effects of pethidine are not pronounced, and, in addition, pethidine use is complicated by unique side effects including serotonergic crisis and normeperidine toxicity. Pethidine's poor efficacy, toxicity, and multiple drug interactions have resulted in a movement to replace pethidine with more efficacious and less toxic opioid analgesics" [142].

The first synthesis of pethidine (**5.3.5**) was described starting with cyclocondensation of bis(2-chloroethyl)methylamine (**5.3.2**) with benzyl cyanide (**5.3.1**) in the presence of sodium amide in toluene, which gave 1-methyl-4-phenylpiperidine-4-carbonitrile (**5.3.3**). The obtained nitrile was hydrolyzed with methanolic potassium hydroxide while heating to 190−200°C. Prepared 1-methyl-4-phenyl-4-piperidinecarboxylic acid (**5.3.4**) on reaction with thionyl chloride easily gave intermediate acid chloride, which on dissolving in ethyl alcohol gave desired pethidine (**5.3.5**) [143] (Scheme 5.15).

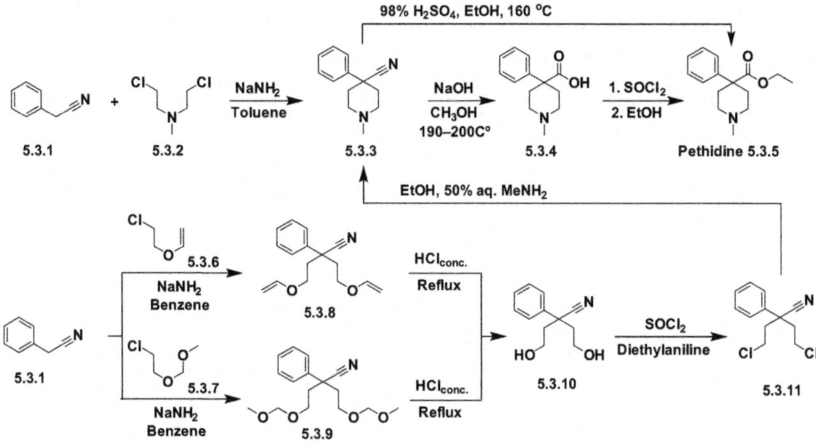

SCHEME 5.15 Synthesis of pethidine.

Another method of preparing 4-arylpiperidine-4-nitriles was described. In this case benzyl cyanide (**5.3.1**) was condensed with β-chloroethyl vinyl ether (**5.3.6**) or 2-(2-methoxy-methoxy)ethyl chloride (**5.3.7**) in the presence of sodium amide in benzene to give α,α-disubstituted benzyl cyanides (**5.3.8**) and (**5.3.9**) correspondingly. Obtained products were hydrolyzed on reflux in concentrated hydrochloric acid to give α,α-bis (2-hydroxyethyl)-benzyl cyanide (**5.3.10**), which on treatment with thionyl chloride in diethylaniline gave corresponding dichloride (**5.3.11**). Heating of the obtained product (**5.3.11**) with 50% aqueous methylamine in ethanol at 145°C gave 4-phenyl-1-methyl-4-piperidinecarbonitrile (**5.3.3**), which was converted to pethidine (**5.3.5**) on heating at 160°C in ethanol containing 98% sulfuric acid and ammonium chloride [144] (Scheme 5.15).

The third proposed method of synthesis of pethidine (**5.3.5**) started from isonicotinic acid (**5.3.12**), which was transferred to isonicotinic acid methochloride (**5.3.14**) in two steps. In the first step isonicotinic acid (**5.3.12**) was quaternized with methyl iodide in the presence of sodium hydroxide in methanol water solution, and then obtained iodide salt (**5.3.13**) was converted to methochloride (**5.3.14**) via passing through Amberlite IRA-400 chloride form column to give product (**5.3.14**). Obtained compound (**5.3.14**) was quantitatively reduced in methanol in the presence of platinum oxide under hydrogen pressure to give 1-methyl-carboxypiperidine (**5.3.15**), which was converted to the intermediate acid chloride under reflux with thionyl chloride, which without separation was condensed with benzene under Friedel–Crafts conditions to give 1-methyl-4-benzoylpiperidine (**5.3.16**). Chlorination of the obtained ketone with chlorine in acetic acid gave the chloroketone (**5.3.17**) in excellent yields.

Obtained chloroketone (**5.3.17**) was refluxed in xylene in presence of powdered sodium hydroxide, which resulted in the quasi-Favorski rearrangement specific to α-haloketones containing no α-hydrogens to give as one of the obtained products 1- methyl-4-phenyl-4-piperidinecarboxylic acid (**5.3.4**), esterification of which on reflux in ethanolic hydrogen chloride solution gave the desired pethidine (**5.3.5**) [145] (Scheme 5.16).

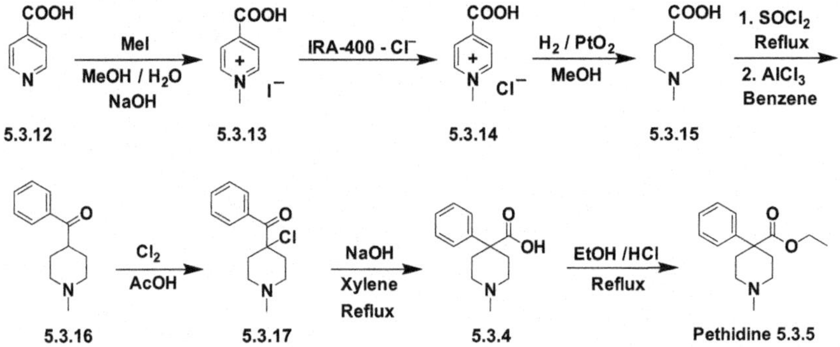

SCHEME 5.16 Synthesis of pethidine.

Phenoperidine (506)

Phenoperidine (**5.3.27**) is another opioid analgesic used in general anesthesia.

Following the discovery of pethidine (**5.3.5**) a considerable number of its *N*-alkyl variations, have been described based on norpethidine (**5.3.22**), which was originally obtained in two ways: from 1-benzyl-4-phenylpiperidine-4-carbonitrile (**5.3.19**) by hydrolysis and esterification followed by debenzylation [146], and from 4-phenyl-1-tosylpiperidine-4-carbonitrile (**5.3.24**) by combined hydrolysis and an esterification procedure using sulphuric acid [146,147].

1-Benzyl-4-phenyl-4-piperidinecarbonitrile (**5.3.19**) was synthesized by the full analogy with the synthesis of 1-methyl-4-phenyl-4-piperidinecarbonitrile (**5.3.3**) via cyclocondensation of bis(2-chloroethyl)benzylamine (**5.3.18**) with benzyl cyanide (**5.3.1**) using sodium amide in toluene. The obtained nitrile (**5.3.19**) was hydrolyzed to the 4-carboxylic acid (**5.3.20**) in two ways: in potassium hydroxide-methanol or in sulfuric acid solutions. Thev acid (**5.3.20**) was esterified directly with ethanol in presence of sulfuric acid, or via intermediate preparation of corresponding acid chloride followed by reaction with ethanol to give ethyl 1-benzyl-4-phenylpiperidine-4-carboxylate (**5.3.21**), which was debenzylated on hydrogenation in ethanol over Pd sponge to give norpethidine (**5.3.22**).

Another method ivolves cyclocondensation of *N,N*-bis(2-chloroethyl)-4-methylbenzenesulfon-amide (**5.3.23**) with with benzyl cyanide (**5.3.1**) by means of sodium amide in toluene which yields 4-phenyl-1-(*p*-tolylsulfonyl)-4-piperidinecarbonitrile (**5.3.24**). Cautious heating of nitrile (**5.3.24**) at 140−150°C with 75% sulfuric acid gave acid (**5.3.25**). The next step was esterification with ethanol acidified with sulfuric acid, accompanied by a spontaneous cleavage of N-Ts, which gave norpethidine (**5.3.22**) (Scheme 5.17).

A huge variety of *N*-alkylations products of norpethidine (**5.3.22**) showed expressed analgesic properties and became commercialized drugs.

One of them is phenoperidine (**5.3.27**) prepared from norpethidine (**5.3.22**) in three ways:

1. Via Mannich reaction on heating under reflux norpethidine (**5.3.22**), acetophenone and paraformaldehyde in iso-propanol in presence of hydrochloric acid;
2. Via condensation of vinyl ketone and norpethidine (**5.3.22**) on heating under reflux in iso-propanol in presence of hydrochloric acid;
3. Via condensation of 2-chloroethyl phenyl ketone and norpethidine (**5.3.22**) by heating reagents in xylene in presence of potassium iodide in a sealed tube at a temperature of 140−146°C. The reduction of obtained product with hydrogen on platinum oxide catalyst in methanol or with sodium borohydride yielded desired phenoperidine (**5.3.27**) [148−151] (Scheme 5.17).

Phenoperidine is classified as a schedule 1 opiate and controlled substance.

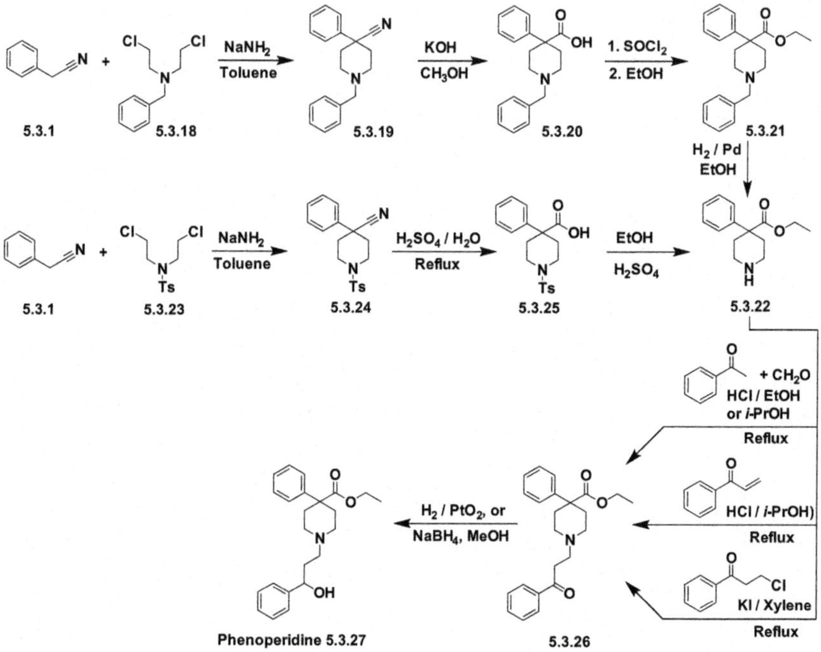

SCHEME 5.17   Synthesis of phenoperidine.

Phenoperidine has a faster and more prolonged analgesic effects and is often used in pediatric intensive care units due to its interesting pharmacologic properties (mild cardiac and respiratory depression) [152–155].

Hundreds of other pethidine analogs were synthesized, some of them entered the pharmaceutical market, but later most of them were withdrawn.

One of them is anileridine (**5.3.28**) (Leritine), a compound where *N*-methyl group of pethidine is replaced by an *N*-aminophenethyl group, which increases its analgesic activity. Anileridine is no longer available in the United States and Canada. It was synthesized by alkylation of norpethidine (**5.3.22**) with 4-(2-chloroethyl)aniline in refluxing ethanol in presence of sodium bicarbonate [156,157].

Benzethidine (**5.3.29**) is another 4-phenylpiperidine derivative prepared by alkylation of norpethidine (**5.3.22**) with ((2-chloroethoxy)methyl)benzene in ethanol in the presence of sodium carbonate on reflux [158] that is related to the opioid analgesic drug pethidine. Benzethidine has side effects comparable to that of other opiates, such as nausea, sedation, and respiratory depression, and is not currently used in medicine and is controlled under UN drug conventions.

Etoxeridine (**5.3.30**) was synthesized by alkylation of norpethidine (**5.3.22**) on reflux witn 2-(2-chloroethoxy)ethan-1-ol in the presence of tri-methylamine but was never commercialised and is not currently used in medicine [159].

Furethidine (**5.3.31**) is another 4-phenylpiperidine derivative that is related to pethidine and was synthesized by alkylation of norpethidine (**5.3.22**) with 2-((2-chloroethoxy)methyl)tetrahydro-furan on reflux in ethanol in presence of sodium carbonate [158]. Furethidine is not currently used in medicine and considered a Class A/Schedule I drug, which is controlled under UN drug conventions.

Piminodine (**5.3.32**), an analog of pethidine, was used in medicine for obstetric analgesia and in dental procedures briefly during the 1960s and 1970s, but has largely fallen out of clinical use.

It was synthesized by alkylation of norpethidine (**5.3.22**) with N-(3-chlor-opropyl)aniline on reflux in buthanol in presence of sodium carbonate [160]. Piminodine produces analgesia, sedation and euphoria and has typical side effects associated with opiods, including potentially serious respiratory depression, which can be life-threatening. Piminodine is currently a Schedule II controlled substance in the United States.

Morpheridine (**5.3.33**) is a strong analgesic prepared by alkylation of nor-pethidine (**5.3.22**) with 4-(2-chloroethyl)morpholine on reflux in ethanol [161–163] and has around four times the potency of pethidine, produces the standard opioid side effects, but unlike pethidine does not cause convulsions. Morpheridine is not currently used in medicine and is a Schedule I drug, which is controlled under UN drug conventions (Fig. 5.4).

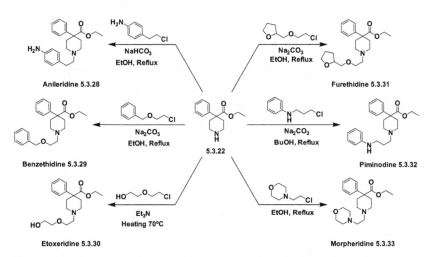

FIGURE 5.4 Analogs of pethidine withdrawn from pharmaceutical market.

Diphenoxylate (878)

Diphenoxylate (**5.3.35**) (Diocalm) is a close analog of loperamide (**5.2.35**) and is commonly used as an effective adjunctive therapy in the management of diarrhea in numerous settings of inflammatory bowel disease [164,165].

It acts by slowing intestinal contractions and peristalsis and allowing the intestines to draw moisture out, thus stopping the formation of loose and liquid stools.

It is often used in combination with with other drugs, usually in combination with atropine to lower the risk of abuse (Lomotil, Nisentil, Dipidolor).

Diphenoxylate act directly on opioid receptors, can cause a sense of false euphoria and is potentially habit-forming and can generate significant tolerance. Common side effects of diphenoxylate are similar to those of loperamide. It produces drowsiness, dizziness and abdominal cramps as well as side effects general to opioids including nausea, vomiting, headache, restlessness, confusion, and changes in mood. Recently it was found that diphenoxylate blocks Kv1.3 potassium channels, which have been proposed as potential therapeutic targets for a range of autoimmune diseases [166].

Diphenoxylate (**5.3.35**) was prepared by interaction of 4-bromo-2,2-diphenylbutanenitrile (**5.3.34**) and norpethidine (**5.3.22**) on reflux in xylene in a sealed tube. In an alternative method 2,2-diphenylacetonitrile (**5.3.37**) was treated with a slurry of sodium amide in benzene and further and obtained 2,2-diphenylacetonitrile anion was reacted with 1-(2-chloroethyl)-4-phenylisonepecotate (**5.3.6**) on reflux in xylene [164,167,168] (Scheme 5.18).

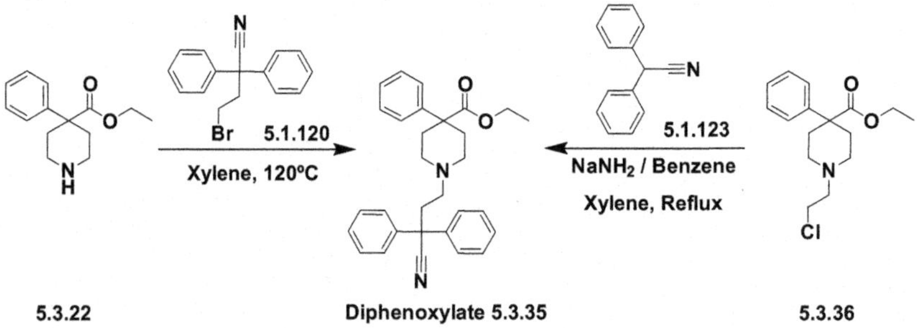

SCHEME 5.18   Synthesis of diphenoxylate.

Levocabastine (854)

Levocabastine (**5.3.41**) (Livostin) is a second-generation selective H1-antihistamine compound, which has been evaluated as a topical treatment (nasal spray and eyedrops) for allergic rhinitis and conjunctivitis used to treat symptoms of eye allergies, such as inflammation, itching, watering, and burning. It was found to be the most potent antihistamine compound available, being 15,000 times more potent than chlorpheniramine. Common side effects include burning, stinging, itching, or watering of the eyes, eye irritation or discomfort, blurred vision, dry or puffy eyes, headache, nausea, or fatigue [169–173]. Levocabastine is no longer available in the United States, but generic versions are still available on the pharmaceutical market.

Levocabastine (**5.3.41**) was synthesized by reductive amination of benzyl (3*S*,4*R*)-3-methyl-4-phenylpiperidine-4-carboxylate (**5.3.38**) with 4-cyano-4-(4-fluorophenyl) cyclohexanone (**5.3.39**) on platinum-on-charcoal catalyst in iso-propanol/thophene mixture, followed by hydrogenolysis of the resulting benzylester (**5.3.40**) on palladium catalyst in iso-propanol [174,175] (Scheme 5.19).

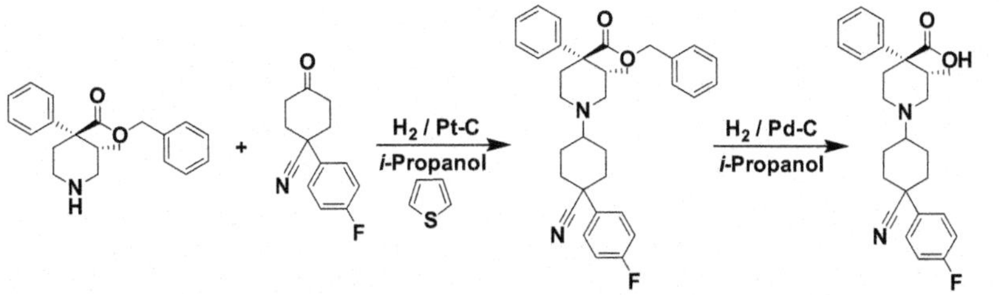

| **5.3.38** | **5.3.39** | **5.3.40** | **Levocabastine 5.3.41** |

SCHEME 5.19   Synthesis of levocabastine.

5.4   DERIVATIVES OF (4-PHENYLPIPERIDIN-4-YL)KETONES

Ketobemidone (722)

Ketobemidone (**5.4.5**) (Ketorax) is a powerful opioid analgesic that plays an important role in the control of the chronic pain of cancer and acute pain, especially postoperative. Ketobemidone is assigned to the Schedule I list of the US's Controlled Substances.

There is no significant differences between the three analgesics — morphine, meperidine and ketobemidone — with respect to efficacy of analgesia or side effects like shivering, nausea, or vomiting. But the binding affinities of ketobemidone to μ, δ, and κ opioid receptors show that they are discriminated better between μ and κ sites than did morphine, but was less μ/δ binding selective. Moreover, ketobemidone also has some *N*-methyl-D-aspartate (NMDA)-antagonist properties. This makes it useful for some types of pain that don't respond well to other opioids [176–178]. Ketobemidone is often used as an alternative to morphine in children in Scandinavian countries [179].

Two synthetic sequences for ketobemidone (**5.4.5**) synthesis have been described. Treatment of *m*-methoxyphenylacetonitrile (**5.4.1**) with bis(2-chloroethyl)methylamine (**5.3.2**) in the presence of sodium amide and heating at 100–105°C in toluene produced nitrile (**5.4.2**). Reaction of obtained nitrile with ethylmagnesium iodide (**5.4.3**) in benzene followed with further hydrolysis of obtained intermediate imine salt converted nitrile (**5.4.2**) into the ketone (**5.4.4**), which was readily demethylated with 48% hydrobromic acid to yield desired product (**5.4.5**) [180,181] (Scheme 5.20).

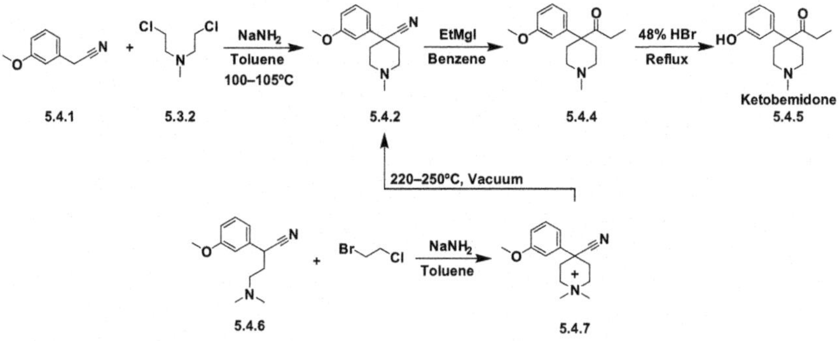

SCHEME 5.20 Synthesis of ketobemidone.

Cyclocondensation reaction of *m*-methoxyphenylaceto-nitrile (**5.4.1**) with bis(2-chloroethyl)-methylamine (**5.3.2**) was carried out also using phase transfer catalysis technique [182].

In the second method 4-dimethylamino-2-(*m*-methoxyphenyl)butyroni-trile (**5.4.6**) and 1-chloro-2-bromoethane reacted in the presence of sodium amide in toluene to yield quaternary salt (**5.4.7**), which was then dequater-nized on heating at 220−250°C under a water-pump vacuum for distlation (**5.4.2**) and further transformed to ketobemidone (**5.4.2**) in the same way [183] (Scheme 5.20).

Ketanserin (9597)

Ketanserin (**5.4.10**) (Sufrexal) is a 5-HT2 receptor antagonist that also pos-sesses weak α1-adrenoceptor antagonistic activity, which may explain its antihypertensive mechanism of action in patients with essential hypertension. It is classified as an antihypertensive drug and is effective in lowering blood pressure in essential hypertension. It also inhibits the effects of serotonin on platelets in cardiovascular disease, inhibits vasoconstriction, and improves some hemorheologic indexes in patients with ischemic diseases. Ketanserin may also be used in the treatment of some peripheral vascular diseases (e.g., Raynaud's phenomenon) as well as of carcinoid syndrome. In addition, it reduces platelet hyperactivity, blood viscosity and total serum cholesterol [184−192].

In addition, ketanserin is also a high affinity antagonist for the H1 receptor and has measurable affinity for other receptors: Ketanserin has a paradoxical effect in wound healing. Topical ketanserin stimulates wound epithelialization and vascularization but inhibits hypertrophic scar formation [193−195].

The synthesis of ketanserin (**5.4.10**) is based on coupling of the key intermediates − (4-fluorophenyl)(piperidin-4-yl)methanone (**5.4.8**) and 3-(2-chloroethyl)quinazoline-2,4(1*H*,3*H*)-dione (**5.4.9**), which was carried out in standard conditions for similar reactions on reflux of reagents in MIBK in presence of sodium carbonate and potassium iodide.

Starting ketone (**5.4.8**), in turn was synthesized from 1-benzylpiperidine-4-carbonitrile (**5.4.11**) which on reaction with (4-fluorophenyl)magnesium bromide in ether gave (1-benzylpiperidin-4-yl)(4-fluorophenyl)methanone (**5.4.12**). Obtained product was debenzylated with ethyl chloroformate on reflux in xylene in presence of sodium carbonate giving the corresponding *N*-carbamate (**5.4.13**) which on heating with 48% HBr yielded the desired secondary amine (**5.4.8**).

For the synthesis of 3-(2-chloroethyl)quinazoline-2,4(1*H*,3*H*)-dione (**5.4.9**), 2-anthranilic acid ethylester (**5.4.14**) was reacted with ethyl chloroformate on reflux in xylene yielding the carbamate (**5.4.15**), which was thermally cyclisized with 2-aminoethanol to the 3-(2-hydroxyethyl)quinazoline-2,4(1*H*,3*H*)-dione (**5.4.16**). The obtained primary alcohol (**5.4.16**) was converted to the chloro derivative (**5.1.9**), used further for coupling with amine (**5.4.8**) [196] (Scheme 5.21).

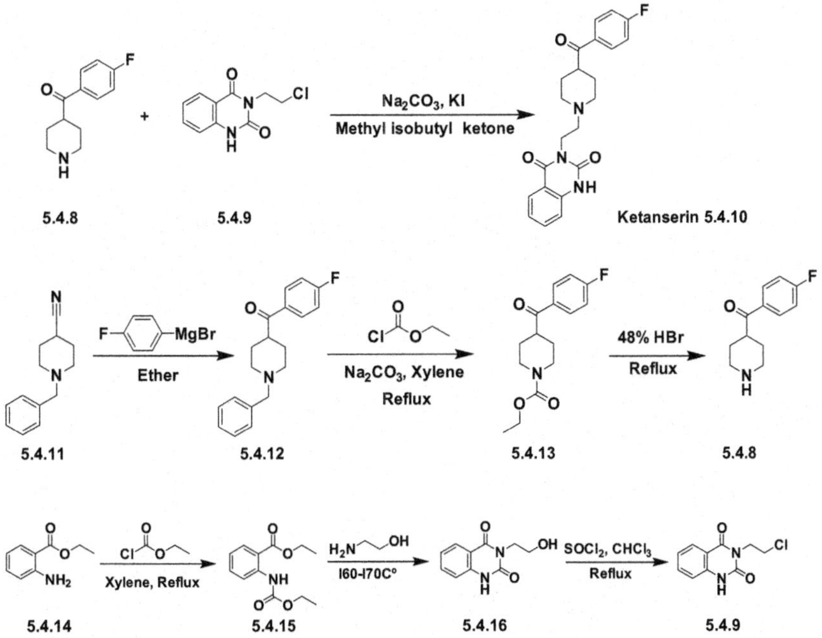

SCHEME 5.21   Synthesis of ketanserin.

Several variations and modifications of this method have been published [197−200].

5.5 DERIVATIVES OF 4-PHENYLPIPERIDINE-4-CARBOXAMIDE

Metopimazine (188)

Metopimazine (**5.5.4**) (Vogalene) is a dopamine receptor antagonist that is used for the prevention and treatment of patients who consider nausea and vomiting as severe adverse events to chemotherapy. It has a high affinity for dopamine D2 receptors (and also α1-adrenoceptors and histamine H1 receptors) but no affinity for serotonin 5-HT3receptors. The most common side effects include constipation, diarrhea, abdominal pain, headache, and asthenia [201−204].

The original process of the synthesis of metopimazine (**5.5.4**) was carried by alkylation of 2-(methylsulfonyl)-10*H*-phenothiazine (**5.5.1**) with 1-bromo-3-chloropropane with use of sodium amide (liquid ammonia, sodium), followed by amination of obtained (**5.5.2**) with piperidine-4-carboxamide (**5.5.3**) on reflux in ethanol in the presence of sodium carbonate [205] (Scheme 5.22).

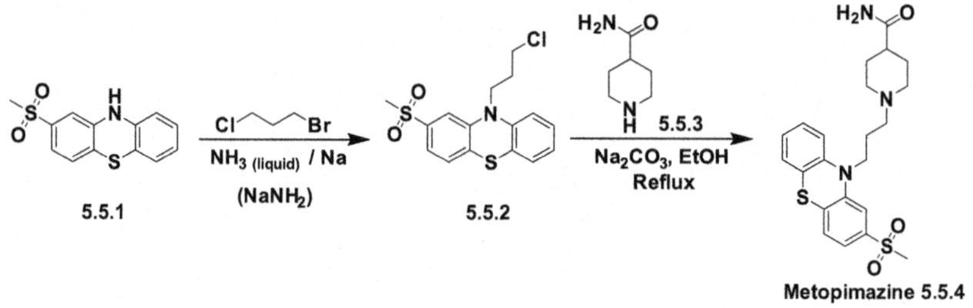

SCHEME 5.22   Synthesis of metopimazine.

Some modified methods of syntheses are proposed starting from 2-(methylthio)-10*H*-phenothiazine and further oxidation of methylthio- group to methylsulfonyl- [206] or alkylation of 2-((2-fluoro-4-(methylsulfonyl)phenyl)thio)aniline with 1-(3-chloropropyl)piperidine-4-carboxamide using NaH in DMSO which is accompanied with simultaneous cyclization to metopimazine (**5.5.4**) [207].

Pipamazine (110)

Pipamazine (**5.5.6**) (Mornidine) [208] chemically related to metopimazine (**5.5.4**) was formerly used as another antiemetic. It was introduced to the US market in 1959 and was eventually withdrawn from the market in 1969, after reports of hepatotoxity and hypotension [209]. There is very little published information on pipamazine.

Pipamazine (**5.5.6**) was synthesized by the *N*-alkylation of piperidine-4-carboxamide (**5.5.3**) with 2-chloro-10-(γ-chloropropyl)phenothiazine (**5.5.5**) on reflux in methyl ethyl ketone or ethyl-acetate in the presence of potassium carbonate and sodium iodide [205,210,211] (Scheme 5.23).

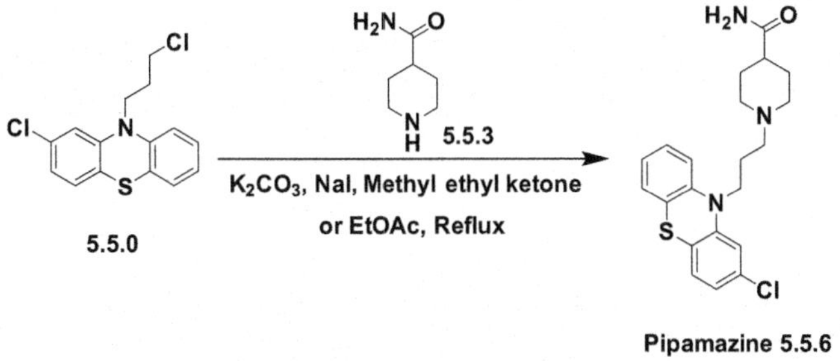

SCHEME 5.23   Synthesis of pipamazine.

5.6 DERIVATIVES OF PIPERIDIN-4-AMINES

Indoramin (700)

Indoramin (**5.6.4**) (Doralese) is a postsynaptic adrenergic antagonist (α1-selective blocker) and at the same time an antagonist of histamine H1 and 5-HT receptors as well. It is used to treat hypertension, benign prostatic hypertrophy, Raynaud's phenomenon, etc. Blood pressure lowering effect results from relaxation of peripheral arterioles as a consequence of a blockade of postsynaptic α1-adrenoceptors. Interestingly, unlike some other α-blockers, lowering of blood pressure by idoramin is rarely associated with reflex tachycardia or postural hypotension. Side effects are as follows: sedation, drowsiness, dizziness, failure of ejaculation, nasal congestion, headache, fatigue, weight gain, diarrhea, nausea, increased need to pass urine [212–218].

A number of synthetic approaches were developed for the synthesis of indoramine (**5.6.4**). The two major synthetic routes are depicted on the Scheme 5.24. The most satisfactory approach to the large scale preparation of indoramine involve alkylation of 4-benzamidopiperidine (**5.6.2**)

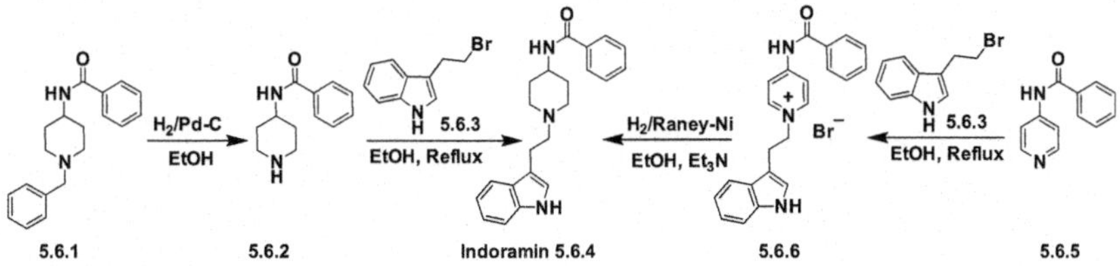

SCHEME 5.24   Synthesis of indoramin.

prepared by different ways, of which the most effective was debenzylation of *N*-(1-benzylpiperidin-4-yl)benzamide (**5.6.1**). The last gave desired indoramine (**5.6.4**) on reflux in ethanol with 3-(2-bromoethyl)indole (**5.6.3**) [219–222].

Another route involved quaternization of 4-benzamidopyridine (**5.6.5**) with 3-(2-bromoethyl)-indole (**5.6.3**), which gave salt (**5.6.6**), and was reduced further on Raney Ni catalyst [220,223–225]. This method proved satisfactory only on a small-scale process.

Clebopride (419)

Clebopride (**5.6.11**) (Cleboril) is a selective D2 receptors antagonist. It bounds to the D2 dopamine receptor with a high affinity and to the α2-adrenoceptor and 5-HT2 serotonin receptor with relatively lower affinity, and not to D1 dopamine, α1-adrenergic, muscarinic acetylcholine, H1 histamine, or opioid receptor.

Clebopride is a dopamine antagonist drug with antiemetic and prokinetic properties and used as an antiemetic, in chemotherapy-induced nausea and vomiting after administration of antitumour drugs, levodopa preparations, or after surgical operations and as an ulcer inhibitor [226–230]. Side effects include somnolence, diarrhea and extrapyramidal-like symptoms. The use of clebopride may be associated with a reversible or persistent parkinsonism syndrome and hemifacial dystonia; therefore, attention must be drawn to these possible side effects [231–233].

Clebopride (**5.6.11**) may be prepared by reacting the mixed anhydride prepared from 4-amino-5-chloro-2-methoxy benzoic acid and ethyl chloroformate – (4-amino-5-chloro-2-methoxybenzoic (ethyl carbonic) anhydride) (**5.6.8**) with 1-benzyl-4-amino-piperidine (**5.6.7**) in THF in the temperature range of 0–10°C. Another method is based on coupling of 4-amino-5-chloro-2-methoxy benzoic acid (**5.6.9**) with 1-benzyl-4-amino-piperidine (**5.6.7**) in the presence of *N*,*N*′-dicyclohexylcarbodiimide (DCC) in methylene chloride at room temperature. The third method employed condensation of methyl-2-methoxy-4-acetamido-5-chloro benzoate (**5.6.10**) with 1-benzyl-4-amino piperidine (**5.6.7**) on reflux in xylene, in the presence of aluminium isopropylate using a Vigreux distillation column until the theoretical quantity of methanol had distilled. Obtained in this case amide was hydrolyzed under reflux with a solution of concentrated hydrochloride acid to yield desired clebopride (**5.6.11**) [234–236].

Another proposed method for the synthesis of clebopride (**5.6.11**) implemented the Hofmann rearrangement of compound (**5.6.13**) initiated with bromine and potassium hydroxide in a water/DMSO mixture. A starting compound (**5.6.13**) was prepared by coupling 1-benzyl-4-amino-piperidine (**5.6.7**) with 4-carbamoyl-5-chloro-2-methoxy-benzoyl chloride (**5.6.12**) in 1,4-dioxane and Et₃N [237,238] (Scheme 5.25).

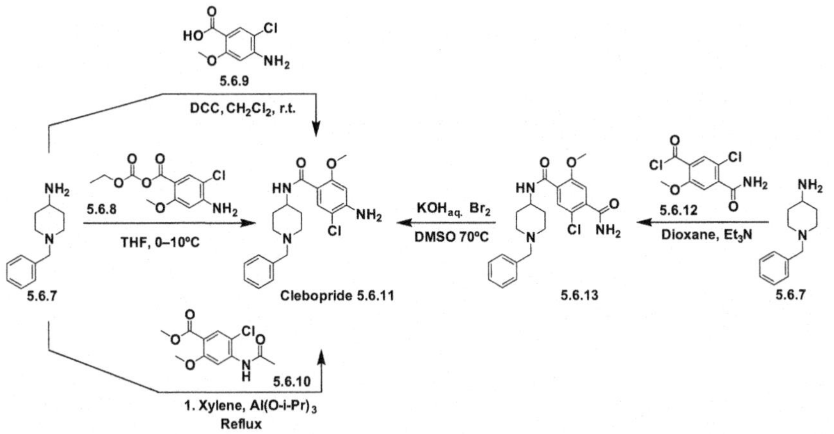

SCHEME 5.25 Synthesis of clebopride.

Cinitapride (115)

Cinitapride (**5.6.17**) has been marketed in Spain (Cidine) and Mexico (Pemix) as an antiemetic medication prescribed to treat conditions caused by an overactive digestive tract. It slows the actions of the muscles to reduce the symptoms of conditions such as acid reflux, ulcer dyspepsia, and delayed gastric emptying. It acts as an agonist of the 5-HT1 and 5-HT4 receptors and as an antagonist of the 5-HT2 and D2 receptors. The drug was effective in minimizing dyspepsia symptoms, and improving the quality of life of patients. It is well tolerated and is almost free of side effects. Nevertheless, side effects such as vomiting, nausea, diarrhea, drowsiness, headache, jaundice, muscle, and joint pain are sometimes observed [239−247].

Cinitapride (**5.6.17**) was prepared by condensation mixed anhydride (**5.6.15**) with 1-(cyclohex-3-en-1-ylmethyl)piperidin-4-amine (**5.6.16**) in DMF/THF mixture. (The mixed anhydride (**5.6.15**) in turn, was prepared by reaction of 2-ethoxy-4-amino-5-nitrobenzoic acid (**5.6.14**) with ethyl chloroformate in DMF in presence of trimethylamine at low temperature.) [248,249]. Another method involves *N*-alkylation of piperidylbenzamide (**5.6.18**) with cyclohexenylmethyl methane-sulfonate (**5.6.19**) on reflux in acetonitrile in the presence of potassium carbonate [250] (Scheme 5.26).

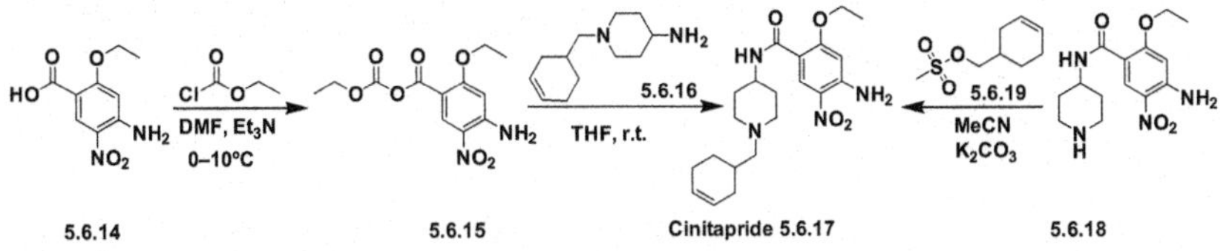

SCHEME 5.26 Synthesis of cinitapride.

Prucalopride (537)

Prucalopride (**5.6.23**) (Resolor) is a selective, high affinity serotonin (5-HT4A and 5-HT4B) receptor agonist which exhibits hundredfold exceeding selectivity against other 5-HT receptor subtypes but without the limiting side effects associated with dopamine D2 receptor antagonism, and has a strong gastrointestinal prokinetic activity mediated through the stimulation of colonic contractions that are closely associated with defecation [251,252]. Prucalopride improves bowel function and was safe and well tolerated and is approved for use in many countries in Europe and in Canada, for the symptomatic treatment of chronic functional constipation in adults in whom laxatives fail to provide adequate relief but is not approved by the US FDA. The most frequently reported adverse reactions are headache and gastrointestinal symptoms (abdominal pain, nausea or diarrhea). Adverse effects are common when initiating treatment [253–265].

Prucalopride (**5.6.23**) was prepared by *N*-acylation of commercial 1-(3-methoxypropyl)-piperidin-4-amine (**5.6.22**) with a mixed anhydride (**5.6.21**), in turn, prepared from 4-amino-5-chloro-2,3-dihydrobenzofuran-7-carboxylic acid (**5.6.20**) with ethyl chloroformate in DMF in the presence of trimethylamine at low temperature.

Another method for the synthesis of prucalopride (**5.6.23**) is that carboxamide (**5.6.24**) is alkylated with 1-chloro-3-methoxypropane (**5.6.25**) or any analogs alkylating reagent where the chloro- group may replaced for a bromo-, mesyl-, or tosyl- groups [266–268] (Scheme 5.27).

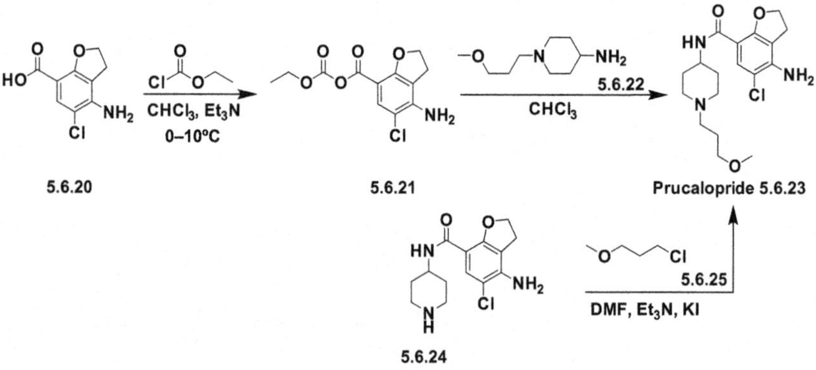

SCHEME 5.27   Synthesis of prucalopride.

A large amount of work dedicated to synthesis of dihydrobenzofuran-7-carboxylic acid (**5.1.20**) [269–272].

Lomitapide (241)

Lomitapide (**5.6.32**) (Juxtapid) is a new, oral inhibitor of microsomal triglyceride transfer protein (MTP), an enzyme located in the lumen of the endoplasmic reticulum responsible for absorbing dietary lipids and transferring triglycerides onto apolipoprotein B (apo-B) in the assembly of very-low-density lipoprotein (VLDL).

It is the first in a new class of drugs indicated as an adjunct to a low-fat diet and other lipid-lowering therapies to improve lipoproteins (total cholesterol, low-density lipoprotein and nonhigh-density lipoprotein cholesterol and (apo B) in patients with homozygous familial hypercholesterolemia (HoFH).

Lomitapide directly binds and inhibits (MTP), thereby preventing the assembly of apo B-containing lipoproteins in enterocytes and hepatocytes. This inhibits the synthesis of chylomicrons and VLDL. The inhibition of the synthesis of VLDL leads to reduced levels of plasma low-density lipoprotein (LDL). Most common adverse reactions of lomitapide are diarrhea, nausea, vomiting, dyspepsia, and abdominal pain. Because of the risk of hepatotoxicity, lomitapide is available only through a restricted program under a Risk Evaluation and Mitigation Strategy [273−280].

The synthesis of lomitapide (**5.6.32**) was carried out starting from 9*H*-fluorene-9-carboxylic acid (**5.6.26**), which was alkylated with 1,4-dibromobutane in the presence of *n*-butyl lithium in THF to give 9-(4-bromobutyl)-9*H*-fluorene-9-carboxylic acid (**5.6.27**). The last dissolved in dichloromethane was converted to intermediate acid chloride using phosgene and catalytic amount of dimethylformamide and then to amide (**5.6.28**) by coupling with (2,2,2-trifluoro-ethylamine). The obtained product was used for alkylation of 4-*N*-Boc-aminopiperidine (**5.6.29**) in a standard way, using potassium carbonate in dimethylformamide followed by removal of the Boc group with 4 N hydrochloric acid in dioxane, which gave amine (**5.6.30**). Amine (**5.6.30**) was then reacted with the acid chloride (**5.6.31**) derived from corresponding acid at 0°C dichloromethane in the presence of trimethylamine and catalytic amount of dimethylaminopyridine to give desired lomitapide (**5.6.32**) [281,282] (Scheme 5.28).

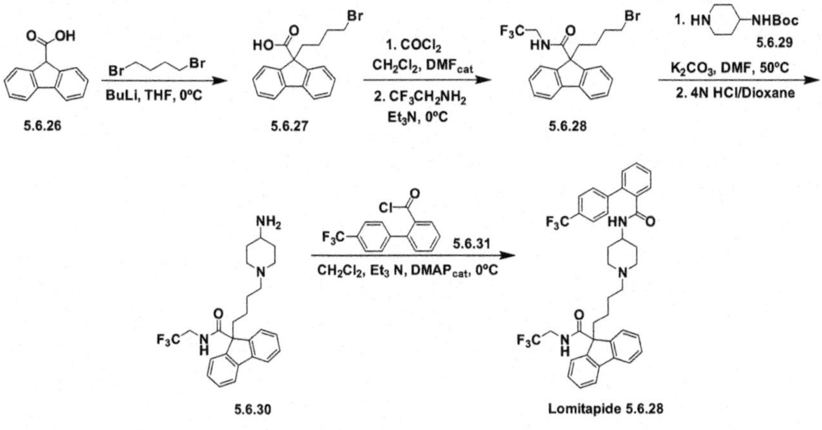

SCHEME 5.28 Synthesis of lomitapide.

Some slightly modified methods for the synthesis of lomitapide (**5.6.32**) are also proposed [283−285].

Cisapride (3766)

Cisapride (**5.6.45**) (Propulsid), an oral prokinetic agent belonging to the pharmacotherapeutic group of propulsives, acts as a serotonin 5-HT4 agonist. It was approved by the US FDA in 1993 for symptomatic relief of nocturnal heartburn due to gastroesophageal reflux disease in adults. It increases lower esophageal sphincter tone, accelerates gastric emptying, and increases small-bowel motility.

It was commonly used for the treatment of conditions that compromise stomach motility, prevent the movement of food and creates delayed gastric emptying, normalizes stomach contractions, and increases the movement of the gastrointestinal tract. It has direct effects on gallbladder and the sphincter of oddi, as well as indirect effects involving gastro-intestinal hormone levels, gastric emptying, gallbladder refilling, interdigestive migrating motor cycle and small intestinal transit [286−294]. Cisapride was used worldwide in the treatment of gastrointestinal motility-related disorders in premature infants, full-term infants, and children. Efficacy data suggest that it is the most effective commonly available prokinetic drug.

Common side effects of cisapride include abdominal pain, nausea, diarrhea, increased frequency of urination, constipation, gas, indigestion, runny or stuffy nose, cough, viral infection, upper respiratory tract infection, pain, fever, urinary tract infection, insomnia, anxiety, nervousness, rash, itching.

Premarketing large-scale trials did not report any serious side effects, but after a long period of availability in several countries it was withdrawn in 2000 because of reports of severe cardiac arrhythmias, and in some cases fatal, cardiac events [295,296].

One of the synthetic strategies involved the preparation of cisapride (**5.6.45**) started from bromination of *N*-benzyl-4-piperidone (**5.2.16**) in acetic acid and the obtained bromide (**5.6.33**) was treated with sodium methoxide. Under these conditions the acetal of *N*-benzyl-3-hydroxy-4-piperidone (**5.6.34**) was formed. The synthesized acetal (**5.6.34**) was converted into the urethane (**5.6.36**) by stepwise debenzylation via hydrogenation over palladium catalyst which gave secondary amine (**5.6.35**). *N*-Acylation of amine (**5.6.35**) with methyl chloroformate gave (**5.6.36**). *O*-Methylation of (**5.6.36**) to compound (**5.6.37**) followed by acidic hydrolysis of the ketal group gave the key methyl 3-methoxy-4-oxopiperidine-1-carboxylate (**5.6.38**). Reductive amination of the last using benzylamine yielded virtually exclusively (98%) the *cis*-aminoether (**5.6.39**) (Scheme 5.29).

Ketal of 3-methoxy-4-oxopiperidine-1-carboxylate (**5.6.37**) was proposed to be synthesized from corresponding piperidin-4-one, first converting it to methyl enol ether using trimethyl orthoformate and then to α-methoxy

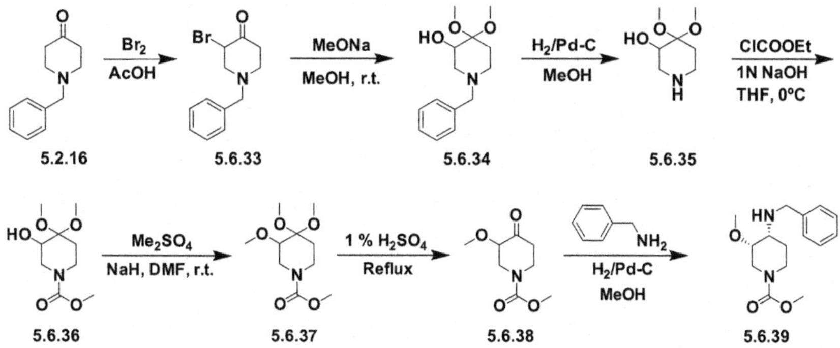

SCHEME 5.29 Synthesis of methyl (3*S*,4*R*)-4-(benzylamino)-3-methoxypiperidine-1-carboxylate.

dimethyl ketal with lead tetraacetate in the presence of boron trifluoride diethyl etherate [297].

At this point, selective deprotection of the two amine functions in (**5.6.39**) became possible and allowed the further development of two complementary synthetic strategies. Catalytic debenzylation of (**5.6.39**) gave 4-amino-3-methoxypiperidine-1-carboxylate (**5.6.40**), and basic hydrolysis gave *N*-benzyl-3-methoxypiperidin-4-amine (**5.6.41**) (Scheme 5.30).

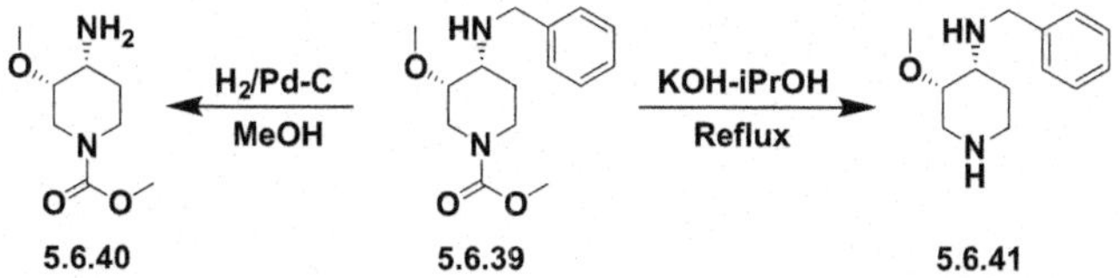

SCHEME 5.30 Selective deprotection of amine functions in methyl 4-(benzylamino)-3-methoxypiperidine-1-carboxylate.

N-Acylation of piperidin-4-amine (**5.6.40**) with the mixed anhydride (**5.6.8**) allowed to synthesize amide (**5.6.42**). Following selective hydrolysis of the urethane group with one equivalent of potassium hydroxide in iso-propanol gave compound (**5.6.43**) *N*-alkylation of which with 1-(3-chloropropoxy)-4-fluorobenzene (**5.6.44**) in dimethylformamide in presence of triethylamine gave desired cisapride (**5.6.45**).

In the other approach, *N*-benzyl-3-methoxypiperidin-4-amine (**5.6.41**) first was *N*-alkylated with 1-(3-chloropropoxy)-4-fluorobenzene (**5.6.44**) in the same conditions (DMF, Et₃N) to give product (**5.6.46**) followed by catalytic debenzylation on palladium catalyst to prepare amine (**5.6.47**) which was converted to desired cisaprid (**5.6.45**) by acylation with the mixed anhydride (**5.6.8**) in the same conditions [298–300] (Scheme 5.31).

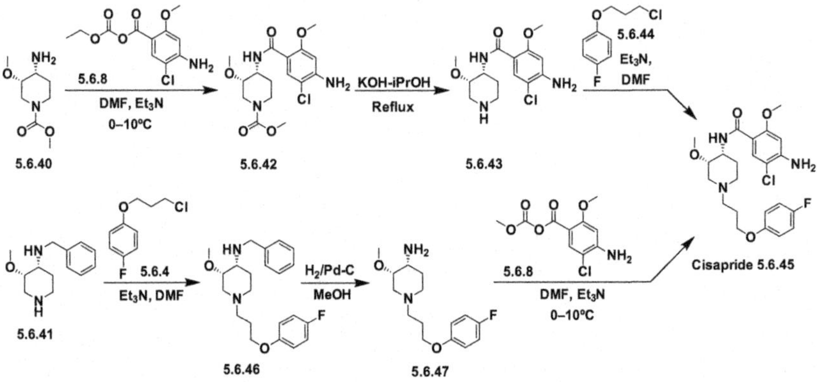

SCHEME 5.31 Synthesis of cisapride.

Another convenient method for preparation of cisapride (**5.6.45**) was proposed, starting with commercially available oxirane (**5.6.48**), which in reaction with sodium azide and almost regioselective, gave ethyl 4-azido-3-hydroxypiperidine-1-carboxylate (**5.6.49**). Hydroxyl group of (**5.6.49**) was methylated with methyl iodide using sodium hydride in dimethylformamide to give (**5.6.50**). After catalytic reduction to amine (**5.6.51**) the product underwent reductive alkylation with benzaldehyde producing the benzyl derivative (**5.6.39**), which was hydrolyzed to known (**5.6.41**) (Scheme 5.32) and then converted to cisapride (**5.6.45**) according to above described sequence (**5.6.41** → **5.6.46** → **5.6.47** → **5.6.45**) (Scheme 5.31) [298−300].

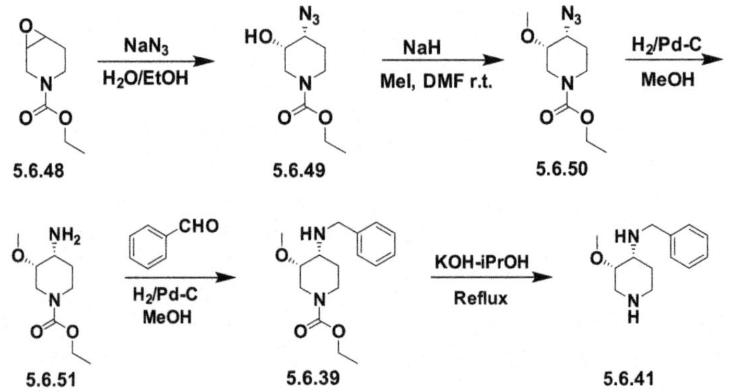

SCHEME 5.32 Synthesis of cisapride.

Modifications of described methods [301−303], as well as implementation of rearrangement reaction of 4-oxopiperidin-3-yl benzoates to *N*-(3-hydroxy-4-methoxypiperidin-4-yl)benzamides and further transformations of obtained product to cisapride are published [304].

Astemizole (2283)

Astemizole (**5.6.59**) (Hismanal), a long-acting, nonsedating H1-receptor antagonist, allowed for once-daily administration, is used to treat allergic disorders such as seasonal and perennial allergic rhinitis, uredo, etc. It has specific properties as easy absorption, fast onset of action, with a standing duration of several hours and with no side effects like drowsiness [305−310]. But due to its cardiotoxicity, an especially rare but potentially fatal arrhythmoginic effect, it has been withdrawn from the market in most countries.

Astemizole, can inhibit tumor cell proliferation ad has gained enormous interest since it also targets important proteins involved in cancer progression, namely, ether à-go-go 1 (Eag1) and Eag-related gene (Erg) potassium channels [311].

The synthesis of astemizole (**5.6.59**) was carried out in two general ways. Thus, the addition of 4-aminopiperidine-1-carboxylate (**5.6.52**) to 1-isothiocyanato-2-nitrobenzene in THF or ethanol at room temperature afforded the (2-nitrophenyl)thiourea (**5.6.53**), which was reduced to thiourea derivative by catalytic hydrogenation of the nitro function with palladium on charcoal. For conversion of obtained (**5.6.54**) to desired 4-((1*H*-benzo[*d*]imidazol-2-yl)amino)piperidine-1-carboxylate (**5.6.56**) two methods − *S*-methylation with methyl iodide followed by reflux in ethanol, or direct cyclodesulfurization using mercury(II) oxide in the presence of catalytic amount of sulfur in THF or ethanol at room temperature − were implemented. Both methods gave the same product (**5.6.56**), but yields were higher on use of mercury(II) oxide. Product (**5.6.54**) was prepared also from 4-isothiocyanatopiperidine-1-carboxylate (**5.6.55**), in turn, synthesized from 4-aminopiperidine-1-carboxylate (**5.6.52**) consistently reacting it with carbon disulfide in sodium hydroxide solution and then with ethyl formate, on reaction with *O*-phenylenediamine. Imidazole ring alkylation of (**5.6.56**) with 1-(chloromethyl)-4-fluorobenzene occurred under weak basic conditions (sodium carbonate, dimethylformamide) to give compound (**5.6.57**), which was deprotected by heating at reflux in 48% aqueous hydrogen bromide solution to give compound (**5.6.58**). Finally, *N*-alkylation of (**5.6.58**) with 1-(2-chloroethyl)-4-methoxybenzene resulted in the title compound astemizole (**5.6.59**) [312−316]. The third method consists of a direct arylation of 4-aminopiperidine-1-carboxylate (**5.6.52**) with 2-chloro-1-(4-fluorobenzyl)-1*H*-benzo[*d*]imidazole (**5.6.60**) at 125°C during 5 hours [317] (Scheme 5.33).

5.7 DERIVATIVES OF 4-ANYLINOPIPERIDINES

The 4-anilidopiperidines are the most potent class of opioid analgesics known to date. The prototype of this class, fentanyl (**5.7.1**), is about 300 times more potent than morphine in mice and rats, compared to 50−100 times in humans [130,318−327]. A very large number of fentanyl analogs −

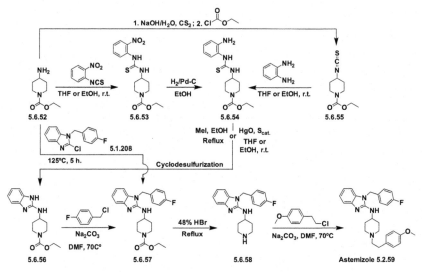

SCHEME 5.33 Synthesis of astemizole.

mefentanyl (**5.7.2**), phenaridine (**5.7.3**), ohmefentanyl (**5.7.4**), ocfentanil (**5.7.5**), brifentanil (**5.7.6**), mirfentanil (**5.7.7**), carfentanil (**5.7.8**), lofentanil (**5.7.9**), thiophentanil (**5.7.10**), remifentanil (**5.1.11**), sufentanyl (**5.7.12**), alfentanyl (**5.7.13**), trefentanil (**5.7.14**) − have been created and are available in the pharmaceutical market (Fig. 5.5).

Fentanyl (31404)

Fentanyl (**5.7.1**) (Fentora) is a potent analgesic that was first synthesized more than 40 years ago. It is still the most popular opioid used in perioperative periods throughout the world. In spite of the development of more potent, faster onset opioids fentanyl remains the mainstay of anesthesiologists. Fentanyl's popularity have resulted in its use in many acute and chronic pain conditions. The clinically useful pharmacologic effects of the interaction of fentanyl are analgesia and sedation. Side effects may include somnolence, hypoventilation, bradycardia, postural hypotension, pruritus, dizziness, nausea, diaphoresis, flushing, euphoria and confusion. Fentanyl is a Schedule II controlled substance that can produce drug dependence pain. The story of fentanyl began in late 1950s/early 1960s [328−331].

The first synthesis of fentanyl [328] started with the condensation of 1-benzypiperidin-4-one (**5.2.16**) with aniline on reflux in toluene forming the corresponding imine (**5.7.15**), which with lithium aluminum hydride was reduced to 1-benzyl-4-anilinopiperidine (**5.7.16**). The resulting product was acylated with propionic anhydride in toluene to give (**5.7.17**). The obtained 1-benzyl-4-*N*-propinoylanilinopiperidine (**5.7.17**) was debenzylated on hydrogenation over a palladium on carbon catalyst to give 4-*N*-

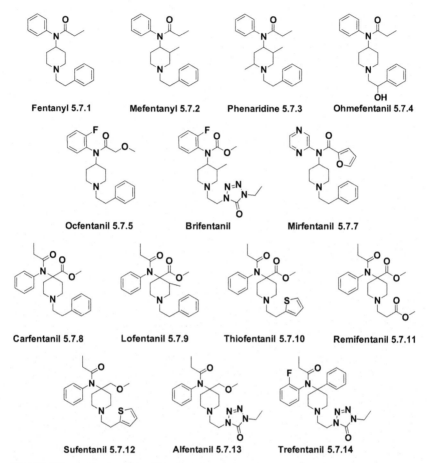

FIGURE 5.5 Analgesics of 4-anilidopiperidines series.

propanoylanilinopiperidine **(5.7.18)**, which was *N*-alkylated by 2-phenylethylchloride or tosylate in MIBK in the presence of potassium iodide, using sodium carbonate as a base to give desired fentanyl **(5.7.1)** [328–331]. Later, a modified and shorter synthesis of fentanyl was derived by the same method, but starting directly from 1-(2-phenethyl)piperidin-4-one **(5.7.19)** → **(5.7.20)** → **(5.7.21)** → **(5.7.1)** has been proposed [332]. Another way, starting from 4-anilinopyridine **(5.7.22)**, was also proposed [333]. According to this method 4-anilinopyridine **(5.7.22)** was acylated with propionic anhydride to give 4-*N*-propinoylanilinopyridine **(5.7.23)**, which was alkylated with 2-phenylethylbromide to give pyridinium salt **(5.7.24)**. Hydrogenation of the last over PtO$_2$ gave 4-*N*-anilinopiperidine derivative **(5.7.21)** which was acylated with propionic anhydride to give desired fentanyl **(5.7.1)** (Scheme 5.34).

Alternative synthesis of fentanyl have been patented [334,335]. A review on fentanyl-related compounds was recently published by us [336].

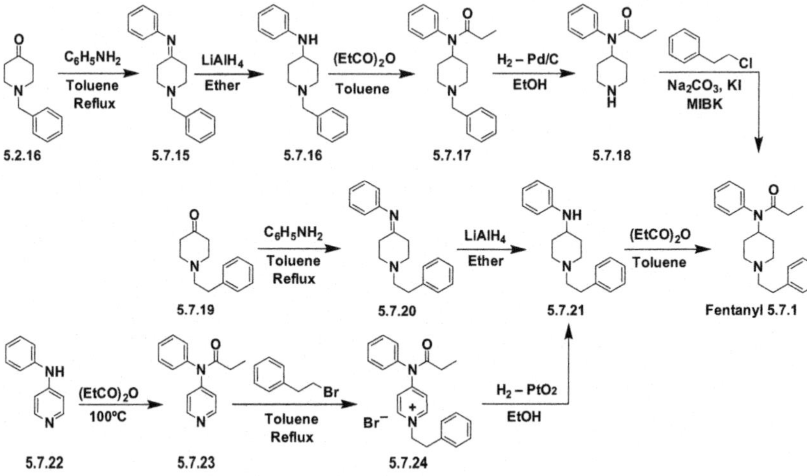

SCHEME 5.34   Synthesis of fentanyl.

The four most popular compounds in medicinal practice are fentanyl (**5.7.1**), remifentanil (**5.7.11**), sufentanil (**5.7.12**), and alfentanil (**5.7.13**). These are considered "tailor made" analgesics designed for specific purposes. The analgesic potency of fentanyl, 300 times higher than that of morphine in the tail withdrawal test in rats, had been enhanced up to 10,000 times than that of morphine in the case of carfentanil (**5.7.8**) one of the most potent known analgesics used as a general anesthesia agent for large animals.

A huge amount of work on structural modifications of fentanyl have been published. Among the changes of practical meaning include the replacement of piperidine ring for other heterocyclic rings as well as synthesis of open chain compounds, replacement of phenyl group in phenethyl- part of molecule for some aromatic heterocyles, insertion of additional methyl groups into the different positions of piperidine ring, and the introduction of fluorine atom into aniline- part of molecule, all of which are reviewed by us [336] and summarized in (Fig. 5.5).

The most sensible change was observed on the attachment of additional methoxymethyl (alfentanil, sufentanil) or metoxycarbonyl groups (lofentanil, carfentanil) into the fourth position of the piperidine ring (Fig. 5.6).

Carfentanil (614)

Carfentanil (**5.7.8**) is an ultra-potent with a clinical potency 10,000 times that of morphine and 100 times that of fentanyl. Its only approved use is by veterinarians as a tranquilizing agent to rapidly incapacitate exotic wildlife for examination and procedures [337,338]. Carfentanil is a Schedule II narcotic controlled substance in the United States.

The first synthesis of carfentanil (**5.7.8**) started from the Strecker reaction of 1-benzypiperidin-4-one (**5.2.16**) with aniline and potassium cyanide in in aqueous acetic acid to give 1-benzyl-4-(phenylamino)piperidine-4-

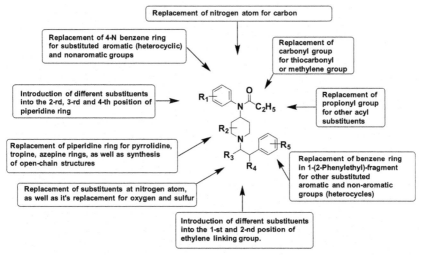

FIGURE 5.6   Variations of fentanyl structure.

carbonitrile (**5.7.25**). Obtained nitrile was hydrolyzed with cold concentric sulgiric acid to the corresponding amide (**5.7.26**). This compound was hydrolyzed further with KOH in refluxing ethylene glycol and coverted to sodium salt (**5.7.27**) and methylated with methyl iodide in dimethyl formamide (DMF) or hexamethylphosphoric triamide (HMPA), or dimethyl sulfoxide (DMSO) to give 4-(phenylamino)piperidine-4-carboxylate (**5.7.28**). Obtained product was acylated with propionic anhydride at reflux temperature to give compound (**5.7.29**). The last was debenzylated by hydrogenation over palladium catalyst and obtained (**5.7.30**) was alkylated with 2-phenylethyl bromide to give desired carfentanil (**5.7.8**) [339−341] (Scheme 5.35).

SCHEME 5.35   Synthesis of carfentanil.

The same method was implemented for the synthesis of carfentanil starting directly from 1-(2-phenethyl)piperidin-4-one (**5.7.19**) avoiding steps of debenzylation and further alkyltion [342].

Another process has been patented for the synthesis of carfentanil. The first step, the Strecker reaction of 1-(2-phenethyl)piperidin-4-one (**5.7.19**) with aniline and potassium cyanide in an aqueous acetic acid, is practicaly identical with the above-described (**5.2.16**) → (**5.7.25**) coversion and gives 1-phenethyl-4-(phenylamino)piperidine-4-carbonitrile (**5.7.31**), which on reaction with formic acid in acetic anhydride gave N-(4-cyano-1-phenethylpiperidin-4-yl)-N-phenyl-formamide (**5.7.32**). Formylation of the aminonitrile was followed by imidate (**5.7.33**) formation, which was carried on reflux of formamide (**5.7.32**) in anhydrous methanol saturated with hydrogen chloride and resulted in methyl 1-phenethyl-4-(phenylamino)piperidine-4-carbimidate (**5.7.33**). The obtained product was converted to aminoamide (**5.7.34**) in dilute sodium hydroxide solution and then after hydrolysis to (**5.7.35**), esterification to (**5.7.36**) and propionylation to carfentanil (**5.7.8**) by direct analogy with the above described sequence (**5.7.28**) → (**5.7.29**) → (**5.7.30**) → (**5.7.8**) on the Scheme 5.35 [343] (Scheme 5.36).

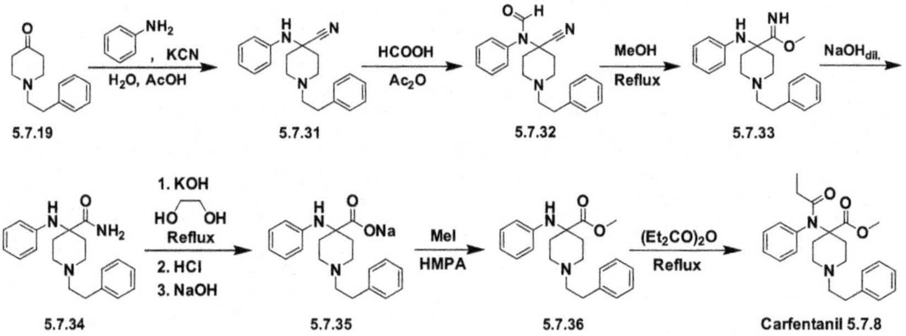

SCHEME 5.36 Synthesis of carfentanil.

The transformation of aminoamides (**5.7.34**) or (**5.7.25**) to corresponding methyl esters (**5.7.36**) or (**5.7.28**) has been proposed to carry out via conversion of aminoamides to an intermediate imidazolone derivative such as (**5.7.37**) using N,N-dimethylformamide dimethyl acetal in methanol and heating the mixture of reagents at 65°C, and further cleavage of obtained imidazolone derivative to methyl ester by heating in methanol with 1.2−1.4 M equivivalents of sulfuric acid at 95°C in a pressure bottle as it is shown on the Scheme 5.37 for nitrile (**5.7.34**) [344] (Scheme 5.37).

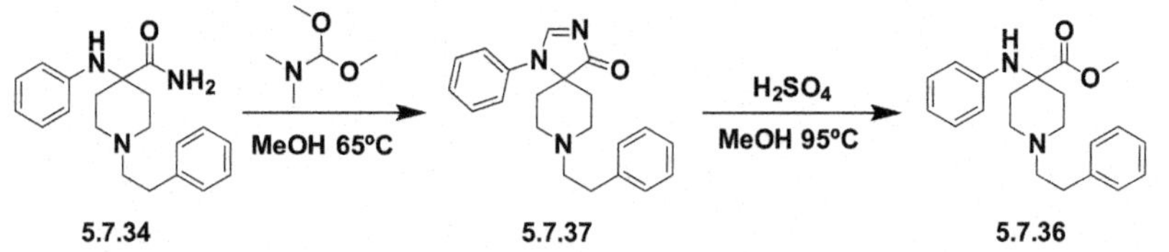

SCHEME 5.37 Synthesis of methyl 1-phenethyl-4-(phenylamino)piperidine-4-carboxylate.

Remifentanil (8289)

Remifentanil (**5.7.11**) (Ultiva), is a relatively new ultrashort action opioid characterized by a rapid onset and ultrashort predictable duration of action providing intense analgesia without prolonged respiratory depression. It is twice as potent as fentanyl and 100−200 times more potent than morphine. It was developed in the early 1990s, introduced into clinical use in 1996, and approved for perioperative use in many countries throughout the world. Remifentanil is a pure agonist at the μ opioid receptor with relatively little binding at the κ, σ, or δ receptors. This is precisely the same profile as the other opioids currently popular in anesthetic practice (fentanyl, alfentanil, sufentanil) and it offers the same advantages (profound analgesia, sedation, attenuation of the stress response). This has led to widespread use of remifentanil as an adjunct to general anesthesia in a variety of clinical settings. It is indicated to provide analgesia and sedation in mechanical ventilated intensive care unit patients and can be recommended even in patients with poor cardiovascular function. Remifentanil has been implicated in the causation of intraoperative bradyarrhythmias and asystole both in adults and in pediatric patients. Remifentanil has protective effects against ischemia-reperfusion injury in vital organs.

A substantial body of evidence supports the use of remifentanil for labor analgesia and cesarean section.

Remifentanil undergoes rapid hydrolysis by non-specific esterases in plasma and tissues. Unlike older opioids, remifentanil clearance is minimally affected by age, gender, body weight, hepatic or renal function, or duration of administration [345−361]. Most common side effects are: blurred vision, difficult or troubled breathing, dizziness, chest pain or discomfort, faintness, confusion, or lightheadedness when getting up from a lying or sitting position suddenly, irregular, fast or slow, or shallow breathing, muscle stiffness or tightness.

Michael-addition reaction was employed for the synthesis of remifentanil (**5.7.11**), which was carried out by the addition of methyl acrylate to a solution of methyl 4-(N-phenylpropion-amido)piperidine-4-carboxylate (**5.7.30**) in acetonitrile at room temperature followed by stirring at 50°C for 2 hours [362,363] (Scheme 5.38).

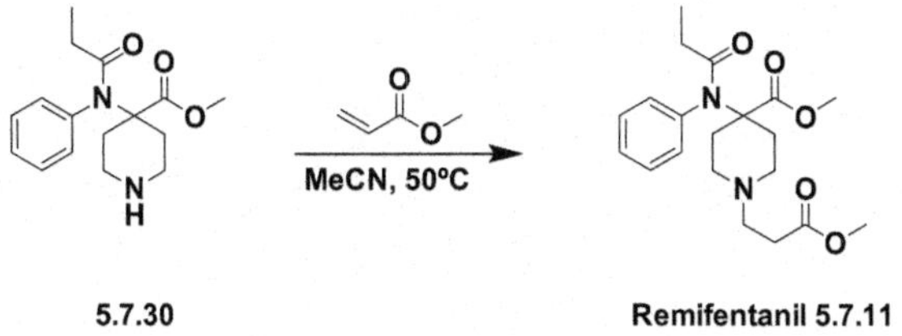

SCHEME 5.38 Synthesis of remifentanil.

Sufentanil (5866)

Sufentanil (**5.7.12**) (Sufenta) is a strong opioid and has been reported to be 6–10 times more potent than fentanyl, depending on the route of administration.

Although morphine and fentanyl remain the predominant epidural opioids, sufentanil offers some unique advantages. Sufentanil has a faster onset of action and longer duration than epidural fentanyl. Compared with morphine, sufentanil has been associated with a lower incidence of side effects, particularly delayed respiratory depression. It is widely used in anesthesia induction, anesthesia maintenance, and pain management. The safety range of sufentanil is large, the adverse reaction and inhibition of respiration is slight, and the analgesic time is longer than the inhibition of respiration. Although sufentanil is not yet approved for chronic pain management, it is being explored for chronic pain therapy [364–373].

Sufentanil (**5.7.12**) was first synthesized starting from the above-described methyl 1-benzyl-4-(phenylamino)piperidine-4-carboxylate (**5.7.28**), which underwent reduction with sodium dihydro-bis(2-methoxyethoxy)aluminate (Red-Al) in benzene to give 4-anilino-1-benzyl-4-piperidinemethanol (**5.7.38**). The obtained product was O-methylated with methyl iodide in hexamethylphosphoric triamide (HMPA) at room temperature using sodium hydride dispersion, which gave 1-benzyl-4-(methoxymethyl)-N-phenylpiperidin-4-amine (**5.7.39**).

Product (**5.7.39**) was acylated with propionic anhydride at reflux and obtained amide (**5.7.40**) hydrogenated on a palladium catalyst to remove N-benzyl group and give N-(4-(methoxymethyl)-piperidin-4-yl)-N-phenylpropionamide (**5.7.41**). The last was coupled with 2-(2-thienyl)ethyl mesylate (**5.7.42**) on reflux in MIBK in the presence of sodium carbonate to give desired sufentanil (**5.7.12**) [374,375] (Scheme 5.39).

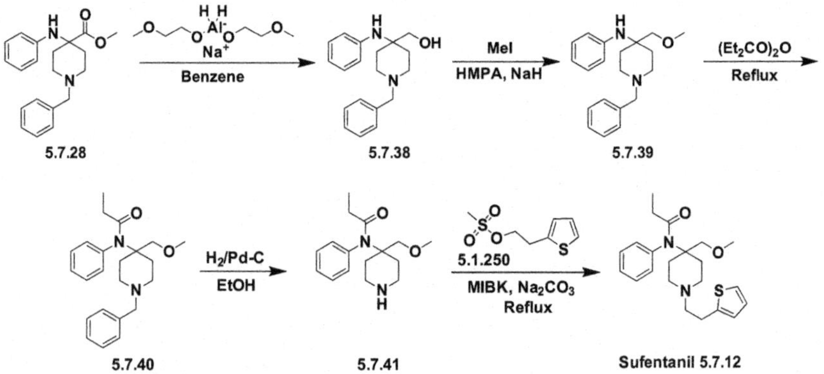

SCHEME 5.39 Synthesis of sufentanil.

Some modifications of this method have been proposed [376−378]. An alternative synthetic route for preparation of sufentanil (**5.7.12**), without describing the details of key steps has been published. The process started from specially prepared, for this purpose, 1-(2-(thiophen-2-yl)ethyl)piperidin-4-one (**5.7.43**), which was converted to oxirane (**5.7.44**) with the use of trimethylsulfonium iodide in the presence of a base − sodium methylsulfinyl-methylide ("dimsyl sodium") in tetrahydrofuran.

Reaction of the last with aniline in methylene chloride in presence of triethyloxonium tetrafluoroborate as Lewis acid turned out to be the best choice for the highest regioselection (1.8:1) in favor of (**5.7.46**). Separation of by column chromatography gave piperidylmethanol (**5.7.45**) with 6.2% yield, and hydroxypiperidine (**5.7.46**) with 4.8% yield (milligram scale). Piperidylmethanol (**5.7.46**) was methylated with diazomethane in ether and the product (**5.7.47**) which was further acylated to desired sufentanil (**5.7.12**) on reflux with with propionic anhydride [379] (Scheme 5.40).

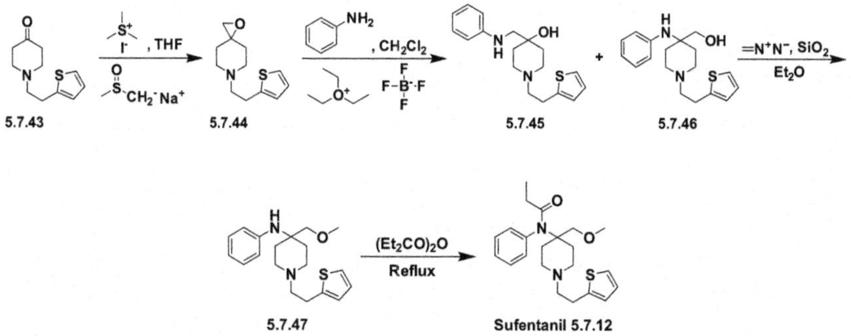

SCHEME 5.40   Synthesis of sufentanil.

Alfentanil (4090)

Alfentanil (**5.7.13**) (Alfenta) is a short- and a fast-acting acting opioid analgesic with around 1/4 to 1/10 the potency of fentanyl and around 1/3 of the duration of action. The onset of action of alfentanil is more rapid than that of an equianalgesic dose of fentanyl and the maximal analgesic and respiratory depressant effect occurs within 1−2 minutes. The duration of action of alfentanil is shorter than that of fentanyl, and is dose-dependent. Alfentanil is effective as an anesthetic during surgery, for supplementation of analgesia during surgical procedures, and as an analgesic for critically ill patients. It produces fast peak analgesic effect and recovery of consciousness [380−385].

Respiratory depression, skeletal muscle rigidity, hypotension, bradycardia, arrhythmia, tachycardia, dizziness, skeletal muscle movements, apnea, chest wall rigidity are characteristic side effects for alfentanil, as well as for entire 4-anilidopiperidine class of analgetics.

A suspension of the above-described *N*-(4-(methoxymethyl)-piperidin-4-yl)-*N*-phenylpropionamide (**5.7.41**) and 1-(2-bromoethyl)-4-ethyl-1,4-dihydro-5*H*-tetrazol-5-one (**5.7.48**) in methyl iso-butyl ketone in the presence sodium carbonate and potassium iodide was stirred and refluxed overnight to give desired alfentanil (**5.7.13**) [386,387] (Scheme 5.41).

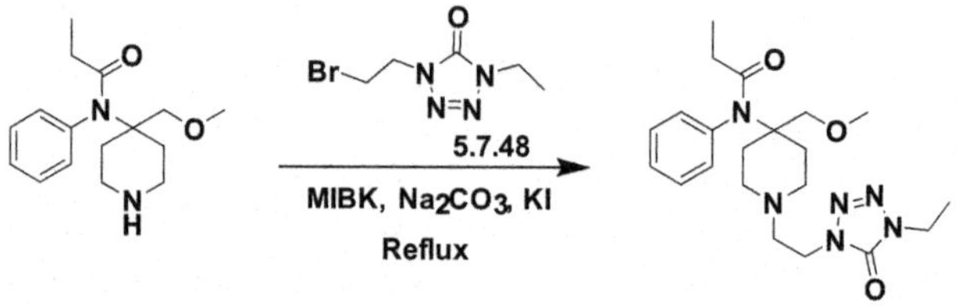

SCHEME 5.41 Synthesis of alfentanil.

New synthetic methods continue to be developed for the creation of new representatives of 4-anilidopiperidine class of analgetics [388].

Lorcainide (358)

Lorcainide (**5.7.53**), a class I antiarrhythmic drug, was reported to be highly efficient for the treatment of ventricular arrhythmias, especially ventricular extrasystoles and recurrent ventricular tachycardia. It is also efficient in the treatment of supraventricular extrasystoles and repetitive auricular tachycardia and ineffective in cases of auricular fibrillation and flutter. Lorcainide is/was used to treat patients with premature ventricular contractions and Wolff–Parkinson–White syndrome. It was reported that lorcainide has a small negative inotropic effect that was not clinically relevant in the patient group that was studied. The antiarrhythmic actions of lorcainide are mediated by an impairment of fast sodium conductance [389–395].

In 1993, a controlled trial report appeared of the antiarrhythmic drug lorcainide in heart attack which showed increased death rate occurred in the lorcainide treated group [396,397].

The development of lorcainide was abandoned.

Two methods for synthesis were proposed for lorcainide (**5.7.53**). In one of them a mixture of 4-oxo-1-piperidinecarboxylate (**5.2.21**) and 4-chloroaniline in toluene and a few crystals of toluenesulfonic acid under reflux gave 4-[(4-chlorophenyl)imino]-1-piperidinecarboxylate (**5.7.49**), which was reduced in methanol with sodium borohydride to give 4-[(4-chlorophenyl)amino]-1-piperidinecarboxylate (**5.7.50**). The last was acylated with phenylacetyl chloride at reflux in benzene to give amide (**5.7.51**). The *N*–protecting group was removed on reflux with 48% hydrobromic acid solution 48% forming (**5.7.52**). The last was alkylated with iso-propyl bromide on reflux in MIBK in the presence of sodium carbonate along with a few crystals of potassium, giving desired lorcainide (**5.7.53**) [398].

According to another patent the synthesis of lorcainide (**5.7.53**) started from 1-iso-propyl-4-piperidinone (**5.7.54**), which was condensed with 4-chloroaniline in toluene on reflux with few drops of acetic acid yielding imine (**5.7.55**). The last was reduced in standard way with sodium borohydride in methanol to give amine (**5.7.56**). Acylation of the last with phenylacetyl chloride at reflux in MIBK gave lorcainide (**5.7.53**) [398−400] (Scheme 5.42).

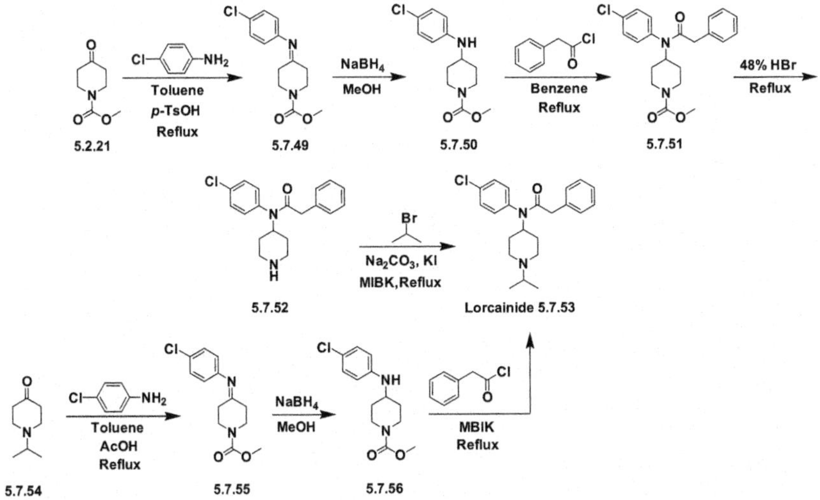

SCHEME 5.42 Synthesis of lorcainide.

5.8   DERIVATIVES OF 4-(*N,N*-DISUBSTITUTED)-PIPERIDINES

Bamipine (189) and Thenalidine (79)

4-Aminopiperidines are substances of sustaining interest, because of their manifold pharmacological activities.

Bamipine (Soventol) (**5.8.4**), a classic antihistaminic that blocks only the histamine H1 receptors, is a 4-aminopiperidine derivative that was and is used as antiallergic ointment in conditions like pruritus and urticaria [401,402]. It's closely related analog, thenalidine (**5.8.5**) [403], has the same pharmacological spectra of action and was withdrawn from the United States, Canadian, and United Kingdom markets in 1963 due to a risk of neutropenia.

Bamipine (**5.8.4**) was synthesized starting from 1-methylpiperidone-4 (**5.8.1**), which was condensed with aniline on reflux in toluene in the presence of several drops of acetic acid to give imine (**5.8.2**), which was reduced to amine (**5.8.3**) using activated borings of aluminium in methanol. The obtained amine was alkylated with benzyl chloride in the presence of sodium amide in boiling benzene to give desired product (**5.8.4**) [404,405]. An analogous method was employed for the synthesis of another antihistaminics, thenalidine (**5.8.5**), using a Raney-Nikel catalyst for reduction of imine (**5.8.2**), and 2-(chloromethyl)thiophene for alkylation of amine (**5.8.3**) [406,407] (Scheme 5.43).

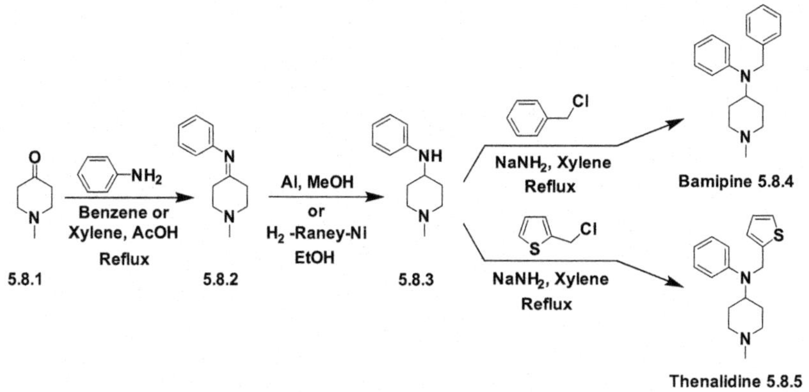

SCHEME 5.43   Synthesis of bamipine and thenalidine.

Mizolastine (573)

Mizolastine (**5.8.12**) (Mizollen) is a second-generation, long-acting H1-anti-histamine indicated for the symptomatic relief of seasonal allergic rhinocon-junctivitis (hay fever), and for seasonal and perennial allergic rhinitis and rhinoconjunctivitis. Mizolastine has a quick and long antihistamine effect and has been widely used in dermatology for treatment of urticarial, eczema, and various dermatoses.

It is a potent H1 receptor antagonist, inhibits 5-lipoxygenase, and has no anticholinergic, antiadrenergic, or antiserotonin activity. Mizolastine inhibits the role of other inflammatory mediators, inhibits leukotriene production, and reduces edema [408−412]. Side effects can include dry mouth and throat, sleepiness, drowsiness, tiredness, stomach ache, nausea and diarrhea.

Three approaches are proposed for the synthesis of mizolastine (**5.8.12**).

According to the first method, piperidin-4-yl carbamate (**5.8.6**) was con-densed with 2-chloro-1-(4-fluorobenzyl)-1*H*-benzo[*d*]imidazole (**5.8.7**) by heating to 140°C in methanol/sodium methylate medium, and the obtained compound (**5.8.8**) was then alkylated with methyl iodide in the presence of sodium hydride in dimethylformamide to give compound (**5.8.9**). The last was hydrolyzed on reflux in hydrobromic acid/acetic acid medium to amine (**5.8.10**), which reacted with *S*-methylthiouracil (**5.8.11**) on heating to 170°C for 10 hours, which gave desired mizolastine (**5.8.12**).

In another method *N*-methylpiperidin-4-amine (**5.8.13**) was alkylated with 2-chloro-1-(4-fluorobenzyl)-1*H*-benzo[*d*]imidazole (**5.8.7**) by refluxing in iso-amyl alcohol in the presence of potassium carbonate for 192 hours to give compound (**5.8.10**), which was converted to mizolastine (**5.8.12**) with methylthiouracil (**5.8.11**) using the above-described method. Finally, the third method was started from 4-(methylamino)piperidine-1-carboxylate (**5.8.14**), which on reaction with methylthiouracil (**5.8.11**) gave compound (**5.8.15**) further hydrolyzed with hydrobromic acid in acetic acid to compound (**5.8.16**) and then converted to mizolastine (**5.8.12**) by alkylation with 2-chloro-1-(4-fluorobenzyl)-1*H*-benzo[*d*]imidazole (**5.8.7**) [413,414] (Scheme 5.44).

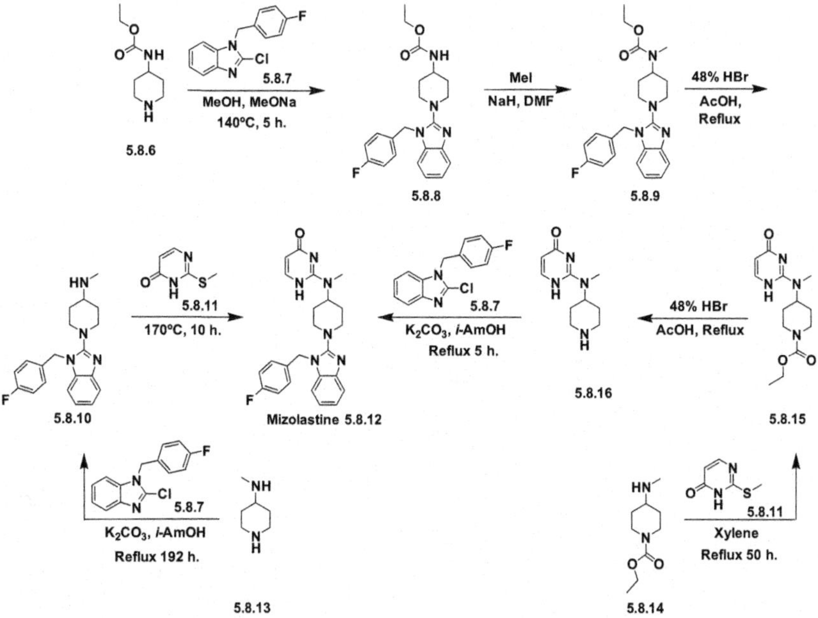

SCHEME 5.44 Synthesis of mizolastine.

Sabeluzole (123)

Sabeluzole (**5.8.22**) is a medication for the treatment of patients suffering from chronic neuro-degenerative diseases such as dementia of the Alzheimer type or Alzheimer's disease, amyotrophic lateral sclerosis, dementia associated with Parkinson's disease, and other central nervous system diseases that are characterized by progressive dementia. Sabeluzole is a drug that seems to protect hippocampal neurons from glutamate- and NMDA–induced toxicity. Except for cognitive-enhancing, it was also described to have antiischemic, antiepileptic properties [415–422].

The compound was well-tolerated and there were no major side effects.

The synthetic strategy for the synthesis of sabeluzole (**5.8.22**) is based on two available compounds: ethyl 4-isothiocyanatopiperidine-1-carboxylate (**5.6.55**) or ethyl-4- methylamino-1-piperidinecarboxylate (**5.8.17**).

A mixture of isothiocyanate (**5.6.55**) and *N*-methylaniline was stirred and refluxed in toluene overnight yielding ethyl 4-[[(methylphenylamino)thioxo-methyl]amino]-1-piperidinecarboxylate (**5.8.18**). The same compound was synthesized via condensation of phenylisothiocyanate with ethyl-4-methylamino-1-piperidinecarboxylate (**5.8.17**) in isopropyl ether at room temperature giving the same compound (**5.8.18**). Oxydative cyclization of obtained arylthiourea derivative (**5.8.18**) with bromine in tetrachloromethane forms a 2-aminobenzothiaozle derivative (**5.8.19**). Further decarbethoxylation in 48% HBr resulted in the formation of the compound (**5.8.20**). Coupling of the compound (**5.8.20**) with 4-fluorophenoxymethyloxyrane (**5.8.21**) in boiling toluene/methanol media gave desired racemic (1:1) sabeluzole (**5.8.22**) [423,424] (Scheme 5.45).

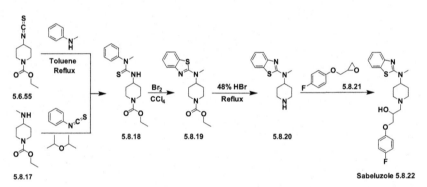

SCHEME 5.45   Synthesis of sabeluzole.

Lubeluzole (198)

Lubeluzole (**5.8.26**) is a novel neuroprotective compound closely related to sabeluzole (**5.8.22**) and its reported effects such as inhibition of glutamate release, inhibition of nitric oxide synthesis, and blockage of voltage-gated sodium and calcium ion channels, suggest a neuroprotective action of this drug [425−427]. Moreover, it was shown to have the efficacy and safety of lubeluzole in the treatment of ischemic stroke. Lubeluzole is safe, effective, and well tolerated at low doses, but unfortunately higher doses can produce a dangerous cardiac side effect [428−432].

Lubeluzole (**5.8.26**) has been synthesized by condensing *N*-methyl-*N*-(piperidin-4-yl)benzo[*d*]thiazol-2-amine (**5.8.20**) with 2-((3,4-difluorophe-noxy)methyl)oxirane (**5.8.24**) on reflux in iso-propyl alcohol in the presence of sodium carbonate [433], or by alkylation of the same compound (**5.8.20**) with 1-chloro(tosyl-, mesyl-)-3-(3,4-difluorophenoxy)propan-2-ol (**5.8.25**) in standard conditions. Amine (**5.8.20**), in turn, can be synthesized by the method proposed for the synthesis of sabeluzole [423], or by alkylation of ethyl 4-(methylamino)piperidine-1-carboxylate (**5.8.17**) with 2-chlorobenzo [d]thiazole (**5.8.23**) [434,435] (Scheme 5.46).

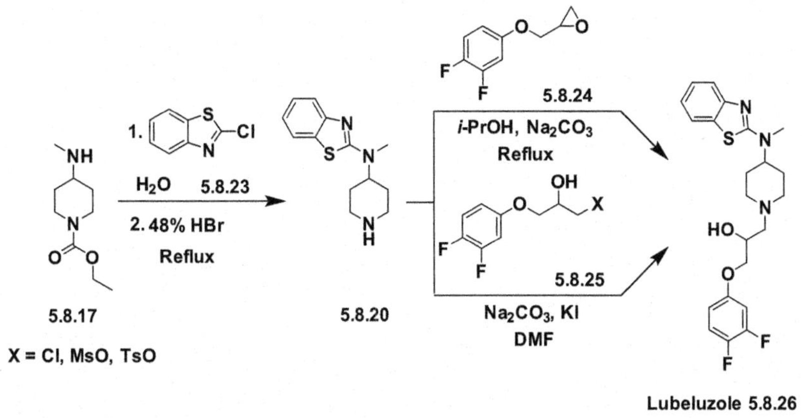

SCHEME 5.46   Synthesis of lubeluzole.

5.9   DERIVATIVES OF 4-(DIALKYLAMINO)PIPERIDINE-4-CARBOXAMIDE

Pipamperone (438)

Pipamperone (**5.9.4**) (Dipiperon) is one of the early butyrophenone derivatives whose pharmacological and clinical profile was distinct from haloperidol and all other known antipsychotic drugs at the time was introduced in 1961. Pipamperone is a typical antipsychotic and a prominent serotonine antagonism demonstrating strong antagonistic affinity at 5-HT2 and D4 receptors, and moderate affinity at D2, a1, and a2 receptors [436,437].

Pipamperone is administered to reduce psychomotor agitation and psychotic conditions in schizophrenic psychoses, to prevent impulsive aggressive behavior in patients with intermittent explosive disorder, personality disorders or mental retardation, symptomatic treatment of serious forms of agitation and anxiety. Pipamperone is known to improve disturbed sleep, social withdrawal, and other symptoms of chronic schizophrenia in the relative absence of extrapyramidal symptoms. It is frequently prescribed for the treatment in children with autism [438–441]. Use of pipamperone is contraindicated in depression.

Despite its "old age," pipamperone remains an intriguing compound in psychopharmacology.

Pipamperone can cause a wide and varied range of side effects because of its affinity for serotonergic, dopaminergic, adrenergic, and histaminergic receptors cases of dyskinesia, hallucinations, loss of visual acuity, color vision, hypothermia, priapism, all of which have observed.

Pipamperone (**5.9.4**) belongs to the class of amino amides, which in general can be prepared starting from 1-benzylpiperidin-4-one (**5.2.16**). The last underwent Strecker synthesis on treatment with potassium cyanide and piperidine hydrochloride to give aminonitrile (**5.9.1**), which was hydrolyzed to amino amide (**5.9.2**) on heating in concentric sulfuric acid. The obtained product was debenzylated on hydrogenation over a palladium catalyst to give compound (**5.9.3**), which was further alkylated with γ-chlorobutyrophenone (**5.2.6**) on reflux in toluene in presence of sodium bicarbonate along with a few crystals of potassium iodide to give desired pipamperone (**5.9.4**) [442,443] (Scheme 5.47).

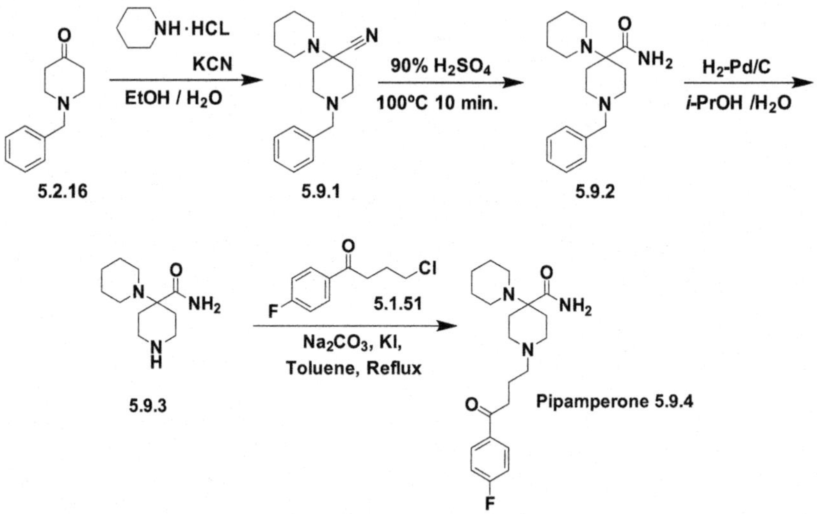

SCHEME 5.47 Synthesis of pipamperone.

Piritramide (704)

Piritramide (**5.9.5**) (Dipidolor) is a synthetic opioid predominantly used for postoperative analgesia and analgosedation in the intensive care unit setting in several European countries in surgical disciplines, such as general surgery, orthopedic surgery, trauma surgery, and gynecological surgery. The onset of analgesia of piritramide is very rapid and as little as 1−2 minutes on intravenous administration. In terms of typical opioid side effects (nausea, vomiting, respiratory depression, constipation) piritramide is believed to be less pronounced [444−449]. Piritramide is a Schedule I narcotic controlled substance.

Synthesis of piritramide has been carried out in full accordance with the synthesis of pipamperone (**5.9.4**) [443], just replacing alkylating agent γ-chlorobutyrophenone on the last stage of the synthesis for 3,3-diphenyl-3-cyanopropyl bromide (**5.3.34**), which gave the desired piritramide (**5.9.5**) [443,450,451] (Scheme 5.48).

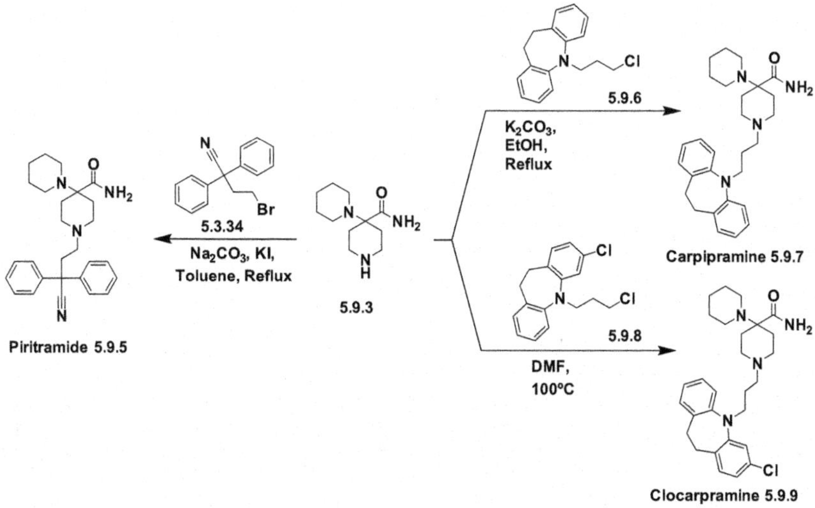

SCHEME 5.48 Synthesis of carpipramine and clocapramine.

Carpipramine (144)

Carpipramine (**5.9.7**) (Defecton) is available in both Japan and France. It is a drug classified as second-generation antipsychotic for treatment of psychoses, and in particular of schizophrenia. It has been found that carpipramine could also been useful in the treatment of depression, anxiety, and sleep disorders. It can induce insomnia, agitation, and extrapyramidal motor side effects, but it displays generally low toxicity. Clinical and biological tolerances seem to be excellent and extrapyramidal side effects are exceptional [452−463].

Carpipramine (**5.9.7**) synthesis is based on alkylation of 4-carbamoyl-4-piperidinopiperidine (**5.9.3**) with 5-(3-chloropropyl)-10,11-dihydro-5H-dibenzo[b,f]azepine (**5.9.6**) after many hours of reflux in ethanol in the presence of potassium carbonate [464] (Scheme 5.48).

Clocarpramine (72)

Clocapramine (**5.9.9**), approved in Japan for the treatment of schizophrenia, was introduced into medicinal practice as a successor to carpipramine (**5.9.7**). Antidopaminergic activity of clocapramine is more potent than that of carpipramine. Clocapramine did not show any imipramine-like action; on the other hand, carpipramine partially did. Clocapramine can induce extrapyramidal motor side effects and insomnia, but it displays generally low toxicity [463,465,466].

Clocapramine (**5.9.9**) was synthesized by reacting 4-carbamoyl-4-piperidinopiperidine (**5.9.3**) with 3-chloro-5-(3-chloropropyl)-10,11-dihydro-5H-dibenz[*b,f*]azepine (**5.9.8**) in dimethylformamide on heating 10 hours at 100°C [467,468] (Scheme 5.48).

5.10   ALVIMOPAN (396)

Alvimopan (**5.10.9**) (Entereg) is a selective, peripherally acting μ-opioid receptor antagonist, with no central nervous system activity, and was approved by the US FDA to accelerate the time to gastrointestinal recovery following bowel resection surgery. It is the first pharmacotherapy to be approved for this application. The polarity of its molecule limits gastrointestinal absorption and central nervous system penetration, so it only has peripheral activity. It works by protecting the bowel from the constipation effects of opioid medications that are used to treat pain after surgery. Alvimopan is only for short-term use for hospitalized patients. Its side effects may include: indigestion, stomach pain, nausea, vomiting, or diarrhea [469–477].

The synthesis of alvimopan (**5.10.9**) started from the methylation of 4-(3-methoxyphenyl)-1-methyl-1,2,3,6-tetrahydropyridine (**5.10.1**) with methyl iodide after initial *n*-butyllithium workup of a starting compound. Obtained 4-(3-methoxyphenyl)-1,4-dimethyl-1,2,3,4-tetrahydropyridine (**5.10.2**) underwent Mannich aminomethylation using 37% aqueous formaldehyde and dimethylamine solution to which was added sulfuric acid to adjust the pH of the mixture to a range of three to four to give compound (**5.10.3**). The last was hydrogenated on platinum on a charcoal catalyst in ethanol to give 1-(4-(3-methoxyphenyl)-1,4-dimethylpiperidin-3-yl)-*N,N*-dimethylmethanamine (**5.10.4**), which was *N*-demethylated with vinyl chloroformate on reflux in dichloroethane in presence of proton sponge [1,8-bis(dimethylamino)naphthalene] to give isomeric (**5.10.5**). After the resolution of optical antipodes of the obtained product, which was accomplished with dibenzoyl D- and L-tartrates, separated 3-((3R,4R)-3,4-dimethylpiperidin-4-yl)phenol (**5.10.5**) was incorporated into the Michael addition reaction with 3-phenyl-2-(ethoxycarbonyl)-L-propene (**5.10.6**), which took place in methanol for 10 days and gave after alkali hydrolysis of ester group (LiOH,THF/MeOH/H$_2$O) product (**5.10.7**). The last was *N*-acylated with glycine *sec*-butyl ester (**5.10.8**) in standard peptide synthesis conditions [1,3-dicyclohexylcarbodiimide (DCC), 1-hydroxybenzotriazole (HOBt), trimethylamine, DMF]. The obtained corresponding *sec*-butyl ester was hydrolyzed in NaOH solution to desired alvimopan (**5.10.9**) [478–482] (Scheme 5.49).

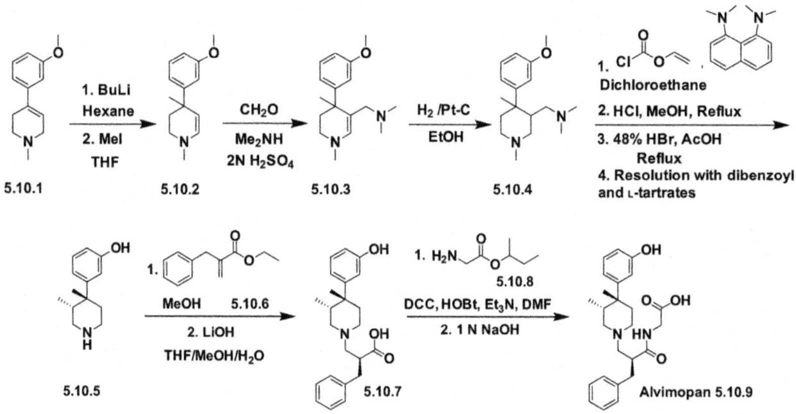

SCHEME 5.49   Synthesis of alvimopan.

REFERENCES

[1]   Matar HE, Almerie MQ, Makhoul S, Xia J, Humphreys P. Periciazine for schizophrenia. Cochrane Database Syst Rev 2014;(5):CD007479.

[2]   Anonymous, Periciazine. Brit Med J 1967;1(5536):352−3.

[3]   Rasch PJ. Treatment of disorders of character and schizophrenia by periciazine (Neulactil). Acta Psychiatr Scand Supplementum 1966;191:200−15.

[4]   By Anonymous. Periciazine (Neulactil). Drug Therapeut Bull 1965;3(21):82−3.

[5]   Morley KC, Haber PS, Morgan ML, Samara F. Periciazine in the treatment of cannabis dependence in general practice: a naturalistic pilot trial. Subst Abuse Rehabil 2012;3:43−7.

[6]   Jacob RM, Robert JG. Phenothiazine derivatives, US 3075976; 1963.

[7]   Robert MJ, Jacques GR. 10-(Heterocyclylalkyl)phenothiazines, FR 1212031; 1960.

[8]   Jacob RM, Robert JG. Preparation of 3-methoxy-10-[3-(4-hydroxypiperidino)-2-methyl-propyl]phenothiazine, DE 1154117; 1963.

[9]   Julou L, Ducrot R, Fournel J, Leau O, Bardone MC, Myon J. Pharmacological properties of 3-methoxy-10-[3-(4-hydroxy-1-piperidyl)-2-methylpropyl]phenothiazine (9.159 R.P.). CR Séances Soc Biol Ses Fil 1966;160(10):1852−8.

[10] Puhakka H, Rantanen T, Virolainen E. Diphenylpyraline (Lergobine) in the treatment of patients suffering from allergic and vasomotor rhinitis. J Int Med Res 1977;5(1):37−41.

[11] Lapa GB, Mathews TA, Harp J, Budygin EA, Jones SR. Diphenylpyraline, a histamine H1 receptor antagonist, has psychostimulant properties. Eur J Pharmacol 2005;506(3):237−40.

[12] Meindl W. Antimycobacterial antihistamines. Arch Pharm 1989;322(8):493−7.

[13] Ohno T, Kobayashi S, Hayashi M, Sakurai M, Kanazawa I. Diphenylpyraline-responsive parkinsonism in cerebrotendinous xanthomatosis: long-term follow up of three patients. J Neurol Sci 2001;182(2):95−7.

[14] Oleson EB, Ferris MJ, Espana RA, Harp J, Jones SR. Effects of the histamine H1 receptor antagonist and benztropine analog diphenylpyraline on dopamine uptake, locomotion and reward. Eur J Pharmacol 2012;683(1-3):161−5.

[15] Knox LH, Kapp R. 4-(1-Alkylpiperidyl) benzhydryl ethers and salts thereof, US 2479843; 1949.

[16] Schuler WA. Basic benzhydryl ethers, DE 934890; 1955.

[17] Satoda I, Kusuda F, Okumura H. N-Alkyl-4-piperidyl benzhydryl ether, JP 37018193; 1962.

[18] Wiseman LR, Faulds D. Ebastine: a review of its pharmacological properties and clinical efficacy in the treatment of allergic disorders. Drugs 1996;51(2):260−77.

[19] Hurst M, Spencer CM. Ebastine: an update of its use in allergic disorders. Drugs 2000;59 (4):981−1006.

[20] Van Cauwenberge P, De Belder T, Sys L. A review of the second-generation antihistamine ebastine for the treatment of allergic disorders. Exp Opin Pharmacother 2004;5(8):1807−13.

[21] Roberts DJ. Towards the optimal antihistamine. Studies with ebastine. Inflam Res 1988;47(Suppl.1):S36−7.

[22] Prieto S, Jose M, Vega NA, Moragues MJ, Spickett RGW. Piperidine derivatives, EP 134124; 1985.

[23] Reddy TM, NC Reddy, Reddy SP, Cheluvaraju PB, Rao MG. Process for preparation of ebastine from 1-[4-(1,1-dimethylethyl)phenyl]-4-(4-hydroxy piperidin-1-yl)butan-1-one and diphenylmethanol, WO 2009157006; 2009.

[24] Bielory L, Duttachoudhury S, McMunn A. Bepotastine besilate for the treatment of pruritus. Exp Opin Pharmacother 2013;14(18):2553−69.

[25] Lyseng-Williamson KA. Oral bepotastine: in allergic disorders. Drugs 2010;70(12): 1579−91.

[26] Graul A, Castaner J. Bepotastine besilate: antihistamine and antiallergic. Drugs Fut 1998;23(3):256−60.

[27] Abelson MB, Torkildsen GL, Williams JI, Gow JA, Gomes PJ, McNamara TR. Time to onset and duration of action of the antihistamine bepotastine besilate ophthalmic solutions 1.0% and 1.5% in allergic conjunctivitis: a phase III, single-center, prospective, randomized, double-masked, placebo-controlled, conjunctival allergen challenge assessment in adults and children. Clin Therapeut 2009;31(9):1908−21.

[28] Kato M, Nishida A, Aga Y, Kita J, Kudo Y, Narita H, et al. Pharmacokinetic and pharmacodynamic evaluation of central effect of the novel antiallergic agent bepotastine besilate. Arzneimit Forsch 1997;47(10):1116−24.

[29] Chen T, Yang H, Chou T, Yao C. Process for the preparation of bepotastine and its benzenesulfonic acid salt, US 20140046068; 2014.

[30] Kita J, Fujiwara H, Takamura S, Yoshioka R, Ozaki Y, Yamada S. Acid - addition salts of optically active piperidine compound and process for producing the same, WO, 9829409; 1988.

[31] Ha T, Park CH, Kim WJ, Cho S, Kim HK, Suh KH. Process for preparing bepotastine and intermediates used therein, WO 9829409; 2008.

[32] Ha TH, Suh K, Lee GS. A novel synthetic method for bepotastine, a histamine H1 receptor antagonist. Bull Korean Chem Soc 2013;34(2):549−52.

[33] Dooley M, Goa KL. Lamifiban. Drugs 1999;57(2):215−21 discussion 222-3

[34] Cases A, Rabasseda X, Castaner J. Lamifiban: platelet antiaggregatory fibrinogen gpIIb/IIIa receptor antagonist. Drugs Fut 1999;24(3):261−8.

[35] Alig L, Edenhofer A, Hadvary P, Huerzeler M, Knopp D, Mueller M, et al. Low molecular weight, non-peptide fibrinogen receptor antagonists. J Med Chem 1992;35(23):4393−407.

[36] Alig L, Hadvary P, Huerzeler M, Mueller M, Steiner B, Weller T. N-acyl-α-amino acids, methods for their preparation and their use for the treatment of diseases associated with binding of adhesive proteins to blood platelets or blood platelet aggregation, EP 505868; 1992.

[37] Chang M, Chen S. Concise synthesis of sibrafiban and lamifiban, two non-peptide fibrinogen receptor (GPIIb/IIIa) antagonists. J Chin Chem Soc 2001;48(2):133−5.

[38] Warren LE. Some observations on eucatropine. Repts Chem Lab Am Med Assoc 1920;13:62−7.

[39] Smith SE. Mydriatic drugs for routine fundal inspection. A reappraisal. Lancet 1971;2 (7729):837−9.

[40] Harries C. About some tropeine of the triacetonamine series. Liebig's Ann 1897;296:328−43.

[41] Harries C. To the knowledge of the euphthalmine. Ber 1898;31:665−6.

[42] Kipping FS. Some derivatives of the vinyldiacetonalkamines. J Chem Soc Transactions 1923;123:3115−19.

[43] Kastning EH, Lischer CF. Vinyldiacetonalkamine, US 2480329; 1949.

[44] Madersbacher H, Murtz G. Efficacy, tolerability and safety profile of propiverine in the treatment of the overactive bladder (non-neurogenic and neurogenic). World J Urol 2001;19(5):324−35.

[45] McKeage K. Propiverine: a review of its use in the treatment of adults and children with overactive bladder associated with idiopathic or neurogenic detrusor overactivity, and in men with lower urinary tract symptoms. Clin Drug Invest 2013;33(1):71−91.

[46] Klosa J, Delmar G. Synthesis of spasmolytic substances. XIV. Synthesis of derivatives of the benzilic acid ester of 1-methyl-4-hydroxypiperidine. J Prakt Chem 1962;16:71−82.

[47] Starke C, Thomas G, Friese J. α,α-Diphenyl-α-alkoxyacetic acid-N-alkylaminoalkyl esters, DD 106643; 1977.

[48] Andagar RR, Roy AK. Process for preparation of propiverine hydrochloride, WO 2011114195; 2011.

[49] Jang SJ, Kong JS, Lee JJ, Lee JM. Method for the preparation of diphenylacetate derivative, KR 2011111782; 2011.

[50] Mishra R, Siddiqui AA, Rashid M, Ramesha AR, Rohini RM, Khaidem S. Synthesis, characterization and pharmacological screening of various impurities present in opipramol, pargeverine and propiverine bulk drugs. Pharma Chemica 2010;2(2):185−94.

[51] Smith VM, Mead Jr. JA, Idea BV, Mallari RP. Clinical appraisal of pentapiperide methyl-sulfate a new anticholinergic drug. Am J Gastroenterol 1963;39:52−60.

[52] Anonymous, Pentapiperide methylsulfate (Quilene). Anticholinergic with antisecretory and antimotility activity on the upper gastrointestinal tract for adjunctive therapy in the management of peptic ulcer. Clin Pharmacol Therapeut 1969;10(6):904−6.

[53] No Inventor data available, Basic esters, their acid salts and quaternary salts, GB 781382; 1957.

[54] Lopez-Munoz F, Alamo C. The consolidation of neuroleptic therapy: Janssen, the discovery of haloperidol and its introduction into clinical practice. Brain Res Bull 2009;79(2):130−41.

[55] Awouters FHL, Lewi PJ. Forty years of antipsychotic drug research - from haloperidol to paliperidone - with Dr. Paul Janssen. Arzneimittl Forsch 2007;57(10):625−32.

[56] Janssen PAJ. Haloperidol and related butyrophenones. Medicinal Chemistry, 4-II. Academic Press; 1967. p. 199−248.

[57] Cookson J. Haloperidol and other first generation antipsychotics in mania, Bipolar Psychopharmacotherapy (2nd Edition) (2011), 109-129. (Edited by Akiskal, Hagop S.; Tohen, Mauricio).

[58] Hassaballa HA, Balk RA. Torsade de pointes associated with the administration of intra-venous haloperidol: a review of the literature and practical guidelines for use. Exp Opin Drug Safety 2003;2(6):543−7.

[59] Granger B. The discovery of haloperidol. Encephale 1999;25(1):59−66.

[60] Settle Jr EC, Ayd Jr. FJ. Haloperidol: a quarter century of experience. J Clin Psych 1983;44(12):440−8.

[61] Beresford R, Ward A. Haloperidol decanoate. A preliminary review of its pharmacodynamics and pharmacokinetic properties and therapeutic use in psychosis. Drugs 1987;33(1):31−49.

[62] Ban TA, Pecknold JC. Haloperidol and the butyrophenones. In: Simpson LL, editor. Drug treatment of mental disorders. Baltimore: Lippincott Williams & Wilkins; 1976. p. 45−60.

[63] Janssen PAJ, van de Westeringh C, Jageneau AHM, Demoen PJA, Hermans BKF, van Daele GHP, et al. Chemistry and pharmacology of CNS depressants related to 4-(4-hydroxy-4-phenyl-piperidino)butyrophenone. I. Synthesis and screening data in mice. J Med Pharm Chem 1959;1:281−97.

[64] Janssen PAJ. 1-Aroylalkyl-4-arylpiperidin-4-ol derivatives, BE 577977; 1959.

[65] Janssen PAJ. Pyrrolidine and piperidine derivatives, GB 895309; 1962.

[66] Janssen PAJ. 1-Aroylalkyl derivatives of arylhydroxypyrrolidines and arylhydroxy-piperidines, US 3438991; 1962.

[67] Janssen PAJ. Phenyl piperidinols, GB 1141664; 1969.

[68] Dryden HL Jr, Erickson RA. Process for the preparation of tertiary amines, US 4086234; 1975.

[69] Kook CS, Reed MF, Digenis GA. Preparation of fluorine-18-labeled haloperidol. J Med Chem 1975;18(5):533−5.

[70] Niemegeers CJE, Janssen PAJ. Bromoperidol, a new potent neuroleptic of the butyrophenone series. Comparative pharmacology of bromoperidol and haloperidol. Arzneimitt -Forsch 1974;24(1):670−84.

[71] Benfield P, Ward A, Clark BG, Jue SG. Bromperidol. A preliminary review of its pharmacodynamic and pharmacokinetic properties, and therapeutic efficacy in psychoses. Drugs 1988;35(6):670−84.

[72] Dubinsky B, McGuire JL, Niemegeers CJE, Janssen PAJ, Weintraub HS, McKenzie BE. Bromperidol, a new butyrophenone neuroleptic: a review. Psychopharmacology 1982;78 (1):1−7.

[73] Purgato M, Adams CE. Bromperidol decanoate (depot) for schizophrenia. Cochrane Database Syst Rev 2012;11:CD001719.

[74] Vincent SH, Shambhu MB, Digenis GA. Synthesis of [82Br] bromperidol and preliminary tissue distribution studies in the rat. J Med Chem 1980;23(1):75−9.

[75] Suehiro M, Yokoi F, Nozaki T, Iwamoto M. No-carrier-added radiobromination via the Gattermann reaction. Synthesis of bromine-75 and bromine-77 bromperidol. J Labelled Comp Radiopharm 1987;24(10):1143−57.

[76] Gross H, Kaltenback E. The clinical position of moperone among the butyrophenones. Nordic J Psychiatry 1959;23(1):4−9.

[77] Janssen PA. The evolution of the butyrophenones, haloperidol and trifluperidol, from meperidine-like 4-phenylpiperidines. Int Rev Neurobiol 1965;8:221−63.

[78] Soudijn W, Van Wijngaarden I, Allewijn FTN. Distribution, excretion, and metabolism of neuroleptics of the butyrophenone type. I., Excretion and metabolism of haloperidol and nine related butyrophenone derivatives in the Wistar rat. Eur J Pharmacol 1967;1(1):47−57.

[79] Janssen PAJ. Arylpiperidine derivatives, GB 881893; 1961.

[80] Janssen PA. Comparative pharmacological data on 6 new basic 4'-fluorobutyrophenone derivatives:haloperidol, haloanisone, triperidol, methylperidide, haloperidide and dipiperone. 2 From. Arzneimittel-Forschung 1961;11:932−8.

[81] Gallant DM, Bishop MP, Timmons E, Steele CA. A controlled evaluation of Trifluperidol: a new potent psychopharmacologic agent. Curr Ther Res Clin Exp 1963;27:463−741.

[82]   Gallant DM, Bishop MP, Timmons E, Steele CA. Trifluperidol: a butyrophenone deriva-
        tive. Am J Psychiatry 1963;120:485−7.
[83]   Janssen PAJ, Niemegeers CJE, Schellekens KHL, Lenaerts FM, Verbruggen FJ, Van
        Nueten JM, et al. Pharmacology of penfluridol (R 16341), a new potent and orally long-
        acting neuroleptic drug. Eur J Pharmacol 1970;11(2):139−54.
[84]   Migdalof BH, Grindel JM, Heykants JJ, Janssen PA. Penfluridol: a neuroleptic drug
        designed for long duration of action. Drug Metabol Rev 1979;9(2):281−99.
[85]   van Praag HM, Schut T, Dols L, van Schilfgaarden R. Controlled trial of penfluridol in
        acute psychosis. Brit Med J 1971;4(5789):710−13.
[86]   Claghorn JL, Mathew RJ, Mirabi M. Penfluridol: a long acting oral antipsychotic drug. J
        Clin Psychiatry 1979;40(2):107−9.
[87]   Bhattacharyya R, Bhadra R, Roy U, Bhattacharyya S, Pal J, Saha SS. Resurgence of pen-
        fluridol: merits and demerits. Am J PharmTech Res 2015;5(5):13−23.
[88]   Soares BGO, Lima MS. Penfluridol for schizophrenia. Cochrane Database Syst Rev
        2006;(2) CD 002923
[89]   Hermans HKF, Niemegeers CJEJ. 4-Phenyl-1-[4-(p-fluorophenyl)-4-phenylbutyl]-4-
        piperidinols, DE 2040231; 1971.
[90]   Protiva M, Rajsner M, Sindelar K, Cervena I, Penfluridol CS. 160865; 1975.
[91]   Sindelar K, Rajsner M, Cervena I, Valenta V, Jilek JO, Kakac B, et al. Neurotropic
        and psychotropic agents. LXVII. 1-[4,4-Bis(4-fluorophenyl)butyl]-4-hydroxy-4-(3-
        trifluoromethyl-4-chlorophenyl)piperidine and related compounds. New synthetic
        approaches, Coll Czech Chem Comm 1973;38(12):3879−901.
[92]   Chen G, Xia H, Cai Y, Ma D, Yuan J, Yuan C. Synthesis and SAR study of di-
        phenylbutylpiperidines as cell autophagy inducers. Bioorg Med Chem Lett 2011;21(1):234−9.
[93]   Awouters F, Megens A, Verlinden M, Schuurkes J, Niemegeers C, Janseen PA J.
        Loperamide: survey of studies on mechanism of its antidiarrheal activity. Dig Dis Sci
        1993;38(6):977−95.
[94]   Ruppin H. Loperamide - a potent antidiarrheal drug with actions along the alimentary
        tract. Aliment Pharmacol Ther 1987;1(3):179−90.
[95]   Ooms LAA, Degryse A-D, Janssen PAJ. Mechanisms of action of loperamide. Scand J
        Gastroenterol, Supplement 1984;19(96):145−55.
[96]   Shriver DA, Rosenthale ME, McKenzie BE, Weintraub HS, McGuire JL. Loperamide.
        Pharmacol Biochem Prop Drug Subst 1981;3:461−76.
[97]   Niemegeers CJE, Colpaert FC, Awouters FHL. Pharmacology and antidiarrheal effect of
        loperamide. Drug Dev Res 1981;1(1):1−20.
[98]   Heel RC, Brogden RN, Speight TM, Avery GS. Loperamide: a review of its pharmaco-
        logical properties and therapeutic efficacy in diarrhea. Drugs 1978;15(1):33−52.
[99]   Baker Daniel E. Loperamide: a pharmacological review. Rev Gastroenterol Disord
        2007;7(Suppl 3):S11−18.
[100]   Ericsson CD, Johnson PC. Safety and efficacy of loperamide. Am J Med 1990;88
        (6A):10S−4S.
[101]   Regnard C, Twycross R, Mihalyo M, Wilcock A. Loperamide. J Pain Sympt Manage
        2011;42(2):319−22.
[102]   Wiles DA, Spiller HA, Russell JL, Casavant MJ. A retrospective assessment of lopera-
        mide toxicity 2000-2011. Int J Pharmacol Toxicol Sci 2014;4(2):1−8.
[103]   Vandenbossche J, Huisman M, Xu Y, Sanderson-Bongiovanni D, Soons P. Loperamide
        and P-glycoprotein inhibition: assessment of the clinical relevance. J Pharm Pharmacol
        2010;62(4):401−12.

[104] Daly JW, Harper J. Loperamide: novel effects on capacitative calcium influx. Cel Mol Life Sci 2000;57(1):149−57.

[105] Hanauer SB. The role of loperamide in gastrointestinal disorders. Rev Gastroenterol Disord 2008;8(1):15−20.

[106] Eggleston W, Clark KH, Marraffa JM. Loperamide abuse associated with cardiac dysrhythmia and death. Ann Emerg Med 2017;69(1):83−6.

[107] Janssen PAJ, Niemegeers CJEJ, Stokbroekx RA, Vandenberk J. 2,2-Diaryl-4-(4-aryl-4-hydroxypiperidino)butyramides, FR 2100711; 1972.

[108] Stokbroekx RA, Vandenberk J, Van Heertum AHMT, Van Laar GMLW, Van der Aa MJMC, Van Bever WFM, et al. Synthetic antidiarrheal agents. 2,2-Diphenyl-4-(4'-aryl-4'-hydroxypiperidino)butyramides. J Med Chem 1973;16(7):782−6.

[109] Ziering A, Berger L. Piperidine derivatives; 4-arylpiperidines. J Org Chem 1947;12(6):894−903.

[110] Carabateas PM, Grumbach L. Strong analgesics. Some 1-substituted-4-phenyl-4-propionoxypiperidines. J Med Pharm Chem 1962;5:913−19.

[111] Janssen PAJ, Eddy NB. J Med Pharm Chem 1960;2:31−45.

[112] Elpern B, Wetterau W, Carabateas P, Grumbach L. Strong analgesics. The preparation of some 4-acyloxy-1-aralkyl-4-phenylpiperidines. J Am Chem Soc 1958;80:4916−18.

[113] Beckett AH, Walker J. The configuration of alphaprodine and betaprodine. J Pharm Pharmacol 1955;7(12):1039−45.

[114] Hartman MM, Schwab JM. The use of alphaprodine (Nisentil) for the production of surgical anesthesia. Surg Clin North Am 1963;43:1229−42.

[115] Dodis A. Use of alphaprodine as an analgesic in surgery & medicine, L'union Med. Can 1957;86(10):1103−8.

[116] Lipson HI, Braford HR. Alphaprodine (nisentil) hydrochloride in anesthesia; its use, with or without antagonists, for supplementation or as sole agent. JAMA 1957;163(14):1244−8.

[117] Bachrach EH, Godholm AN, Betcher AM. Alphaprodine (nisentil) hydrochloride in anesthesia; its use, with or without antagonists, for supplementation or as sole agent. Surgery 1955;37(3):440−5.

[118] Burnett RG, White CA. Alphaprodine for continuous intravenous obstetric analgesia. Obstet Gynecol 1966;27(4):472−7.

[119] Carter WJ, Bogert JA. An effective pre-medication procedure for dental patients. J Mo Dent Assoc 1966;46(6):8−9.

[120] Fuller JD, Crombleholme WR. Respiratory arrest and prolonged respiratory depression after one low, subcutaneous dose of alphaprodine for obstetric analgesia. A case report. J Reprod Med 1987;32(2):149−51.

[121] Ziering A, Motchane A, Lee J. Piperidine derivatives. IV. 1,3-Disubstituted-4-aryl-4-acyloxypiperidines. , J Org Chem 1957;22:1521−8.

[122] McElvain SM, Barnett MD. Piperidine derivatives. XXVIII. 1-Methyl-3-alkyl-4-phenyl-4-acyloxypiperidines. J Am Chem Soc 1956;78:3140−3.

[123] Nazarov IN, Prostakov NS, Shvetsov NI. Heterocyclic compounds. XXXIX. Synthetic analgesic substances. 4. Esters of 1,2,5-trimethyl-4-phenyl-4-piperidol with aliphatic acids. Synthesis of Promedol and Isopromedol. Zh Obshch Khim 1956;26:2798−811.

[124] Casy AF, McErlane K. Analgesic potency and stereochemistry of trimeperidine and its isomers and analogs. J Pharm Pharmacol 1971;23(1):68−9.

[125] Fries D, Portoghese PS. Stereochemical studies on medicinal agents. 18. Absolute configuration and analgetic potency of trimeperidine enantiomers. J Med Chem 1974;17(9):990−3.

[126] Van Bever, W.; Stokbroeckx, R.; Vandenberk, J.; Wouters, M., Chemistry of modern antidiarrheals. III. Synthesis of diphenoxylate, difenoxin, loperamide, and related compounds. From Modern Pharmacology-Toxicology (1976), 7(Synth. Antidiarrheal Drugs: Synth. - Preclin. Clin. Pharmacol.), 23-36, 61-63. (M. Dekker).

[127] Lee J, Ziering A. 1,3 dimethyl-4-propionoxy-4-phenyl-piperidine and acid addition salts thereof. US 1950;2498433.

[128] Ziering A, Lee J. Piperidine derivatives. V. 1,3-Dialkyl-4-aryl-4-acyloxypiperidines. J Org chem 1947;12:911−14.

[129] Abdel-Monem MM, Larson DL, Kupferberg HJ, Portoghese PS. Stereochemical studies on medicinal agents. 11. Metabolism and distribution of prodine isomers in mice. J Med Chem 1972;15(5):494−500.

[130] Casy AF. Analgesics and their antagonists: recent developments. Progr Drug Res 1978;22:149−227.

[131] Iorio MA, Casy AF, May EL. 3-Alkyl and 3-alkenyl diastereoisomers related to the reversed ester of pethidine. Eur J Med Chem 1975;10(2):178−81.

[132] Ahmed FR, Barnes WH, Kartha G. Configuration of the α-prodine molecule, Chem Ind, 485.

[133] Ahmed FR, Barnes WH, Masironi, Liliana A. Configuration of the β-prodine molecule in DL-β- prodine-HBr. Chem Ind 1962;97−8.

[134] Kartha G, Ahmed FR, Barnes WH. The crystal and molecular structure of dl-alphaprodine hydrochloride. Acta Crystallograph 1960;13:525−31.

[135] Ahmed FR, Barnes WH. The crystal and molecular structure of dl-betaprodine hydrochloride. Acta Crystallograph 1963;16(12):1249−52.

[136] Portoghese PS, Larson DL. Absolute stereochemistry and analgesic potency of prodine enantiomers. J Pharm Sci 1968;57(4):711−13.

[137] Portoghese PS, Shefter E. Stereochemical studies on medicinal agents. 19. The x-ray crystal structures of two (+-)-allylprodine diastereomers. Role of the allyl group in conferring high stereoselectivity and potency at analgetic receptors. J Med, Chem 1976;19 (1):55−7.

[138] Fries DS, Portoghese PS. Stereochemical studies on medicinal agents. 20. Absolute configuration and analgetic potency of α-promedol enantiomers. The role of the C-4 chiral center in conferring stereoselectivity in axial- and equatorial-phenylprodine congeners. J Med Chem 1976;19(9):1155−8.

[139] Ngan KWD. Intrathecal pethidine: pharmacology and clinical applications. Anaesth Intensive Care Med 1998;26(2):137−46.

[140] Ngan KWD. Epidural pethidine: pharmacology and clinical experience. Anaesth Intensive Care Med 1998;26(3):247−55.

[141] Clark RF, Wei EM, Anderson PO. Meperidine: therapeutic use and toxicity. J Emerg Med 1955;13(6):797−802.

[142] Latta KS, Ginsberg B, Barkin RL. Meperidine: a critical review. Am J Therapeut 2002;9 (1):53−68.

[143] Eisleb O. Piperidine compounds (therapeutics and intermediates), US 2167351; 1939.

[144] Bergel F, Morrison AL, Rinderknecht H. Synthetic analgesics. II. New synthesis of pethidine and similar compounds. J Chem Soc 1944;265−77.

[145] Hite G. Quasi-Favorski rearrangement. I. Preparation of Demerol and β-pethidine. J Am Chem Soc 1959;81:1201−3.

[146] No Inventor data available, Piperidine derivatives, GB 501135; 1939.

[147] Eisleb O. New syntheses with sodium amide. Chem Ber 1941;74B:1433−50.

[148] Pohland A. Mannich bases derived from 4-carbethoxy-4-phenylpiperidines, US 2951080; 1960.

[149] Cutler FA Jr, Fisher JF. 1-Substituted propylpiperidines, US 2962501; 1960.

[150] Janssen PAJ, Jageneau AH, Demoen PJA, van de Westeringh C, Raeymaekers AHM, Wouters MSJ, et al. Compounds related to Pethidine. I. Mannich bases derived from nor-pethidine and acetophenones. J Med Pharm Chem 1959;1:105−20.

[151] Janssen PAJ, Eddy NB. Compounds related to pethidine. IV. General chemical methods of increasing the analgesic activity of pethidine. J Med Pharm Chem 1960;2:31−45.

[152] Claris O, Bertrix L. Phenoperidine pharmacology and use in pediatric resuscitation. Pediatrie 1988;43(6):509−13.

[153] Rollason WN, Sutherland JS. Phenoperidine (R 1406), a new analgesic. Anaesthesia 1963;18:16−22.

[154] Viars P, Gaveau T. Clinical use of phenoperidine in aged persons and adolescents, Anesthesia, analgesie. reanimation 1968;25(3):319−33.

[155] de Castro J, Van de Water A, Wouters L, Xhonneux R, Reneman R, Kay B. Comparative study of cardiovascular, neurological and metabolic side effects of 8 narcotics in dogs. Pethidine, piritramide, morphine phenoperidine, fentanyl, R 39 209, sufentanil, R 34 995. III. Comparative study of the acute metabolic toxicity of the narcotics used in high and massive doses in curarised and mechanically ventilated dogs. Acta Anaesthesiol Belg 1979;30(1):71−90.

[156] No Inventor data available, Substituted piperidine-4-carboxylic acid esters, GB 793010; 1958.

[157] Weijlard J, Orahovats PD, Sullivan Jr. AP, Purdue G, Heath FK, Pfister III K. A new synthetic analgesic. J Am Chem Soc 1956;78:2342−3.

[158] Frearson PM, Stern ES. Analgesic N-substituted 4-phenylpiperidino-4-carboxylic ethyl esters, DE 1256219; 1967.

[159] Morren HG. Piperidine derivatives, BE 558883; 1957.

[160] Elpern B, Carabateas P, Soria AE, Gardner LN, Grumbach L. Strong analgesics. The preparation of some ethyl 1-anilinoalkyl-4-phenylpiperidine-4-carborylates. J Am Chem Soc 1959;81:3784−9.

[161] Thorp RH, Walton E. Search for new analgesics; further homologues of pethidine and the pharmacology of these and other compounds. J Chem Soc 1948;2:559−61.

[162] Anderson RJ, Frearson PM, Stern ES. Some new analogs of pethidine. J Chem Soc 1956;4088−91.

[163] Stern ES, Anderson RJ. Piperidine compounds, US 2795581; 1957.

[164] Janssen PAJ, Jageneau AH, Huygens J. Synthetic anti-diarrhoeal agents. I. Some pharmacological properties of R 1132 and related compounds. J Med Pharm Chem 1959;1:299−308.

[165] Lustman F, Walters EG, Shroff NE, Akbar FA. Diphenoxylate hydrochloride (Lomotil) in the treatment of acute diarrhea. Brit J Clin Pract 1987;41(3):648−51.

[166] Nguyen W, Howard BL, Jenkins DP, Wulff H, Thompson PE, Manallack DT. Structure-activity relationship exploration of Kv1.3 blockers based on diphenoxylate. Bioorg Med Chem Lett 2012;22(23):7106−9.

[167] Janssen PAJ. 2,2-Diaryl-ω-(4-phenylpiperidino)alkanenitriles, US 2898340; 1959.

[168] Van Bever W, Stokbroeckx R, Vandenberk J, Wouters M. Chemistry of modern antidiarrheals. III. Synthesis of diphenoxylate, difenoxin, loperamide, and related compounds. Modern Pharmacol -Toxicol 1976;7(23-36):61−3.

[169] Noble S, McTavish D. Levocabastine: an update of its pharmacology, clinical efficacy and tolerability in the topical treatment of allergic rhinitis and conjunctivitis. Drugs 1995;50(6):1032–49.

[170] Heykants J, Van Peer A, Van de Velde V, Snoeck E, Meuldermans W, Woestenborghs R. The pharmacokinetic properties of topical levocabastine: a review. Clin Pharmacokinet 1995;29(4):221–30.

[171] Dechant KL, Goa KL. Levocabastine. A review of its pharmacological properties and therapeutic potential as a topical antihistamine in allergic rhinitis and conjunctivitis. Drugs 1991;41(2):202–24.

[172] Abelson MB, Weintraub D. Levocabastine eye drops: a new approach for the treatment of acute allergic conjunctivitis. Eur J Ophtamol 1994;4(2):91–101.

[173] Janssens MM, Vanden BG. Levocabastine: an effective topical treatment of allergic rhinoconjunctivitis. Clin Exp Allergy 1991;21(Suppl. 2):29–36.

[174] Stokbroekx RA, Luyckx MGM, Willems JJM. 1-Cyclohexyl-4-aryl-4-piperidinecarboxylic acid derivatives, US 4369184; 1983.

[175] Stokbroekx RA, Luyckx MGM, Willems JJM, Janssen M, Bracke JOMM, Joosen RLP, et al. Levocabastine (R 50547): the prototype of a chemical series of compounds with specific H1-antihistaminic activity. Drug Devel Res 1986;8(1-4):87–93.

[176] Bondesson U, Arner S, Anderson P, Boreus LO, Hartvig P. Clinical pharmacokinetics and oral bioavailability of ketobemidone. Eur J Clin Pharmacol 1980;17 (1):45–50.

[177] Christensen CB. The opioid receptor binding profiles of ketobemodone and morphine. Pharmacol Toxicol 1993;73(6):344–5.

[178] Andersen S, Dickenson AH, Kohn M, Reeve A, Rahman W, Ebert B. The opioid ketobemidone has a NMDA blocking effect. Pain 1996;67(2,3):369–74.

[179] Lundeberg S, Stephanson N, Stiller C-O, Eksborg S. Pharmacokinetics after a single intravenous dose of the opioid ketobemidone in neonates. Acta Anaesthesiol Scand 2012;56(8):1026–31.

[180] Eisleb O. 1-Alkyl-4-(m-hydroxyphenyl)piperidine compounds, DE 752755; 1952.

[181] Avison AWD, Morrison AL. Synthetic analgesics. VI. The synthesis of ketobemidon. J Che Soc 1950;1469–71.

[182] Cammack T, Reeves PC. Synthesis of ketobemidoneprecursors via phase-transfer catalysis. J Het Chem 1986;23(1):73–5.

[183] Kagi H, Miescher K. New synthesis of 4-phenyl-4-piperidyl alkyl ketones and related compounds with morphinelike action. Helv Chim Acta 1949;32:2489–507.

[184] Leysen JE, Niemegeers CJE, Van Nueten JM, Laduron PM. [3H]Ketanserin (R 41 468), a selective 3H-ligand for serotonin2 receptor binding sites. Binding properties, brain distribution, and functional role. Mol Pharmacol 1982;21(2):301–14.

[185] Brogden RN, Sorkin EM. Ketanserin: a review of its pharmacodynamic and pharmacokinetic properties, and therapeutic potential in hypertension and peripheral vascular disease. Drugs 1990;40(6):903–49.

[186] Vanhoutte P, Amery A, Birkenhaeger W, Breckenridge A, Buehler F, Distler A, et al. Serotoninergic mechanisms in hypertension, Focus on the effect of ketanserin. Hypertension 1988;11(2):111–33.

[187] Awouters F. The pharmacology of ketanserin, the first selective serotonin S2-antagonist. Drug Dev Res 1985;6(4):263–300.

[188] Matsumoto M, Yoshioka M, Togashi H, Minami M, Saito H. Serotonin-2 receptor antagonist: Ketanserin. Focus on the sympathoinhiitory action. Progr Hypertens 1992;2:189–205.

[189] Frishman WH, Okin S, Huberfeld S. Serotonin antagonism in the treatment of systemic hypertension: the role of ketanserin. Med Clin North Am 1988;72(2):501—22.

[190] Robertson JIS, Stott DJ, Ball SG. Antihypertensive mode of action of ketanserin. J Cardiovasc Pharmacol 1987;10(Suppl. 3):S45—7.

[191] Hedner T, Persson B. Am J Hypertens 1988;1(3 Pt 3):317S—23S.

[192] Vanhoutte PM, Ball SG, Berdeaux A, Cohen ML, Hedner T, McCall R, et al. Mechanism of action of ketanserin in hypertension. Trends Pharmacol Sci 1986;7(2): 58—9.

[193] Malinin A, Oshrine B, Serebruany V. Treatment with selective serotonin reuptake inhibitors for enhancing wound healing. Med Hypotheses 2004;63(1):103—9.

[194] Kim M, Ustuner ET, Schuschke D, Morsing A, Kjolseth D, Fingar V, et al. Ketanserin accelerates wound epithelialization and neovascularization. Wound Repair Regen 1995; 3(4):506—11.

[195] Rooman RP, Moeremans M, De Wever B, Daneels G, Geuens G, Aerts F, et al. Ketanserin in wound healing and fibrosis: investigations into its mechanism of action, in serotonin. In: Paoletti R, Vanhoutte PM, Brunello N, Maggi FM, editors. Cell biology to pharmacology and therapeutics. New York: Springer; 1990.

[196] Vandenberk J, Kennis L, Van der Aa M, Van Heertum A. Piperidinylalkylquinazoline compounds, composition and method of use, US 4335127; 1982.

[197] Fakhraian H, Heydary M. Reinvestigation of the synthesis of ketanserin and its hydrochloride salt via 3-(2-chloroethyl)-2,4-(1H,3H)-quinazolinedione or dihydro-5H-oxazolo [2,3-b]quinazolin-5-one. J Het Chem 2014;51(1):151—6.

[198] Berridge M, Comar D, Crouzel C, Baron JC. Carbon-11-labeled ketanserin: a selective serotonin S2 antagonist. J Labell Comp Radiopharmaceut 1983;20(1):73—8.

[199] Wouters W, Janssen CGM, Van Dun J, Thijssen JBA, Laduron PM. In vitro labeling of serotonin-S2 receptors. Synthesis and binding characteristics of [3H]-7-aminoketanserin. J Med Chem 1986;29(9):1663—8.

[200] Janssen CGM, Lenoir HAC, Thijssen JBA, Knaeps AG, Verluyten WLM, Heykants JJP. Synthesis of 3H- and 14C- ketanserin. J Labell Comp Radiopharmaceut 1988;25 (7):783—92.

[201] Croom KF, Keating GM. Metopimazine: a review of its use in the treatment of chemotherapy-induced nausea and vomiting. Am J Cancer 2006;5(2):123—36.

[202] Herrstedt J. Chemotherapy-induced nausea and vomiting with special emphasis on metopimazine. Danish Med Bull 1998;45(4):412—22.

[203] Herrstedt J, Sigsgaard TC, Nielsen HA, Handberg J, Langer SW, Ottesen S, et al. Randomized, double-blind trial comparing the antiemetic effect of tropisetron plys metopimazine with tropisetron plys placebo in patients receiving multiple cycles of multiple-day cisplatin-based chemotherapy. Supportive Care Cancer 2007;15: 417—26.

[204] Khamales S, Bethune-Volters A, Chidiac J, Bensaoula O, Delgado A, Di Palma M. A randomized, double-blind trial assessing the efficacy and safety of sublingualmetopimazine and ondansetron in the prophylaxis of chemotherapy-induced delayedemesis. Anticancer Drugs 2006;17(2):217—24.

[205] Jacob RM, Robert JG. Phenothiazine derivatives, DE 1092476; 1959.

[206] Reddy MS, Eswaraiah S, Satyanarayana K. Process for the preparation of metopimazine, IN 2010CH00360; 2010.

[207] Sindelar K, Holubek J, Koruna I, Hrubantova M, Protiva M. Modified syntheses of 2-(methylthio)-10-[2-(1-methyl-2-piperidinyl)ethyl]phenothiazine (thioridazine) and 1-[3-

[2-(methylsulfonyl)-10-phenothiazinyl]propyl]piperidine-4-carboxamide (metopimazine). Coll Czech Chem Comm 1990;55(6):1586−601.

[208]   Dobkin AB, Purkin N. The antisialogogue effect of phenothiazine derivatives: comparison of pecazine, perphenazine, fluphenazine, thiopropazate, pipamazine and triflupromazine. Brit J Anaest 1960;32:57−9.

[209]   Blatchford E. Studies of anti-emetic drugs: a comparative study of cyclizine (Marzine®), pipamazine (Mornidine®), trimethobenzamide (Tigan®), and hyoscine. Can Anaesth Soc J 1961;8:159−65.

[210]   Cusic JW, Sause HW. N-Phenothiazinylalkylpiperidinecarboxamides, US 2957870; 1960.

[211]   No Inventor data available, Piperidinecarboxamide derivatives, GB 830709; 1960.

[212]   Archibald JL. In: Scriabine A, editor. Indoramin, from pharmacology of antihypertensive drugs. New York: Raven Press; 1980. p. 161−77.

[213]   Pierce DM. A review of the clinical pharmacokinetics and metabolism of the α1-adrenoceptor antagonist indoramin. Xenobiotica 1990;20(12):1357−67.

[214]   Marmo E. Indoramin, a drug with mainly antihypertensive activity. Riforma Medica 1979;94(4):145−51.

[215]   Holmes B, Sorkin EM. Indoramin. A review of its pharmacodynamic and pharmacokinetic properties, and therapeutic efficacy in hypertension and related vascular, cardiovascular and airway diseases. Drugs 1986;31(6):467−99.

[216]   Archibald JL. Recent developments in the pharmacology and pharmacokinetics of indoramin. J Cardiovasc Pharmacol 1986;8(Suppl. 2):S16−19.

[217]   Archibald JL. Medicinal chemistry and animal pharmacology of indoramin. Brit J Clin Pharmacol 1981;12(Suppl. 1):45−7.

[218]   Shanks RG. The clinical pharmacology of indoramin - α-adrenoceptors. Brit J Clin Pharmacol 1981;12(Suppl. 1):43−4.

[219]   Neumeyer JL, Moyer UV, Leonard JE. Pharmacologically active acetylene compounds. II. Propynyl-substituted indole derivatives. J Med Chem 1969;12(3):450−2.

[220]   Archibald JL, Alps BJ, Cavalla JF, Jackson JL. Synthesis and hypotensive activity of benzamidopiperidylethylindoles. J Med Chem 1971;14(11):1054−9.

[221]   Archibald JL, Fairbrother P, Jackson JL. Benzamidopiperidines. 3. Carbocyclic derivatives related to indoramin. J Med Chem 1974;17(7):739−44.

[222]   Archibald JL, Benke GA. Benzamidopiperidines. 2. Heterocyclic compounds related to indoramin. J Med Chem 1974;17(7):736−9.

[223]   Gluchowski C, Forray CC, Chiu G, Branchek TA, Wetzel JM, Hartig PR. Preparation of α1C adrenergic receptor antagonists for treatment of benign prostatic hyperplasia, US 6015819; 2000.

[224]   Archibald JL. Indole derivatives, US 3527761; 1975.

[225]   Archibald JL, Jackson JL. Antiinflammatory 3-[2-[4-(substituted-benzamido)piperidino] ethyl]indoles, ZA 6803204; 1969.

[226]   Lin L. In: Lee PW, editor. *Clebopride*, from handbook of metabolic pathways of xenobiotics, vol. 3. New York: Wiley; 2014. p. 1139−41.

[227]   Fernandez AG, Massingham R, Roberts DJ. Potentiation of the gastric antisecretory activity of histamine H2-receptor antagonists by, Methods Find. Exp Clin Pharmacol 1988;10(5):285−93.

[228]   Takeda K, Taniyama K, Kuno T, Sano I, Ishikawa T, Ohmura I, et al. Clebopride enhances contractility of the guinea pig stomach by blocking peripheral D2 dopamine receptor and α2-adrenoceptor. J Pharmacol Exp Theraput 1991;257(2):806−11.

[229] Roberts DJ. The pharmacological basis of the therapeutic activity of clebopride and related substituted benzamides. Curr Therapeut Res 1982;31(Suppl. 1):1−44.

[230] Roberts DJ, Salazar W, Beckett PR, Nahorski SR. Clebopride, a new orthopramide with central antidopaminergic properties, dosage and selectivity. Arch Farmacol Toxicol 1978;4(1):102−4.

[231] Sempere AP, Duarte J, Palomares JM, Coria F, Claveria LE. Parkinsonism and tardive dyskinesia after chronic use of clebopride. Mov Disord 1994;9(1):114−15.

[232] Montagna P, Gabellini AS, Monari L, Lugaresi E. Parkinsonian syndrome after long-term treatment with clebopride. Mov Disord 1992;7(1):89−90.

[233] Bosco D, Plastino M, Marcello MG, Mungari P, Fava A. Acute hemifacial dystonia possibly induced by clebopride. Clin Neuropharmacol 2009;32(2):107−8.

[234] Prieto J, Moragues J, Spickett RG, Vega A, Colombo M, Salazar W, et al. Synthesis and pharmacological properties of a series of antidopaminergic piperidyl benzamides. J Pharm Pharmacol 1977;29(3):147−52.

[235] Spickett RGW, Noverola AV, Soto JP, Mauri JM. N-(1-benzylpiperid-4-yl)benzamides, methods for their preparation and pharmaceutical compositions containing them, GB 1507463; 1978.

[236] Spickett RGW, Moragues MJ, Prieto SJ. Benzamido-substituted nitrogen heterocycles, DE 2513136; 1975.

[237] Takahashi W, Yamane H. Preparation of clebopride, JP 63295557; 1988.

[238] De Knaep AGM, Moens LJR, Rey M. A new process for preparing cisapride, WO 9816511; 1988.

[239] Portincasa P, Mearin F, Robert M, Plazas MJ, Mas M, Heras J. Efficacy and tolerability of cinitapride in the treatment of functional dyspepsia and delayed gastric emptying, Gastroenterol. Hepatol 2009;32(10):669−76.

[240] Du Y, Su T, Song X, Gao J, Zou D, Zuo C, et al. Efficacy and safety of cinitapride in the treatment of mild to moderate postprandial distress syndrome-predominant functional dyspepsia. J Clin Gastroenterol 2014;48(4):328−35.

[241] Baqai MT, Malik MN, Ziauddin F. Efficacy and safety of cinitapride in functional dyspepsia. J Pak Med Assoc 2013;63(6):747−51.

[242] Suros A, Adell F, De Novoa V, Castellarnau J, Diosdado D, Carrasquer J, et al. Cinitapride in the treatment of gastroesophageal reflux. Comparative study with metoclopramide and placebo. Rev Med Univ Navarra 1992;37(1):18−23.

[243] Gallego SJ, Fombuena FJ, Martinez LJ. Efficacy and tolerance of cinitapride on the disturbances of gastrointestinal transit. Rev Med Univ Navarra 1991;36(3):12−18.

[244] Robert M, Salva M, Segarra R, Pavesi M, Esbri R, Roberts D, et al. The prokinetic cinitapride has no clinically relevant pharmacokinetic interaction and effect on QT during coadministration with ketoconazole. Drug Metab Dispos 2007;35(7):1149−56.

[245] Alarcon de la Lastra C, La Casa C, Martin MJ, Motilva V. Effects of cinitapride on gastric ulceration and secretion in rats. J Inflamm Rese 1998;47(3):131−6.

[246] Fernandez AG, Massingham R. Peripheral receptor populations involved in the regulation of gastrointestinal motility and the pharmacological actions of metoclopramide-like drugs. Life Sci 1985;36:1−14.

[247] Fernandez AG, Roberts DJ. Cinitapride hydrogen tartrate. Drugs Fut 1991;16:885−92.

[248] Iglesias JB, Soto JP, Noverola AV, Mauri JM. Piperidine compounds, GB 1574419; 1980.

[249] Noverola AV, Iglesias JB, Mauri JM, Spickett RGW. Therapeutic piperidine derivatives, DE 2751139; 1978.

[250]   Mauri JM, Spickett RGW. Process for the preparation of N-(piperid-4-yl)benzamides and their salts, useful as antidopaminergics and gastrointestinal stimulants, ES 2001458; 1988.

[251]   Briejer MR, Bosmans J-P, Van Daele P, Jurzak M, Heylen L, Leysen JE, et al. The in vitro pharmacological profile of prucalopride, a novel enterokinetic compound. Eur J Pharmacol 2001;423(1):71–83.

[252]   Camilleri M, Kerstens R, Rykx A, Vandeplassche L. A placebo-controlled trial of prucalopride for severe chronic constipation. N Eng J Med 2008;358(22):2344–54.

[253]   Garnock-Jones KP. Prucalopride: a review in chronic idiopathic constipation. Drugs 2016;76(1):99–110.

[254]   Keating GM. Prucalopride: a review of its use in the management of chronic constipation. Drugs 2013;73(17):1935–50.

[255]   Tack J, Camilleri M, Chang L, Chey WD, Galligan JJ, Lacy BE, et al. Systematic review: cardiovascular safety profile of 5-HT4 agonists developed for gastrointestinal disorders. Aliment Pharmacol Therapeut 2012;35(7):745–67.

[256]   Quigley EMM. Prucalopride: safety, efficacy and potential applications. Therapeut Adv Gastroenterol 2012;5(1):23–30.

[257]   Sanger GJ, Quigley EMM. Constipation, IBS and the 5-HT4 receptor: what role for prucalopride? Clin Med Insights Gastroenterol 2010;3:21–33.

[258]   Wong BS, Manabe N, Camilleri M. Role of prucalopride, a serotonin (5-HT4) receptor agonist, for the treatment of chronic constipation. Clin Exp Gastroenterol 2010;3:49–56.

[259]   Camilleri M, Deiteren A. Prucalopride for constipation. Exp Opin Pharmacother 2010;11 (3):451–61.

[260]   Frampton JE. Prucalopride. Drugs 2009;69(17):2463–76.

[261]   Sanger GJ. Translating 5-HT4 receptor pharmacology. Neurogastroenterol Motil 2009;21 (12):1235–8.

[262]   Bassotti G, Villanacci V. Prucalopride for chronic constipation. Nat Rev Gastroenterol Hepatol 2009;6(6):324–5.

[263]   Lacy BE, Loew B, Crowell MD. Prucalopride for chronic constipation. Drugs Today 2009;45(12):843–53.

[264]   Yiannakou Y, Piessevaux H, Bouchoucha M, Schiefke I, Filip R, Gabalec L, et al. A randomized, double-blind, placebo-controlled, phase 3 trial to evaluate the efficacy, safety, and tolerability of prucalopride in men with chronic constipation,. Am J Gastroenterol 2015;110(5):741–8.

[265]   Piessevaux H, Corazziari E, Rey E, Simren M, Wiechowska-Kozlowska A, Kerstens R, et al. A randomized, double-blind, placebo-controlled trial to evaluate the efficacy, safety, and tolerability of long-term treatment with prucalopride. Neurogastroenterol Motil 2015;27(6):805–15.

[266]   Van Daele GHP, Bosmans JRMA, Schuurkes JAJ. Preparation of an N-piperidiylbenzofurancarboxamide as an enterokinetic agent, WO 9616060; 1996.

[267]   Castaner J. Prucalopride: treatment of irritable bowel syndrome; 5-HT4 agonist. Drugs Fut 1999;24(7):729–34.

[268]   Liu KK-C, Sakya SM, O'Donnell CJ, Flick AC, Ding HX. Synthetic approaches to the 2010 new drugs. Bioorg Med Chem 2012;20(3):1155–74.

[269]   Baba Y, Usui T, Iwata N. Preparation of benzo[b]furancarboxamide gastrointestinal mobility-enhancing agent 5HT4-receptor agonists, EP 640602; 1995.

[270]   Fancelli D, Caccia C, Severino D, Vaghi F, Varasi M. Preparation of substituted dihydrobenzo[b]furan derivatives as 5-HT4 agonists, WO 9633186; 1996.

[271] Van Daele GHP, Van den Keybus FMA. Preparation of N-(3-methoxy-4-piperidinyl) dihydrobenzofuran-, -dihydro-2H-benzopyran-, or -dihydrobenzodioxincarboxamides for enhancing gastrointestinal motility, EP 389037; 1990.

[272] Van Daele GHP, Bosmans JPRMA, De Cleyn MAJ. Preparation of N-(4-piperidinyl) dihydrobenzofuran- or dihydro-2H-benzopyrancarboxamides as gastrointestinal motility stimulants, EP 445862; 1991.

[273] Rader DJ, Kastelein JJP. Lomitapide and mipomersen: two first-in-class drugs for reducing low-density lipoprotein cholesterol in patients with homozygous familial hypercholesterolemia. Circulation 2014;129(9):1022–32.

[274] Neef D, Berthold HK, Gouni-Berthold I. Lomitapide for use in patients with homozygous familial hypercholesterolemia: a narrative review. Exp Rev Clin Pharmacol 2016;9 (5):655–63.

[275] Davis KA, Miyares MA. Lomitapide: a novel agent for the treatment of homozygous familial hypercholesterolemia. Am J Health Syst Pharm 2014;71(12):1001–8.

[276] Panno MD, Cefalu AB, Averna MR. Lomitapide: a novel drug for homozygous familial hypercholesterolemia. Clin Lipidol 2014;9(1):19–32.

[277] By Anonymous. Lomitapide. Am J Cardiovasc Drugs 2011;11(5):347–52.

[278] Rizzo M. Lomitapide, a microsomal triglyceride transfer protein inhibitor for the treatment of hypercholesterolemia. IDrugs 2010;13(2):103–11.

[279] Gouni-Berthold I, Berthold HK. Mipomersen and lomitapide: two new drugs for the treatment of homozygous familial hypercholesterolemia, Atherosclerosis. Supplements 2015;18:28–34.

[280] deGoma EM. Lomitapide for the management of homozygous familial hypercholesterolemia. Rev Cardiovasc Med 2014;15(2):109–18.

[281] Biller SA, Dickson JK, Lawrence RM, Magnin DR, Poss MA, Sulsky RB, et al. Preparation of heterocyclic inhibitors of microsomal triglyceride transfer protein, US 5739135; 1998.

[282] Biller SA, Dickson JK, Lawrence RM, Magnin DR, Poss MA, Robl JA, et al. Preparation of 9-thioxanthenecarboxamides and 9-fluorenecarboxamides as inhibitors of microsomal triglyceride transfer protein, US 5712279; 1998.

[283] Parthasaradhi RB, Rathnakar RK, Vamsi KB, Mukunda RJ. Process for the preparation of Lomitapide, WO 2016071849; 2016.

[284] Desai SJ, Khera B, Patel JM, Shah HB, Upadhyay ASN, Agravat SN. Processes for the preparation of polymorphic forms of Lomitapide and its salts, US 20160083345.

[285] Vadali LR, Yerva ER, Vemavarapu GPS, Rao PB. Process for preparation of Lomitapide mesylate, WO 2016012934; 2016.

[286] Schuurkes JAJ, Van Nueten JM, Van Daele PGH, Reyntjens AJ, Janssen PAJ. Motor-stimulating properties of cisapride on isolated gastrointestinal preparations of the guinea pig. J Pharmacol Exp Ther 1985;234(3):775–83.

[287] Wiseman LR, Faulds D. Cisapride, An updated review of its pharmacology and therapeutic efficacy as a prokinetic agent in gastrointestinal motility disorders. Drugs 1994;47 (1):116–52.

[288] Cucchiara S. Cisapride therapy for gastrointestinal disease. J Pediatr Gastroenterol Nutr 1966;22(3):259–69.

[289] Barone JA, Jessen LM, Colaizzi JL, Bierman RH. Cisapride: a gastrointestinal prokinetic drug. Ann Pharmacother 1994;28(4):488–500.

[290] Quigley EMM. Cisapride: what can we learn from the rise and fall of a prokinetic? J Dig Dis 2011;12(3):147–56.

[291] Aboumarzouk OM, Agarwal T, Antakia R, Shariff U, Nelson RL. Cisapride for intestinal constipation. Cochrane Database Syst Rev 2011;(1)):CD007780.

[292] Vandenplas Y, Belli DC, Benatar A, Cadranel S, Cucchiara S, Dupont C, et al. The role of cisapride in the treatment of pediatric gastroesophageal reflux. J Pediatr Gastroenterol Nutr 1999;28(5):518−28.

[293] Bedford TA, Rowbotham DJ. Cisapride: drug interactions of clinical significance. Drug Safety 1966;15(3):167−75.

[294] von Kiedrowski R, Huijghebaer S, Raedsch R. Mechanisms of cisapride affecting gall-bladder motility. Dig Dis Sci 2001;46(5):939−44.

[295] Layton D, Key C, Shakir SAW. Prolongation of the QT interval and cardiac arrhythmias associated with cisapride: limitations of the pharmacoepidemiological studies conducted and proposals for the future. Pharmacoepidemiol Drug Safety 2003;12(1):31−40.

[296] Barbey JT, Lazzara R, Zipes DP. Spontaneous adverse event reports of serious ventricular arrhythmias, QT prolongation, syncope, and sudden death in patients treated with cisapride. J Cardiovasc Pharmacol Therapeut 2002;7(2):65−76.

[297] Singh VS, Singh C, Dikshit DK. Reaction of enol ethers with lead tetraacetate: an improved method for the synthesis of α-methoxy ketones. Synth Commun 1998;28 (1):45−9.

[298] Van Daele G. N-(3-Hydroxy-4-piperidinyl)benzamide derivatives, EP 76530; 1983.

[299] Van Daele GHP, De Bruyn MFL, Sommen FM, Janssen ML, Van Nueten JM, Schuurkes JAJ, et al. Synthesis of cisapride, a gastrointestinal stimulant derived from cis-4-amino-3-methoxypiperidine. Drug Dev Res 1986;8(1-4):225−32.

[300] De Knaep AGM, Moens LJR, Rey M. A new process for preparing cisapride. WO 1998;9816511.

[301] Cossy J, Molina JL, Desmurs J-R. A short synthesis of cisapride: a gastrointestinal stimulant derived from cis-4-amino-3-methoxypiperidine. Tetrahedron Leyy 2001;42 (33):5713−15.

[302] Davies SG, Huckvale R, Lorkin TJA, Roberts PM, Thomson JE. Concise, efficient and highly selective asymmetric synthesis of (+)-(3S,4R)-cisapride. Tatrahedron: Assymetry 2011;22(14-15):1591−3.

[303] Kim BJ, Pyun DK, Jung HJ, Kwak HJ, Kim JH, Kim EJ, et al. Practical synthesis of ethyl cis-4-amino-3-methoxy-1-piperidine carboxylate, a key intermediate of cisapride. Synth Commun 2001;31(7):1081−9.

[304] Lu Y, So R, Slemon C, Oudenes J, Ngooi T. Novel oxopiperidines, their preparation and preparation of cisapride from them, WO 9611186; 1996.

[305] Emanuel MB. Towards complete histamine blockade: the role of astemizole. Drugs Today 1986;22(1):39−51.

[306] Awouters FHL, Niemegeers CJE, Janssen PAJ. Pharmacology of the specific histamine H1-antagonist astemizole. Arzneimitt -Forsch 1983;33(3):381−8.

[307] Krstenansky PM, Cluxton Jr. RJ. Astemizole: a long-acting, nonsedating antihistamine. Drug Intell Clin Pharm 1987;21(12):947−53.

[308] Ray JA. The development of a new antihistamine: astemizole. New Eng Reg Allergy Proceed 1985;6(1):71−7.

[309] Richards DM, Brogden RN, Heel RC, Speight TM, Avery GS. Astemizole. A review of its pharmacodynamic properties and therapeutic efficacy. Drugs 1984;28(1):38−61.

[310] Van Wauwe J, Awouters F, Niemegeers CJE, Janssens F, Van Nueten JM, Janssen PAJ. In vivo pharmacology of astemizole, a new type of H1-antihistaminic compound. Arch Int Pharmacodyn Ther 1981;251(1):39−51.

[311] Garcia-Quiroz J, Camacho J. Astemizole: an old anti-histamine as a new promising anti-cancer drug. Anti-cancer Agents Med Chem 2011;11(3):307–14.

[312] Janssens F, Luyckx M, Stokbroekx R, Torremans J. N-Heterocyclyl-4-piperidinamines, US 4219559; 1980.

[313] Janssens F, Torremans J, Janssen M, Stokbroekx RA, Luyckx M, Janssen PAJ. New antihistaminic N-heterocyclic 4-piperidinamines. 1. Synthesis and antihistaminic activity of N-(4-piperidinyl)-1H-benzimidazol-2-amines. J Med Chem 1985;28(12):1925–33.

[314] Janssens F, Torremans J, Janssen M, Stokbroekx RA, Luyckx M, Janssen PAJ. New antihistaminic N-heterocyclic 4-piperidinamines. 2. Synthesis and antihistaminic activity of 1-[(4-fluorophenyl)methyl]-N-(4-piperidinyl)-1H-benzimidazol-2-amines. J Med Chem 1985;28(12):1934–43.

[315] Janssens F, Janssen MAC, Awouters F, Niemegeers CJE, Vanden Bussche G. Chemical development of astemizole-like compounds. Drug Dev Res 1986;8(1-4):27–36.

[316] Thijssen JBA, Knaeps AG, Heykants JJP. Synthesis of tritium- and carbon-14-labeled astemizole (R 43 512). J Labell Comp Radiopharmaceut 1983;20(7):861–8.

[317] Anaya de Parrodi C, Quintero-Cortes L, Sandoval-Ramirez J. A short synthesis of astemizole. Synth Commun 1996;26(17):3323–9.

[318] Bagley JR, Kudzma LV, Lalinde NL, Colapret JA, Huang B, Lin B, et al. Evolution of the 4-anilidopiperidine class of opioid analgesics. Med Res Rev 1991;11(4):403–36.

[319] Stanley TH, Fentanyl J. Pain Sympt. Manage 2005;29(5S):S67–71.

[320] Davis MP. Fentanyl for breakthrough pain: a systematic review. Exp Rev Neurotherapeut 2011;11(8):1197–216.

[321] Stanley TH. The history and development of the fentanyl series. J Pain Sympt Manage 1992;7(3 Suppl):S3–7.

[322] Casy AF. Opioid receptors and their ligands: recent developments. Adv Drug Res 1989;18:177–289.

[323] Casy AF, Huckstep MR. Structure-activity studies of fentanyl. J Pharm Pharmacol 1988;40(9):606–8.

[324] Casy AF, Hassan MMA, Simmonds AB, Staniforth D. Structure-activity relations in analgesics based on 4-anilinopiperidine. J Pharm Pharmacol 1969;21(7):434–40.

[325] Famini GR, Ashman WP, Mickiewicz AP, Wilson LY. Using theoretical descriptors in quantitative structure-activity relationships: opiate receptor activity by fentanyl-like compounds. Quant Struct-Act Relat 1992;11(2):162–70.

[326] Vuckovic S, Prostran M, Ivanovic M, Dosen-Micovic Lj, Todorovic Z, Nesic Z, et al. Fentanyl analogs: structure-activity-relationship study. Curr Med Chem 2009;16 (19):2468–74.

[327] Dosen-Micovic Lj. Molecular modeling of fentanyl analogs. J Serb Chem Soc 2004;69 (11):834–54.

[328] Janssen PAJ, Gardocki JF. Method for producing analgesia, US 3141823; 1964.

[329] Janssen PAJ. N-(1-Aralkyl-4-piperidyl)alkanoic acid anilides, FR M2430; 1964.

[330] Janssen PAJ, Niemegeers CJ, Dony JG. The inhibitory effect of fentanyl and other morphine-like analgesics on the warm water induced tail withdrawal reflex in rats. Arzneimit -Forsch 1963;13:502–7.

[331] Janssen PAJ. 1-(g-Aroylpropyl)-4-(N-arylacylamino)piperidines. FR 1344366; 1963.

[332] Jonczyk A, Jawdosiuk M, Makosza M, Czyzewski J. Search for a new method for synthesis of the analgesic agent "Fentanyl". Przem Chem 1978;57(3):131–4.

[333] Zee S-H, Wang W-K. A new process for the synthesis of fentanyl. J Chin Chem Soc 1980;27(4):147–9.

[334] Pease JP, Lepine AJ, Smith CM. Process for preparation fentanyl and fentanyl intermediates, US 20130281702; 2013.

[335] Gupta PK, Manral L, Ganesan K, Malhotra RC, Sekhar K. Process for preparation of fentanyl, WO 2009116084; 2009.

[336] Vardanyan RS, Hruby VJ. Fentanyl-related compounds and derivatives: current status and future prospects for pharmaceutical applications. Fut Med Chem 2014;6 (4):385−412.

[337] George AV, Lu JJ, Pisano MV, Metz J, Erickson TB. Carfentanil - an ultra potent opioid. Am J Emerg Med 2010;28(4):530−2.

[338] Cookson RF. Carfentanil and lofentanil. Clinics Anaesthesiol 1983;1(1):172−9 156-8

[339] Van Daele PGH, De Bruyn MFL, Boey JM, Sanczuk S, Agten JTM, Janssen PAJ. Synthetic analgesics: N-(1-[2-arylethyl]-4-substituted 4-piperidinyl)-N-arylalkanamides. Arzneimitt -Forsch 1976;26(8):1521−31.

[340] Bagley JR, Thomas SA, Rudo FG, Spencer HK, Doorley BM, Ossipov MH, et al. New 1-(heterocyclylalkyl)-4-(propionanilido)-4-piperidinyl methyl ester and methylene methyl ether analgesics. J Med Chem 1991;34(2):827−41.

[341] Feldman PL, Brackeen MF. A novel route to the 4-anilido-4-(methoxycarbonyl)piperidine class of analgetics. J Org Chem 1990;55(13):4207−9.

[342] Janssen PAJ, Van Daele GHP. 4-Anilinopiperidines, US 4179569; 1979.

[343] Reiff LP, Sollman PB. Carfentanil and related analgesics, US 5106983; 1992.

[344] Walz AJ, Hsu F. Synthesis of of 4-anilinopiperidine methyl esters, intermediates in the production of carfentanil, sufentanil, and remifentanil. Tetrahedron Lett 2014;55(2):501−2.

[345] Feldman PL. Discovery and development of the ultrashort-acting analgesic remifentanil. In: Chorghade MS, editor. From drug discovery and development, 1. New York: Wiley; 2006. p. 339−51.

[346] Scott LJ, Perry CM. Remifentanil. A review of its use during the induction and maintenance of general anaesthesia. Drugs 2005;65(13):1793−823.

[347] Beers R, Camporesi E. Remifentanil update: clinical science and utility. CNS Drugs 2004;18(15):1085−104.

[348] Videira RLR, Cruz JRS. Remifentanil in the clinical practice. Rev Bras Anesteziol 2004;54(1):114−28.

[349] Wilhelm W, Wrobel M, Kreuer S, Larsen R. Remifentanil. An update. Anaesthesist 2003;52(6):473−94.

[350] Sivak EL, Davis PJ. Review of the efficacy and safety of remifentanil for the prevention and treatment of pain during and after procedures and surgery. Loc Reg Anesth 2010;3:35−43.

[351] Stroumpos C, Manolaraki M, Paspatis GA. Remifentanil, a different opioid: potential clinical applications and safety aspects. Exp Opin Drug Safety 2010;9(2):355−64.

[352] Battershill AJ, Keating GM. Remifentanil: a review of its analgesic and sedative use in the intensive care unit. Drugs 2006;66(3):365−85.

[353] James MK. Remifentanil and anesthesia for the future. Exp Opin Invest Drugs 1994;3 (4):331−40.

[354] Patel SS, Spencer CM. Remifentanil. Drugs 1996;52(3):417−27.

[355] Thompson JP, Rowbotham DJ. Remifentanil-an opioid for the 21st century. Brit J Anaesth 1996;76(3):341−3.

[356] Egan TD. Remifentanil pharmacokinetics and pharmacodynamics. A preliminary appraisal. Clin Pharmacokinet 1995;29(2):80−94.

[357] Glass PSA, Gan TJ, Howell S. A review of the pharmacokinetics and pharmacodynamics of remifentanil. Anesth Analg 1999;89(Suppl. 4):S7−14.

[358] Servin F, Desmonts JM, Watkins WD. Remifentanil as an analgesic adjunct in local/regional anesthesia and in monitored anesthesia care. Anesth Analg 1999;89(Suppl. 4): S28−32.

[359] Rivosecchi RM, Rice MJ, Smithburger PL, Buckley MS, Coons JC, Kane-Gill SL. An evidence based systematic review of remifentanil associated opioid-induced hyperalgesia. Exp Opin Drug Safety 2014;13(5):587−603.

[360] Hinova A, Fernando R. Systemic remifentanil for labor analgesia. Anesth Analg 2009;109(6):1925−9.

[361] Muchatuta NA, Kinsella SM. Remifentanil for labour analgesia: time to draw breath? Anaesthesia 2013;68(3):231−5.

[362] Feldman PL, James MK, Brackeen MF, Johnson MR, Leighton HJ. Preparation of N-phenyl-N-(4-piperidinyl)amides useful as analgesics, EP 383579; 1990.

[363] Feldman PL, James MK, Brackeen MF, Bilotta JM, Schuster SV, Lahey AP, et al. Design, synthesis, and pharmacological evaluation of ultrashort- to long-acting opioid analgesics. J Med Chem 1991;34(7):2202−6.

[364] Monk JP, Beresford R, Ward A. Sufentanyl. A review of its pharmacological properties and therapeutic use. Drugs 1988;36(3):286−313.

[365] Sanford Jr. TJ. A review of sufentanyl Semin. Anesth 1988;7(2):127−36.

[366] Maciejewski D. Sufentanyl in anaesthesiology and intensive therapy. Anaesthesiol Intens Ther 2012;44(1):35−41.

[367] Rosow CE. Sufentanyl citrate: a new opioid analgesic for use in anesthesia. Pharmacother 1984;4(1):11−19.

[368] Grass JA. Sufentanyl: clinical use as postoperative analgesic−epidural/intrathecal route. J Pain Sympt Manage 1992;7(5):271−86.

[369] Frampton JE. Sublingual sufentanyl: a review in acute postoperative pain. Drugs 2016;76 (6):719−29.

[370] Minkowitz HS. A review of sufentanyl and the sufentanyl sublingual tablet system for acute moderate to severe pain. Pain Manag 2015;5(4):237−50.

[371] Savoia G, Loreto M, Gravino E. Sufentanyl: an overview of its use for acute pain management. Minerva Anestesiol 2001;67(9 Suppl 1):206−16.

[372] Joshi GP. Sufentanyl for chronic pain management. Future Neurol 2010;5(6):791−6.

[373] Celleno D. Spinal sufentanyl. Anaesthesia 1998;53(Suppl. 2):49−50.

[374] Janssen PAJ, Van Daele, Georges HP. 4-Anilinopiperidines, US 4179569; 1979.

[375] Niemegeers CJE, Schellekens KHL, Van Bever WFM, Janssen PAJ. Sufentanil, a very potent and extremely safe intravenous morphine-like compound in mice, rats and dogs. Arzneimitt -Forsch 1976;26(8):1551−6.

[376] Jacob M, Killgore JK. New methods for the syntheses of alfentanil, sufentanil and remifentanil, WO 2001040184; 2001.

[377] Mathew J. Process for the preparation of sufentanil derivatives by carbene addition/aminolysis of 4-piperidone, WO 9509152; 1995.

[378] Puthuparampil PK, Eturi SR, Carroll R. Process for making sufentanil, WO 2008005423; 2008. From PCT Int. Appl., 2008, WO 2008005423 A1 20080110.

[379] Shin D, Ryu J, Hyun S, Park H, Jeon R, Suh Y. Total synthesis of sufentanil, rch. Pharm Res 1999;22(4):398−400.

[380] Rosow Cl. Afetanil. Semin Anesth 1988;7(2):107−12.

[381] Larijani GE, Goldberg ME. Alfentanil hydrochloride: a new short-acting narcotic analgesic for surgical procedures. Clin Pharm 1987;6(4):275–82.

[382] Cookson RF, Niemegeers CJE, Vanden Bussche G. The development of alfentanil. Brit J Anaesth 1983;55(Suppl. 2):147–55.

[383] Kay B. Alfentanil. Clin Anaesthesiol 1983;1(1):172–9 143-6

[384] Chrubasik S, Chrubasik J, Friedrich G. Clinical use of alfentanil. Anaesthesiol Reanim 1994;19(3):60–6.

[385] Reitz JA. Alfentanil in anesthesia and analgesia. Drug Intel Clin Pharm 1986;20 (5):335–41.

[386] Janssens F. N-Phenyl-N-(4-piperidinyl)-amides, DE 2819873; 1978.

[387] Janssens F, Torremans J, Janssen PAJ. Synthetic 1,4-disubstituted 1,4-dihydro-5H-tetrazol-5-one derivatives of fentanyl: Alfentanil (R 39209), a potent, extremely short-acting narcotic analgesic. J Med Chem 1986;29(11):2290–7.

[388] Jacob M, Killgore JK. New methods for the syntheses of alfentanil, sufentanil and remifentanil, WO 2001040184; 2001.

[389] Carmeliet E, Janssen PA, Marsboom R, Van Nueten JM, Xhonneux R. Antiarrhythmic, electrophysiologic and hemodynamic effects of lorcainide. Arch Int Pharmacodyn Ther 1978;231(1):104–30.

[390] Keefe DL. Pharmacology of lorcainide. Am J Cardiol 1984;54(4):18–21.

[391] Eiriksson CE, Brogden RN. Lorcainide. A preliminary review of its pharmacodynamic properties and therapeutic efficacy. Drugs 1984;27(4):279–300.

[392] Amery WK, Aerts T. Lorcainide, new drugs annual: cardiovasc. Drugs 1983;1:109–32.

[393] Koch H. Lorcainide hydrochloride. Drugs Today 1982;18(1):12–22.

[394] Amery WK, Heykants JJ, Xhonneux R, Towse G, Oettel P, Gough DA, et al. Lorcainide (R 15 889), a first review. Acta Cardiol 1981;36(3):207–34.

[395] Alkhatib S, Pritchett E. Clinical features of wolff-parkinson-white syndrome. Am Heart J 1999;138(3):403–13.

[396] Cowley AJ, Skene A, Stainer K, Hampton JR. The effect of lorcainide on arrhythmias and survival in patients with acute myocardial infarction: an example of publication bias. Int J Cardiol 1993;40:161–6.

[397] Chalmers I. The true lorcainide story. BMJ 2015;350:h287.

[398] Hermans HKF, Sanczuk S. N-Aryl-N-(4-piperidinyl)arylacetamides, US 4197303; 1980.

[399] Sanczuk S, Hermans HKF. N-Aryl-N-(1-alkyl-4-piperidinyl)arylacetamides, US 4197303; 1978.

[400] Thijssen JBA, Knaeps AG, Heykants JJP. Synthesis of tritium-labeled lorcainide monohydrochloride (R 15889). J Labell Comp Radiopharm 1981;18(9):1379–86.

[401] Kraushaar A. A new antihistamine: 4-(N-benzylanilino)-1-methylpiperidine. Deutsch Med Wochenschr 1950;75:1148–9.

[402] Haas H. Antihistamine-combined preparations; experimental trials with soventol. Arzneimitt -Forsch 1953;3(7):328–33.

[403] Getzler NA, Ereaux LP. Evaluation of thenalidine tratrate (sandostene) in dermatological disorders. Can Med Assoc J 1959;80(6):445–8.

[404] Kallischnigg R. 4-(N-Phenyl-N-benzylamino)-1-alkylpiperidine and derivatives, DE 891547; 1953.

[405] Kallischnigg R. 4-(N-Phenylbenzylamino)-1-alkylpiperidines substituted in the aromatic radicals, US 2683714; 1954.

[406] Stoll A, Bourquin JP. Piperidine derivatives, US 2717251; 1955.

[407] Stoll A, Bourquin JP. Piperidine derivatives, US 2757175; 1956.

[408] Mealy N, Ngo J, Castaner J. Mizolastine. Drugs Fut 1996;21(8):799–804.

[409] Simons FER. Mizolastine: antihistaminic activity from preclinical data to clinical evaluation. Clin Exp Allergy 1999;29(Suppl. 1):3–8.

[410] Selve N, Pichat PH, Goldhill J, Depoortere H, Arbilla S. Pharmacological profile of mizolastine, a novel histamine H1 receptor antagonist. In: Marone G, Lichtenstein LM, Galli SJ, editors. From mast cells and basophils. Academic Press; 2000. p. 625–40.

[411] Triggiani M, Palumbo C, Gentile M, Granata F, Marone G. Antihistaminic and anti-inflammatory effects of mizolastine, Mast Cells and BasophilsIn: Marone G, Lichtenstein LM, Galli SJ, editors. Academic Press; 2000. p. 665–72.

[412] Prakash A, Lamb HM. Mizolastine: a review of its use in allergic rhinitis and chronic idiopathic urticarial. BioDrugs 1998;10(1):41–63.

[413] Manoury P, Binet J, DeFosse G. 2-[4-Pyrimidin-2-yl-amino)piperidin-1-yl]benz-imidazole compound, US 4912219; 1990.

[414] No Inventor data available, 2-[4-(Pyrimidin-2-ylamino)piperidin-1-yl]benzimidazole derivatives as allergy inhibitors, JP 62061979; 1987.

[415] Mohr E, Nair NPV, Sampson M, Murtha S, Belanger G, Pappas B, et al. Treatment of Alzheimer's disease with sabeluzole: functional and structural correlates. Clin Neuropharmacol 1997;20(4):338–45.

[416] Uberti D, Rizzini C, Galli P, Pizzi M, Grilli M, Lesage A, et al. Priming of cultured neurons with sabeluzole results in long-lasting inhibition of neurotoxin-induced tau expression and cell death. Synapse 1997;26(2):95–103.

[417] Aldenkamp AP, Overweg J, Smakinan J, Beun AM, Diepman L, Edelbroek P, et al. Effect of sabeluzole (R 58 735) on memory functions in patients with epilepsy. Neuropsychobiol 1995;32(1):37–44.

[418] Geerts H, Nuydens R, De Jong M, Cornelissen F, Nuyens R, Wouters L. Sabeluzole stabilizes the neuronal cytoskeleton. Neurobiol Aging 1996;17(4):573–81.

[419] Geerts H, Nuydens R, Cornelissen F, De Brabander M, Pauwels P, Janssen PAJ, et al. Sabeluzole, a memory-enhancing molecule, increases fast axonal transport in neuronal cell cultures. Exp Neurol 1992;117(1):36–43.

[420] Clincke GHC, Tritsmans L. Sabeluzole (R 58 735) increases consistent retrieval during serial learning and relearning of nonsense syllables. Psychopharmacol 1988;96(3):309–10.

[421] Tritsmans L, Clincke G, Amery WK. The effect of sabeluzole (R 58735) on memory retrieval functions. Psychopharmacol 1988;94(4):527–31.

[422] Clincke GHC, Tritsmans L, Idzikowski C, Amery WK, Janssen PAJ. The effect of R 58 735 (sabeluzole) on memory functions in healthy elderly volunteers. Psychopharmacol 1988;94(1):52–7.

[423] Stokbroekx RA, Luyckx MGM, Janssens FE. Benzoxazol- and benzothiazolamine derivatives, EP 184257; 1986.

[424] Werbrouck L, Megens AAHP, Stokbroekx RA, Niemegeers CJE. Comparison of the in vivo pharmacological profiles of sabeluzole and its enantiomers. Drug Dev Res 1991;24(1):41–51.

[425] Lesage AS, Peeters L, Leysen JE. Lubeluzole, a novel long-term neuroprotectant, inhibits the glutamate-activated nitric oxide synthase pathway. J Pharmacol Exp Ther 1996;279(2):759–66.

[426] Diener HC, Hacke W, Hennerici M, Radberg J, Hantson L, Keyser J. De, Lubeluzole in acute ischemic stroke: a double-blind, placebo-controlled phase II trial. Stroke 1996;27(1):76–81.

[427]   Culmsee C, Junker V, Wolz P, Semkova I, Krieglstein J. Lubeluzole protects hippocampal neurons from excitotoxicity in vitro and reduces brain damage caused by ischemia. Eur J Pharmacol 1998;342(2/3):193−201.

[428]   Grotta J. Lubeluzole treatment of acute ischemic stroke. Stroke 1997;28(12):2338−46.

[429]   Diener HC, Cortens M, Ford G, Grotta J, Hacke W, Kaste M, et al. Lubeluzole in acute ischemic stroke treatment: a double-blind study with an 8-hour inclusion window comparing a 10-mg daily dose of lubeluzole with placebo. Stroke 2000;31(11):2543−51.

[430]   Mueller RN, Deyo DJ, Brantley DR, Disterhoft JF, Zornow MH. Lubeluzole and conditioned learning after cerebral ischemia. Exp Brain Res 2003;152(3):329−34.

[431]   Koinig H, Vornik V, Rueda C, Zornow MH. Lubeluzole inhibits accumulation of extracellular glutamate in the hippocampus during transient global cerebral ischemia. Brain Res 2001;898(2):297−302.

[432]   Maiese K, TenBroeke M, Kue I. Neuroprotection of lubeluzole is mediated through the signal transduction pathways of nitric oxide. J Neurochem 1997;68(2):710−14.

[433]   Stokbroekx RA, Grauwels GAJ. Preparation of 4-[(2-benzothiazolyl)methylamino]-α-[(3, 4-difluorophenoxy)methyl]-1-piperidineethanol as stroke inhibitor, EP 501552; 1992.

[434]   Cavalluzzi MM, Viale M, Bruno C, Carocci A, Catalano A, Carrieri A, et al. A convenient synthesis of lubeluzole and its enantiomer: evaluation as chemosensitizing agents on human ovarian adenocarcinoma and lung carcinoma cells. Bioorg Med Chem Lett 2013;23(17):4820−3.

[435]   Kommi DN, Kumar D, Seth K, Chakraborti AK. Protecting group-Free Concise Synthesis of (RS)/(S)-Lubeluzole. Org Lett 2013;15(6):1158−61.

[436]   Schotte A, Janssen PFM, Gommeren W, Luyte WHML, Van Gompel P, Lesage AS, et al. Psychopharmacol 1996;124(1/2):57−73.

[437]   Van Craenenbroeck K, Gellynck E, Lintermans B, Leysen JE, Van Tol HHM, Haegeman G, et al. Influence of the antipsychotic drug pipamperone on the expression of the dopamine D4 receptor. Life Sci 2006;80(1):74−81.

[438]   Haegeman J, Duyck F. A retrospective evaluation of pipamperone (Dipiperon) in the treatment of behavioural deviations in severely mentally handicapped. Acta Psych Belg 1978;78(2):392−8.

[439]   Noordhuizen GJ. Treating severe maladjustment with pipamperone (Dipiperon). Acta Psych Belg 1977;77(6):754−60.

[440]   Squelart P, Saravia J. Pipamperone (Dipiperon), a useful sedative neuroleptic drug in troublesome chronic psychotic patients. Acta psychiatrica Belgica 1977;77(2):284−93.

[441]   Van Renynghe de Voxvrie G, De Bie M. Character neuroses and behavioural disorders in children: their treatment with pipamperone (Dipiperon). A clinical study. Acta Psych Belg 1976;76(4):688−95.

[442]   Janssen PAJ. 1-(Arylalkyl)-4-piperidinecarboxamides, BE 610830; 1962.

[443]   van de Westeringh C, van Daele P, Hermans B, van der Eycken C, Boey J, Janssen PAJ. 4-Substituted piperidines. I. Derivatives of 4-tertiaryamino-4-piperidinecarboxamides. J Med Chem 1964;7(5):619−23.

[444]   Cathelin M, Rosemblatt JM, Conseiller C, Viars P. Analgesic activity of piritramide. Anesth Analg Reanim 1976;33(3):443−56.

[445]   Dopfmer UR, Schenk MR, Kuscic S, Beck DH, Dopfmer S, Kox WJ. A randomized controlled double-blind trial comparing piritramide and morphine for analgesia after hysterectomy. Eur J Anaesthesiol 2001;18(6):389−93.

[446]   Feifel G, Thurmayr R. Clinical evaluation of the analgesic effect of drugs in postoperative pains. (Comparative clinical trial of piritramide). Der Anaesthesist 1970;19(10):369−73.

[447] De Castro J, Van de Water A, Wouters L, Xhonneux R, Reneman R, Kay B. Comparative study of cardiovascular, neurological and metabolic side-effects of eight narcotics in dogs: pethidine, piritramide, morphine, phenoperidine, fentanyl, R39209, sufentanil, R34995. Acta Anaesthesiol Belg 1979;30(1):5−99.

[448] Bouillon T, Kietzmann D, Port R, Meineke I, Hoeft A. Population pharmacokinetics of piritramide in surgical patients. Anesthesiol 1999;90(1):7−15.

[449] Kietzmann D, Bouillon T, Hamm C, Schwabe K, Schenk H, Gundert-Remy U, et al. Pharmacodynamic modelling of the analgesic effects of piritramide in postoperative patients. Acta Anaesthesiol Belg 1997;41(7):888−94.

[450] Janssen PAJ. Preparation of 1-ω,ω-diphenylalkyl)-4-aminopiperidine- or 1-(ω,ω-diphenylalkyl)-3-aminopyrrolidine-3-carboxamides, DE 1238472; 1967.

[451] Janssen PAJ. 1-(ω,ω-Diphenylalkyl)-4-amino-4-piperidinecarboxamide derivatives, BE 606850; 1961.

[452] Kishi T, Matsunaga S, Matsuda Y, Iwata N. Iminodibenzyl class antipsychotics for schizophrenia: a systematic review and meta-analysis of carpipramine, clocapramine, and mosapramine. Neuropsychiatr Dis Treat 2014;10:2339−51.

[453] Deniker P, Loo H, Zarifian E, Verdeaux G, Garreau G. A new psychotropic drug: carpipramine, intermediate compound between 2 therapeutic classes. Encephale 1977;3 (2):133−48.

[454] Nakanishi M, Tsumagari T, Okada T, Kase Y. Pharmacological studies on 5-[3-(4-piperidino-4-carbamoylpiperidino)propyl]-10,11-dihydro-5(H)-dibenz[b,f]azepine (carpipramine) dihydro-chloride, a new psychotropic agent. Arzneimitt -Forsch 1968;18(11):1435−41.

[455] des Lauriers A, Clery-Melin P, Allilaire JF, Baruch P, Ammar S, Vilamot B. Carpipramine. Encephale 1990;16(1):53.

[456] Deniker P, Poirier MF. Position of carpipramine among psychotropic drugs. Encephale 1978;4(5):553−67.

[457] Feline A, Pilate C. Carpipramine, a new psychotropic drug. Ann Medico-psychol 1979;137(3-4):230−6.

[458] Deniker P, Loo H, Zarifian E, Garreau G, Benyacoub A, Roux JM. Carpipramine, a specific psychotropic drug between neuroleptics and anti-depressive drugs, Ann. Medico-psychol 1978;136(9):1069−80.

[459] Benyacoub A, Roux JM. Carpipramine and drug addiction. Encephale 1978;4(Suppl. 5):596−600.

[460] Peyrouzet JM, Perier M. Carpipramine in neuroses. Encephale 1978;4(Suppl. 5):577−85.

[461] Garreau G, Sechter D. Carpipramine in psychoses. Encephale 1978;4(Suppl. 5):569−76.

[462] Clavelou P, Fialip J, Monange P, Lavarenne J, Dordain G, Tournilhac M. Visual hallucinations and carpipramine. Therapie 1988;43(3):239−40.

[463] Setoguchi M, Sakamori M, Takehara S, Fukuda T. Effects of iminodibenzyl antipsychotic drugs on cerebral dopamine and α-adrenergic receptors. Eur J Pharmacol 1985;112(3):313−22.

[464] Nakanishi M, Tashiro C, Munakata T, Araki K, Tsumagari T, Imamura H. Piperidine derivatives. 1. J Med Chem 1970;13(4):644−8.

[465] Nakanishi M, Tsumagari T, Nakanishi A. Psychotropic drugs. XI. Pharmacology of 3-chloro-5-[3-(4-piperidino-4-carbamoylpiperidino)propyl]-10,11-dihydro-(5H)-dibenz-[b,f]-azepine dihydrochloride monohydrate. Arzneimitt -Forsch 1971;21(3):391−5.

[466] Kurihara M, Tsumagari T, Setoguchi M, Fukuda T. A study on the pharmacological and biochemical profile of clocapramine. Int Pharmacopsych 1982;17(2):73−90.

[467] Nakanishi M, Tashiro C. Psychotropic dibenzazepine derivatives, DE 1905765; 1969.

[468] Hasegawa G, Tashiro C. 5-(Piperidinoalkyl)-10,11-dihydro-5H-dibenz[b,f]azepines, JP 50111087; 1975.

[469] Curran MP, Robyns GW, Scott LJ, Perry CM. Alvimopan. Drugs 2008;68(14):2011−19.

[470] Delaney CP, Yasothan U, Kirkpatrick P. Alvimopan. Nat Rev Drug Disc 2008;7 (9):727−8.

[471] Leslie JB. Alvimopan: a peripherally acting Mu-opioid receptor antagonist. Drugs Today 2007;43(9):611−25.

[472] Leslie JB. Alvimopan for the management of postoperative ileus. Ann Pharmacother 2005;39(9):1502−15-10.

[473] Neary P, Delaney CP. Alvimopan. Exp Opin Invest Drugs 2005;14(4):479−88.

[474] Schmidt WK. Alvimopan (ADL 8-2698) Is a Novel Peripheral Opioid Antagonist. Am J Surg 2001;182(5A):27S−38S.

[475] Bream-Rouwenhorst HR, Cantrell MA. Alvimopan for postoperative ileus. Am J Health-Syst Pharm 2009;66(14):1267−77.

[476] Rodriguez RW. Off-label uses of alvimopan and methylnaltrexone. Am J Health-Syst Pharm 2014;71(17):1450−5.

[477] Marderstein EL, Delaney CP. Management of postoperative ileus: focus on alvimopan, Therapeut. Clin Risk Manage 2008;4(5):965−73.

[478] Cantrell BE, Zimmerman DM. Preparation of ω-(4-phenylpiperidino)alkanoates as peripheral opioid receptor receptor antagonists, EP 506478; 1992.

[479] Cantrell BE, Zimmerman DM. Preparation of phenylpiperidine derivatives as peripheral opioid antagonists, CA 2064373; 1992.

[480] Cantrell BE, Zimmerman DM. Peripherally selective piperidine carboxylate opioid antagonists, US 5250542; 1993.

[481] Zimmerman DM, Gidda JS, Cantrell BE, Schoepp DD, Johnson BG, Leander JD. Discovery of a potent, peripherally selective trans-3,4-dimethyl-4-(3-hydroxyphenyl) piperidine opioid antagonist for the treatment of gastrointestinal motility disorders. J Med Chem 1994;37(15):2262−5.

[482] Werner JA, Cerbone LR, Frank SA, Ward JA, Labib P, Tharp-Taylor RW, et al. Synthesis of trans-3,4-dimethyl-4-(3-hydroxyphenyl)piperidine opioid antagonists: application of the cis-thermal elimination of carbonates to alkaloid synthesis. J Org Chem 1996;61(2):587−97.

Chapter 6

Piperidin-4-Ylidene Substituted Tricyclic Compounds

Antihistamine and antiserotonin as well as central depressant or antidepressant actions were likewise reported for a series of piperidylidene-substituted tricyclic compounds, including (5*H*-dibenzo [*a,d*]cyclohepten; 6,11-dihydro-5*H*-benzo[5,6]cyclohepta[1,2-*b*]pyridine; 9,10-dihydro -4*H*-benzo-[4,5]cyclohepta [1,2]-thiophenexanthene; thioxanthene; dibenzo[*b,e*]thiepin; perithiadene, etc.; and piperidylidene derivatives. Some of them, which became widespread antihistamine drugs, are presented on Fig. 6.1.

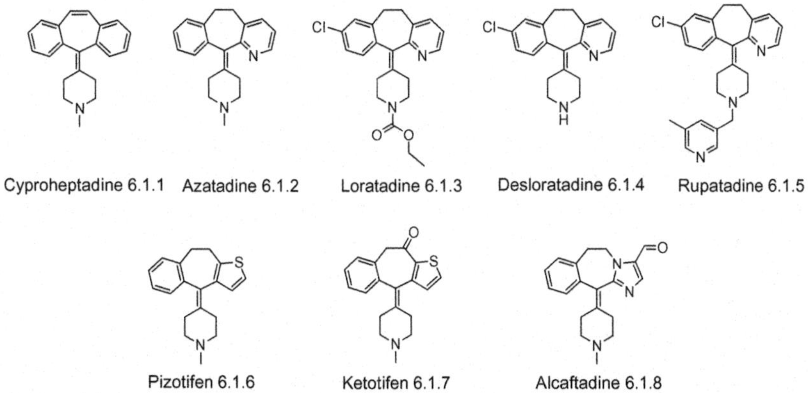

Cyproheptadine 6.1.1   Azatadine 6.1.2   Loratadine 6.1.3   Desloratadine 6.1.4   Rupatadine 6.1.5

Pizotifen 6.1.6   Ketotifen 6.1.7   Alcaftadine 6.1.8

FIGURE 6.1 Piperidylidene-substituted tricyclic compounds representing antihistamine drugs.

6.1 CYPROHEPTADINE (6511)

Cyproheptadine (**6.1.4**) (Periactin) is a first-generation antihistamine available in over-the-counter drugs and marketed for the treatment of various allergic problems caused by seasonal allergies, food, and as a relief from the symptoms of hay fever. Cyproheptadine is a medication used to relieve allergy symptoms such as watery eyes, runny nose, itching eyes/nose, sneezing, hives, and itching. It is also used to treat vasomotor mucosal edema and edema of the throat, and to relieve itching associated with some skin conditions.

Piperidine-Based Drug Discovery. DOI: http://dx.doi.org/10.1016/B978-0-12-805157-3.00006-5
223

Cyproheptadine can be used in chronic allergic and pruritic conditions, such as dermatitis, including neurodermatitis and neurodermatitis circumscripta, eczema, eczematoid dermatitis, dermatographism, conjunctivitis due to inhalant allergens, urticarial, angioneurotic edema, drug and serum reactions, and pruritus of chicken pox.

It may also be used also for the treatment of migraine and related headache, a hormone disorder (Cushing's syndrome), sexual function problems due to certain drugs, and eating disorders (anorexia nervosa).

Cyproheptadine is a histamine H1 blocker and a serotonin antagonist and anticholinergic that can be used as an antipruritic, appetite stimulant, and for the postgastrectomy dumping syndrome, etc. Cyproheptadine was noted to cause unexpected weight gain. It has been marketed in Europe as a drug for weight gain and evaluated for anorexia. Cyproheptadine causes drowsiness, as is common for first-generation antihistamines. Other common side effects include: dizziness, blurred vision, constipation, dry mouth, throat, or nose, excitability, nausea, nervousness, and restlessness [1].

A breakthrough finding was reported recently in two advanced hepatocellular carcinoma patients, of whom one achieved complete remission of liver tumors and the other a normalized α-fetoprotein level, along with complete remission of their lung metastases, after the use cyproheptadine [2].

Cyproheptadine (**6.1.1**) synthesis started from a Grignard reaction of (1-methylpiperidin-4-yl)magnesium chloride (**6.1.9**) with dibenzosuberone (**6.1.10**) in tetrahydrofuran, which gave tertiary alcohol (**6.1.11**). The last was dissolved in acetic acid, the solution saturated with dry hydrochloric acid while cooling, treated with acetic anhydride and heated on a steam bath to give desired cyproheptadine (**6.1.1**) [3,4]. Later, a serious improvement, employment of in situ-prepared homogeneous catalyst – $ZnCl_2$, Me_3SiCH_2MgCl, and LiCl – effectively increased yields of Grignard reaction up to 99% [5].

Another approach for the synthesis of cyproheptadine (**6.1.1**) was demonstrated on implementation of McMurry's method for titanium reductive mixed carbonyl coupling for generation of exocyclic double bond (**6.1.13**) using ethyl 4-oxopiperidine-1-carboxylate (**6.1.12**) and dibenzosuberone (**6.1.10**) in dimethoxy ethane in the presence of low-valent titanium ($TiCl_3$/Li) [6–8] (Scheme 6.1).

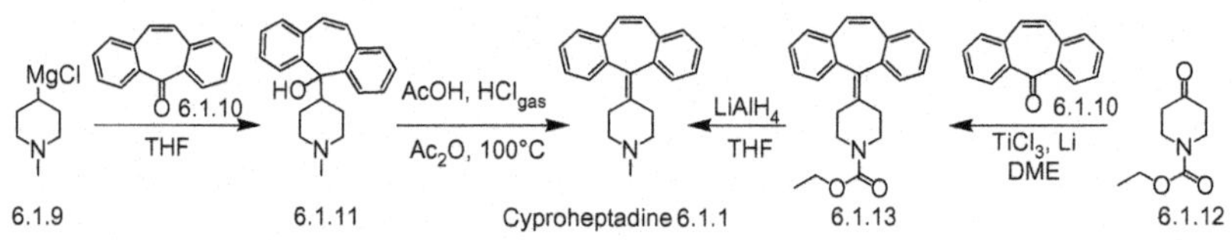

SCHEME 6.1 Synthesis of cyproheptadine.

6.2 AZATADINE (395)

Azatadine (**6.1.2**) (Optimine) is an analog of cyproheptadine that bears a structural resemblance to cyproheptadine and differs from cyproheptadine (**6.1.1**) by the replacement of the 4-carbon by nitrogen and saturation of the 10,11-olefinic linkage.

It is a first-generation antihistamine and anticholinergic and possesses a long acting, potent antihistaminic, anticholinergic, antiserotinin, and antiana- phylactic properties. Antihistamine potency of azatadine's is equal to cypro- heptadine, but it is a more potent antianaphylactic agent and has greater therapeutic indices than cyproheptadine.

Azatadine is also indicated for the relief of symptoms associated with upper respiratory mucosal congestion in perennial and allergic rhinitis, and for the relief of nasal congestion, and it can also be found as an ingredient in some cough and cold preparations. General adverse effects are drowsiness, sedation, dizziness, disturbed coordination, fatigue, confusion, euphoria, excitation, nervousness, restlessness, insomnia, tremor, irritability, dryness of nose and throat, tinnitus, and blurred vision. Azatadine has been succeeded by other anti- histamines, and marketing approvals have been widely withdrawn [9−14].

Azatadine (**6.1.2**) was proposed to synthesize by the method implemented to cyproheptadine synthesis.

The Grignard reagent derived from 1-methyl-4-chloropiperidine (**6.1.9**) was added to the tricyclic ketone − 5,6-dihydro-11H-benzo[5,6]cyclohepta [1,2-b]pyridin-11-one (**6.1.14**), producing the tertiary carbinol (**6.1.15**). Dehydration of the obtained carbinol to the exocyclic piperidylidene deriva- tive was accomplished in a solution of warm sulfuric acid in an acetic anhy- dride giving desired azatadine (**6.1.2**) [15,16]. The second method consists in reaction of the same Grignard reagent (**6.1.9**) with 3-phenethylpicolinonitrile (**6.1.16**). Obtained intermediate (1-methylpiperidin-4-yl)(3-phenethylpyridin- 2-yl)methanimine was hydrolyzed to corresponding ketone (**6.1.17**), which was cyclized to desired azatadine (**6.1.2**) on heating at 160°C for 6 hours in polyphosphoric acid [17] (Scheme 6.2).

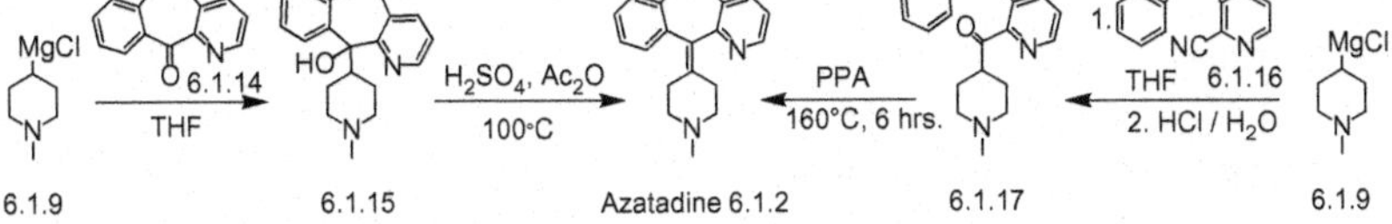

SCHEME 6.2 Synthesis of azatadine.

6.3 LORATADINE (3853) AND DESLORATADINE (1609)

The evolution of azatadine (**6.1.2**) to loratadine (**6.1.3**) (Claritin) and its active metabolite, desloratadine (**6.1.4**) (Clarinex), is an excellent example that shows that a minor change in the molecular structure can have an enormous impact on the pharmacological profile. Further optimization of azatadine afforded the 8-chloroderivative with a longer duration of action. Its active metabolite desloratadine (**6.1.4**) was launched in 2001.

Insertion of a chloro atom into 8- position of 6,11-dihydro-5*H*-benzo[5,6] cyclohepta[1,2-*b*]pyridine ring, and replacement of *N*-methyl roup in piperidine ring for ethyl carbamate group resulted in the creation of loratadine (**6.1.3**), a second-generation H1-receptor antagonist, which is a potent, nonsedating, long-acting antihistamine drug, exhibiting partial selectivity for peripheral histamine H1-receptors. To date, loratadine has been evaluated in allergic rhinitis, urticaria and, to a limited extent, in asthma. Loratadine is superior and faster acting than azatadine in patients with allergic rhinitis and chronic urticarial and is well tolerated. It has demonstrated effective control of asthma symptoms, improved pulmonary function, and long duration of action in patients with allergic bronchial asthma. Loratadine exhibits a dose-related inhibition of the histamine-induced skin wheal and flare response. The introduction of a relatively nonsedating H1-receptor antagonists ushered in a new era in the symptomatic treatment of allergic disorders. Unlike first-generation H1-receptor antagonists, the second-generation compounds do not cross the blood-brain barrier readily and are thus comparatively free of central nervous system effects. Efficacy of the H1 antagonists is maintained during chronic therapy. The second-generation H1-receptor antagonists are appropriate for use as first-line treatment of allergic rhinoconjunctivitis and urticaria. Loratadine's commonly reported adverse events are somnolence, fatigue, and headache. Sedation occurred less frequently with loratadine than with azatadine. Serious ventricular arrhythmias, as reported with some other second-generation histamine H1-receptor antagonists, have not been observed with loratadine to date [18−23].

Desloratadine (**6.1.4**) (Clarinex) is a major and active metabolite of loratadine (**6.1.3**) and a selective peripheral histamine H1 receptor antagonist that has a 24-hour duration of action, enabling once-daily dosing. The binding affinity of desloratadine for H1-receptors is the highest among all antihistamines. It is a currently approved nonsedating, nonimpairing antihistamine that is effective in relieving nasal and nonnasal symptoms of allergic rhinitis, including nasal congestion safely and effectively [24−34].

Loratadine (**6.1.3**) and the product of its carbamate group cleavage, desloratadine (**6.1.4**), were proposed to be synthesized in different ways. According to previously published methods described above for cyproheptadine and azatadine, the 8-chloro-5,6-dihydro-11*H*-benzo[5,6]cyclohepta[1,2-*b*]pyridin-11-one − ketone (**6.1.18**), was reacted with a Grignard reagent (**6.1.9**) to give the corresponding tertiary carbinol, which was dehydrated in acidic media affording the 8-chloro-11-piperidinylidene derivative (**6.1.21**) − 8-chloroazatadine. The last reacted with ethyl chloroformate on reflux in benzene to give the desired compound, loratadine (**6.1.3**) [35,36].

Another method was developed based on building central seven-membered rings via cyclizing intermediate ketone obtained from the reaction of the same Grignard reagent (**6.1.9**) with tailor-made 3-(3-chlorophenethyl) picolinonitrile (**6.1.19**) using a different super acid systems media, in particular, a system composed of HF and BF3 [37−41]. The loratadine synthesis was accomplished by conversion of the *N*-methylpiperidine to the corresponding ethyl carbamate (**6.1.3**) − loratadine, by the same way: refluxing with chloroformate in toluene.

The third process is based on the low-valent titanium promoted reductive coupling reaction between the two ketones: 8-chloro-5,6-dihydro-11*H*-benzo [5,6]cyclohepta[1,2-*b*]pyridin-11-one (**6.1.18**) and ethyl 4-oxopiperidine-1-carboxylate (**6.1.20**), which gave 8-chloroazatadine (**6.1.21**) [42].

Another method is based on the Wittig reaction between ethyl 4-(diethoxyphosphoryl)piperidine-1-carboxylate (**6.1.22**) and ketone (**6.1.18**) in presence of lithium diisopropylamide in xylene-THF media to give the β-hydroxy phosphonate (**6.1.23**), which was further refluxed in xylene to give on thermal decomposition loratadine (**6.1.3**) [43].

For the synthesis of desloratadine (**6.1.4**), the carbamate group of loratadine (**6.1.3**) was further cleaved under alkaline or acidic conditions to release the desired product (**6.1.4**). Essentially demethylation of key intermediate (**6.1.21**) could be and was carried out by treatment with cyanogen bromide (von Braun reaction) affording cyanamide (**6.1.24**), which was hydrolyzed and decarboxylated in one step by refluxing in a mixture of acetic acid and concentrated hydrogen chloride [26,37] (Scheme 6.3).

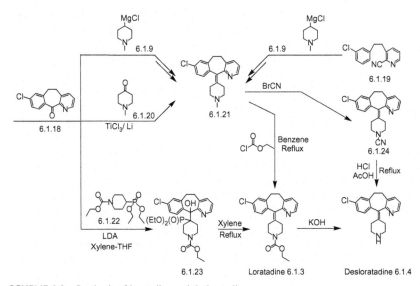

SCHEME 6.3 Synthesis of loratadine and desloratadine.

6.4 RUPATADINE (368)

Rupatadine (**6.1.5**) (Rupafin), is a novel, nonsedating, second-generation antihistamine approved recently in Europe for the treatment of allergic rhinitis and chronic idiopathic urticarial. It has been shown to be highly efficacious, and is safe and well tolerated as other commonly employed antihistamines in the treatment of allergic disease. Rupatadine has a dual affinity for histamine H1-receptors and of platelet-activating factor receptors. Both of these receptors have been shown to play an important role in common allergic inflammatory conditions. Rupatadine also suppresses the release of several inflammatory mediators in response to allergens. It inhibits the degranulation of mast cells, and reduces the release of cytokines [5,6], including tumor necrosis factor, from mast cells and monocytes. Rupatadine is indicated for the treatment of allergic rhinitis and urticaria. It does not present the side effects of first-generation H1-antihistamines, such as somnolence, fatigue, headache, impaired memory and learning, sedation, increased appetite, dry mouth, dry eyes, visual disturbances, constipation, urinary retention, and erectile dysfunction [44−55].

Several approaches were reported for the synthesis of rupatadine (**6.1.5**). It was synthesized by the reaction of desloratadine (**6.1.4**) with 3-(bromo-, chloromethyl)-5-methyl pyridine (**6.1.25**) in dichloromethane in the presence of trimethylamine (TEA) [56]; in carbon tetrachloride in the presence of 4-dimethylaminopyridine [57]; in dimethylformamide and powdered potassium carbonate [58]; in phase transfer catalysis conditions (dichloromethane, tetrabutylammonium bromide and NaOH aqueous solution) [59]; by Grignard coupling 8-chloro-6,11-dihydro-5*H*-benzo[5,6]cyclohepta[1,2-*b*]pyridin-11-one (**6.1.18**) with *N*-[(5-methyl-3-pyridinyl)methyl]-4-chloromagnesiumpiperidine (**6.1.26**), which gave carbinol (**6.1.27**) dehydrated with sulfuric acid [60]; by *N*-acylation of desloratadine (**6.1.4**) with 5-methylnicotinic acid (**6.1.28**) using *N*,*N'*-dicyclohexylcarbodiimide (DCC) and hydroxybenzotriazole (HOBt), and by chlorination/reduction of the obtained intermediate amide (**6.1.29**) using phosphorus oxychloride followed by sodium borohydride reduction [61], summarized in [62,63] (Scheme 6.4).

6.5 KETOTIFEN (3522)

Ketotifen (**6.1.7**) (Zaditor) is a second-generation selective, noncompetitive histamine antagonist (H1-receptor) and a release mast cell stabilizer. Ketotifen inhibits the release of inflammatory mediators from cells involved in hypersensitivity reactions and prevents eosinophil accumulation. These multiple pharmacological actions provided the rationale for assessing the efficacy and safety of ketotifen, the action of which occurs rapidly with an effect seen within minutes after administration. It has been proposed for the treatment of asthma, rhinitis, skin allergies, and

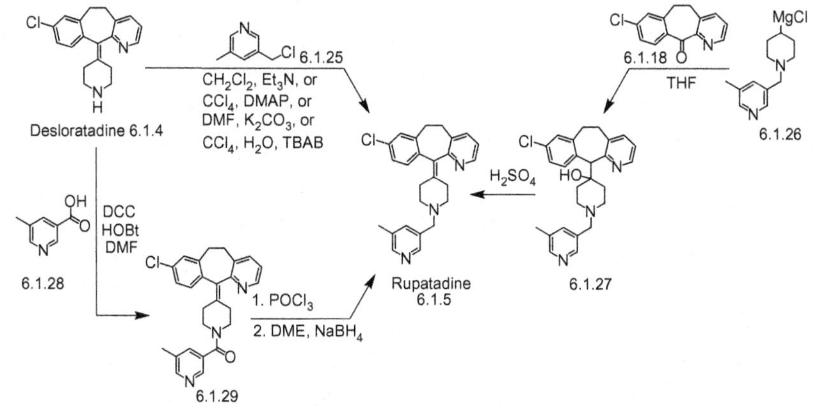

SCHEME 6.4   Synthesis of rupatadine.

anaphylaxis and antiallergic eye drops preventing eye itchiness and irritation associated with allergies [64−74].

The synthesis of ketotifen (**6.1.7**) started with *N*-bromosuccinimide bromination of 9,10-dihydro-4*H*-benzo[4,5]cyclohepta[1,2-*b*]thiophen-4-one (**6.1.30**) on reflux in carbon tetrachloride in the presence of dibenzoyl peroxide, which gave a dibromo- compound (**6.1.31**). Dehydrobromination of the last item mentioned with potassium hydroxide in refluxing methanol gave a pair of isomeric vinyl bromides (**6.1.32a**, **6.1.32b**). Prepared compounds were involved in the Grignard reaction with (1-methylpiperidin-4-yl)magnesium chloride (**6.1.9**) in THF to give a mixture of carbinols (**6.1.33a**, **6.1.33b**) which were dehydrated on boiling in 3 N hydrochloric acid to give piperidilidene derivatives (**6.1.34a**, **6.1.34b**). The obtained mixture of compounds was reacted with piperidine on reflux in dioxane containing potassium tert-butylate to give a mixture of enamines (**6.1.35a**, **6.1.35b**), which was hydrolyzed on reflux in hydrochloric acid to give compound (**6.1.36**) and desired ketotifen (**6.1.6**), which were separated chromatographically [75−78] (Scheme 6.5).

6.6   PIZOTIFEN (623)

Pizotifen (**6.1.6**) (Sandomigran) is an antimigraine drug, acting principally as an antagonist of serotonin 5-HT2A and 5-HT2C receptors used in the preventive treatment of migraine and eating disorders. Pizotifen has strong antiserotonergic and antihistaminic effects, and a weak anticholinergic and antidepressant activity. The mechanism of action is not fully understood. It is used for prophylactic treatment of recurrent vascular headaches, such as typical or atypical migraine, vasomotor headache, cluster headache (Horton's syndrome), and is also indicated in eating disorders. Pizotifen is less

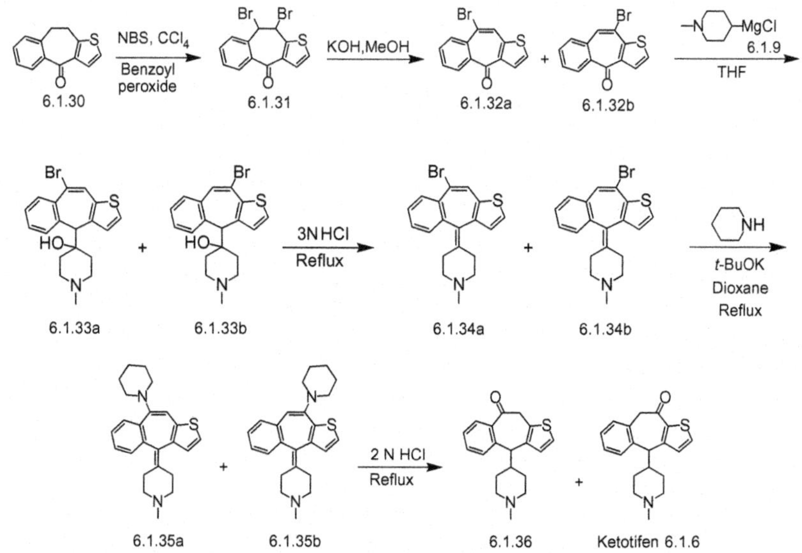

SCHEME 6.5 Synthesis of ketotifen.

effective in tension headache and in psychogenic and posttraumatic headaches. It is not effective in relieving migraine attacks once in progress [64−68,79−83]. The most common side effects are an appetite stimulating effect, an increase in body weight, and sedation (including somnolence and fatigue).

Pizotifen was ineffective in classic tests for antidepressant activity. It neither antagonized the effects of reserpine in rats (hypothermia, ptosis) nor potentiated the effects of amphetamine, nialamide, or L-dopa on locomotor activity. However, its antidepressant activity was found in the "despair test" in rats. The mechanisms of the central action of pizotifen may involve inhibition of monoamine release, and its effects on mood are probably due to a number of mechanisms [80,84]. It was supposed that clinical efficacy of pizotifen in the treatment of migraine may be related to its calcium channel blocking ability [85−94].

In addition to its use in migraine prophylaxis, pizotifen has been proposed for use in several other conditions (anorexia nervosa, depression, pruritus of polycythemia vera, the treatment of underweight patients), but the FDA has approved none of these uses.

The preparation of pizotifen (**6.1.6**) was based on Grignard coupling of (1-methylpiperidin-4-yl)magnesium chloride (**6.1.9**) with specially synthesized 9,10-dihydro-4H-benzo[4,5]cyclohepta[1,2-b]thiophen-4-one (**6.1.37**), which gave carbinol (**6.1.38**), which was dehydrated on reflux in an acetic and concentrated hydrochloric acid [95,96] (Scheme 6.6).

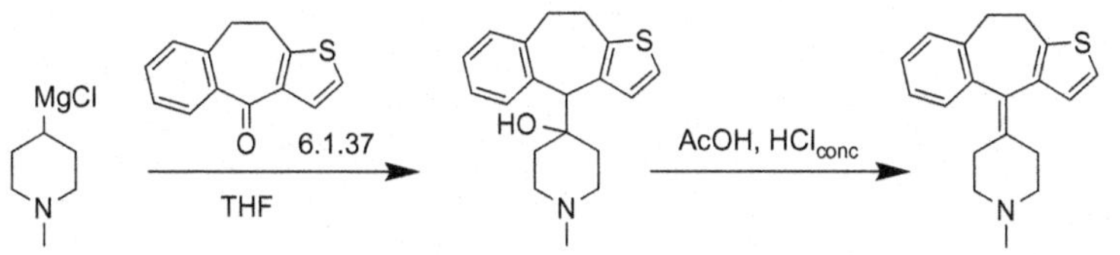

SCHEME 6.6   Synthesis of pizotifen.

6.7   ALCAFTADINE (74)

Alcaftadine (**6.1.8**) (Lastacaft) is an ophthalmic dual-acting H1-antihistamine and mast cell stabilizer approved for the prevention of itching associated with allergic conjunctivitis. Decreased chemotaxis and inhibition of eosinophil activation has also been demonstrated. Indicated for prevention of itching associated with allergic conjunctivitis. The most common eye-related side effects that were reported were eye irritation, burning and/or stinging in the eyes after use, eye redness, and eye itching [97−99].

Two basic principles for the synthesis of alcaftadine (**6.1.8**) were disclosed in the first published on this topic patent [100].

According to the first approach the synthesis started from ethyl 4-(chlorocarbonyl)piperidine-1-carboxylate (**6.1.37**), which was used to acylate 1-phenethyl-1*H*-imidazole (**6.1.38**) in acetonitrile in presence of TEA to give ketone (**6.1.39**). The protecting carbamate group was hydrolyzed on reflux in 12 N hydrochloric acid to give piperidine compound (**6.1.40**), which was methylated on reflux with formaldehyde and formic acid to give methylpiperidin derivative (**6.1.41**). The obtained product was then added to trifluorosulfonic acid at 25−50°C and heated to 140−145°C to give 11-(1-methylpiperidin-4-ylidene)-6,11-dihydro-5*H*-benzo[*d*]imidazo[1,2-*a*]azepine (**6.1.42**). Hydroxymethylation of imidazole ring was simply carried out with 37% aqueous formaldehyde solution in the presence of acetic acid (50 mL), 37% aqueous formaldehyde (500 mL) and potassium acetate on heating the reaction mixture to 100°C, which gave compound (**6.1.43**). The obtained compound was oxidized with manganese dioxide in chloroform at 55−65°C to give desired alcaftadine (**6.1.8**).

Another method described in the same patent comprises Grignard reaction of (1-methylpiperidin-4-yl)magnesium chloride (**6.1.9**) with especially prepared 11*H*-benzo[*d*]imidazo[1,2-*a*]azepin-11-one (**6.1.44**), which gave carbinol (**6.1.45**). The double bond in the 11*H*-benzo[*d*]imidazo[1,2-*a*]azepin-11-ol part of (**6.1.45**) was hydrogenated on palladium catalyst to give compound (**6.1.46**) and was further dehydrated to key compound (**6.1.42**) followed by transformation to alcaftadine (**6.1.8**) the same way (**6.1.42** → **6.1.43** → **6.1.8**).

Alcaftadine key intermediate (**6.1.42**) was also proposed to be prepared simply by reaction of *N*-methyl-4-piperidine carboxylic acid chloride (**6.1.47**) with 1-phenethyl-1*H*-imidazole (**6.1.38**) in acetonitrile to give (**6.1.48**). Follow up cyclodehydration with polyphosphoric acid gave required (**6.1.42**) [101]. For the oxidation step of the synthesis of alcaftadine (**6.1.43** → **6.1.8**), various other oxidizing agents were claimed [102–106] (Scheme 6.7).

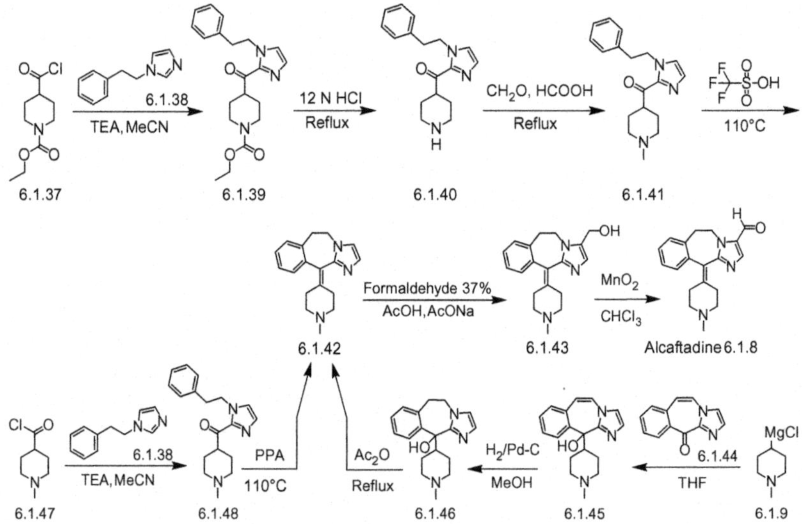

SCHEME 6.7 Synthesis of alcaftadine.

6.8 RITANSERIN (2840)

Ritanserin (**6.1.4**) was never marketed for clinical use but is widely used as a scientific tool as a ritanserin, has very high affinity for all three 5-HT2 receptor sites (5-HT2A-2B-2C) acting as receptor antagonist, and a potency of about 100-fold greater than that for the dopamine D2, the adrenoceptors α1, α2, and the serotonin 5-HT1A receptors. In a clinical pilot study, ritanserin was effective as an add-on medication to neuroleptic treatment in patients with schizophrenia, as an adjunctive therapy to haloperidol, and as an effective medication in the treatment of a variety of syndromes related to anxiety and depression with a rapid onset of action. It appears to improve mood and subjective well being of depressive schizophrenic patients as a supplementary therapy to a highly potent neuroleptic. Ritanserin produces significant therapeutic effects on the negative symptoms of schizophrenia (apathy, absent, blunted, or incongruous emotional responses, reductions in speech, social withdrawal, sexual problems, lethargy) and improves extrapyramidal side effects induced by neuroleptics [107–119]. Ritanserin may be of value in treating alcohol abuse and drug dependence in addicts [120].

The only adverse events reported for ritanserin were headache and preexisting sleep/wake problems, numbness to touch, impaired coordination, constipation, and dry mouth.

Ritanserin (**6.1.54**) was proposed to synthesize by reaction of ethyl 1-benzylpiperidine-4-carboxylate (**6.1.49**) in tetrahydrofuran with Grignard reagent, previously prepared from 1-bromo-4-fluorobenzene, which resulted in (1-benzylpiperidin-4-yl)bis(4-fluorophenyl)methanol (**6.1.50**). The last was dehydrated on reflux in solution of hydrochloric and acetic acids for 2 hours giving 1-benzyl-4-(bis(4-fluorophenyl)methylene)piperidine (**6.1.51**). The obtained product was debenzylated on hydrogenation on palladium- or rhodium-on-charcoal catalyst in methanol to give 4-(bis(4-fluorophenyl)methylene)-piperidine (**6.1.52**), which was alkylated with a specially prepared 6-(2-chloroethyl)-7-methyl-5*H*-thiazolo[3,2-*a*]pyrimidin-5-one (**6.1.53**) in standard conditions (sodium carbonate, potassium methyl isobutyl ketone, reflux) to give desired ritanserin (**6.1.54**) [121] (Scheme 6.8).

An alternative approach mentioned in the same sources envisages a reaction of ethyl 4-(4-fluorobenzoyl)piperidine-1-carboxylate (**6.1.55**) with the same (4-fluorophenyl)magnesium bromide, which gave ethyl 4-(bis(4-fluorophenyl)(hydroxy)methyl)piperidine-1-carboxylate (**6.1.56**), which was simultaneously dehydrated and deprotected on reflux in a solution of concentric hydrochloric acid and ethanol on a constant gaseous hydrogen chloride introduction to give the key compound (**6.1.52**), which was further alkylated with (**6.1.53**) in the same conditions to give ritanserin (**6.1.54**) [121,122,123] (Scheme 6.8).

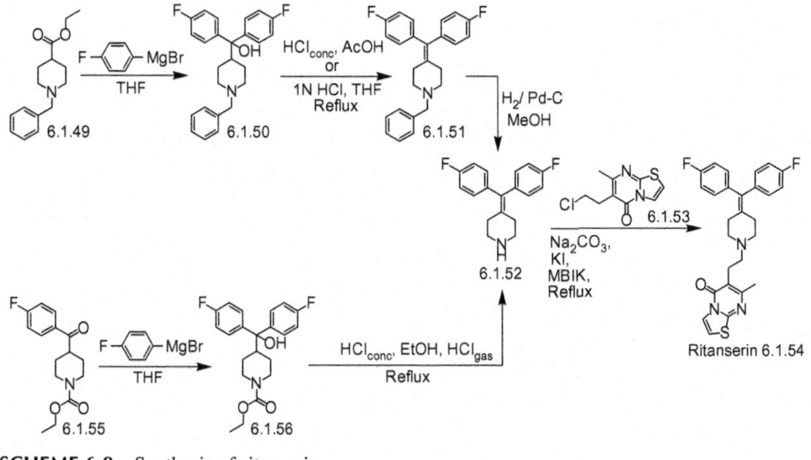

SCHEME 6.8   Synthesis of ritanserin.

REFERENCES

[1] Aboul-Enein HY, Al-Badr AA. Cyproheptadine. Anal Profiles Drug Subst 1980;9: 155–79.

[2] Feng Y, Feng C, Chen S, Hsieh H, Chen Y, Hsu C. Cyproheptadine, an antihistaminic drug, inhibits proliferation of hepatocellular carcinoma cells by blocking cell cycle progression through the activation of P38 MAP kinase. BMC Cancer 2015;15:134.

[3] Engelhardt EL. Derivatives of dibenzo[a,e]cycloheptatriene, US 3014911; 1961.

[4] Engelhardt EL, Zell HC, Saari WS, Christy ME, Colton CD, Stone CA, et al. Structure-activity relations in the cyproheptadine series. J Med Chem 1965;8(6):829–35.

[5] Hatano M, Ito O, Suzuki S, Ishihara K. Zinc(II)-catalyzed addition of grignard reagents to ketones. J Org Chem 2010;75(15):5008–16.

[6] Cid MM, Seijas JA, Villaverde MC, Castedo L. New synthesis of cyproheptadine and related compounds using low valent titanium. Tetrahedron 1988;44(19):6197–200.

[7] Honrubia MA, Rodriguez J, Dominguez R, Lozoya E, Manaut F, Seijas JA, et al. Synthesis, affinity at 5-HT2A, 5-HT2B and 5-HT2C serotonin receptors and structure-activity relationships of a series of cyproheptadine analogs. Chem Pharm Bull 1997;45 (5):842–8.

[8] Loza MI, Sanz F, Cadavid MI, Honrubia M, Orallo F, Fontenla JA, et al. Cyproheptadine analogs: synthesis, antiserotoninergic activity, and structure-activity relationships. J Pharm Sci 1993;82(11):1090–3.

[9] Tozzi S, Roth FE, Tabachnick IIA. Pharmacology of azatadine, a potential antiallergy drug. Agents Actions 1974;4(4):264–70.

[10] Lehrer JF. Inhibition of mediator release by azatadine. JAMA 1986;255(24):3366–7.

[11] Ahmed AR, Moy R. Treatment of acquired cold urticaria with azatadine. Austr J Dermatol 1981;22(2):53–5.

[12] Wilson JD, Hillas JL, Somerfield SD. Azatadine maleate (Zadine): evaluation in the management of allergic rhinitis. N Z Med J 1981;94(689):79–81.

[13] Luscombe DK, Nicholls PJ, Spencer PS. Effect of azatadine on human performance. Brit J Clin Pract 1980;34(3):75–9.

[14] By Anonymous. Azatadine (optimine) - a new antihistamine. Med Letter Drugs Therapeut 1977;19(19):77–9.

[15] Villani FJ. Amino derivatives of phenethylpyridines, US 3301863; 1967.

[16] Villani FJ, Daniels PJL, Ellis CA, Mann TA, Wang K, Wefer EA. Derivatives of 10,11-dihydro-5H-dibenzo[a,d]cycloheptene and related compounds. 6. Aminoalkyl derivatives of the aza isosteres. J Med Chem 1972;15(7):750–4.

[17] Dagger, Raymond Ellery; Motyka, Linda Ann., Process and intermediates for the production of the pyridobenzocycloheptene antihistaminic azatadine, EP 319254; 1989.

[18] Kay GG, Harris AG. Loratadine: a nonsedating antihistamine. Review of its effects on cognition, psychomotor performance, mood and sedation. Clin Exp Allergy 1999;29 (Suppl. 3):147–50.

[19] Menardo J, Horak F, Danzig MR, Czarlewski W. A review of loratadine in the treatment of patients with allergic bronchial asthma. Clin Therapeut 1997;19(6):1278–93.

[20] Monroe EW. Loratadine in the treatment of urticarial. Clin Therapeut 1997;19(2):232–42.

[21] Clissold SP, Sorkin EM, Goa KL. Loratadine. A preliminary review of its pharmacodynamic properties and therapeutic efficacy. Drugs 1989;37(1):42–57.

[22] Haria M, Fitton A, Peters DH. Loratadine. A reappraisal of its pharmacological properties and therapeutic use in allergic disorders. Drugs 1994;48(4):617–37.

[23] Katelaris C. Comparative effects of loratadine and aatadine in the treatment of seasonal allergic rhinitis. Asian Pac J Allergy Immunol 1990;8(2):103−7.

[24] McClellan K, Jarvis B. Desloratadine. Drugs 2001;61(6):789−96.

[25] Agrawal DK. Pharmacology and clinical efficacy of desloratadine as an anti-allergic and anti-inflammatory drug. Exp Opin Invest Drugs 2001;10(3):547−60.

[26] Graul A, Leeson PA, Castaner J. Desloratadine: treatment of allergic rhinitis histamine H1 antagonist. Drugs Fut 2000;25(4):339−46.

[27] Murdoch D, Goa KL, Keam SJ. Desloratadine: an update of its efficacy in the management of allergic disorders. Drugs 2003;63(19):2051−77.

[28] Limon L, Kockler DR. Desloratadine: a nonsedating antihistamine. Ann Pharmacother 2003;37(2):237−46.

[29] DuBuske LM. Pharmacology of desloratadine: special characteristics. Clin Drug Invest 2002;22(Suppl. 2):1−11.

[30] Dizdar EA, Sekerel BE, Tuncer A. Desloratadine: a review of pharmacology and clinical efficacy in allergic rhinitis and urticarial. Therapy 2008;5(6):817−28.

[31] Bachert C, van Cauwenberge P. Desloratadine treatment for intermittent and persistent allergic rhinitis: a review. Clin Therapeut 2007;29(9):1795−802.

[32] Wilken JA, Daly AF, Sullivan CL, Kim H. Desloratadine for allergic rhinitis. Exp Rev Clin Immunol 2006;2(2):209−24.

[33] DuBuske LM. Review of desloratadine for the treatment of allergic rhinitis, chronic idiopathic urticaria and allergic inflammatory disorders. Exp Opin Pharmacother 2005;6 (14):2511−23.

[34] Agrawal DK. Anti-inflammatory properties of Desloratadine. Clin Exp Allergy 2004;34 (9):1342−8.

[35] Villani FJ, Daniels PJL, Ellis CA, Mann TA, Wang K, Wefer EA. Derivatives of 10, 11-dihydro-5H-dibenzo[a,d]cycloheptene and related compounds. Aminoalkyl derivatives of the aza isosteres. J Med Chem 1972;15(7):750−4.

[36] Vilani FJ. Antihistaminic 11-(4-piperidylidene)-5H-benzo-(5,6)-cyclohepta-[1,2-b]-pyridines, US 4282233; 1981.

[37] Villani FJ, Wong JK. Preparation of antihistaminic 8-(halo)-substituted 6,11-dihydro-11-(4-piperidylidene)-5H-benzo[5,6]cyclohepta[1,2-b]pyridines and pharmaceutical compositions containing them, US 4659716; 1987.

[38] Schumacher DP, Murphy BL, Clark JE. US. Process for preparing piperidylidene dihydro-dibenzo(a,d)-cycloheptenes or aza-derivatives thereof, US 4 731 447; 1988.

[39] Inventor data unavailable, Piperidylidenedihydrodibenzo[a,d]cycloheptane analogs, JP 61289087; 1986.

[40] Schumacher DP, Murphy BL, Clark JE, Tahbaz P, Mann TA. Superacid cyclodehydration of ketones in the production of tricyclic antihistamines. J Org Chem 1989;54(9):2242−4.

[41] Piwinski JJ, Wong JK, Green MJ, Ganguly AK, Billah MM, West RE, et al. J Med Chem 1991;34:457−61.

[42] Stampa A, Camps P, Rodriguez G, Bosch J, Onrubia MC. Process for the preparation of loratadine, US 6084100; 2000.

[43] Doran HJ, O'Neill Pat M. Process for preparing tricyclic compounds having antihistaminic activity; 2000.

[44] Keam SJ, Plosker GL. Rupatadine a review of its use in the management of allergic disorders. Drugs 2007;67(3):457−74.

[45] Gonzalez-Nunez V, Bachert C, Mullol J. Rupatadine: global safety evaluation in allergic rhinitis and urticarial. Exp Opin Drug Safety 2016;15(10):1439−48.

[46] Mullol J, Bousquet J, Bachert C, Canonica GW, Gimenez-Arnau A, Kowalski ML, et al. Update on rupatadine in the management of allergic disorders. Allergy 2015;70(Suppl. 100):1−24.

[47] Ridolo E, Montagni M, Fassio F, Massaro I, Rossi O, Incorvaia C, et al. Rupatadine for the treatment of allergic rhinitis and urticaria: a look at the clinical data. Clin Invest 2014;4(5):453−61.

[48] Nettis E, Delle Donne P, Di Leo E, Calogiuri G, Ferrannini A, Vacca A. Rupatadine for the treatment of urticarial. Exp Opin Pharmacother 2013;14(13):1807−13.

[49] Picado C. Rupatadine: pharmacological profile and its use in the treatment of allergic disorders. Exp Opin Pharmacother 2006;7(14):1989−2001.

[50] Mullol J, Bousquet J, Bachert C, Canonica WG, Gimenez-Arnau A, Kowalski ML, et al. Rupatadine in allergic rhinitis and chronic urticarial. Allergy 2008;63(Suppl. 87):5−28.

[51] Katiyar S, Parkash S. Pharmacological profile, efficacy and safety of rupatadine in allergic rhinitis. Prim. Care Respir. J. 2009;18(2):57−68.

[52] Merlos M, Giral M, Balsa D, Ferrando R, Queralt M, Puigdemont A, et al. Rupatadine, a new potent, orally active dual antagonist of histamine and platelet-activating factor (PAF). J Pharm Exp Ther 1997;280(1):114−21.

[53] Metz M, Maurer M. Rupatadine for the treatment of allergic rhinitis and urticarial. Exp Rev Clin Immunol 2011;7(1):15−20.

[54] Izquierdo I, Merlos M, Garcia-Rafanell J. Rupatadine: a new selective histamine H1 receptor and platelet-activating factor (PAF) antagonist. A review of pharmacological profile and clinical management of allergic rhinitis. Drugs Today 2003;39(6):451−68.

[55] Van Den Anker-Rakhmanina NY. Rupatadine (J Uriach & Cia). Curr Opin Anti-Inflamm Immunomodul Invest Drugs 2000;2(2):127−32.

[56] Carceller E, Merlos M, Giral M, Balsa D, Almansa C, Bartroli J, et al. [(3-pyridylalkyl) piperidylidene]benzocycloheptapyridine derivatives as dual antagonists of PAF and histamine. J Med Chem 1994;37(17):2697−703.

[57] Carceller E, Recasens N, Almansa C, Almansa J, Merlos M, Giral M, et al. Process for preparation of 8-chloro-11-[1-[(5-methyl-3-pyridyl)methyl]-4-piperidylidene]-6,11-dihydro-5H-benzo[5,6]cyclohepta[1,2-b]pyridine and analogs as antihistaminics and PAF antagonists, ES 2042421; 1993.

[58] Agarwal R, Bhirud SB, Bijukumar G, Khude GD. Expedient synthesis of rupatadine. Synth Commun 2008;38(1):122−7.

[59] Khamar BM, Modi IA, Chandrakant SM, Kashyapbhai PK, Prabhakar DS, Ravi P, et al. Process for the preparation of rupatadine by PTC-catalyzed N-alkylation of desloratadine, WO 2006114676; 2006.

[60] Carceller E, Jimenez PJ, Salas J. Preparation of 8-chloro-6,11-dihydro-11[1-[(5-methyl-3-pyridinyl)methyl]-5H-benzo[5,6]cyclohepta[1,2-b]pyridine, ES 2120899; 1998.

[61] Carceller E, Recasens N, Almansa C, Bartroli J, Merlos M, Giral M. 8-Chloro-11-[1-[(5-methyl-3-pyridyl)methyl]-4-piperidylidene]-6,11-dihydro-5H-benzo[5,6]cyclohepta[1,2-b] pyridine fumarate and its preparation and use as a PAF antagonist and antihistaminic, ES 2087818; 1996.

[62] Garcia-Rafanell J. Rupatadine fumarate. UR-12592 fumarate. Antiallergic. Histamine and PAF antagonist. Drugs Fut 1996;21(10):1032−6.

[63] Liu G, Xu H, Chen G, Wang PWY, Liu H, Yu D. Stereoselective synthesis of desloratadine derivatives as antagonist of histamine. Bioorg Med Chem 2010;18(4):1626−32.

[64] Alhadeff M. Ketotifen. From Drugs of Today 1978;14(8):367−73.

[65] Grant SM, Goa KL, Fitton A, Sorkin EM. Ketotifen. A review of its pharmacodynamic and pharmacokinetic properties, and therapeutic use in asthma and allergic disorders. Drugs 1990;40(3):412−48.

[66] Craps LP, Ney UM. Ketotifen: current views on its mechanism of action and their therapeutic implications. Respiration 1984;45(4):411−21.

[67] Martin U, Greenwood C, Craps LP, Baggiolini M. Ketotifen. Pharm Biochem Prop Drug Subst 1981;3:424−60.

[68] Greenwood C. The pharmacology of ketotifen. Chest 1982;82(Suppl. 1):45−8.

[69] Ney UM, Bretz U, Martin U, Mazzoni L. Pharmacology of ketotifen. Res Clin Forums 1982;4(1):9−16.

[70] Sokol KC, Amar NK, Starkey J, Grant JA. Ketotifen in the management of chronic urticaria: resurrection of an old drug. Ann Allergy Asthma Immunol 2013;111(6):433−6.

[71] Craps L. Ketotifen in the oral prophylaxis of bronchial asthma: a review. Pharmacotherapeut 1981;3(1):18−35.

[72] Mikotic-Mihun Z, Kuftinec J, Hofman H, Zinic M, Kajfez F, Meic Z. Ketotifen. Analy Profiles Drug Subst 1984;13:239−63.

[73] Martin U, Roemer D. The pharmacological properties of a new, orally active antianaphylactic compound: ketotifen, a benzocycloheptathiophene. Arzneimitt -Forsch 1978;28(5):770−82.

[74] Herzog H. Asthma, antihistamines and ketotifen. Allergol Immunopathol Suppl 1986;10:168−78.

[75] Bourquin JP, Schwarb G, Waldvogel E. 4-(4-Piperidylidene)-4H-benzo[4,5]cyclohepta [1,2-b]thiophenones, DE 2111071; 1971.

[76] Bourquin J, Schwarb G, Waldvogel E. (1-Alkyl-4-piperidylidine)-4H-benzo. nu.,5]cyclohepta[1,2-6], US 3682930; 1972.

[77] Bourquin JP, Schwarb G, Waldvogel E. 4-(4-Piperidylidene)-4H-benzo[4,5]cyclohepta [1,2-b]thiophen-10(9H)-ones, DE 2302970; 1973.

[78] Waldvogel E, Schwarb G, Bastian JM, Bourquin JP. Studies on synthetic drugs. The 9- and 10-oxo derivatives of 9,10-dihydro-4H-benzo[4,5]-cyclohepta[1,2-b]thiophenes. Helv Chim Acta 1976;59(3):866−77.

[79] Speight TM, Avery GS. Pizotifen (BC-105). Review of its pharmacological properties and its therapeutic efficacy in vascular headaches. Drugs 1972;3(3−4):159−203.

[80] Dixon AK, Hill RC, Roemer D, Scholtysik G. Pharmacological properties of 4(1-methyl-4-piperidylidine)-9,10-dihydro-4H-benzo[4,5]cyclohepta[1,2]-thiophene hydrogen maleate (pizotifen). Arzneimitt -Forsch 1977;27(10):1968−79.

[81] Carrol JD, Maclay WP. Pizotifen (BC-105) in migraine prophylaxis. Curr Med Res Opin 1975;3:68−71.

[82] Crowder D, Maclay WP. Pizotifen once daily in the prophylaxis of migraine; results of a multicentre general practice study. Curr Med Res Opin 1984;9:280−5.

[83] Mueller-Schweinitzer E. Pizotifen, an antimigraine drug with venoconstrictor activity in vivo. J Cardiovasc Pharmacol 1986;8(4):805−10.

[84] Przegalinski E, Baran L, Palider W, Siwanowicz J. The central action of pizotifen. Psychopharmaol 1979;62(3):295−300.

[85] Peroutka SJ, Banghart SB, Allen GS. Calcium channel antagonism by pizotifen. J Neurol Neurosurg Psychiatry 1985;48(4):381−3.

[86] Standal JE. Pizotifen as an antidepressant. Acta Psychiatr Scand 1977;56(4):276−9.

[87] Weischer ML, Opitz K. Effect of pizotifen, ketotifen and MK-711 on food intake in female tree shrews (Tupaia belangeri). IRCS Med Sci Library Compend 1983;11(4):332.

[88] Weischer ML, Opitz K. Orexigenic effects of pizotifen and cyproheptadine in cats. From IRCS Med Sci Library Compend 1979;7(11):555.

[89] Swiergiel AH, Peters G. Failure of serotonin antagonist pizotifen to stimulate feeding or weight gain in free-feeding rats. Pharmacol Biochem Behavior 1990;35(1):61−7.

[90] Jowett NI. Severe weight loss after withdrawal of chronic pizotifen treatment. J Neurol Neurosurg Psychiatry 1998;65(1):137.

[91] Lin OA, Karim ZA, Vemana HP, Espinosa EVP, Khasawneh FT. The antidepressant 5-HT2A receptor antagonists pizotifen and cyproheptadine inhibit serotonin-enhanced platelet function. PLoS One 2014;9(1) e87026/1-e87026/12

[92] Richards MH. Effects of cyproheptadine and pizotifen on central muscarinic receptors. Eur J Pharmacol 1991;195(3):403−5.

[93] Eltze M, Mutschler E, Lambrecht G. Affinity profiles of pizotifen, ketotifen and other tricyclic antimuscarinics at muscarinic receptor subtypes M1, M2 and M3. Eur J Pharmacol 1992;211(3):283−93.

[94] Minnema DJ, Hendry JS, Rosecrans, John A. Discriminative stimulus properties of pizotifen maleate (BC105): a putative serotonin antagonist. Psychopharmacol 1984;83 (2):200−4.

[95] Jucker E, Ebnoether A, Stoll A, Bastian J, Rissi E. Substituted 4[piperidylidene(4')]-9, 10-dihydro-4h-benzo [4, 5] cyclohepta [1, 2-b] thiophene and the 4 piperidyl 4 ol compounds, US 3272826; 1966.

[96] Jucker E, Ebnoether A, Stoll A. New heterocyclic derivatives, BE 636717; 1964.

[97] Mahvan TD, Buckley WA, Hornecker JR. Alcaftadine for the prevention of itching associated with allergic conjunctivitis. Ann Pharmacother 2012;46(7−8):1025−32.

[98] Namdar R, Valdez C. Alcaftadine: a topical antihistamine for use in allergic conjunctivitis. Drugs Today 2011;47(12):883−90.

[99] Chigbu DeGaulle I, Coyne AM. Update and clinical utility of alcaftadine ophthalmic solution 0.25% in the treatment of allergic conjunctivitis. Clin Ophthalmol 2015;9:1215−25.

[100] Janssens FE, Diels GSM, Leenaerts JE. Anti-allergic imidazo[2,1-b][3]benzazepine derivatives, compositions and method of use, US 5468743; 1995.

[101] Wang X, Lixin J, Li Y, Gao Y, Lv X, Shen X, et al. Method for preparing alcaftadine intermediate 6,11-dihydro-11-(1-methyl-4-piperidylidene)-5H-imidazole[2,1-b][3]benzazepine, CN 103408549; 2013.

[102] Ponnaiah R, Hashmi AM, Neela PK, Seemala DN, Mokka R, Lahoti AM, et al. A process for the preparation of alcaftadine, WO 2014083571; 2014.

[103] Sivakumar BV, Rao KE, Patel GB, Vaidya SD, Tripathi AP. Preparation of polymorphic forms of alcaftadine, IN 2012MU03334; 2014.

[104] Sivakumar BV, Rao KE, Patel GB, Vaidya SD, Tripathi AP. A process of preparing alcaftadine, IN 2012MU03449; 2014.

[105] Janssens FE, Diels GS, Leenaerts JF, Imidazo[2,1-b]benzazepine derivatives, compositions and method of use, US 5468743; 1995.

[106] Pansuriya PB, Maguire GEM, Friedrich HB. Synthesis and structural elucidation of a novel polymorph of alcaftadine, Spectrochim. Acta, Part A: Mol Biomol Spectr 2015;142:311−19.

[107] Shi W, Nathaniel P, Bunney BS. Ritanserin, a 5-HT1A/2C antagonist, reverses direct dopamine agonist-induced inhibition of midbrain dopamine neurons. J Pharmacol Exp Ther 1995;274(2):735−40.

[108] Strauss WH, Klieser E. Psychotropic effects of ritanserin, a selective S2 antagonist: an open study. Eur Neuropsycopharmacol 1991;1(2):101−5.

[109] Klieser E, Strauss WH. Study to establish the indication for the selective S2 antagonist ritanserin. Pharmacopsychiatry 1988;21(6):391−3.

[110] Den Boer JA, Vahlne J-O, Post P, Heck AH, Daubenton F, Olbrich R. Ritanserin as add-on medication to neuroleptic therapy for patients with chronic or subchronic schizophrenia. Human Psychopharmacol 2000;15(3):179−89.

[111] Lammers GJ, Arends J, Declerck AC, Kamphuisen HA, Schouwink G, Troost J. Ritanserin, a 5-HT2 receptor blocker, as add-on treatment in narcolepsy. Sleep 1991;14 (2):130−2.

[112] Slassi A. Recent advances in 5-HT1B/1D receptor antagonists and agonists and their potential therapeutic applications. Curr Top Med Chem 2002;2(6):559−74.

[113] Den Boer JA, Westenberg HGM. Serotonin function in panic disorder: a double blind placebo controlled study with fluvoxamine and ritanserin. Psychopharmacology 1990;102(1):85−94.

[114] Hensman R, Guimaraes FS, Wang M, Deakin JFW. The effects of ritanserin on aversive classical conditioning in humans. Psychopharmacology 1991;104(2):220−4.

[115] Leysen JE, Gommeren W, Van Gompel P, Wynants J, Janssen PFM, Laduron PM. Receptor-binding properties in vitro and in vivo of ritanserin: a very potent and long acting serotonin-S2 antagonist. Mol Pharmacol 1985;27(6):600−11.

[116] Gelders YG. Thymostenic agents, a novel approach in the treatment of schizophrenia. From Br J Psychiatry Supplement 1989;5:33−6.

[117] Bersani G, Grispini A, Marini S, Pasini A, Valducci M, Ciani N. 5-HT2 antagonist ritanserin in neuroleptic-induced parkinsonism: a double-blind comparison with orphenadrine and placebo. Clin Neuropharmacol 1990;13(6):500−6.

[118] Bersani G, Grispini A, Marini S, Pasini A, Valducci M, Ciani N. Neuroleptic-induced extrapyramidal side-effects: clinical perspectives with ritanserin (R-55667), a new selective 5-HT2 receptor blocking agent. Curr Ther Res 1986;40:492−9.

[119] Ruiu S, Marchese G, Saba PL, Gessa GL, Pani L. The 5-HT2 antagonist ritanserin blocks dopamine reuptake in the rat frontal cortex. Mol Psychiatry 2000;5(6):673−7.

[120] Meert TF. Ritanserin and alcohol abuse and dependence. Alcohol Alcoholism 1996;29 (Suppl. 2):523−30 Volume date 1994

[121] Kennis LEJ, Vandenberk J, Mertens JC. [[Bis(aryl)methylene]-1-piperidinyl]alkyl-pyrimidinones, US 4533665 A; 1985.

[122] Kennis LEJ, Vandenberk J, Mertens JC. [Bis(aryl)methylene]-1-piperidinyl)alkyl pyrimidinones, EP 110435; 1984.

[123] Kennis LEJ, Vandenberk J, Boey JM, Mertens JC, Van Heertum AHM, Janssen M, et al. The chemical development of selective and specific serotonin S2-antagonists. Drug Dev Res 1986;8(1−4):133−40.

Chapter 7

Piperidine-Based Nonfused Biheterocycles With C–N and C–C Coupling

7.1 PIPERIDINE-BASED NONFUSED BIHETEROCYCLES WITH C–N COUPLING

(1-(1,2,3,6-tetrahydropyridin-4-yl)-1,3-dihydro-2*H*-benzo[*d*]imidazol-2-ones and 1-(piperidin-4-yl)-1,3-dihydro-2*H*-benzo[*d*]imidazol-2-ones (Droperidol analogues and derivatives).

Presented below are a series of drugs such as droperidol (**7.1.1**), benperidol (**7.1.2**), timiperone (**7.1.3**), pimozide (**7.1.4**), zaldaride (**7.1.5**), domperidone (**7.1.6**), and bezitramide (**7.1.7**) as well as many other analogs structurally related compounds that are effective as antipsychotic agents (Fig. 7.1).

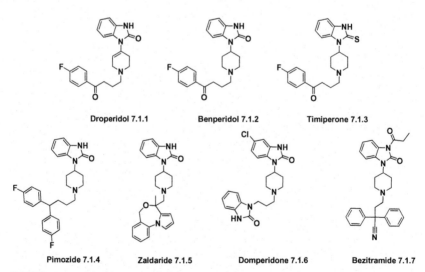

Droperidol 7.1.1   Benperidol 7.1.2   Timiperone 7.1.3

Pimozide 7.1.4   Zaldaride 7.1.5   Domperidone 7.1.6   Bezitramide 7.1.7

FIGURE 7.1   (1-(1,2,3,6-tetrahydropyridin-4-yl)-1,3-dihydro-2*H*-benzo[*d*]imidazol-2-ones and 1-(piperidin-4-yl)-1,3-dihydro-2*H*-benzo[*d*]imidazol-2-ones.

Piperidine-Based Drug Discovery. DOI: http://dx.doi.org/10.1016/B978-0-12-805157-3.00007-7

Droperidol (4475)

Droperidol (**7.1.1**) (Inapsine) is a dopamine D2 receptor antagonist and one of the first compounds of the butyrophenone series, an effective medication used in emergency medicine practice for control of a variety of uses, such as psychosis/agitation [1−5], as an antiemetic [6,7], for vertigo [8,9], and as a migraine remedy [10]. The most common side effects of droperidol are drowsiness, dizziness, or feeling restless or anxious [11−13].

Droperidol's general properties are similar to those of haloperidol. It produces marked tranquilization and sedation; it provides a state of mental detachment and indifference while maintaining a state of reflex alertness. Droperidol potentiates other central nervous system (CNS) depressants, produces mild alpha-adrenergic blockade, peripheral vascular dilation, and provides a reduction of the pressor effect of epinephrine. The onset of action is from 3 to 10 minutes following administration; it generally continues for 2−4 hours.

Droperidol is used in conjunction with an opioid analgesic such as fentanyl to maintain the patient in a calm state of neuroleptanalgesia with an indifference to surroundings. It is also used for the control of agitation in acute psychoses and as a premedicant, as an antiemetic.

In 2001, the US Food and Drug Administration issued a black box warning for droperidol over concerns of QT prolongation (The QT interval is a measure of the time between the start of the Q wave and the end of the T wave in the heart's electrical cycle.) − a heart rhythm condition that can potentially cause fast, chaotic heartbeats and the potential for "torsades de pointes" recommending specific electrocardiographic monitoring during the time of use of the drug [14]. Recently it was shown that droperidol is safe and effective for calming violent and aggressive emergency patients, and the negative effects that garnered a black box warning are actually quite rare. A new study was published [15].

The preparation of *N*-alkenylbenzimidazol-2-ones was developed in 1960 and involves the condensation of *o*-phenylenediamines 2 with *b*-keto esters either in the presence, or absence, of an acid catalyst and currently represents the only known method for the construction of this functionality [16,17].

Droperidol (**7.1.1**) was prepared using the methodology described, by a portionwise addition of a solution of L-benzyl-3-carbethoxy-4- piperidone (**7.1.8**) in xylene to the boiling solution of *o*-phenylenediamine (**7.1.9**) in the same solvent. During the addition, an equal quantity of xylene was distilled off. Obtained 1-(1-benzyl-1,2,3,6-tetrahydropyridin-4-yl)-1,3-dihydro-2*H*-benzo[*d*]imidazol-2-one (**7.1.10**) was hydrogenated in ethanol at an atmospheric pressure of about 50°C in the presence of palladium on a charcoal catalyst. The reduction stopped after one equivalent hydrogen was absorbed to give debenzylated product (**7.1.11**). The last was alkylated with 4-chloro-1-(4-fluorophenyl)butan-1-one (**7.1.12**) in refluxing methyl isobutyl ketone (MBIK) in the presence of sodium carbonate and potassium iodide for 64 hours to give desired droperidol (**7.1.1**) [18−20] (Scheme 7.1).

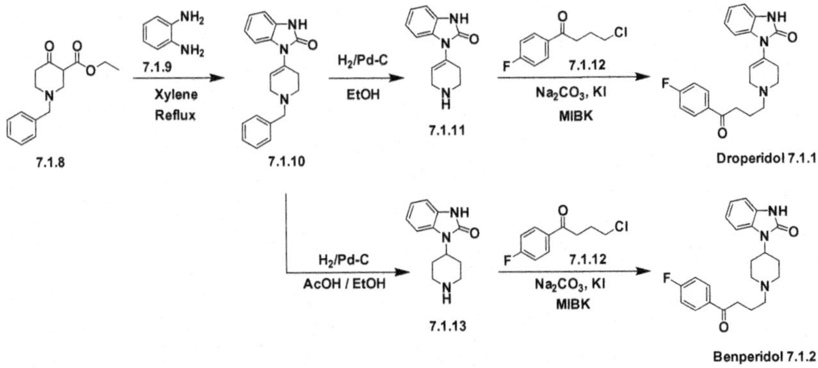

SCHEME 7.1 Synthesis of droperidol and benperidol.

Benperidol (1114)

Benperidol (**7.1.2**) (Anquil) is structurally similar to droperidol and differs only in the absence of a double bond in the 4,5 position in the piperidine ring. It also belongs to the D2 antagonist pharmacological group and is also classified as an antipsychotic agent. Benperidol exhibited sedative activity, serotonin- and *N*-cholinoblocking activity as well as some *M*-cholinomimetic activity, and is used to treat schizophrenia and control deviant antisocial sexual behavior [21−24]. Side effects were observed in more than half of the cases under treatment with all neuroleptics, such as extrapyramidal symptoms, e.g., tremor, rigidity, hypersalivation, bradykinesia, akathisia, acute dystonia, oculogyric crisis, and laryngeal dystonia.

Benperidol (**7.1.2**) was synthesized by the same methods and its synthesis is described in the same patents [19,20] that revealed the synthesis of droperidol (**7.1.1**) and lies in the hydrogenation of 4,5- double bond, and debenzylation (**7.1.12** → **7.1.13**), and is carried out in one stage in acetic acid − ethanolmedia at 50°C and atmospheric pressure until two equivalents of hydrogen are taken up. An additional step − alkylation of obtained 1-(piperidin-4-yl)-1,3-dihydro-2*H*-benzo[*d*]imidazol-2-one (**7.1.13**) to desired benperidol (**7.1.2**) with 4-chloro-1-(4-fluorophenyl)butan-1-one (**7.1.12**) − was done in a standard manner on refluxing reactants in MBIK in the presence of sodium carbonate and potassium iodide (Scheme 7.1).

Timiperone (108)

The next change in the structure of benperidol (**7.1.2**) was the replacement of carbonyl oxygene in the benzimidazolone ring for sulfur thereby resulting in a new antipsychotic drug that belongs to the butyrophenone series: timiperone (**7.1.3**).

Timiperone is a first-generation (typical) antipsychotic drug approved in Japan and available in a number of countries worldwide for the treatment of schizophrenia. It has general properties similar to those of haloperidol, but with higher affinity (as an antagonist) for D2 and 5-HT2A receptors. It also has modest binding affinity for sigma receptors. It can induce extrapyramidal side effects, drowsiness, and constipation, but it displays low toxicity. Relatively many appearances of akathisia, dyskinesia, Parkinsonian syndrome, and insomnia were noted, but these were controlled [25–27].

Benperidol (**7.1.2**) was synthesized by the routes outlined in Scheme 7.2. According to the first of them *N*-(2-nitrophenyl)piperidin-4-amine (**7.1.14**) was alkylated in butanol in the presence of sodium carbonate and potassium iodide with 2-(3-chloropropyl)-2-(4-fluorophenyl)-1,3-dioxolane (**7.1.15**) to give compound (**7.1.16**), hydrogenateed over Raney nickel catalyst to an *o*-penylendiamine derivative (**7.1.17**). Cyclization of the last with carbon disulfide in the presence of potassium hydroxide in water/ethanol media on heating at 80°C in a sealed tube after acidic hydrolysis of protecting dioxolane group gave desired benperidol (**7.1.2**) [28,29].

According to another method published by the same authors, *N*1-(pyridin-4-yl)benzene-1,2-diamine (**7.1.19**) prepared by coupling with 4-chloropyridine (**7.1.18**) in boiling xylene was alkylated with 2-(3-chloropropyl)-2-(4-fluorophenyl)-1,3-dioxolane (**7.1.15**) to give pyridinium salt (**7.1.20**), which was reduced with sodium borohydride to give *o*-penylendiamine derivative (**7.1.17**) converted to benperidol (**7.1.2**) the same way described above [30].

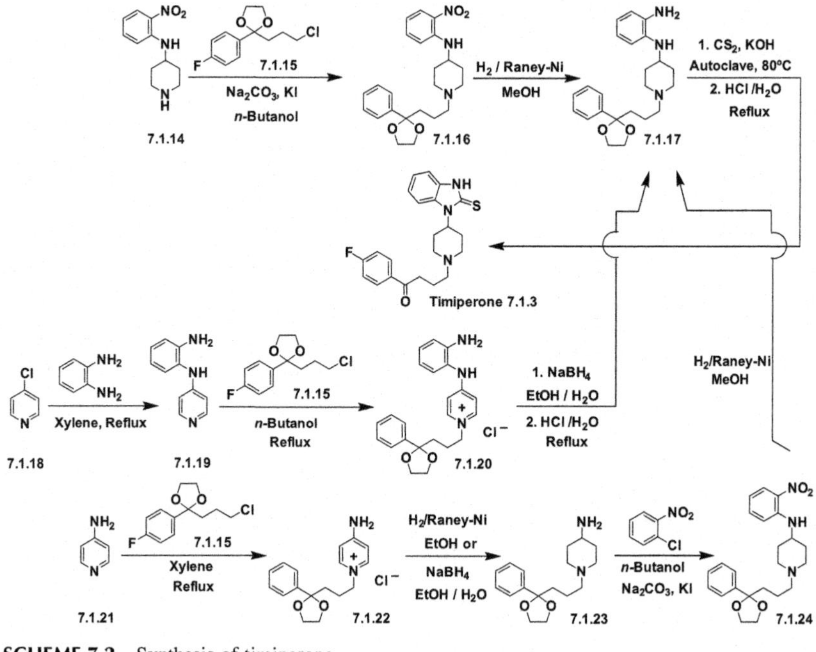

SCHEME 7.2 Synthesis of timiperone.

Lastly, alkylation of 4-aminopyridine (**7.1.21**) with 2-(3-chloropropyl)-2-(4-fluorophenyl)-1,3-dioxolane (**7.1.15**) gave pyridinium salt (**7.1.22**) further hydrogenated on Raney nickel catalyst or sodium borohydride gave 1-(3-(2-phenyl-1,3-dioxolan-2-yl)propyl)piperidin-4-amine (**7.1.23**). The obtained compound was arylated with 1-chloro-2-nitrobenzene on reflux in butanol in the presence of sodium carbonate and potassium iodide to give *N*-(2-nitrophenyl)-1-(3-(2-phenyl-1,3-dioxolan-2-yl)propyl)piperidin-4-amine (**7.1.24**) nitro group of which was reduced further to known *o*-penylendiamine derivative (**7.1.17**) and converted to benperidol (**7.1.2**) (Scheme 7.2).

Pimozide (5441)

Pimozide (**7.1.4**) (Orap) is a first-generation antipsychotic drug which shares with other antipsychotics the ability to blockade dopaminergic receptors in the CNS. Although its exact mode of action has not been established it is proved that pimozide inhibits D2 receptors, thereby decreasing dopamine neurotransmission and reducing the occurrence of motor and vocal tics. In addition, pimozide antagonizes a-adrenergic and 5-HT2 receptors [31−33].

The ability of pimozide to suppress motor and phonic tics in Tourette's disorder is thought to be a function of its dopaminergic blocking activity. It was approved by the FDA for treatment of Tourette's disorder, an inherited tic disorder that begins with simple facial tics and may progress to more complex tics, including grunting and compulsive utterances that can be publicly embarrassing and disabling for the individual [34,35].

In clinical practice, however, physicians often prescribe drugs for "off-label" uses, e.g., psychotic disorders, such as schizophrenia, paranoidal personality disorder, deflusional disorder, delusions of parasitosis. In addition, pimozide has been found to be efficacious in the treatment of some body dysmorphic disorder, metastatic melanoma, trichotillomania, and trigeminal and postherpetic neuralgia [36−38], in dermatology [39,40].

Pimozide (**7.1.4**) was synthesized by direct alkylation of described above 1-(piperidin-4-yl)-1,3-dihydro-2*H*-benzo[*d*]imidazol-2-one (**7.1.13**) with 4,4-bis(4-fluorophenyl)butyl chloride (**7.1.25**) on reflux in MBIK in presence sodium carbonate and potassium iodide [41,42] (Scheme 7.3).

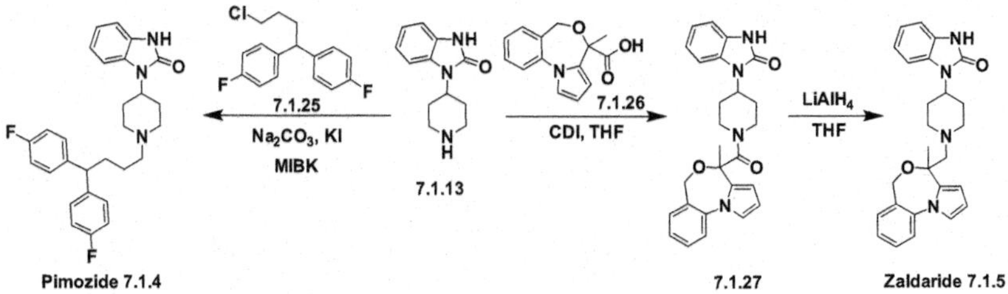

SCHEME 7.3 Synthesis of pimozide and zaldaride.

Zaldaride (39)

Zaldaride (**7.1.5**) is a novel, potent, and selective inhibitor of calmodulin, which plays an important role in the regulation of processes, such as the assembly and disassembly of microtubules by controlling protein kinase activities, by exerting an indirect influence upon a wide variety of cellular processes, and decreases the severity and duration of traveler's diarrhea mechanism of action, the nature of which is not clear.

Zaldaride has been shown to be an effective antidiarrheal due to its antisecretory properties, which does not interfere with gastrointestinal transit time. It is effective for the symptomatic relief of acute diarrhea. Interestingly, a structural analog of benperidol, pimozide, and other representatives of the series of compounds discussed in this chapter does not possess potential antidopaminergic activity [43−48].

Zaldaride (**7.1.5**) was synthesized starting from the same 1-(piperidin-4-yl)-1,3-dihydro-2*H*-benzo[*d*]imidazol-2-one (**7.1.13**), which was acylated with 4-methyl-4*H*,6*H*-benzo[*e*]pyrrolo[2,1-*c*][1,4]oxazepine-4-carboxylic acid (**7.1.26**) in tetrahydrofuran with the use of *N*,*N*-carbonyldiimidazole to prepare compound (**7.1.27**) amide carbonyl group of which was further reduced with lithium aluminum hydride to give the desired zaldaride (**7.1.5**) [49−51] (Scheme 7.3).

Domperidone (5565)

Domperidone (**7.1.6**) (Motilium), a peripherally selective D2-like receptor antagonist, regulates the motility of the gastric and small intestinal smooth muscles and has been shown to have some effects on the motor function of the esophagus. It effectively prevents bile reflux but does not affect gastric secretion. As a result of the blockade of dopamine receptors in the chemoreceptor trigger zone it also has an antiemetic activity. Domperiodone provided relief of such symptoms as anorexia, nausea, vomiting, abdominal pain, early satiety, bloating, and distension in patients with symptoms of diabetic gastropathy. It also provided short-term relief of symptoms in patients with dyspepsia or gastroesophageal reflux, prevented nausea and vomiting associated with emetogenic chemotherapy, and prevented the gastrointestinal and emetic adverse effects of antiparkinsonian drugs. Because domperidone does not readily cross the blood brain barrier and does not inhibit dopamine receptors in the brain, reports of adverse effects on the CNS, such as dystonic reactions, are rare [52−61]. Domperidone is widely used in many countries and can now be officially prescribed to patients in the United States. There are very few treatment options currently available for patients with gastrointestinal motility disorders, especially for patients with gastroparesis. Domperidone has been successfully used in the United States and in many countries as a second-line treatment option for the treatment of gastroparesis.

Synthesis of domperidone (**7.1.6**) started with arylation of ethyl 4-aminopiperidine-1-carboxylate (**7.1.28**) with 1,4-dichloro-2-nitrobenzene (**7.1.29**) on heating at 150°C in cyclohexanol in the presence of sodium carbonate and potassium iodide (in a later disclosure in toluene in presence of sodium carbonate [62]) to give compound (**7.1.30**), which on reflux in 48% hydrobromic acid solution yielded *N*-(4-chloro-2-nitrophenyl)piperidin-4-amine (**7.1.31**). The obtained product was alkylated with 1-(3-chloropropyl)-1,3-dihydro-2*H*-benzo[*d*]imidazol-2-one (**7.1.32**) on reflux in MBIK in the presence of sodium carbonate and potassium iodide to give compound (**7.1.33**). The ring closure could be effected by heating *o*-phenylene diamine (**7.1.33**) with an appropriate cyclizing agent, such as phosgene, urea, potassium isocyanate [63], and the like. In this patent potassium isocyanate dissolved in water was carefully added to a solution of compound (**7.1.34**) in 10 N hydrochloric acid solution (exothermic reaction) to give desired domperidone (**7.1.6**) [64,65] (Scheme 7.4).

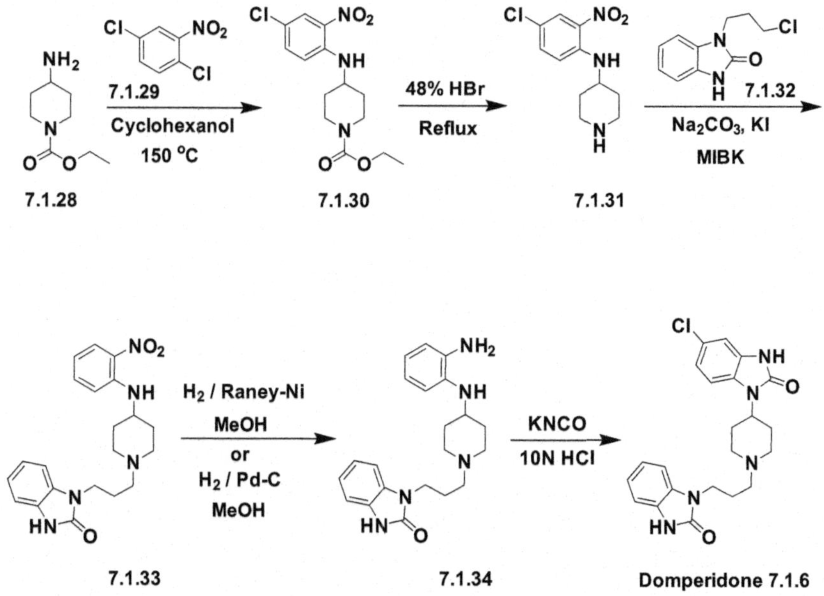

SCHEME 7.4   Synthesis of domperidone.

Bezitramide (204)

Bezitramide (**7.1.7**) (Burgodin) is an effective, long-acting opioid analgesic that was introduced and clinically tested for the treatment of severe chronic pain. It has a peculiar combination of characteristics: potency about 20-times that of methadone, relatively long duration of action − up to 12 hours after chronic oral administration very poor water solubility, which restricts administration to the oral route. Its onset of action is slow, with a peak in analgesic effect noted between 2.5 and 3.5 hours after dosing. It is noted to illicit a strong antitussive effect [66−73]. The drug was withdrawn from the market in 2004 after cases of fatal overdose. Bezitramide was never approved for clinical use in the United States. It is presently an illegal substance classified under Schedule II of the Controlled Substances Act.

Synthesis of bezitramide (**7.1.7**) was carried out via coupling of 1-(piperidin-4-yl)-1,3-dihydro-2*H*-benzo[*d*]imidazol-2-one (**7.1.13**) with 4-bromo-2,2-diphenylbutyronitrile (**7.1.35**) in standard conditions (sodium carbonate, a few crystals of potassium iodide, MBIK, reflux) which gave an intermediate compound (**7.1.36**) further acylated with propionic acid anhydride in refluxing benzene to give desired bezitramide (**7.1.7**) [74,75] (Scheme 7.5).

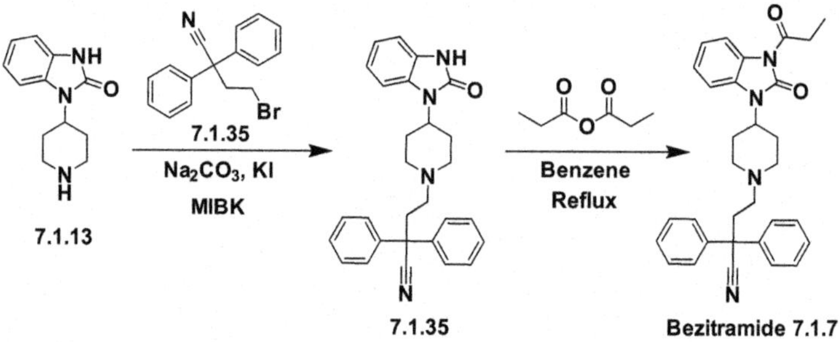

SCHEME 7.5 Synthesis of bezitramide.

Benzpiperylon (30) and Piperylon (1)

Two 4-(1*H*-pyrazol-1-yl)piperidine derivatives, benzpiperylon (**7.1.43**) and piperylon (**7.1.44**), that had good analgetic and antipyretic properties coupled with very low toxicity and were proposed for the pharmaceutical market at the beginning of 1960s, did not get wide application and dissemination [76,77].

Benzpiperylon (**7.1.43**) and piperylon (**7.1.44**) were prepared starting from 1-methylpiperidin-4-one (**7.1.36**), which was converted to benzoylhydrazone (**7.1.38**) on reflux with benzohydrazide (**7.1.37**) in ethanol. Hydrogenation of the obtained product in acetic acid on Adams's catalyst (PtO$_2$) followed by hydrolysis of obtained (**7.1.39**) on boiling with hydrochloric acid, which gave *N*-methylpiperidyl-4-hydrazine (**7.1.40**). The last was cyclized to pyrazolone derivatives benzpiperylon (**7.1.43**) and piperylon (**7.1.44**) on reaction with benzoylacetic acid ethyl ester derivatives − ethyl 2-benzyl-3-oxo-3-phenylpropanoate (**7.1.41**) or ethyl 2-benzoylbutanoate (**7.1.42**), which gave benzpiperylon (**7.1.43**) and piperylon (**7.1.44**), respectively [76,78−83] (Scheme 7.6).

Interestingly, 4-(1*H*-pyrazol-1-yl)piperidine matrix is involved in the structures of very interesting compounds (Fig. 7.2) with a fantastic plethora

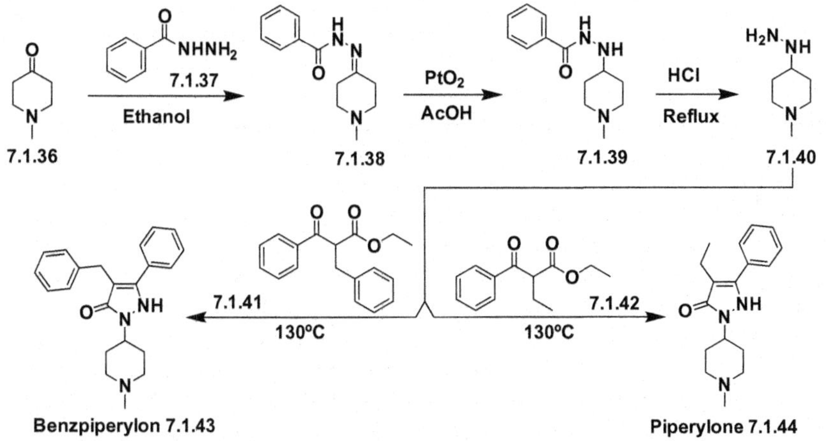

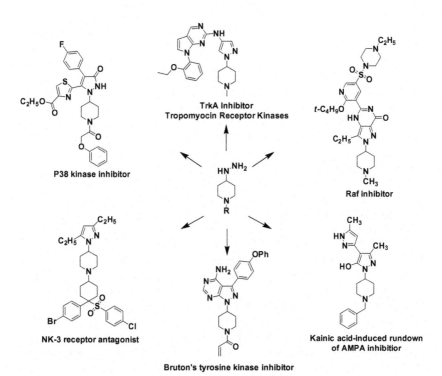

SCHEME 7.6 Synthesis of benzpiperylon and piperylon.

FIGURE 7.2 Plethora of piperidine-based nonfused 4-(1*H*-pyrazol-1-yl)piperidine bihetero-cycles with C−N coupling with diverse biological activities.

of diverse biological activity disclosed in 17,224 publications listed in SciFinder, but has not yet reached the pharmaceutical market.

7.2 PIPERIDINE-BASED NONFUSED BIHETEROCYCLES WITH C−C COUPLING

(3-(Piperidin-4-yl)benzo[*d*]isoxazoles and 3-(piperidin-4-yl)-1*H*-indoles)

   3-(Piperidin-4-yl)benzo[*d*]isoxazoles such as risperidone (**7.2.1**), paliperidone (**7.2.2**) and iloperidone (**7.2.3**) are atypical antipsychotic drugs with high affinity for 5-hydrotryptamine (5-HT) and dopamine D2 receptors for the treatment of schizophrenia symptoms. 3-(Piperidin-4-yl)-1*H*-indole derivatives sertindole (**7.2.42**), one of the newer antipsychotic medications, is also while it's structural analog naratriptan (**7.2.22**) is a selective 5-hydroxytryptamine agonist used for the treatment of migraine headaches (Fig. 7.3).

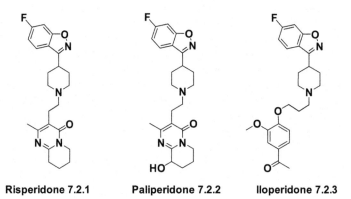

Risperidone 7.2.1   Paliperidone 7.2.2   Iloperidone 7.2.3

FIGURE 7.3 Series of 3-(piperidin-4-yl)benzo[*d*]isoxazoles risperidone, paliperidone and iloperidone.

Risperidone (15970)

Risperidone (**7.2.1**) (Risperdal) is the first second-generation antipsychotic that was specifically designed as a combined D2 and serotonin 5-HT(2A) receptor antagonist, thus following the pharmacological mechanism thought to be responsible for the antipsychotic effects. After its advent in the 1990s as the first novel second-generation antipsychotic, risperidone has achieved worldwide acceptance. It was initially approved for use in schizophrenia, mania of bipolar disorder, and irritability and aggression of autism. But it is also effectively used in other instances of psychosis, including schizoaffective disorder, depression with psychotic features, and psychosis secondary to general medical conditions. Risperidone may be effective in other conditions such as major depression, various anxiety disorders, delirium, dementia, for Alzheimer's dementia, which occurs in 6–8% of persons older than 65 and increases to 30% among those 85 years or older, and substance abuse disorders [84–113].

Risperidone is proposed for inclusion in the WHO Model List of Essential Medications for treatment of schizophrenia, mania, and autism.

Risperidone (**7.2.1**) was synthesized starting from 1-acetyl-4-piperidine-carbonyl chloride (**7.2.4**), which was used to acylate 1,3-difluorobenzene (**7.2.5**) in dichloromethane using aluminum chloride as Lewis acid. The reaction gave 1-(4-(2,4-difluorobenzoyl)piperidin-1-yl)ethan-1-one (**7.2.6**). The protecting acetyl group of the last was removed off by hydrolysis in 6 N hydrochloric acid on reflux, which gave (2,4-difluorophenyl)(piperidin-4-yl) methanone (**7.2.7**). The obtained product was converted further to corresponding oxime (**7.2.8**) on reaction with hydroxylamine hydrochloride in ethanol in the presence of *N,N*-diethylenethanamine. Synthesized oxime (**7.2.8**) was cyclized to 6-fluoro-3-(piperidin-4-yl)benzo[*d*]isoxazole (**7.2.9**) on reflux with 50% potassium hydroxide solution in water. At the final stage the obtained product (**7.2.9**) was alkylated with 3-(2-chloroethyl)-2-methyl-6,7,8,9-tetrahydro-4*H*-pyrido[1,2-*a*]pyrimidin-4-one (**7.2.10**) on heating at 85–90°C in dimethylformamide in the presence of sodium carbonate and potassium iodide, which gave the desired product, risperidone (**7.2.1**) [114,115]. Later, another method of (**7.2.7**) → (**7.2.1**) transformation was proposed, which involved the reductive alkylation of (2,4-difluorophenyl)(piperidin-4-yl) methanone (**7.2.7**) with aldehyde (**7.2.11**) and sodium cyanoborohydride, which gave compound (**7.2.12**), coherently converted to oxime (**7.2.13**) and further to the desired compound, risperidone (**7.2.1**) [116] (Scheme 7.7).

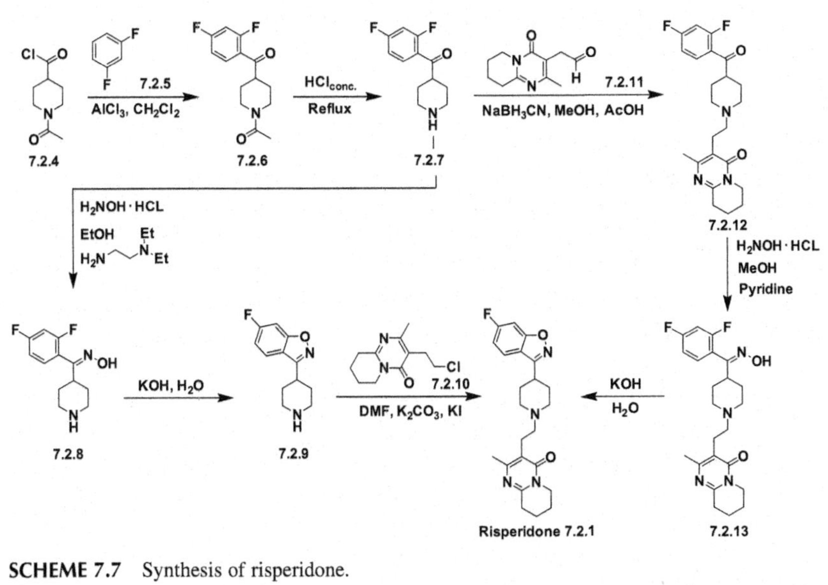

SCHEME 7.7   Synthesis of risperidone.

Paliperidone (1968)

Paliperidone (**7.2.2**) (Invega), 9-hydroxy-risperidone, is the major active metabolite of the atypical antipsychotic risperidone and is one of the most recent antipsychotic medications on the market and is widely prescribed. The mechanism of action of this second-generation antipsychotic, as with other drugs having efficacy in schizophrenia, is unknown, but it has been proposed that the drug's therapeutic activity in schizophrenia is mediated through a combination of central dopamine D2 receptors in the mesolimbic pathway and 5-HT2A receptors in the prefrontal cortex. Differences in receptor binding, 5-HT2A /D2 binding ratios, and mitochondrial proteomics suggest that the effects of risperidone (**7.2.1**) and paliperidone (**7.2.2**) on neuronal firing, regulation of mitochondrial function, and movement are different. Paliperidone is also active as an antagonist at a1 and a2 adrenergic and H1-histaminergic receptors. Paliperidone has no affinity for cholinergic muscarinic or β1- and β2-adrenergic receptors, which may explain some of the other effects of the drug.

Paliperidone is indicated for the acute and maintenance treatment of schizophrenia, acute treatment of schizoaffective disorder, and as a mono-therapy acute treatment of schizoaffective disorder as an adjunct to mood stabilizers and/or antidepressants. It is also used in combating autism, hypertension, vascular disorders, obesity, and the withdrawal symptoms associated with cessation of drinking and smoking [117−124].

Once-daily paliperidone extended-release formulation provides stable plasma drug concentrations over a 24-hour period [125−127].

The most common adverse reactions were extrapyramidal symptoms, tachycardia, and akathisia, and extrapyramidal symptoms, somnolence, dyspepsia, constipation, weight increase, but it is safer and more tolerable than many other antipsychotics and was found to overcome some of the problems associated with their use.

Paliperidone palmitate is a prodrug of paliperidone. Paliperidone palmitate is formulated as an aqueous nansuspension with low solubility leading to extended-release properties for once-a-month intramuscular injection. Injected paliperidone palmitate is gradually hydrolyzed to paliperidone with low rate [128−132].

The first synthetic routes described for paliperidone are very similar to those previously employed for risperidone, namely, by *N*-alkylating 6-fluoro-3-piperi-din-4-yl-1,2-benzisoxazole (**7.2.9**) with specially prepared 3-(2-chloroethyl)-9-hydroxy-2-methyl-6,7,8,9-tetrahydro-4*H*-pyrido[1,2-*a*]pyrimidin-4-one (**7.2.14**) on reflux in methanol in the presence of diisopropylamine (DIPA) [133]. Later this method was employed following art-known *N*-alkylation procedures such as coupling of mentioned regents in DMF in the presence of sodium carbonate and potassium iodide [134], or using 1,8-diazabicyclo[5.4.0]undec-7-ene (DBU) in methanol [135] as well as engaging many other specific chemical tricks [136−147] (Scheme 7.8).

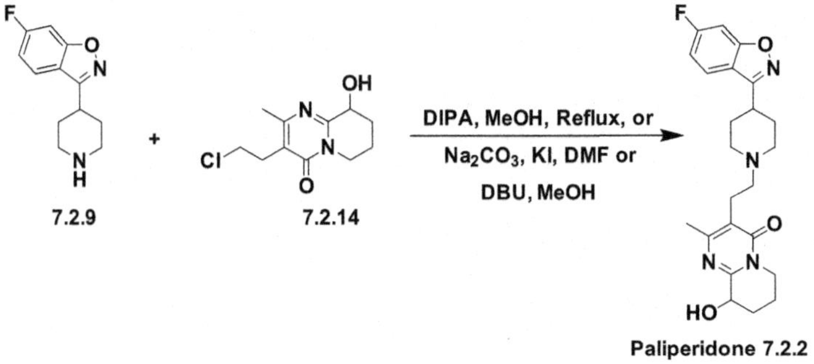

SCHEME 7.8 Synthesis of paliperidone.

Iloperidone (595)

Iloperidone (**7.2.3**) (Fanapt) is an atypical antipsychotic for the treatment of schizophrenia symptoms whose primary mechanism of action is unknown. However, it is proposed that the efficacy of iloperidone is mediated through a combination of D2/5-HT2A antagonism, with greater affinity for the 5-HT2A receptor than D2 receptors. Iloperidone has other binding affinities that give it a unique receptor binding "fingerprint." In particular, iloperidone has a high affinity for the dopamine D3 receptor and very strong affinity for the noradrenergic A1 receptor. Iloperidone has low affinity to serotonin 5HT1A, dopamine D1 and histamine H1, and muscarinic receptors.

The characteristics of iloperidone are few extrapyramidal adverse reactions, lower incidence of metabolic abnormalities, and increased body mass. However, it cut can cause QT prolongation — a heart rhythm disorder that can cause serious arrhythmias.

Iloperidone is the most recent addition to the current group of second-generation antipsychotics. While it may share many qualities with other agents in this class, its unique neuroreceptor signature and adverse-effect profile may prove beneficial in clinical practice [148−167]. The most common adverse events associated with iloperidone treatment include dizziness, dry mouth, dyspepsia, sedation, orthostatic hypotension, and weight gain.

A number of syntheses of iloperidone (**7.2.3**) were reported in the literature. According to initial publications, a mixture of 6-fluoro-3-(piperidin-4-yl)benzo[*d*]isoxazole (**7.2.9**) and 1-(4-(3-chloropropoxy)-3-methoxyphenyl)ethan-1-one (**7.2.15**) was heated in dimethylformamide in presence of potassium carbonate at 90°C for 16 hours to give desired product, iloperidone (**7.2.3**) [168−170]. Different bases and solvents including phase transfer catalysis procedure were reported for this process [171−178]. According to another approach benzoisoxazole derivative (**7.2.9**) was first alkylated with 1-chloro-3-bromo propane in dietylformamide in the presence of potassium carbonate, which provided the chloride intermediate (**7.2.16**). Subsequent reaction with acetovanillone (**7.2.17**) under basic conditions (acetone, potassium carbonate, reflux) gave the desired iloperidone (**7.2.3**) [179] (Scheme 7.9).

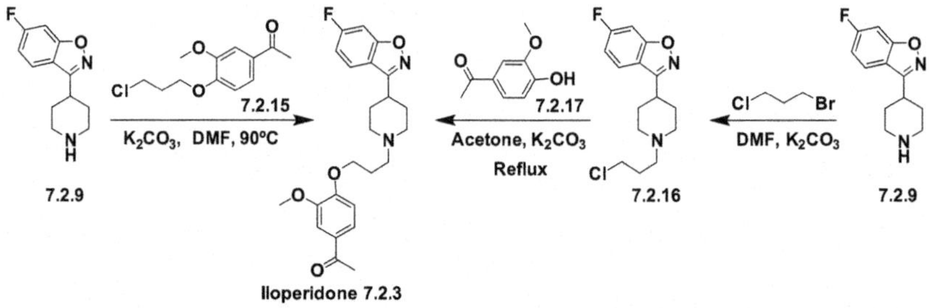

SCHEME 7.9 Synthesis of iloperidone.

Naratriptan (903)

Naratriptan (**7.2.22**) (Amerge) is a selective 5-HT(1B/1D) receptor agonist, with a high affinity at the 5-HT(1B), 5-HT(1D) and 5-HT(1F) receptor subtypes, which belongs to a family of tryptamine-based drugs (triptans) that are indicated for the acute treatment of migraine headaches. Naratriptan is not intended for the prophylactic therapy of migraine or for use in the management of hemiplegic, basilar, or ophthalmoplegic migraine. Triptans, in general, are a rescue treatment for immediate pain relief of an acute migraine headache attack. The biological and pharmacokinetic profile of naratriptan, excellent tolerability profile, and long duration of action, differs significantly from other selective 5-HT(1B/1D) agonists [180–190].

Most common adverse reactions are paresthesias, nausea, dizziness, drowsiness, malaise/fatigue, and throat/neck symptoms.

Literature study reveals several methods of synthesis for naratriptan (**7.2.22**). The first patent [191] and related publications [192–195] disclose several processes for its preparation. One of them comprises reacting 1-methylpiperidin-4-one (**7.1.36**) with 5-bromoindole (**7.2.18**) by means of potassium hydroxide in methanol at room temperature to give 5-bromo-3-(4-(hydroxy-1-methylpiperidin-4-yl)-1*H*-indole (**7.2.19**), which was condensed with *N*-methylvinylsulfonamide (**7.2.20**) using palladium acetate and tri-*p*-tolyl phosphine (P(*p*-Tolyl)₃) in hot DMF in sealed vessels to afford the Heck reaction product (*E*)-*N*-methyl-2-[3-(1-methyl-1,2,3,6-tetrahydropyridin-4-yl)-1*H*-indol-5-yl]vinylsulfonamide (**7.2.21**). Finally, this compound was hydrogenated over palladium on a charcoal catalyst to give naratriptan (**7.2.22**).

In another process described in the same patent, the reaction of 5-bromoindole (**7.2.18**) with 1-methylpiperidin-4-one (**7.1.36**) in methanol in the presence of potassium hydroxide was carried out at reflux the resulting product was 5-bromo-3-(1-methyl-1,2,3,6-tetrahydropyridin-4-yl)-1*H*-indole (**7.2.23**), which was further condensed with *N*-methylvinylsulfonamide

(**7.2.20**), as described above, to obtain compound (**7.2.21**) converted to naratriptan (**7.2.22**) on hydrgenation.

The same approach and same reaction conditions was demonstrated in [194] reacting of 2-(1*H*-indol-5-yl)-*N*-methylethanesulfonamide (**7.2.24**) with 1-methylpiperidin-4-one (**7.1.36**) using potassium hydroxide in refluxing methanol, which directly yielded *N*-methyl-2-[3-(1-methyl-1,2,3,6- tetrahydropyridin-4-yl)-1*H*-indol-5-yl]ethanesulfonamide (**7.2.21**), reducing the C = C double bond to give naratriptan (**7.1.22**).

Another method disclosed in [192] comprises Fischer cyclization of intermediate hydrazone obtained from reaction of 2-(1-methylpiperidin-4-yl)acetaldehyde (**7.2.25**) with 2-(4-hydrazinophenyl)-*N*-methylethanesulfonamide (**7.2.26**) in 2 N hydrochloric acid using polyphosphate ester (PE) on heating at reflux in chloroform for eight minutes to give naratriptan (**7.2.22**) (Scheme 7.10).

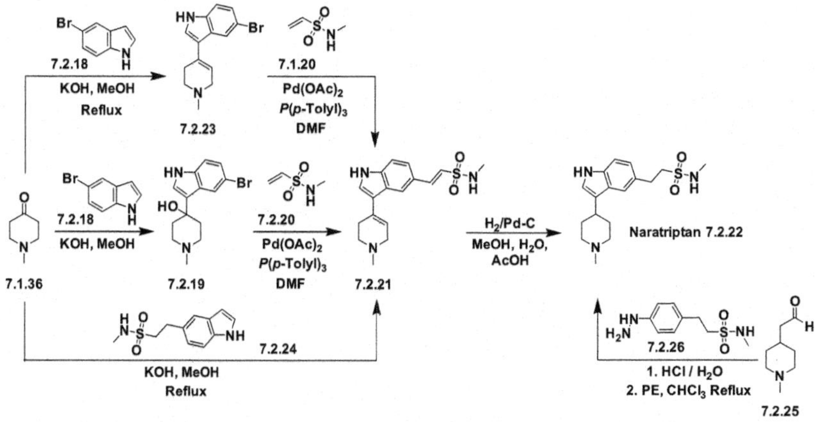

SCHEME 7.10 Synthesis of naratriptan.

Another approach for the preparation of naratriptan (**7.2.22**) is based on the Japp—Klingemann reaction as a key step in building the indole moiety. The method is disclosed in patents [196,197]. The process comprises diazotizing *N*-methyl-2-(4-aminophenyl)-ethane sulfonamide (**7.2.27**) with nitrous acid prepared on treating sodium nitrite solution in water at −5 to 0°C. To the obtained diazonium salt (**7.2.28**) specially synthesized methyl 3-oxo-2-(pyridin-4-ylmethyl)butanoatemethyl-2-acetyl-3-pyridyl propanoate (**7.2.29**) was added to afford the corresponding intermediate hydrazone compound (**7.2.30**), which was cyclized in the presence of hydrogene chloride in methanol (AcCl + MeOH) giving methyl-5-methyl sulfamoylethyl-3-(4-pyridyl)-1*H*-2-indole carboxylate (**7.2.31**). It was quaternized using methyl iodide in dimethylsulfoxide to obtain 1-methyl-4-(2-methoxy carbonyl-5-methyl sulfamoylethyl-1*H*-3-indolyl)pyridinium iodide (**7.2.32**) followed by reduction with sodium borohydride in methanol to give methyl 3-(1-methyl-1,2,3,6-tetrahydropyridin-4-yl)-5-(2-(N-methylsulfamoyl)ethyl)-1H-indole-2-carboxylate (**7.2.33**). The obtained product was hydrogenated over a Raney nickel catalyst in methanol under hydrogen pressure at 50°C for 12 hours to give methyl 3-(1-methylpiperidin-4-yl)-5-(2-(N-methylsulfamoyl)ethyl)-1H-indole-2-carboxylate (**7.2.34**). The ester group in obtained (**7.2.34**) was hydrolyzed in refluxing solution of sodium carbonate in water and then decarboxylated on reflux in concentrated hydrochloric acid to give naratriptan (**7.2.22**). An analogs sequence of reactions was carried out that was specially synthesized from (**7.2.29**) methyl 2-((1-methylpiperidin-4-yl)methyl)-3-oxo-butanoate (**7.2.35**) to prepare naratriptan (**7.2.22**) [196,197] (Scheme 7.11).

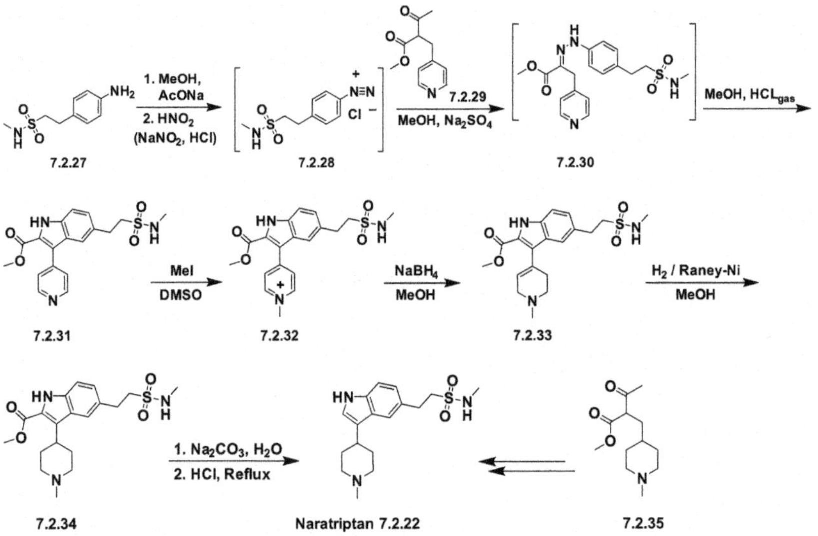

SCHEME 7.11 Synthesis of naratriptan.

Sertindole (1242)

Sertindole (**7.2.42**) (Serdolect) is a second-generation atypical antipsychotic that is thought to give a lower incidence of extrapyramidal side effects than typical antipsychotic drugs. It displays a broad pharmacological profile and has a high affinity as an antagonist at dopamine D2, serotonin 5-HT2A, 5-HT2C, and α1-adrenergic receptors and shows modest affinity for H1-histaminergic and muscarinic receptors. It was introduced into the market for the treatment of schizophrenia. It also finds application in the treatment of anxiety, cognitive disorders, drug abuse, and hypertension. In contrast to other antipsychotics, sertindole has no associated sedative effects [198–209].

Sertindole was taken off the market by its manufacturer in 1998 due to concerns of its association with prolongation of QT intervals, serious cardiac arrhythmia, and sudden cardiac death associated with its use. But after an extensive postmarketing analysis and epidemiological studies regarding its safety and a reevaluation of its risks and benefits, sertindole was relaunched in Europe in 2005 [210−214].

Synthesis of sertindole (**7.2.42**) involves reaction of piperidin-4-one (**7.2.36**) with 5-chloro-1-(4- fluorophenyl)indole (**7.2.37**) prepared by copper catalyzed *N*-arylation of 5-chloroindole with 4-fluorobromobenzene. Contrary to previous reports describing alkaline conditions for the reaction of condensation of 4-piperidones to indoles, the authors propose acidic conditions (a mixture of trifluoroacetic acid in acetic acid media, gentle reflux), which affords 5-chloro-1-(4-fluorophenyl)-3-(1,2,3,6-tetrahydropyridin-4-yl)-1*H*-indole (**7.2.38**) in good yields. Alkylation of the obtained compound with 1-(2-chloroethyl)-2-imidazolidinone (**7.2.39**) yielded product (**7.2.40**). Catalytic hydrogenation of the last on Adams catalyst gave sertindole (**7.2.42**).

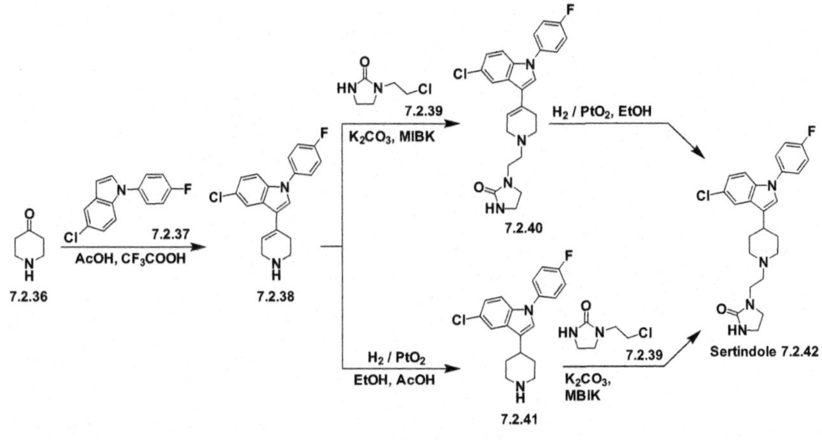

SCHEME 7.12 Synthesis of sertindole.

The alternative route − catalytic hydrogenation of compound (**7.2.38**) over the same Adams catalyst in ethanol/acetic acid − gave 5-chloro-1-(4-fluorophenyl)-3-(piperidin-4-yl)-1*H*-indole (**7.2.41**). The obtained product was alkylated with 1-(2-chloroethyl)-2-imidazolidinone (**7.2.39**) to give desired sertindole (**7.2.42**), which was found to be the preferred way to synthesize sertindole [215−221] (Scheme 7.12).

REFERENCES

[1] Cure S, Rathbone J, Carpenter S. Droperidol for acute psychosis. Cochrane Database Syst Rev 2004;(4):CD002830.

[2] Shale JH, Shale CM, Mastin WD. Safety of droperidol in behavioural emergencies. Exp Opin Drug Safety 2004;3(4):369−78.

[3] Shale JH, Shale CM, Mastin WD. A review of the safety and efficacy of droperidol for the rapid sedation of severely agitated and violent patients. J Clin Psych 2003;64 (5):500–5.

[4] Chambers RA, Druss BG. Droperidol: efficacy and side effects in psychiatric emergencies. J Clin Psych 1999;60(10):664–7.

[5] Brown ES, Dilsaver SC, Bowers TC, Swann AC. Droperidol in the interim management of severe mania: case reports and literature review. Clin Neuropharmacol 1998;21 (5):316–18.

[6] McKeage K, Simpson D, Wagstaff AJ. Intravenous droperidol: a review of its use in the management of postoperative nausea and vomiting. Drugs 2006;66(16):2123–47.

[7] Storrar J, Hitchens M, Platt T, Dorman S. Droperidol for treatment of nausea and vomiting in palliative care patients. Cochrane Database Syst Rev 2014;(11):CD006938.

[8] Richards John R, Richards Irina N. Ozery Gal; Derlet Robert W., Droperidol analgesia for opioid-tolerant patients. Cochrane Database Syst Rev 2011;41(4):389–96.

[9] Richards JR, Schneir AB. Droperidol in the emergency department: is it safe? J Emerg Med 2003;24(4):441–7.

[10] Thomas MC, Musselman ME, Shewmaker J. Droperidol for the treatment of acute migraine headaches. Ann Pharmacother 2015;49(2):233–40.

[11] Cozanitis DA, Rosenberg PH. 'Intense inner agitation': an overlooked side effect of droperidol. Acta Anaesthesiol Scand 2012;56(2):261–2.

[12] Dershwitz M. Is droperidol safe? Probably. Semin Anesth 2004;23(4):291–301.

[13] Janicki CA, Gilpin RK. Droperidol. Anal Profiles Drug Subst 1978;7:171–92.

[14] Jackson CW, Sheehan AH, Reddan JG. Evidence-based review of the black-box warning for droperidol. Am J Health-Syst Pharm 2007;64(11):1174–85.

[15] Calver L, Page CB, Downes MA, Chan B, Kinnear F, Wheatley L, et al. The safety and effectiveness of droperidol for sedation of acute behavioral disturbance in the emergency department. Ann Emerg Med 2015;66(3):230–238.e1.

[16] Davoll J. Reaction of o-phenylenediamine with α,β-unsaturated acids and with β-oxo esters. J Chem Soc 1960;308–14.

[17] Rossi A, Hunger A, Kebrle J, Hoffmann K. Benzimidazole derivatives. IV. Condensation of o-phenylenediamine with α-aryl- and γ-arylacetoacetates Helv. Chim Acta 1960;43:1046–56.

[18] Janssen PAJ. 1-(1-Aroylpropyl-4-piperidyl)-2-benzimidazolinones and related compounds, US 3161645; 1964.

[19] Janssen C. 1-(1-Aroylpropyl-4-piperidyl)-2-benzimidazolinones, BE 626307; 1963.

[20] Janssen PAJ, Gardocki JF. Method for producing analgesia, US 3141823; 1964.

[21] Fluegel KA, Pfeiffer, Wolfgang M. Clinical experience with the butyrophenone benperidol. Arzneimitt -Forsch 1967;17(4):483–5.

[22] Takla PG, James KC, Gassim AEH. Benperidol. Anal Profiles Drug Subst 1985; 14:245–72.

[23] Germane S, Veselova SV. Benperidol (review). Eksperiment Klin Farmakoter 1982;11:6–15.

[24] Leucht S, Hartung B. Benperidol for schizophrenia. Cochrane Database Syst Rev 2005; (2):CD003083.

[25] Yamasaki T, Kojima H, Sakurai T, Kasahara A, Watanabe S, Fujiwara M, et al. Pharmacological studies on timiperone, a new neuroleptic drug. Part I: behavior effects. Arzneimitt -Forsch 1981;31(4):701–7.

[26] Yamasaki T, Kojima H, Tanaka M, Aibara S, Hirohashi M, Kasai Y, et al. Pharmacological studies on timiperone, a new neuroleptic drug Part II: general pharmacological properties. Arzneimitt -Forsch 1981;31(4):707–15.

[27] Nakazawa T, Ohara K, Sawa Y, Edakubo T, Matsui H, Sawa J, et al. Comparison of efficacy of timiperone, a new butyrophenone derivative, and clocapramine in schizophrenia: a multiclinic double-blind study. J Int Med Res 1983;11(5):247−58.

[28] Ueno K, Sato M, Arimoto M, Kojima H, Yamasaki T, Sakurai T. 1,2-Disubstituted benzimidazole derivatives, US 3963727; 1976.

[29] Sato M, Arimoto M, Ueno K, Kojima H, Yamasaki T, Sakurai T, et al. Psychotropic agents. 3. 4-(4-Substituted piperidinyl)-1-(4-fluorophenyl)-1-butanones with potent neuroleptic activity. J Med Chem 1978;21(11):1116−20.

[30] Sato M, Arimoto M. Psychotropic agents. VI. An improved synthetic method for 4′-fluoro-4-[4-(2-thioxo-1-benzimidazolinyl)piperidino] butirophenone. Chem Pharm Bull 1982;30(2):719−22.

[31] Pinder RM, Brogden RN, Sawyer PR, Speight TM, Spencer R, Avery GS. Pimozide: a review of its pharmacological properties and therapeutic uses in psychiatry. Drugs 1976;12(1):1−40.

[32] Smyj R, Wang X, Han F. Pimozide. Profiles Drug Subst Excip Relat Methodol 2012; 37:287−311.

[33] Tueth MJ, Cheong JA. Clinical uses of pimozide. South Med J 1993;86(3):344−9.

[34] Pringsheim T, Marras C. Pimozide for tics in Tourette's syndrome. Cochrane Database Syst Rev 2009;(2):CD006996.

[35] Colvin CL, Tankanow RM. Pimozide: use in Tourette's syndrome. Drug Intell Clin Harm 1985;19(6):421−4.

[36] Opler LA, Feinberg SS. The role of pimozide in clinical psychiatry: a review. J Clin Psych 1991;53(5):221−33.

[37] Mothi M, Sampson S. Pimozide for schizophrenia or related psychoses. Cochrane Database Syst Rev 2013;(11):CD001949.

[38] Sultana A, McMonagle T. Pimozide for schizophrenia or related psychoses. Cochrane Database Syst Rev 2000;(3):CD001949.

[39] Lorenzo CR, Koo J. Pimozide in dermatologic practice: a comprehensive review. Am J Clin Derm 2004;5(5):339−49.

[40] van Vloten Willem A. Pimozide: use in dermatology. Dermatol Online J 2003;9(2):3.

[41] Janssen C. Benzimidazolinylpiperidines and −tetrahydropyridines, BE 633495; 1963.

[42] Janssen PAJ. 1-(4,4′-Diarylbutyl)-4-(2-oxo-1-benzimidazolyl)piperidines, FR M3695; 1964.

[43] Okhuysen PC, DuPont HL, Ericsson CD, Marani S, Martinez-Sandoval FG, Olesen MA, et al. Zaldaride maleate (a new calmodulin antagonist) versus loperamide in the treatment of traveler's diarrhea: randomized, placebo-controlled trial. Clin Infect Dis 1995;21 (2):341−4.

[44] Norman JA, Ansell J, Stone GA, Wennogle LP, Wasley JW. CGS 9343B, a novel, potent, and selective inhibitor of calmodulin activity. Mol Pharmacol 1987;31(5):535−40.

[45] Silberschmidt G, Schick MT, Steffen R, Kilpatrick ME, Murphy JR, Oyofo BA, et al. Treatment of travellers' diarrhea: zaldaride compared with loperamide and placebo. Eur J Gastroenterol Hepatol 1995;7(9):871−5.

[46] DuPont HL, Ericsson CD, Mathewson JJ, Marani S, Knellwolf-Cousin AL, Martinez-Sandoval FG. Zaldaride maleate, an intestinal calmodulin inhibitor, in the therapy of travelers' diarrhea. Gastroenterol 1993;104(3):709−15.

[47] Aikawa N, Kishibayashi N, Karasawa A, Ohmori K. The effect of zaldaride maleate, an antidiarrheal compound, on acetylcholine-induced intestinal electrolyte secretion. Biol Pharmaceut Bull 2000;23(11):1377−8.

[48]   Aikawa N, Karasawa A, Ohmori K. Effect of zaldaride maleate, an antidiarrheal compound, on 16,16-dimethyl prostaglandin E2-induced intestinal ion secretion in rats. Jap J Pharmacol 2000;83(3):269−72.

[49]   Wasley JWF, Norman J. Pyrrolo[1,2-a] [4,1]benzoxazepine derivatives useful as calmodulin and histamine inhibitors, US 4758559; 1988.

[50]   Inventor data available. Preparation of 4-(aminoalkyl)-4H,6H-pyrrolo[1,2-a][4,1]benzoxazepines as antidiarrheal and antiulcer agents, JP 62169791; 1987.

[51]   Boyer S, Blazier E, Barabi M, Long G, Zaunius G, Wasley JWF, et al. The synthesis of 1,3-dihydro-1-[1-[(4-methyl-4H,6H-pyrrolo[1,2-a][4,1]benzoxazepin-4-yl)methyl]-4-piperidinyl]-2H-benzimidazol-2-one (1:1) maleate (CGS 9343 B, potent calmodulin inhibitor). J Het Chem 1988;25(3):1003−5.

[52]   Brogden RN, Carmine AA, Heel RC, Speight TM, Avery GS. Domperidone. A review of its pharmacological activity, pharmacokinetics and therapeutic efficacy in the symptomatic treatment of chronic dyspepsia and as an antiemetic. Drugs 1982;24 (5):360−400.

[53]   Barone JA. Domperidone: a peripherally acting dopamine2-receptor antagonist. Ann Pharmacother 1999;33(4):429−40.

[54]   Reddymasu SC, Soykan I, McCallum RW. Domperidone: review of pharmacology and clinical applications in gastroenterology. Am J Gastroenterol 2007;102(9):2036−45.

[55]   Ahmad N, Keith-Ferris J, Gooden E, Abell T. Making a case for domperidone in the treatment of gastrointestinal motility disorders. Curr Opin Pharmacol 2006;6(6):571−6.

[56]   Phan H, DeReese A, Day AJ, Carvalho M. The dual role of domperidone in gastroparesis and lactation. Int J Pharmaceut Comp 2014;18(3):203−7.

[57]   Prakash A, Wagstaff AJ. Domperidone: a review of its use in diabetic gastropathy. Drugs 1998;56(3):429−45.

[58]   Champion MC. Domperidone. Gen Pharmacol 1988;19(4):499−505.

[59]   Hopkins SJ. Domperidone. Drugs Today 1981;17(1):19−23.

[60]   Albright LM. Use of domperidone as a prokinetic and antiemetic. Int J Pharmaceut Comp 2005;9(2):120−5.

[61]   Rossi M, Giorgi G. Domperidone and long QT syndrome. Curr Drug Safety 2010;5 (3):257−62.

[62]   Li Z, Liu C, Li W, Cao F. Process for preparation of domperidone maleate, CN 1810805; 2006.

[63]   Janssen PAJ, Van Wijngaarden I, Soudijn W. 1-{1-[2-(1,4-Benzodioxan-2-yl)-2-hydroxyethyl]-4-piperidyl}-2-benzimidazolinones, DE 2400094; 1974.

[64]   Vanderberk J, Kennis LEJ, Van der Aa MJMC, Van Hertum AHMT. 1,3-Dihydro-1-[3-(1-piperidinyl)propyl]-2H-benzimidazol-2-ones and related compounds, US 4066772 A; 1978.

[65]   Vandenberk J., Kennis LEJ, Van der Aa MJMC, Van Heertum AAM Th. 1-(Benzazolylalkyl)piperidines and their salts with acids, DE 2632870; 1977.

[66]   Knape H. Bezitramide, an orally active analgesic. An investigation on pain following operations for lumbar disc protrusion (preliminary report). Brit J Anaesth 1970;42(4):325−8.

[67]   Knape H. Further experiences with bezitramide. Its analgesic action and side effects in patients operated upon for lumbar disc protrusion. Brit J Anaesth 1971;43(1):76−83.

[68]   Janssen PAJ, Niemegeers CJE, Schellekens KHL, Marsboom RHM, Herin VV, Amery WKP, et al. Bezitramide (R 4845), a new potent and orally long-acting analgesic compound. Arzneimitt -Forsch 1971;21(6):862−7.

[69] Amery WK, Admiraal PV, Beck PH, Bosker JT, Crul JF, Feikema JJ, et al. Peroral management of chronic pain by means of bezitramide (R 4845), a long-acting analgesic, and droperidol (R 4749), a neuroleptic. A multicentric pilot-study. Arzneimitt -Forsch 1971;21 (6):868−71.

[70] Vaerenberg C. Clinical experiences with bezitramide (R 4845), a potent orally long-acting analgesic. Acta Clin Belg 1971;26(1):11−20.

[71] Meijer DKF, Hovinga G, Versluis A, Broering J, Van Aken K, Moolenaar F, et al. Pharmacokinetics of the oral narcotic analgesic bezitramide and preliminary observations on its effect on experimentally induced pain. Eur J Clin Pharmacol 1984;27(5):615−18.

[72] Admiraal PV, Knape H, Zegveld C. Experience with bezitramide and droperidol in the treatment of severe chronic pain. Brit J Anaesth 1972;44(11):1191−6.

[73] Kay B. A study of strong oral analgesics: the relief of postoperative pain using dextromoramide, pentazocine and bezitramide. Brit J Anaesth 1973;45(6):623−8.

[74] Janssen C. Benzimidazolinylpiperidines and −tetrahydropyridines, BE 633495; 1963.

[75] Janssen PAJ. Benzimidazolinyl piperidines, US 3196157; 1965.

[76] Ebnother A, Jucker E, Lindenmann A. Synthetic drugs. III. New basically substituted pyrazolone derivatives. Helv Chim Acta 1959;42:1201−14.

[77] Cerletti A, Berde B, Neubold K, Taeschler M. Pharmacology of a new serotonin antagonist with analgesic antiphlogistic properties. Boll Chim Farm 1963;102(9):602−15.

[78] Jucker E, Ebnother A, Lindenmann A. Pyrazolone derivatives, US 2903460; 1959.

[79] Jucker E, Ebnoether A, Lindenmann AJ. Pyrazolone derivatives, CH 346885; 1960.

[80] Jucker E, Ebnoether A, Lindenmann AJ. Pyrazolone derivatives, CH 346886; 1960.

[81] Jucker E, Lindenmann AJ. Pyrazolone derivatives, DE 1116674; 1961.

[82] Leemann HG, Antenen K. Determination of the structure of an isomer occurring in the preparation of 1-(N-methyl-4-piperidyl)-3-phenyl-4-benzyl-5-pyrazolone as 1-(N-methyl-4-piperidyl)-4-benzyl-5-phenyl-3-pyrazolone. I. Communication on x-ray structure analysis. Helv Chim Acta 1962;45:177−9.

[83] Jucker E. New basic substituted hydrazines and their application in the synthesis of pharmaceuticals. Angew Chem 1959;71:321−33.

[84] Corena-McLeod M. Comparative pharmacology of risperidone and paliperidone. Drugs in R&D 2015;15(2):163−74.

[85] Germann D, Kurylo N, Han F. Risperidone. Profiles Drug Subst Excip Relat Methodol 2012;37:313−61.

[86] Megens AAHP, Awouters FHL, Schotte A, Meert TF, Dugovic C, Niemegeers CJE, et al. Survey on the pharmacodynamics of the new antipsychotic risperidone. Psychopharmacol 1994;114(1):9−23.

[87] Cohen LJ. Risperidone. Pharmacother 1994;14(3):253−65.

[88] Bhana N, Spencer CM. Risperidone: a review of its use in the management of the behavioural and psychological symptoms of dementia. Drugs Aging 2000;16(6):451−71.

[89] Lane H, Lee C, Liu Y, Chang W. Pharmacogenetic studies of response to risperidone and other newer atypical antipsychotics. Pharmacogenom 2005;6(2):139−49.

[90] Harrison TS, Goa KL. Long-acting risperidone: a review of its use in schizophrenia. CNS Drugs 2004;18(2):113−32.

[91] Colpaert FC. Timeline: discovering risperidone: the LSD model of psychopathology. Nat Rev Drug Disc 2003;2(4):315−20.

[92] Madaan V, Bestha DP, Kolli V, Jauhari S, Burket RC. Clinical utility of the risperidone formulatio in the management of schizophrenia. Neuropsych Dis Treat 2011;7:611−20.

[93] Madaan V. Risperidone: a review of efficacy studies in adolescents with schizophrenia. Drugs Today 2009;45(1):55–62.

[94] Seto K, Dumontet J, Ensom MHH. Risperidone in schizophrenia: is there a role for therapeutic drug monitoring? Therapeut Drug Monitor 2011;33(3):275–83.

[95] Bobo WV, Shelton RC. Risperidone long-acting injectable (Risperdal Consta) for maintenance treatment in patients with bipolar disorder. Exp Rev Neurotherapeut 2010;10 (11):1637–58.

[96] Deeks ED. Risperidone long-acting injection: in bipolar I disorder. Drugs 2010;70 (8):1001–12.

[97] de Leon J, Wynn G, Sandson NB. The pharmacokinetics of paliperidone versus risperidone. Psychosomat 2010;51(1):80–8.

[98] Raja M. Pharmacotherapy update: risperidone in the treatment of schizophrenia. Clin Med Ther 2009;1:1199–214.

[99] Kemp DE, Canan F, Goldstein BI, McIntyre RS. Long-acting Risperidone: a review of its role in the treatment of bipolar disorder. Adv Ther 2009;26(6):588–99.

[100] Keith S. Use of long-acting risperidone in psychiatric disorders: focus on efficacy, safety and cost-effectiveness. Exp Rev Neurotherapeut 2009;9(1):9–31.

[101] McNeal KM, Meyer RP, Lukacs K, Senseney A, Mintzer J. Using risperidone for Alzheimer's dementia-associated psychosis. Exp Opin Pharmacother 2008;9(14):2537–43.

[102] Ravindran AV, Bradbury C, McKay M, da Silva TL. Novel uses for Risperidone: focus on depressive, anxiety and behavioral disorders. Exp Opin Pharmacother 2007;8 (11):1693–710.

[103] Burns A, De Deyn PP. Risperidone for the treatment of neuropsychiatric features in dementia. Drugs Aging 2006;23(11):887–96.

[104] Sajatovic M, Madhusoodanan S, Fuller MA. Risperidone in the treatment of bipolar mania. Neuropsychiat Dis Treat 2006;2(2):127–38.

[105] Fenton C, Scott LJ. Risperidone. A review of its use in the treatment of bipolar mania. CNS Drugs 2005;19(5):429–44.

[106] Khanna S, Eduard E, Lyons B, Grossman F, Eerdekens M, Kramer M. Risperidone in the treatment of acute mania, double-blind, placebo-controlled study. Brit J Psychiatry 2005;187(3):229–34.

[107] Moeller H. Risperidone: a review. Exp Opin Pharmacother 2005;6(5):803–18.

[108] Owens DG. Extrapyramidal side effects and tolerability of risperidone: a review. J Clin Ppsychiat 1994;55(Suppl.):29–35.

[109] Correia C, Vicente AM. Pharmacogenetics of risperidone response and induced side effects. Per Med 2007;4(3):271–93.

[110] Conley RR. Risperidone side effects. J Clin Psych 2000;61(Suppl. 8):20–5.

[111] Grant S, Fitton A. Risperidone. A review of its pharmacology and therapeutic potential in the treatment of schizophrenia. Drugs 1994;48(2):253–73.

[112] Doh J. Risperidone beyond psychoses. Clin Psychopharmacol Neurosci 2003;1(Suppl. 1):147–56.

[113] Green B. Focus on risperidone. Curr Med Res Opin 2000;16(2):57–65.

[114] Kennis L.E.J., Vandenberk J. 3-piperidinyl-substituted 1,2-benzisoxazoles and 1,2-benzisothiazoles, US 4804663 A; 1989.

[115] Kennis L.E.J., Vandenberk J. Preparation of 1,2-benzisoxazol-3-yl and 1,2-benzisothiazol-3-yl derivatives as antipsychotics, EP 196132; 1986.

[116] Kim D, Kang M, Kim JS, Jeong J. An efficient synthesis of risperidone via Stille reaction: antipsychotic, 5-HT2, and dopamine-D2-antagonist. Arch Pharm Res 2005;28(9):1019–22.

[117]   Capasso A, Milano W. Paliperidone in the treatment of schizophrenia: an overview. Curr Neurobiol 2012;3(2):133–50.

[118]   Alphs L, Fu D, Turkoz I. Paliperidone for the treatment of schizoaffective disorder. Exp Opin Pharmacother 2016;17(6):871–83.

[119]   Corena-McLeod M. Comparative pharmacology of risperidone and paliperidone. Drugs R&D 2015;15(2):163–74.

[120]   Wang S, Han C, Lee S, Patkar AA, Pae C, Fleischhacker WW. Paliperidone: a review of clinical trial data and clinical implications. Clin Drug Invest 2012;32(8):497–512.

[121]   Yang LPH. Oral paliperidone: a review of its use in the management of schizoaffective disorder. CNS Drugs 2011;25(6):523–38.

[122]   Green B. Paliperidone: a clinical review. Curr Drug Ther 2009;4(1):7–11.

[123]   Rao PV, Valli SM, Prabhakar T, Suneetha S, Panda J. Clinical and pharmacological review on novel atypical antipsychotic drug: paliperidone. Biomed Pharmacol J 2008;1(1):167–72.

[124]   Dolder C, Nelson M, Deyo Z. Paliperidone for schizophrenia. Am J Health-Sys Pharm 2008;65(5):403–13.

[125]   Chwieduk CM, Keating GM. Paliperidone extended release: a review of its use in the management of schizophrenia. Drugs 2010;70(10):1295–317.

[126]   Gahr M, Koelle MA, Schoenfeldt-Lecuona C, Lepping P, Freudenmann RW. Paliperidone extended-release: does it have a place in antipsychotic therapy? Drug Des Devel Ther 2011;5:125–46.

[127]   Yang LPH, Plosker GL. Paliperidone extended release. CNS Drugs 2007;21(5):125–46.

[128]   Bishara D. Once-monthly paliperidone injection for the treatment of schizophrenia. Neuropsychiatr Dis Treat 2010;6:561–72.

[129]   Carter NJ. Extended-release intramuscular paliperidone palmitate: a review of its use in the treatment of schizophrenia. Drugs 2012;72(8):1137–60.

[130]   Yin J, Collier AC, Barr AM, Honer WG, Procyshyn RM. Paliperidone palmitate long-acting injectable given intramuscularly in the deltoid versus the gluteal muscle: are they therapeutically equivalent?. J Clin Psychopharmacol 2015;35(4):447–9.

[131]   Zhang LL, Li JT, Zhao YJ, Su YA, Si T. Critical evaluation of paliperidone in the treatment of schizophrenia in Chinese patients: a systematic literature review. Neuropsychiatr Dis Treat 2016;12:113–31.

[132]   Chue P, Chue J. A review of paliperidone palmitate. Exp Rev Neurotherapeut 2012;12(12):1383–97.

[133]   Janssen CGM, Knaeps AG, Kennis LEJ, Vandenberk J. Pharmaceuticals containing antipsychotic 3-piperidinyl-1,2-benzisoxazoles, US 5158952; 1992.

[134]   Vandenberk J, Kennis LEJ. 9-Hydroxy-pyrido 1,2-a!pyrimidin-4-one ether derivatives, US 5688799; 1997.

[135]   Solanki PV, Uppelli SB, Pandit BS, Mathad VT. An improved and efficient process for the production of highly pure paliperidone, a psychotropic agent, via DBU catalyzed N-alkylation, ACS sustain. Chem Eng 2013;1(2):243–8.

[136]   Vandenberk J, Kennis LEJ. Preparation of [(benzisoxazolylpiperidinyl)alkyl]pyrido[1,2-a]pyrimidinones as neurotransmitter antagonists useful as antipsychotics. WO 9514691; 1995.

[137]   Ruzic M, Pecavar A, Prudic D, Plaper I, Klobcar A, Hvala J, et al. Process for the synthesis of paliperidone and purification thereof in the presence of metal cations, EP 2275423; 2011.

[138] Dolitzky B. Process for preparation of paliperidone by reaction of 3-(2-chloroethyl)-6,7,8,9-tetrahydro-9-hydroxy-2-methyl-4H-pyrido[1,2-a]-pyrimidin-4-one with 6-fluoro-3-piperidino-1,2-benzisoxazole, WO 2008021345; 2008.

[139] Modi IA, Sodagar KR, Vineet M, Jain SH, Parikh SN, Sharma AO, et al. Process of synthesis of paliperidone, WO 2010064134; 2010.

[140] Ini S, Shmuely Y. Process for the synthesis of 3-(2-chloroethyl)-6,7,8,9-tetrahydro-9-hydroxy-2-methyl-4H-pyrido[1,2-a]pyrimidin-4-one, a paliperidone intermediate, WO 2009045489; 2010.

[141] Riva R, Banfi L, Castaldi G, Ghislieri D, Malpezzi L, Musumeci F, et al. Eur J Org Chem 2011;12:2319−25; S2319/1-S2319/11.

[142] Rameshchandra SK, Vineet M, Hukamchand JS, Natvarlal PS, Omprakash SA, Rajaram BU, et al. An Improved Process for the Preparation of Paliperidone, WO2010/089643A1; 2010.

[143] Narayanrao KR, Ramachandra RD, Laxminarayan PS. Processes for the Preparation of Paliperidone and Pharmaceutically Acceptable Salts Thereof and Intermediates for Use in the Processes, WO2009/047499A2; 2009.

[144] Reddy RB, Muthulingam A, Saravanakumar KG, Kondalarao P. An Improved Process for the Preparation of Pure Paliperidone, WO2011/015936A2; 2011.

[145] Reddy MS, Eswaraiah S, Satyanarayana R. Processes for the preparation of paliperidone, US 8481729; 2013.

[146] Bartl J, Kraicovic J, Benovsky P. Synthesis of paliperidone, US 7977480 B2; 2011.

[147] Bartl J, Benovsky P. WO 2010003702A1; 2010.

[148] Arif SA, Mitchell MM. Iloperidone: a new drug for the treatment of schizophrenia. Am J Health-Syst Pharm 2011;68(4):301−8.

[149] Citrome L. Iloperidone: a clinical overview. J Clin Psych 2011;72(Suppl. 1):19−23.

[150] Bishop JR, Bishop DL. Iloperidone for the treatment of schizophrenia. Drugs Today 2010;46(8):567−79.

[151] Citrome L. Iloperidone: chemistry, pharmacodynamics, pharmacokinetics and metabolism, clinical efficacy, safety and tolerability, regulatory affairs, and an opinion. Exp Opin Drug Metabol Toxicol 2010;6(12):1551−64.

[152] Hale KS. Iloperidone-a second-generation antipsychotic for the treatment of acute schizophrenia. J Pharm Technol 2010;26(4):193−202.

[153] Rado J, Janicak PG. Iloperidone for schizophrenia. Exp Opin Pharmacother 2010;11 (12):2087−93.

[154] Marino J, Caballero J. Iloperidone for the treatment of schizophrenia. Ann Pharmacother 2010;44(5):863−70.

[155] Citrome L. Iloperidone for schizophrenia: a review of the efficacy and safety profile for this newly commercialised second-generation antipsychotic. Int J Clin Pract 2009;63 (8):1237−48.

[156] Caccia S, Pasina L, Nobili A. New atypical antipsychotics for schizophrenia: iloperidone. Drug Design Dev Ther 2010;4:33−48.

[157] Cutler AJ. Iloperidone: a new option for the treatment of schizophrenia. Exp Rev Neurotherapeut 2009;9(12):1727−41.

[158] Scott LJ. Iloperidone in schizophrenia. CNS Drugs 2009;23(10):867−80.

[159] Ehret MJ, Sopko Jr MA, Levine A. Iloperidone: a novel atypical antipsychotic for the treatment of schizophrenia. Formulary 2008;43(6):190−2 194-196, 203.

[160] Albers LJ, Musenga A, Raggi MA. Iloperidone: a new benzisoxazole atypical antipsychotic drug. Is it novel enough to impact the crowded atypical antipsychotic market? Exp Opin Invest Drugs 2008;17(1):61−75.

[161] Keppel HJM. Iloperidone (Novartis). IDrugs 2002;5(1):84–90.

[162] Jain KK. An assessment of Iloperidone for the treatment of schizophrenia. Exp Opin Invest Drugs 2000;9(12):2935–43.

[163] Mucke HAM, Castaner J. Iloperidone: antipsychotic, dopamine D2 antagonist, 5-HT2A antagonist. Drugs Fut 2000;25(1):29–40.

[164] Corbett R, Griffiths L, Shipley JE, Shukla U, Strupczewski JT, Szczepanik AM, et al. Iloperidone: preclinical profile and early clinical evaluation. CNS Drug Rev 1997;3 (2):120–47.

[165] Nnadi CU, Malhotra AK. Clinical and pharmacogenetic studies of iloperidone. Personal Med 2008;5(4):367–75.

[166] Rado JT, Janicak PG. Long-term efficacy and safety of iloperidone: an update. Neuropsych Dis Treat 2014;10:409–15.

[167] Weiden PJ. Iloperidone for the treatment of schizophrenia: an updated clinical review. Clin Schizoph Relat Psychoses 2012;6(1):34–44.

[168] Strupczewski JT, Helsley GC, Chiang Y, Bordeau KJ. EP 402644; 1990.

[169] Strupczewski JT, Helsley GC, Chiang Y, Bordeau KJ, Glamkowski EJ. US 5364866; 1994.

[170] Strupczewski JT, Bordeau KJ, Chiang Y, Glamkowski EJ, Conway PG, Corbett R, et al. 3-[[(Aryloxy)alkyl]piperidinyl]-1,2-Benzisoxazoles as D2/5-HT2 Antagonists with Potential Atypical Antipsychotic Activity: Antipsychotic Profile of Iloperidone (HP 873). J Med Chem 1995;38(7):1119–31.

[171] Dwivedi SD, Patel DJ, Shah AP. Process for preparing iloperidone and its salts, IN 2010MU03103; 2013.

[172] Ansari SA, Hirpara HM, Yadav AK, Gianchandani JP. Process for the preparation of ilo-peridone, IN 2011KO00760; 2016.

[173] Azad MAK, Pandey G, Singh K, Prasad M, Arora SK. Processes for the preparation of iloperidone, WO 2012090138; 2012.

[174] Bettoni P, Roletto J, Paissoni P. One-pot process for the preparation of iloperidone, EP 2644608; 2013.

[175] Reguri BR, Arunagiri M, Yarroju PCi, Gurusamy SK, Ponnapalli K. An improved pro-cess for the preparation of Iloperidone, IN 2009CH02695; 2012.

[176] Raman JV, Rane D, Kevat J, Patil D. An improved process for preparing iloperidone, IN 2010MU01752; 2012.

[177] Athalye SS, Parghi KD, Ranbhan KJ, Sarjekar PB. A process for the preparation of ilo-peridone, IN 2011MU01458; 2012.

[178] Solanki PV, Uppelli SB, Pandit BS, Mathad VT. Improved and efficient process for the production of highly pure iloperidone: a psychotropic agent. Org Proc Res Dev 2014;18 (2):342–8.

[179] Liu KK-C, Sakya SM, O'Donnell CJ, Flick AC, Ding HX. Synthetic approaches to the 2010 new drugs. Bioorg Med Chem 2012;20(3):1155–74.

[180] Salonen R. Naratriptan: the gentle triptan. Front Headache Res 2001;10:228–35.

[181] Connor HE, Beattie DT. Naratriptan – pharmacology, monographs in clinical neurosci-ence. In: Diener HC, editor. Drug treatment of migraine and other headaches, 17. Basel: Karger; 2000. p. 124–33.

[182] Goadsby PJ. Treatment of acute migraine attacks with naratriptan from monographs in clinical neuroscience. In: Diener HC, editor. Drug treatment of migraine and other head-aches, 17. Basel: Karger; 2000. p. 134–40.

[183] Salonen R. Naratriptan. Int J Clin Practice 1999;53(7):552–6.

[184] Tfelt-Hansen PC. Published and not fully published double-blind, randomised, controlled trials with oral naratriptan in the treatment of migraine: a review based on the GSK Trial Register. J Headache Pain 2011;12(4):399−403.

[185] Dulli DA. Naratriptan: an alternative for migraine. Ann Pharmacother 1999;33(6):704−11.

[186] Mathew NT. Naratriptan: a review. Exp Opin Invest Drugs 1999;8(5):687−95.

[187] Reddy P, Lee N. Focus on naratriptan: an oral 5-HT1 receptor agonist for acute treatment of migraine. Formulary 1998;33(6):521−4 527−528, 530, 533.

[188] Massiou H. Naratriptan. Curr Med Res Opin 2001;17(Suppl. 1):s51−53.

[189] Lambert GA. Preclinical neuropharmacology of naratriptan. CNS Drug Rev 2005;11 (3):289−316.

[190] Mealy N, Castaner J. Drugs Fut 1996;21(5):476−9.

[191] Oxford AW, Butina D, Owen MR. Preparation and formulation of 3-(4-piperidinyl) indole-5-ethanesulfonamides for treatment of headache, EP 303507; 1989.

[192] Oxford AW, Butina D, Owen MR. Indole derivatives U.S. Patent 4997841; 1991.

[193] Blatcher P, Carter M, Hornby R, Owen MR. Process for preparation of N-methyl-3-(1-methyl-4-piperidinyl)-1H-indole-5-ethanesulfonamide, WO 9509166; 1995.

[194] Kumar US, Sankar VR, Kumar SB, Prabhu MP, Rao SM. An investigation into the formation of impurity b during the optimization of naratriptan hydrochlorid. Org Proc Res Dev 2009;13(3):468−70.

[195] Shashikumar ND, Krishnamurthy GN, Rao KSR, Shridhara K, Naik HSB, Nagarajan K. An improved process for the synthesis of 5-bromo-3-(1-methylpiperidin-4-yl)-1H-indole: a key intermediate in the synthesis of naratriptan hydrochloride. Org Proc Res Dev 2010;14(4):918−20.

[196] Islam A, Sahadev K, Reddy MV, Kulkarni RV, Layek MM, Bhar C. Process for preparing naratriptan hydrochloride via Fischer indole synthesis, WO 2006010079; 2006.

[197] Aggarwal AK, Sarin GS, Srinivasan CV, Wadhwa L. Preparation of naratriptan and indole derivatives, WO 2008072257; 2008.

[198] Zoccali RA, Bruno A, Muscatello MRA. Efficacy and safety of sertindole in schizophrenia: a clinical review. J Clin Psychopharmacol 2015;35(3):286−95.

[199] Juruena MF, Ponde de Sena E, Reis de Oliveira I. Sertindole in the management of schizophrenia. J Centr Nerv Syst Dis 2011;3:75−85.

[200] Muscatello MRA, Bruno A, Pandolfo G, Mico U, Settineri S, Zoccali R. Emerging treatments in the management of schizophrenia - focus on sertindole. Drug Des Dev Ther 2010;4:187−201.

[201] Cincotta SL, Rodefer JS. Emerging role of sertindole in the management of schizophrenia. Neuropsychiatr Dis Treat 2010;6:429−41.

[202] Spina E, Zoccali R. Sertindole: pharmacological and clinical profile and role in the treatment of schizophrenia. Exp Opin Drug Metabol Toxicol 2008;4(5):629−38.

[203] Murdoch D, Keating GM. Sertindole: a review of its use in schizophrenia. CNS Drugs 2006;20(3):233−55.

[204] Lewis R, Bagnall A, Leitner M. Sertindole for schizophrenia. Cochrane Database Syst Rev 2000;(2):CD001715.

[205] Hale A. Sertindole: a clinical efficacy profile. Int J Psych, Clin Practice 2002;6(Suppl. 1):S21−6.

[206] Brown LA, Levin GM. Sertindole, a new atypical antipsychotic for the treatment of schizophrenia. Pharmacother 1998;18(1):69−83.

[207] Kane JM, Tamminga CA. Sertindole (serdolect): preclinical and clinical findings of a new atypical antipsychotic. Exp Opin Invest Drugs 1997;6(11):1729−41.

[208] Cardoni AA, Myer S. Sertindole: an atypical antipsychotic for the treatment of schizophrenia. Formulary 1997;32(9):907–10 913-914, 922, 925.

[209] Dunn CJ, Fitton A. Sertindole. CNS Drugs 1996;5(3):224–30.

[210] Kasper S, Quiner S, Pezawas L. A review of the benefit:risk profile of sertindole. Int J Psychiatr Clin Pract 1998;2(Suppl. 2):S59–64.

[211] Kasper S. Sertindole: safety and tolerability profile. Int J Psychiatr Clin Pract 2002;6 (Suppl. 1):S27–32.

[212] Lindstroem E, Levander S. Sertindole: efficacy and safety in schizophrenia. Exp Opin Pharmacother 2006;7(13):1825–34.

[213] Muscatello MRA, Bruno A, Micali BP, Pandolfo G, Zoccali RA. Sertindole in schizophrenia: efficacy and safety issues. Exp Opin Pharmacother 2014;15(13):1943–53.

[214] Pae C. Sertindole: dilemmas for its use in clinical practice. Exp Opin Drug Safety 2013;12(3):321–6.

[215] Perregaard JK, Arnt J, Boegesoe KP, Hyttel J, Sanchez C. Noncataleptogenic, centrally acting dopamine D-2 and serotonin 5-HT2 antagonists within a series of 3-substituted 1-(4-fluorophenyl)-1H-indoles. J Med Chem 1992;35(6):1092–101.

[216] Perregaard JK. Indole derivatives and their antipsychotic activity, EP 200322; 1986.

[217] Perregaard JK, Costall B. Preparation of 3-(4-piperidinyl)-N-arylindoles as anxiolytics, WO 9200070; 1992.

[218] Perregaard JK. 1-(4′-fluorophenyl)-3,5-substituted indoles useful in the treatment of psychic disorders and pharmaceutical compositions thereof, US 4710500; 1987.

[219] Perregaard JK Skarsfeldt T. Use of sertindole for the treatment of schizophrenia, EP 0392959; 1990.

[220] Zanon J, Villa M, Ciardella F. Method for manufacture of sertindole, WO03/080597; 2003.

[221] Kumar VS, Anjaneyulu SR, Bindu VH. Identification and synthesis of impurities formed during sertindole preparation. Beilstein J Org Chem 2011;7:29–33.

Chapter 8

Piperidine-Based Fused Biheterocycles

Piperidine-based fused biheterocycles are represented on the pharmaceutical market as thienopyridine-derivatives. Ticlopidine (**8.1.1**), clopidogrel (**8.1.2**), and prasugrel (**8.1.3**) are the primary platelet inhibitors. There are also γ-carbolines such as antihistaminic mebhydrolin (**8.1.34**), atypical antipsychotics gevotroline (**8.1.40**) and carvotroline (**8.1.42**); and 5,6,7,8-tetrahydropyrido[4,3-*c*]pyridazine — a hypotensive drug endralazine (**8.1.52**) (Fig. 8.1).

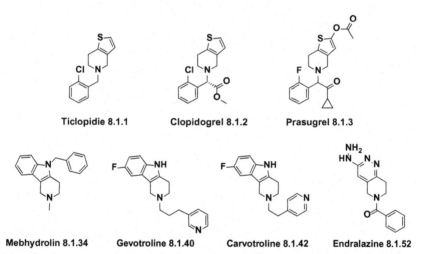

Ticlopidie 8.1.1   Clopidogrel 8.1.2   Prasugrel 8.1.3

Mebhydrolin 8.1.34   Gevotroline 8.1.40   Carvotroline 8.1.42   Endralazine 8.1.52

FIGURE 8.1 Piperidine-based fused biheterocycles represented on the pharmaceutical market.

Platelet inhibitors play a major role in the management of cardiovascular, cerebrovascular, and peripheral vascular diseases. They are approved for a variety of indications including treatment and/or prevention of acute coronary syndromes, stroke/transient ischemic attack, and thrombocythemia. In general the platelet inhibitors are also indicated to prevent thrombosis in patients undergoing cardiovascular procedures and/or surgery.

Piperidine-Based Drug Discovery. DOI: http://dx.doi.org/10.1016/B978-0-12-805157-3.00008-9

8.1 THIENOPYRIDINE DERIVATIVES

Ticlopidine (12218)

Ticlopidine (**8.1.1**) (Ticlid) is an antiplatelet agent that interferes with platelet membrane function by inhibiting adenosine diphosphate (ADP)—induced platelet activation. Ticlopidine behave in vivo as specific antagonists of the P2Y12 purinergic receptor, one of the ADP receptors on platelets.

It is used to reduce the occurrence of atherothrombotic arterial events: stroke, peripheral arterial disease, and unstable angina. Ticlopidine has proved effective in the treatment of atherosclerosis and for the prevention of atherothrombosis [1−16].

Ticlopidine has rare but serious adverse reactions, including thrombotic thrombocytopenic purpura, which is a life-threatening drug-associated disease characterized by Moschcowitz's pentad: thrombocytopenia, microangiopathic hemolytic anemia, fluctuating neurological signs, renal failure, and fever [17−20].

A variety of synthetic approaches for the synthesis of ticlopidine have been described, including improvements on the different steps of synthetic processes.

Evidently the first synthesis of ticlopidine that was considered was a method comprising condensation of a thieno[3,2-c]pyridine (**8.1.3**) with o-chlorobenzyl chloride (**8.1.4**), which took place in boiling acetonitrile. Obtained salt − 5-(2-chloro-benzyl)-thieno[3,2-c]pyridinium chloride (**8.1.5**) − was hydrogenated with sodium borohydride in ethanol water solution to give desired ticlopidine (**8.1.1**) [21,22].

According to another approach 2-(2-thienyl)ethyl amine (**8.1.6**) was added to 37% aqueous formaldehyde solution, and the formed intermediate formimine of 2-(2′-thienyl)ethylamine was cyclized to 4,5,6,7-tetrahydrothieno[3,2-c]pyridine (**8.1.7**) just on shaking with hydrochloric acid. Alkylation of the obtained product with o-chlorobenzyl chloride (**8.1.4**) in the presence of potassium carbonate on reflux in ethanol or in tetrahydrofuran in the presence of sodium hydride gave ticlopidine (**8.1.1**) [23,24].

Benzoylation of 4,5,6,7-tetrahydrothieno[3,2-c]pyridine (**8.1.7**) with 2-chlorobenzoyl chloride (**8.1.8**) gave amide (**8.1.9**), which on LiAlH$_4$ reduction, was converted to ticlopidine (**8.1.1**) [25,26].

Another synthetic route involved a sequence of reactions, including preparation of a 2-(2-thienyl)ethanol (**8.1.11**), synthesized via interaction of thiophen-2-yllithium, prepared on reaction thiophen (**8.1.10**) with butyl lithium. Its conversion to corresponding tosylate (**8.1.12**) in diisopropyl ether was followed by benzylation with o-chlorobenzylamine (**8.1.13**) in reluxing acetonitrile, which gave N-(2-chloro-benzyl)-2-(2-thienyl)ethylamine (**8.1.14**). The last was cyclized to sought ticlopidine (**8.1.1**) on addition of formalin, followed by 2 N hydrochloric acid treatment [27].

The key compound *N*-(2-chloro-benzyl)-2-(2-thienyl)ethylamine (**8.1.14**) was used in the same Pictet-Spengler—type reactions: acid-catalyzed intra-molecular cyclization of 2-arylethylimines to desired ticlopidine (**8.1.1**) using dioxolane instead of formaldehyde, paraformaldehyde, or 1,3,5-trioxane [28], as well as chloromethylmethylether [29,30] (Scheme 8.1).

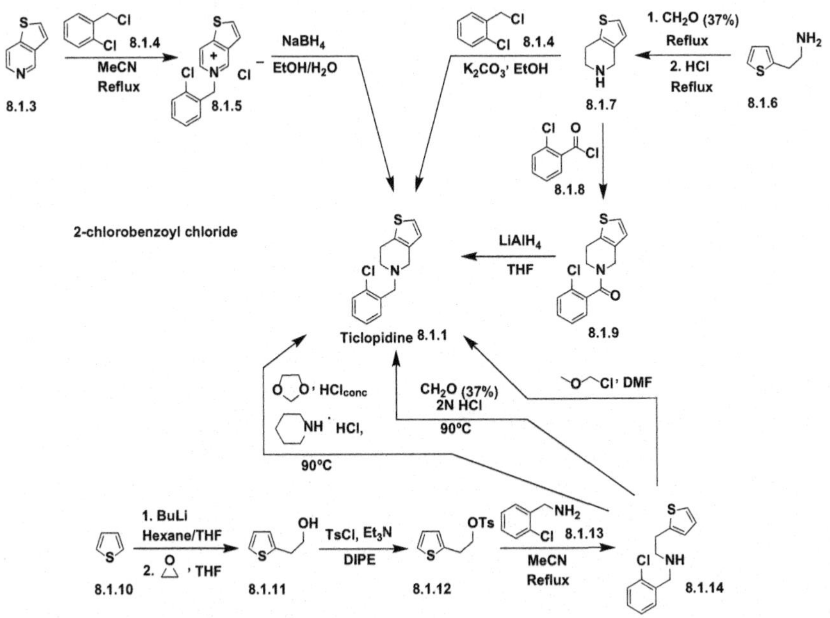

SCHEME 8.1   Synthesis of ticlopidine.

Clopidogrel (20379)

Clopidogrel (**8.1.2**) (Plavix) is one of the most widely clinically used platelet receptor inhibitors used in patients with acute coronary syndromes and/or undergoing percutaneous coronary interventions. More than 40 million patients in the world receive clopidogrel, which is indicated for the reduction of atherothrombotic events in cardiovascular patients with recent myocardial infarction, stroke or established peripheral arterial disease. In 2010, clopido-grel was the second-best selling drug with $ 9.4 billion in global sales.

It is also used in combination with aspirin to treat chest pain, heart attack, unstable angina and to keep blood vessels open and prevent blood clots in patients with acute coronary syndromes and/or undergoing percutaneous coronary interventions and after certain procedures such as cardiac stent, coronary bypass graft balloon angioplasty.

Clopidogrel is a prodrug that has to be metabolized in the liver to generate the active metabolite, which selectively and irreversibly inhibits ADP − induced platelet aggregation targeting the ADP P2Y12 receptor on platelets. Clopidogrel is very efficient in reducing ischemic cardiovascular events but exposes patients to an increased risk of bleeding.

Clopidogrel resistance (failure of clopidogrel to achieve its antiaggregatory effect), even adverse clinical events, could be presented in some patients taking clopidogrel.

Proton pump inhibitors often are prescribed in combination with clopidogrel to decrease the risk of gastrointestinal bleeding after an acute coronary syndrome. But both medications are metabolized largely by the same CYP2C19 enzyme. Therefore concerns exist that a drug−drug interaction during concomitant treatment may result in a reduction of platelet inhibition [3,31−64].

Currently a variety of synthetic methods for preparation of clopidogrel (**8.1.2**) are reported.

One of the first reports disclosed synthesis of racemic (±) clopidogrel (**8.1.2**) described reaction of 4,5,6,7-tetrahydrothieno[3,2-*c*]pyridine (**8.1.7**) with methyl 2-chloro-2-(2-chlorophenyl)-acetate (**8.1.15**) in dimethylformamide in presence of potassium carbonate [65,66]. Process for the separation of (+)-(*S*) clopidogrel (**8.1.2**) from its racemic compound is based on crystallization of the salts prepared from racemic compound (±) (**8.1.2**) with (+)-10-camphor sulphonic acid (CSA) as resolution agent in acetone.

Implementing an analog reaction − methyl (*R*)-2-(2-chlorophenyl)-2-(tosyloxy)acetate tolenesulfonate ester (**8.1.16**) − was coupled with thieno [3,2-*c*]pyridine (**8.1.7**) in the presence of potassium carbonate in dichloromethane/water media. Which allowed to obtain directly desired enantiomerically pure (+)-(*S*) isomer of clopidogrel (**8.1.2**) by S_N2 displacement [67].

According to another method, racemic methyl 2-(2-chlorophenyl)-2-((2-(thiophen-2-yl)ethyl)amino)acetate (**8.1.18**), prepared by reaction of methyl 2-amino-2-(2-chlorophenyl)-acetate (**8.1.17**) with tosylate of 2-(thiophen-2-yl)ethan-1-ol (**8.1.12**) was suspended in a 30% aqueous solution of formaldehyde and kept at 50°C to give (±) racemic clopidogrel (**8.1.2**). The desired (+)-(*S*) enantiomer of (**8.1.2**) was similarly obtained by recrystallization of the salt of the racemic compound with (+)-10-camphorsulphonic acid in acetone. The same sequence of reactions with (+)-methyl 2-amino-2-(2-chlorophenyl)acetate (**8.1.17a**) prepared via resolution of (**8.1.17**) with (L)-(+)-tartaric acid in the media of three solvents − methanol, acetonitrile and methyl ethyl ketone − at about 60°C, which gave pure (+)-(*S*) isomer of clopidogrel (**8.1.2**) [68,69].

An interesting method that implemented a Strecker reaction of *o*-chlorobenzaldehyde (**8.1.19**) with sodium cyanide and 4,5,6,7-tetrahydrothieno [3,2-*c*]pyridine (**8.1.7**) as a secondary amine component in aqueous sodium bisulfite media gave nitrile (**8.1.20**). Treatment of this nitrile with potassium hydroxide in *t*-butanol gave amide (**8.1.21**). Resolution of the obtained racemic amide (**8.1.21**) using (+)-10-CSA gave product (**8.1.22**) subjected directly to methanolysis with dimethyl sulfate in refluxing methanol to give (*S*)-(+)-clopidogrel (**8.1.21**) in high yield [70].

A variety of synthetic modifications to making clopidogrel [71−85] are generalized on the Scheme 8.2.

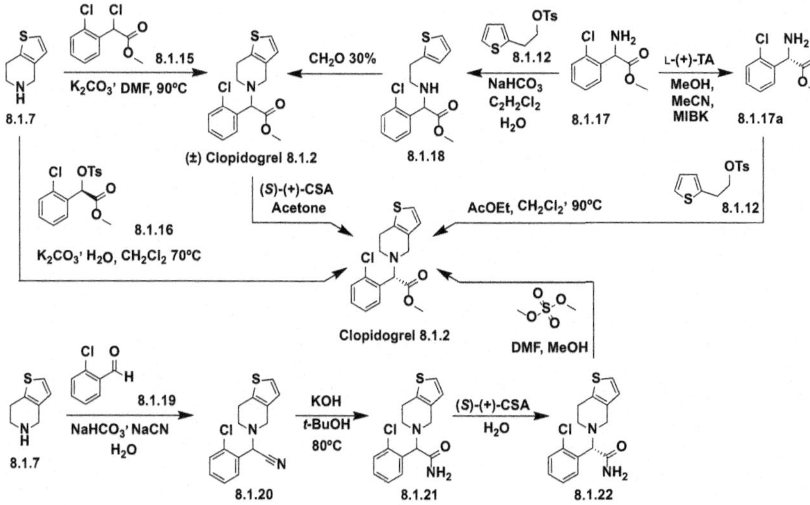

SCHEME 8.2 Synthesis of clopidogrel.

Prasugrel (3208)

Prasugrel (**8.1.3**) (Effient) is the most recent, third-generation development of thienopyridine-type antiplatelet drugs structurally and pharmacologically related to clopidogrel (**8.1.1**) and ticlopidine (**8.1.2**). Prasugrel is also an inactive prodrug that requires metabolic processing in vivo to generate the active antiplatelet metabolite. The efficacy of this bioactivation is the key determinant for the pharmacodynamic potency of the compound, i.e., the irreversible blockade of the platelet P2Y12-ADP receptor.

This newest agent, currently available for use in the United States is similar to clopidogrel, but it is about 10 times more potent and has a quicker onset of action.

Prasugrel is used with aspirin to prevent thrombosis by patients with heart disease (recent heart attack, unstable angina) who undergo a certain heart procedure (angioplasty). This medication helps to prevent other serious heart/blood vessel problems (such as heart attacks, strokes, blood clots in stents).

Prasugrel may cause unwanted side effects. Some of the more common ones are blurred vision, dizziness, headache, nervousness, pounding in the ears, and slow or fast heartbeat [45,86−105].

The first patent that disclosed synthesis of prasugrel (**8.1.3**) described it as starting from the Grignard reaction of cyclopropyl cyanide (**8.1.23**) with (2-fluorobenzyl)magnesium bromide (**8.1.24**) in ether to provide cyclopropyl 2-fluorobenzyl ketone (**8.1.25**), which is then reacted with bromine in carbon tetrachloride solution to provide 2-bromo-1-cyclopropyl-2-(2-fluorophenyl) ethan-1-one (**8.1.26**). Coupling of the last with 5,6,7,7a-tetrahydro-thieno [3,2-*c*]pyridin-2(4*H*)-one (**8.1.27**) in the presence of potassium carbonate in dimethylformamide afforded 5-[2-cyclopropyl-1-(2-fluorophenyl)-2-oxoethyl]-5,6,7,7a-tetrahydro-thieno[3,2-*c*]pyridin-2(4*H*)-one (**8.1.28**), which was finally reacted with acetic anhydride and sodium hydride in dimethylformamide to yield prasugrel (**8.1.3**) [106,107] (Scheme 8.3).

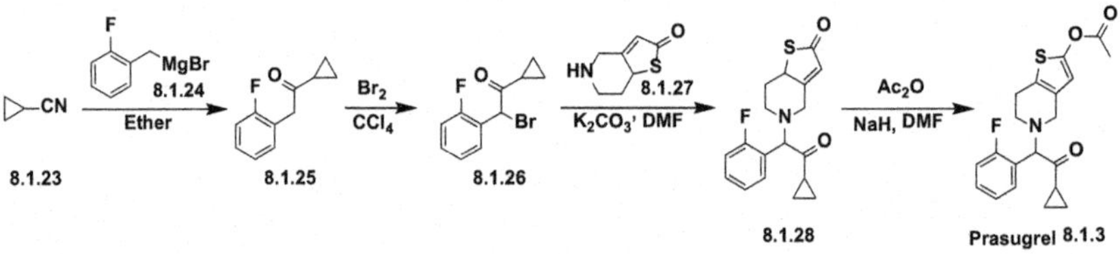

SCHEME 8.3 Synthesis of prasugrel.

The 5,6,7,7a-tetrahydro-thieno[3,2-*c*]pyridin-2(4*H*)-one (**8.1.27**) used in this synthesis was prepared starting from above-described 4,5,6,7-tetrahydrothieno[3,2-*c*]pyridine (**8.1.7**), which was coupled with trityl chloride in dichloromethane in the presence of trimethylamine to give 5-trityl-4,5,6,7-tetrahydrothieno[3,2-*c*]pyridine (**8.1.29**). To a solution of obtained (**8.1.29**) in tetrahydrofuran, butyl lithium solution in hexane was added and then tri-n-butyl borate was dissolved in tetrahydrofuran. After cooling of the reaction medium to −40°C, 30% aqueous hydrogen peroxide was added to give 5-trityl-5,6,7,7a-tetrahydrothieno[3,2-*c*]pyridin-2(4*H*)-one (**8.1.30**). Desired compound (**8.1.27**) was prepared on heating to 90°C for 1 hour, obtained (**8.1.30**) in 98% formic acid [108,109] (Scheme 8.4).

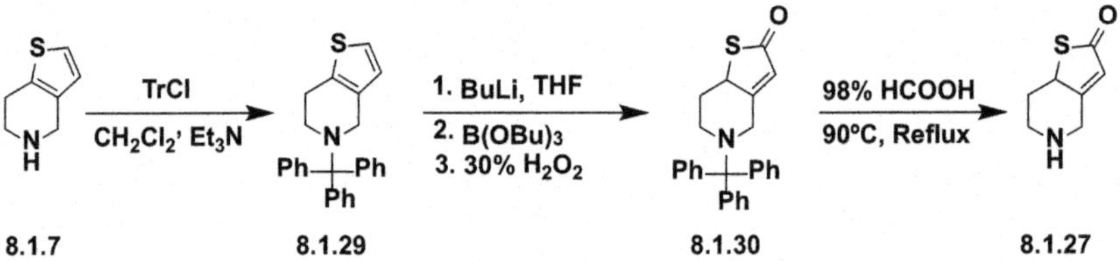

SCHEME 8.4 Synthesis of 5,6,7,7a-tetrahydro-thieno[3,2-*c*]pyridin-2(4*H*)-one.

Several modified and improved methods for synthesizing prasugrel (**8.1.3**) were reported [106,110−125].

8.2 γ-CARBOLINES

Mebhydrolin (74)

Mebhydrolin (**8.1.34**) (Diazolin) is a somewhat out of fashion antihistamine, used for symptomatic relief of allergic symptoms caused by histamine release, including nasal allergies and allergic dermatosis, allergic conditions including urticaria, eczema, pruritic skin disorders, and in various dermato-logic diseases.

Mebhydrolin is an antihistamine with antimuscarinic and sedative proper-ties [126,127].

Mebhydrolin (**8.1.34**) was synthesized in one step using a Fischer reac-tion. To this purpose 1-methylpiperidin-4-one (**8.1.31**) hydrochloride and 1-benzyl-1-phenylhydrazine (**8.1.32**) were converted to the intermediate hydrazone (**8.1.33**) on mixing together in saturated ethanolic hydrochloric acid solution [128,129], or in diluted (5−10%) sulfuric acid [130,131] and then refluxed in the same media for a short time to give desired mebhydrolin (**8.1.34**) (Scheme 8.5).

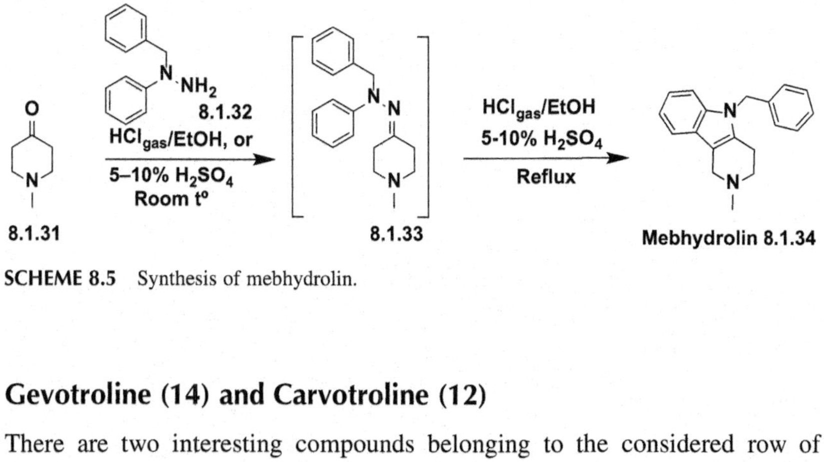

SCHEME 8.5 Synthesis of mebhydrolin.

Gevotroline (14) and Carvotroline (12)

There are two interesting compounds belonging to the considered row of piperidine-based fused biheterocycles: gevotroline (**8.1.40**) and carvotroline (**8.1.42**), which have not yet entered the category of the vogue pharmaceuti-cals. Gevotroline and carvotroline are atypical antipsychotics under develop-ment for the treatment of schizophrenia. They act as balanced, modest affinity D2 and 5-HT2 receptor antagonists and also possess a high affinity for the sigma receptor. Gevotroline is well-tolerated and showed efficacy in phase II clinical trials but was never marketed [132,133].

Scheme 8.6 describes the synthesis of gevotroline (**8.1.40**) and carvotroline (**8.1.42**) implementing on the first step standard Fisher synthesis conditions and starting with ethyl 4-oxopiperidine-1-carboxylate (**8.1.35**) and 4-Fluorophenylhydrazine hydrochloride (**8.1.36**), which were heated at reflux in ethanol to give ethyl 8-fluoro-1,3,4,5-tetrahydro-2H-pyrido[4,3-b]indole-2-carboxylate (**8.1.37**). The carbethoxy-group in the first position of piperidine ring was removed on reflux in Claisen's alkali (KOH dissolved in water and methanol (35:25:100)) and obtained key compound − 8-fluoro-2,3,4,5-tetrahydro-1H-pyrido[4,3-b]indole (**8.1.38**), which was alkylated with 3-(3-bromopropyl)pyridine (**8.1.39**) in dimethylformamide in the presence of potassium carbonate and cesium carbonate to give gevotroline (**8.1.40**). The same reaction conditions with the use of 4-(2-bromoethyl)pyridine (**8.1.41**) gave carvotroline (**8.1.42**) [134,135] (Scheme 8.6).

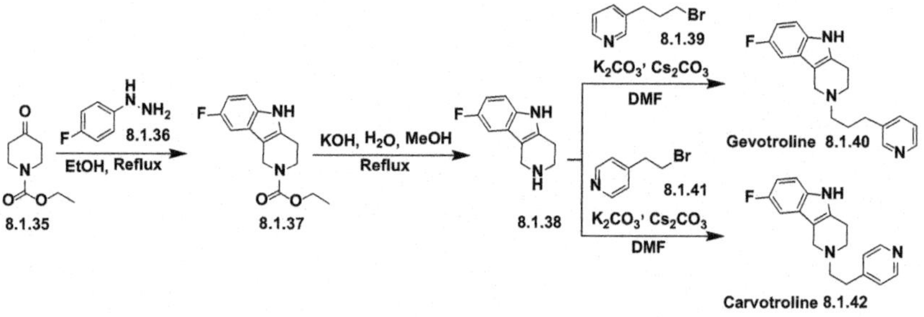

SCHEME 8.6 Synthesis of gevotroline and carvotroline.

8.3 5,6,7,8-TETRAHYDROPYRIDO[4,3-C]PYRIDAZINES

Endralazine (114)

Endralazine (**8.1.52**) is a hypotensive drug, a peripheral vasodilating drug related to the hydralazine series that reduces peripheral vascular resistance, resulting in substantial falls in blood pressure and increase in heart rate with no significant additional orthostatic effect. Endralazine chemically and pharmacologically related to hydralazine, exhibits a longer half-life and is only minimally influenced by metabolic acetylation.

Endralazine is used as a drug for treating severe and moderately severe hypertension commonly in combined therapy with beta-receptor blockers and diuretics, angiotensin converting enzyme inhibitors (ACE inhibitors). Common side effects of endralazine include: headache, dizziness, anxiety, muscle or joint pain, runny or stuffy nose, or mild itching or skin rash, and edema that disappeared spontaneously [136−144].

The synthesis of endralazine (**8.1.52**) was based on the product of cyclcondensation of ethyl 3-(2-ethoxy-2-oxoethyl)-4-oxopiperidine-1-carboxylate (**8.1.45**) with hydrazine − ethyl 3-oxo-3,4,4a,5,7,8-hexahydropyrido[4,3-c]pyridazine-6(2H)-carboxylate (**8.1.46**). The starting compound

(**8.1.45**) for this series of reactions was synthesized via alkylation of enamine (**8.1.42**) with ethyl 2-bromoacetate (**8.1.43**). Enamine (**8.1.42**), in turn was prepared via reaction of ethyl 4-oxopiperidine-1-carboxylate (**8.1.35**) with pyrrolidine in boiling benzene. Hydrolysis of obtained after akylation enaminoester (**8.1.44**) with hydrochloric acid and subequent cyclization of prepared keto ester (**8.1.45**) with hydrazine hydrate took place at reflux in the mixture of absolute ethanol and glacial acetic acid gave dihydropyridazin-3-one derivative (**8.1.46**), which was oxidized with bromine in chloroform giving 3-hydroxy-pyridazine derivative (**8.1.47**). The carbethoxy- protective group in piperidine ring of the last was removed at reflux in concentrated hydrochloric acid. The obtained intermediate 5,6,7,8-tetrahydropyrido[4,3-*c*]pyridazin-3-ol (**8.1.48**) was suspended in phosphorus oxychloride and heated to the boil to give 3-chloro-5,6,7,8-tetrahydropyrido[4,3-*c*]pyridazine (**8.1.49**). Synthesized product was acylated with benzoyl chloride (**8.1.50**) in ethylene chloride in to presence of triethylamine at room temperature to give (3-chloro-7,8-dihydro-pyrido[4,3-*c*]pyridazin-6(5*H*)-yl)(phenyl)methanone (**8.1.51**). Finally, a suspension of obtained product (**8.1.51**) was boiled at reflux at temperature of 110°C for 1 hour in hydrazine hydrate to give desired endralazine (**8.1.52**) [145−149] (Scheme 8.7).

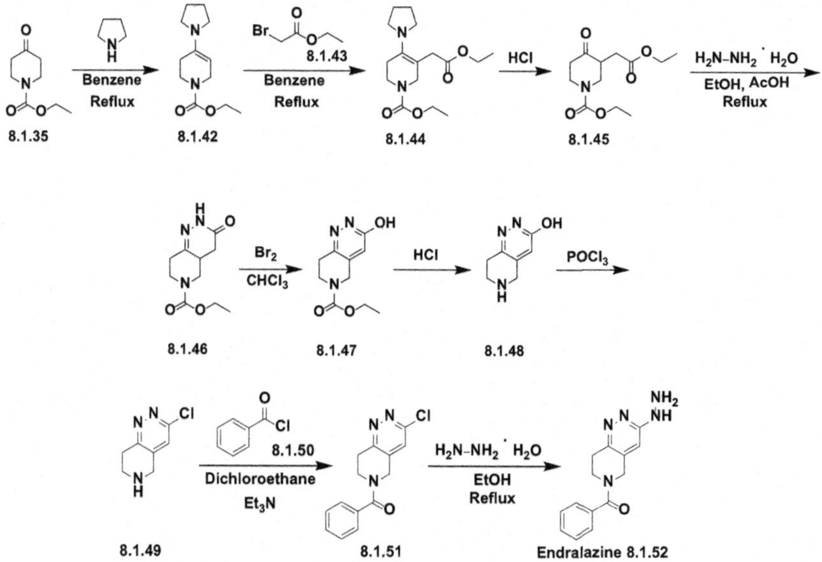

SCHEME 8.7 Synthesis of endralazine.

Phenindamine (413)

Phenindamine (**8.1.57**) does not belong to piperidine-based fused biheterocycles, but its structure is more suited for description in this chapter.

Phenindamine (**8.1.57**) (Thephorin) is a potent H1-receptor antagonist that was developed almost 50 years ago. The chemical structure antihistaminic agent is different from all other known antihistamines.

It was used to treat perennial and seasonal allergic rhinitis and chronic urticarial sneezing, runny nose, itching, watery eyes, hives, rashes, itching, and other symptoms of allergies [150−153]. Common side effects are sleepiness, fatigue, or dizziness, headache, dry mouth. The product is no longer available.

Two methods have been proposed for the synthesis of phenindamine (8.1.57). According the first of them, Mannich reaction of acetophenone (**8.1.53**), formaldehyde, and methylamine in boiling ethanol, gave bis(3-phenyl-3-oxo-propyl)methylamine (**8.1.54**), which was cyclized using sodium hydroxide solution forming β-hydroxy ketone (**8.1.55**). The last subjected cyclodehydration on reflux with 48% hydrobromic acid to give 2-methyl-9-phenyl-2,3-dihydro-1*H*-indeno[2,1-*c*]pyridine (**8.1.56**) and further hydrogenated at 48−50°C with a Raney nickel catalyst to give desired phenindamine (**8.1.57**) [154] (Scheme 8.8).

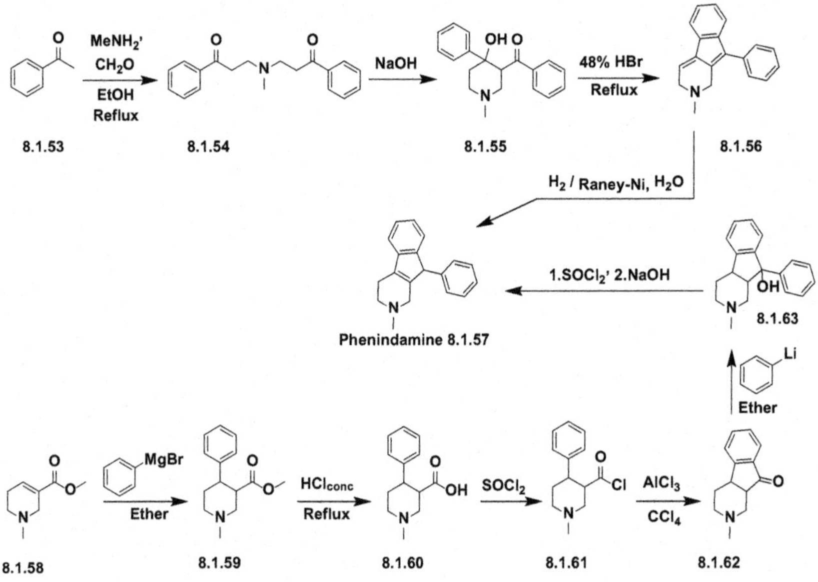

SCHEME 8.8 Synthesis of phenindamine.

The second method started from arecoline (**8.1.58**), which at 10°C, was reacted with phenylmagnesium bromide to give methyl 1-methyl-4-phenylpiperidine-3-carboxylate (**8.1.59**), which on reflux in concentric hydrochloric acid hydrolyzed to the corresponding acid (**8.1.60**). The last was converted to congruent acid chloride by the use of thionyl chloride (**8.1.61**). After adding anhydrous aluminum chloride to the solution of synthesized compound (**8.1.61**) an intramolecular cyclization reaction occurred to give 2-methyl-1,2,3,4,4a,9a-hexahydro-9*H*-indeno[2,1-*c*]pyridin-9-one (**8.1.62**). Obtained ketone was reacted in ether with phenyllithium and the formed 2-methyl-9-phenyl-2,3,4,4a,9,9a-hexahydro-1*H*-indeno[2,1-*c*]pyridin-9-ol (**8.1.63**) was converted to corresponding chloride with thionyl chloride and further dehydrohalogenated with 10% sodium hydroxide solution to give phenindamine (**8.1.57**) [155] (Scheme 8.8).

REFERENCES

[1]   Quinn MJ, Fitzgerald DJ. Ticlopidine and clopidogrel. Circulation 1999;100 (15):1667–72.

[2]   Sharis PJ, Cannon CP, Loscalzo J. The antiplatelet effects of ticlopidine and clopidogrel. Ann Intern Med 1998;129(5):394–405.

[3]   Savi P, Herbert J. Clopidogrel and ticlopidine: P2Y12 adenosine diphosphate-receptor antagonists for the prevention of atherothrombosis. Semin Thromb Hemost 2005;31 (2):174–83.

[4]   Cattaneo M. Aspirin and clopidogrel. Efficacy, safety, and the issue of drug resistance. Arterioscler Throm Vasc Biol 2004;24(11):1980–7.

[5]   McTavish D, Faulds D, Goa KL. Ticlopidine: an updated review of its pharmacology and therapeutic use in platelet-dependent disorders. Drugs 1990;40(2):238–59.

[6]   Saltiel E, Ward A. Ticlopidine. A review of its pharmacodynamic and pharmacokinetic properties, and therapeutic efficacy in platelet-dependent disease states. Drugs 1987;34 (2):222–62.

[7]   Jacobson AK. Platelet ADP receptor antagonists: ticlopidine and clopidogrel. Best Pract Res Clin Haematol 2004;17(1):55–64.

[8]   Paciaroni M, Bogousslavsky J, Gallai V. Ticlopidine and clopidogrel. In: Bogousslavsky J, editor. Drug therapy for stroke prevention. Boca Raton, FL: CRC Press; 2001. p. 49–78.

[9]   Sharis PJ, Loscalzo J. 2nd Edition Thienopyridines: ticlopidine and clopidogrel in Fundamental and Clinical Cardiology, 46. M. Dekker; 2003. p. 431–49.

[10] Curtin R, Cox D, Fitzgerald D. Clopidogrel and ticlopidine. In: Michelson AD, editor. Platelets. New York: Academic Press; 2002. p. 787–801.

[11] Noble S, Goa KL. Ticlopidine. A review of its pharmacology, clinical efficacy, and tolerability in the prevention of cerebral ischemia and stroke. Drugs Aging 1996;8(3):214–32.

[12] Feuerstein GZ, Ruffolo Jr. RR. Ticlopidine: a novel antiplatelet drug for prevention of thrombotic disorders. Exp Opin Invest Drugs 1994;3(11):1163–9.

[13] Desager J. Clinical pharmacokinetics of ticlopidine. Clin Pharmacokinet 1994;26 (5):347–55.

[14] Schroer K. The basic pharmacology of Ticlopidine and Clopidogrel. Platelets 1993;4 (5):252–61.

[15] Rollini F, Franchi F, Muniz-Lozano A, Angiolillo DJ. In: Waksman R, Gurbel PA, Gaglia Jr MA, editors. Ticlopidine, antiplatelet therapy in cardiovascular disease. Wiley; 2014. p. 150–9.

[16] Dusitanond P, Hankey G. Ticlopidine. J Drug Eval 2004;2(6):163–76.

[17] Jacob S, Dunn BL, Qureshi ZP, Bandarenko N, Kwaan HC, Pandey DK, et al. Ticlopidine-, clopidogrel-, and prasugrel-associated thrombotic thrombocytopenic

purpura: a 20-year review from the southern network on adverse reactions (SONAR). Semin Thromb Hemost 2012;38(8):845−53.

[18] Black C, Paterson KR. New antiplatelet agents: ticlopidine and clopidogrel. Antiplatelet therapy but at what cost? Adverse Drug React Toxicol Rev 2001;20(4):277−303.

[19] Love BB, Biller J, Gent M. Adverse hematological effects of ticlopidine. Prevention, recognition and management. Drug Safety 1998;19(2):89−98.

[20] Ebihara A. Safety of Panaldine (ticlopidine hydrochloride). Med Pharm 1981;15 (8):272−6.

[21] Castaigne ARJ. Thieno[3,2-c]pyridine derivatives, US 4051141; 1977.

[22] No Inventor data available, Pharmaceutical 4,5,6,7-tetrahydrofuro- and -thieno[3,2-c]pyridines, DE 2404308; 1974.

[23] Gronowitz S, Sandberg E. Thiophene isosteres of isoquinoline. I. Synthesis of thieno[2,3-c]pyridines and thieno[3,2-c]pyridines. Arkiv Kemi 1970;32(19):217−27.

[24] DeHoff BS. Preparation of 2-(2'-thienyl)ethylamine derivatives and synthesis of thieno [3,2-c]pyridine derivatives therefrom, US 5191090; 1993.

[25] Maffrand JP, Eloy F. Synthesis of thienopyridines and furopyridines of therapeutic interest. Eur J Med Chem 1974;9(5):483−6.

[26] Maffrand JP, Eloy F. New syntheses of thieno[3,2-c]- and thieno[2,3-c]pyridines. J Het Chem 1976;13(6):1347−9.

[27] Braye E. Process for the preparation of thieno-pyridine derivatives, US 4127580; 1978.

[28] Sumita K, Koumori M, Ohno S. A modified Mannich reaction using 1,3-dioxolane. Chem Pharm Bull 1994;42(8):1676−8.

[29] Bousquet A, Braye E. Process for the preparation of thienopyridine derivatives, US 4174448 A; 1979.

[30] No inventor data available. Thienopyridines, JP 54019994; 1979.

[31] Herbert JM, Frehel D, Vallee E, Kieffer G, Gouy D, Berger Y, et al. Clopidogrel, a novel antiplatelet and antithrombotic agent. Cardiovasc Drug Rev 1993;11(2):180−98.

[32] Coukell AJ, Markham A. Clopidogrel. Drugs 1997;54(5):745−50.

[33] Herbert JM. Clopidogrel and antiplatelet therapy. Exp Opin Invest Drugs 1994;3 (5):449−55.

[34] Angiolillo DJ, Fernandez-Ortiz A, Bernardo E, Alfonso F, Macaya C, Bass TA, et al. Variability in Individual Responsiveness to Clopidogrel: clinical implications, management, and future perspectives. J Am Coll Cardiol 2007;49(14):1505−16.

[35] Lerner RG, Frishman WH, Mohan KT. Clopidogrel: a new antiplatelet drug. Heart Dis 2000;2(2):168−73.

[36] Savi P, Nurden P, Nurden AT, Levy-Toledano S, Herbert J-M. Clopidogrel: a review of its mechanism of action. Platelets 1998;9(3/4):251−5.

[37] Bezerra DC, Bogousslavky J. Clopidogrel: cardiologists' panacea or neurologists' headache? Fut Cardiol 2005;1(5):579−90.

[38] Gurbel PA, Antonino MJ, Tantry US. Recent developments in clopidogrel pharmacology and their relation to clinical outcomes. Exp Opin Drug Metabol Toxicol 2009;5 (8):989−1004.

[39] Fox KAA, Chelliah R. Clopidogrel: an updated and comprehensive review. Exp Opin Drug Metabol Toxicol 2007;3(4):621−31.

[40] Diener H, Ringleb PA, Savi P. Clopidogrel for the secondary prevention of stroke. Exp Opin Pharmacother 2005;6(5):755−64.

[41] Plosker GL, Lyseng-Williamson KA. Clopidogrel: a review of its use in the prevention of thrombosis. Drugs 2007;67(4):613−46.

[42] Shah Bhukhanwala K, Godbole R. Clopidogrel bisulfate (Plavix Sanofi and Bristol-Myers Squibb): another billion dollar drug under attack by the generics. Exp Opin Therapeut Pat 2006;16(12):1609−11.

[43] Scott SA, Sangkuhl K, Stein CM, Hulot J-S, Mega JL, Roden DM, et al. Clinical pharmacogenetics implementation consortium guidelines for CYP2C19 genotype and clopidogrel therapy: 2013 update. Clin Pharmacol Therapeut 2013;94(3):317−23.

[44] Sarafoff N, Byrne RA, Sibbing D. Clinical use of clopidogrel. Curr Pharm Design 2012;18(33):5224−39.

[45] Farid NA, Kurihara A, Wrighton SA. Metabolism and disposition of the thienopyridine antiplatelet drugs ticlopidine, clopidogrel, and prasugrel in humans. J Clin Pharmacol 2010;50(2):126−42.

[46] Mega JL, Simon T, Collet J, Anderson JL, Antman EM, Bliden K, et al. Reduced-function CYP2C19 genotype and risk of adverse clinical outcomes among patients treated with clopidogrel predominantly for PCI. A meta analysis. JAMA 2010;304(16):1821−30.

[47] Johnson JA, Roden DM, Lesko LJ, Ashley E, Klein TE, Shuldiner AR. Clopidogrel: a case for indication-specific pharmacogenetics. Clin Pharmacol Ther 2012;91(5):774−6.

[48] Maffrand J. The story of clopidogrel and its predecessor, ticlopidine: could these major antiplatelet and antithrombotic drugs be discovered and developed today? Compt Rend Chim 2012;15(8):737−43.

[49] Laine L, Hennekens C. Proton pump inhibitor and clopidogrel interaction: fact or fiction? Am J Gastroenterol 2010;105(1):34−41.

[50] Momary K, Cavallari LH. Clopidogrel and proton pump inhibitors: between a rock and a hard place. Pharmacother 2010;30(8):762−5.

[51] Bainey KR, Lai TF, Mehta SR. Clopidogrel in acute coronary syndromes: where are we now? Thromb Haemost 2011;105(5):766−73.

[52] Campo G, Fileti L, Valgimigli M, Tebaldi M, Cangiano E, Cavazza C, et al. Poor response to clopidogrel: current and future options for its management. J Thromb Trombolysis 2010;30(3):319−31.

[53] Nguyen TA, Diodati JG, Pharand C. Resistance to clopidogrel: a review of the evidence. J Am Coll Cardiol 2005;45(8):1157−64.

[54] Jaruis B, Simpson K. Clopidogrel: a review of its use in the prevention of atherothrombosis. Drugs 2000;60(2):347−77.

[55] Budaj A. Clopidogrel. In: Waksman R, Gurbel PA, Gaglia Jr MA, editors. Antiplatelet therapy in cardiovascular disease. Wiley; 2014. p. 160−5.

[56] Bates ER, Lau WC, Angiolillo DJ. Clopidogrel-drug interactions. J Am Coll Cardiol 2011;57(11):1251−63.

[57] Sambu N, Warner T, Curzen N. Clopidogrel withdrawal: is there a rebound phenomenon? Thromb. Haemostas 2011;105(2):211−20.

[58] Sadanandan S, Singh IM. Clopidogrel: the data, the experience, and the controversies. Am J Cardiovasc Drugs 2012;12(6):361−74.

[59] Bertino Jr. JS. The clopidogrel conundrum. J Clin Pharmacol 2013;53(8):841−2.

[60] Ford NF, Taubert D. Clopidogrel, CYP2C19, and a black box. J Clin Pharmacol 2013;53 (3):241−8.

[61] Juel J, Pareek M, Jensen SE. The clopidogrel-PPI interaction: an updated mini-review. Curr Vasc Pharmacol 2014;12(5):751−7.

[62] Jiang X, Samant S, Lesko LJ, Schmidt S. Clinical pharmacokinetics and pharmacodynamics of clopidogrel. Clin Pharmacokinet 2015;54(2):147−66.

[63] Howell LA, Stouffer GA, Polasek M, Rossi JS. Review of clopidogrel dose escalation in the current era of potent P2Y12 inhibitors. Exp Rev Clin Pharmacol 2015;8(4):411−21.

[64] Qureshi Z, Hobson AR. Clopidogrel "resistance": where are we now? Cardiovasc Therapeut 2013;31(1):3−11.

[65] Aubert D, Ferrand C, Maffrand JP. Thieno[3,2-c]pyridine derivatives and their therapeutical use, EP 99802; 1984.

[66] Aubert D, Ferrand C, Maffrand PJ. Thieno [3,2-c] pyridine derivatives and their therapeutic application, US 4529596 A; 1985.

[67] Bousquet A, Musolino A. Hydroxyacetic ester derivatives, namely (R)-methyl 2-(sulfony-loxy)-2-(chlorophenyl) acetates, preparation method, and use as synthesis intermediates for clopidogrel, WO 9918110; 1999.

[68] Descamps M, Radisson J. Preparation of methyl α-[4,5,6,7-tetrahydrothieno[3,2-c]pyrid-5-yl]-2'-chlorophenylacetate, EP 466569; 1992.

[69] Descamps M, Radisson J. Process for the preparation of an n-phenylacetic derivative of tetrahydrothieno(3,2-c)pyridine and its chemical intermediate, US 5204469 A; 1993.

[70] Pandey B, Lohray VB, Lohray BB. Process for preparing clopidogrel and analogs via synthesis and/or resolution of corresponding acetamide and acetonitrile derivatives, WO 2002059128; 2002.

[71] Bousquet A, Calet S, Heymes A. Isopropyl 2-thienylglycidate, process for its preparation, and its use as synthetic intermediate for ticlopidine and clopidogrel, EP 465358; 1992.

[72] Bouisset M, Radisson J. Process for preparing phenylacetic derivatives of thienopyridines and intermediate α-bromophenylacetic acids, EP 420706; 1991.

[73] No Inventor data available. Preparation of d-α-5-(4,5,6,7-tetrahydro[3,2-c]thienopyridyl)-2-(chlorophenyl) acetic acid methyl ester as an antithrombotic, JP 63203684; 1988.

[74] Tarur VR, Srivastava RP, Srivastava AR, Somani SK. Preparation of acid addition salts of 4, 5, 6, 7 - tetrahydrothieno (3,2-c) pyridine derivatives having antithrombotic activity, WO 2002018357; 2002.

[75] Castro B, Dormoy J, Previero A. Improved method for preparing 2-thienylethylamine derivatives, including an intermediate for clopidogrel, WO 9839322; 1998.

[76] Kumar A, Vyas KD, Barve SG, Bhayani PJ, Nandavadekar S, Shah CH, et al. Industrial process for preparation of clopidogrel hydrogen sulfate, WO 2005104663; 2005.

[77] Castaldi G, Barreca G, Bologna A. A process for the preparation of clopidogrel, WO 2003093276; 2003.

[78] Valeriano M, Daverio P, Bianchi S. Racemization and enantiomer separation of clopido-grel, US 20040024012; 2004.

[79] Venkat R, Vyakaranam KR, Sirigiri AK, Bodapati SR, Billa RR, Gudibandi SR, et al. Process for preparation of clopidogrel bisulfate Form-1, US 20040024012; 2007.

[80] Pandey B, Lohray VB, Lohray BB. Process for preparing clopidogrel and analogs via synthesis and/or resolution of corresponding acetamide and acetonitrile derivatives, US 20040024012; 2002.

[81] Simonic IR, Benkic P, Zupet R, Smrkolj M, Stukelj M. Process for the synthesis of clopidogrel and new forms of pharmaceutically acceptable salts thereof, WO 2008034912; 2008.

[82] Alla VR, Vyakaranam KR, Sirigiri AK, Bodapati SR, Billa RR, Gudibandi SR, et al. Process for preparation of clopidogrel bisulfate Form-1, US 20070191609; 2007.

[83] Wang L, Shen J, Tang Y, Chen Y, Wang W, Cai Z, et al. Synthetic improvements in the preparation of clopidogrel. Org Proc Res Dev 2007;11(3):487−9.

[84] Jung K, Kim J, Kim T, Kim J. A Facile Solid-Phase Synthesis of (+)-(S)-Clopidogrel. Helv Chim Acta 2013;96(2):326−9.

[85] Li JJ, Johnson DS, Sliskovic DR, Roth BD. Contemporary Drug Synthesis. Wiley-Interscience; 2004. p. 4−10.

[86] Shan J, Sun H. The discovery and development of prasugrel. Exp Opin Drug Disc 2013;8(7):897−905.

[87] Jakubowski JA, Winters KJ, Naganuma H, Wallentin L. Prasugrel: a novel thienopyridine antiplatelet agent. A review of preclinical and clinical studies and the mechanistic basis for its district antiplatelet profile. Cardiovasc Drug Rev 2007;25(4):357−74.

[88] Niitsu Y, Jakubowski JA, Sugidachi A, Asai F. Pharmacology of CS-747 (prasugrel, LY640315), a novel, potent antiplatelet agent with in vivo P2Y12 receptor antagonist activity. Sem Thromb Hemost 2005;31(2):184−94.

[89] Tantry US, Bliden KP, Gurbel PA. Prasugrel. Exp Opin Ivest Drugs 2006;15 (12):1627−33.

[90] Tello-Montoliu A, Tomasello SD, Angiolillo DJ. Prasugrel. Adv Cardiol 2012;47:39−63.

[91] Riley AB, Tafreshi MJ, Haber SL. Prasugrel: a novel antiplatelet agent. Am J Health-Syst Pharm 2008;65(11):1019−28.

[92] Gurbel PA, Tantry US. Prasugrel, a third generation thienopyridine and potent platelet inhibitor. Curr Opin Ivest Drugs 2008;9(3):324−36.

[93] Duggan ST, Keating GM. Prasugrel: a review of its use in patients with acute coronary syndromes undergoing percutaneous coronary intervention. Drugs 2009;69(12):1707−26.

[94] Reinhart KM, White CM, Baker WL. Prasugrel: a critical comparison with clopidogrel. Pharmacother 2009;29(12):1441−51.

[95] Huber K, Yasothan U, Hamad B, Kirkpatrick P. Prasugrel. Nat Rev Drug Disc 2009;8 (6):449−50.

[96] Angiolillo DJ, Suryadevara S, Capranzano P, Bass TA. Prasugrel: a novel platelet ADP P2Y12 receptor antagonist. A review on its mechanism of action and clinical development. Exp Opin Pharmacother 2008;9(16):2893−900.

[97] Serebruany V, Shalito I, Kopyleva Ol. Prasugrel development - claims and achievements. Thromb Haemost 2009;101(1):14−22.

[98] Dobesh PP. Pharmacokinetics and pharmacodynamics of prasugrel, a thienopyridine P2Y12 inhibitor. Pharmacother 2009;29(9):1089−102.

[99] Varenhorst C, Oskarsson A, James S. Prasugrel. In: Waksman R, Gurbel PA, Gaglia Jr MA, editors. Antiplatelet Therapy in Cardiovascular Disease. Wiley; 2014. p. 166−72.

[100] Tomasello SD, Tello-Montoliu A, Angiolillo DJ. Prasugrel for the treatment of coronary thrombosis: a review of pharmacological properties, indications for use and future development. Exp Opin Ivest Drugs 2011;20(1):119−33.

[101] Nanau RM, Delzor F, Neuman MG. Efficacy and safety of prasugrel in acute coronary syndrome patients. Clin Biochem 2014;47(7−8):516−28.

[102] Morici N, Colombo P, Mafrici A, Oreglia JA, Klugmann S, Savonitto S. Prasugrel and ticagrelor: is there a winner? J Cardiovasc Med 2014;15(1):8−18.

[103] Mousa SA, Jeske WP, Fareed J. Prasugrel: a novel platelet ADP P2Y12 receptor antagonist, in Methods in Molecular Biology. 2nd Edition Anticoagulants, Antiplatelets, and Thrombolytics, 663. Humana Press, a part of Springer Science + Business Media; 2010. p. 221−8.

[104] Jacob S, Dunn BL, Qureshi ZP, Bandarenko N, Kwaan HC, Pandey DK, et al. Ticlopidine-, clopidogrel-, and prasugrel-associated thrombotic thrombocytopenic

purpura: a 20-year review from the southern network on adverse reactions (SONAR). Sem Thromb Hemostas 2012;38(8):845−53.

[105] Alexopoulos D. Prasugrel resistance: fact or fiction. Platelets 2012;23(2):83−90.

[106] Koike H, Asai F, Sugidachi A, Kimura T, Inoue T, Nishino S, et al. Preparation of substituted hydrothienopyridine derivatives having antithrombotic activity, CA 2077695; 1993.

[107] Koike H, Asai F, Sugidachi A, Kimura T, Inoue T, Nishino S, et al. Tetrahydrothienopyridine derivatives, furo and pyrrolo analogs thereof and their preparation and uses for inhibiting blood platelet aggregation, US 5288726 A; 1994.

[108] Badorc A, Frehel D, Maffrand JP, Vallee E. Derivatives of alpha-(2-oxo 2,4,5,6,7,7a-hexahydro thieno[3,2-c]5-pyridyl) phenyl acetic acid, and their use as platelet and thrombotic aggregation inhibitors, US 4740510 A; 1988.

[109] Badorc A, Frehel D, Maffrand JP, Vallee E. Preparation and therapeutic application of derivatives of α-(2-oxo-2,4,5,6,7,7a-hexahydro-5-thieno[3,2-c]pyridyl)phenylacetic acid, Fr 2576901; 1986.

[110] Rao AVVS, Jaware J, Goud S, Bijukumar GP, Nadkarni SS. Process for the preparation of 2-acetoxy-5-(α-cyclopropylcarbonyl-2-fluorobenzyl)-4,5,6,7-tetrahydrothieno[3,2-c] pyridine, WO 2009122440; 2009.

[111] Padi PR, Peri SRS, Ganta MR, Polavarapu S, Cherukupally P, Ireni B, et al. Processes for the preparation of prasugrel, and its salts and polymorphs, US 20100261908 A1; 2010.

[112] Padi PR, Peri SRS, Ganta MR, Polavarapu S, Cherukupally P, Ireni B, et al. A process for preparing prasugrel and its salts and polymorphs, WO 2009062044; 2009.

[113] Ahmed KM, Sanikommu SR, Antyakula P, Bhaskar R, Sanganabhatla S. Preparation of prasugrel and its hydrochloride salt and crystal forms, WO 2010070677; 2012.

[114] Asai F, Ogawa T, Naganuma H, Yamamura N, Inoue T, Nakamura K. Tetrahydrothienopyridine derivative acid addition salts, WO 2002004461; 2002.

[115] Inoue T, Nakamura K, Hagihara M, Miyata H, Wada Y, Yokota N. Process for producing high-purity prasugrel and acid addition salts thereof, WO 2007114526; 2007.

[116] Satyanarayana RM, Eswaraiah S, Venkat RG. Process for the preparation of prasugrel from 4,5,6,7-tetrahydrothieno[3,2-c]pyridine, α-cyclopropylcarbonyl-2-fluorobenzyl bromide, and acetic anhydride, WO 2009066326; 2009.

[117] Satyanarayana RM, Thirumalai RS, Eswaraiah S, Rama SRK, Kondal Reddy B, Venkat Reddy G. Processes for preparing prasugrel and pharmaceutically acceptable salts thereof, US 20120202066; 2012.

[118] Stepankova H, Kaminska K, Hajicek J. Method of producing highly pure prasugrel and pharmaceutically acceptable salts thereof, WO 2011057592; 2011.

[119] Stepankova H, Hajicek J. Process for the manufacturing of 5-[2-cyclopropyl-1-(2-fluoro-phenyl)-2-oxoethyl]-4,5,6,7-tetrahydrothieno[3,2-c]pyridin-2-yl acetate (prasugrel), WO 2009006859; 2009.

[120] Porcs-Makkay M, Gregor T, Volk B, Nemeth G, Barkoczy J, Nyulasi B, et al. A process for preparing prasugrel and its intermediates, 2011077173; 2011.

[121] Porcs-Makkay M, Volk B, Gregor T, Barkoczy J, Mezei T, Broda J, et al. A process for preparing intermediates of prasugrel, their crystal forms, and their application to synthesize prasugrel, WO 2011077174; 2011.

[122] Aalla S, Gilla G, Metil DS, Anumula RR, Vummenthala PR, Padi PR. Process improvements of prasugrel hydrochloride: an adenosine diphosphate receptor antagonist. Org Proc Res Dev 2012;16(2):240−3.

[123] Ou W, Yi W, Liu F, Pan X, Peng X. An improvement to the preparation of prasugrel hydrochlorid. J Chem Res 2013;37(6):369−71.

[124] Sastry TU, Rao KN, Reddy TA, Gandhi P. Identification and synthesis of impurities formed during prasugrel hydrochloride preparation. Asian J Chem 2013;25(14):7783−9.

[125] Pan X, Huang R, Zhang J, Ding L, Li W, Zhang Q, et al. Efficient synthesis of prasugrel, a novel P2Y12 receptor inhibitor. Tetrahedron Lett 2012;53(40):5364−6.

[126] Kharkevich DA. Pharmacology of diazoline. Farmakologiya i Toksikologiya (Moscow) 1957;20(6):46−51.

[127] Franks HM, Lawrie M, Schabinsky VV, Starmer GA, Teo RK. Interaction between ethanol and antihistamines: 3. Mebhydrolin. Med J Australia 1981;2(9):477−9.

[128] No Inventor data available, Derivatives of β- and γ-carbolines, GB 721171; 1954.

[129] Horlein U. Derivatives of 2-N-methyl-1, 2, 3, 4-tetrahydro-gamma-carbolines, US 2786059; 1957.

[130] Kucherova NF, Kochetkov NK. Derivatives of indole. II. Synthesis of some derivatives of 1,2,3,4-tetrahydro-γ-carboline. Zh Obshch Khim 1956;26:3149−54.

[131] Horlein U. Tetrahydrocarboline compounds. I Chem Berichte 1954;87:463−72.

[132] Abou-Gharbia M, Patel UR, Webb MB, Moyer JA, Andree TH, Muth EA. Antipsychotic activity of substituted γ-carbolines. J Med Chem 1987;30(10):1818−23.

[133] Moyer JA, Abou-Gharbia M, Muth EA. Behavioral pharmacology of the gamma carboline WY 47,384: a potential antipsychotic agent. Drug Dev Res 1988;13(1):11−28.

[134] Abou-Gharbia M, Patel U, Moyer J, Muth E. Psychotropic agents: synthesis and antipsychotic activity of substituted g-carbolines. J Med Chem 1987;30:1100−5.

[135] Abou-Gharbia M, Marquis K, Andree T. WY-47,791. Antipsychotic Drugs Future 1991;16:1008−13.

[136] Thien T, Hoffmann JJML. Endralazine, a new vasodilator. Pharm Weekbl 1983;118 (45):941−4.

[137] Reece PA, Cozamanis I, Zacest R. Endralazine - a new hydralazine-like antihypertensive with high systemic bioavailability. Eur J Clin Pharmacol 1983;25(4):553−6.

[138] Wu R, Spence JD, Carruthers SG. Evaluation of once daily endralazine in hypertension. Eur J Clin Pharmacol 1986;30(5):553−7.

[139] Hoffmann JJ, Thien T, van T'Laar A. Effects of intravenous endralazine in essential hypertension. Brit J Clin Pharmacol 1983;16(1):39−44.

[140] Elliott HL, Meredith PA, Howden CW, Lawrie CB, Reid JL. Clinical pharmacological studies with the vasodilator endralazine in normotensive subjects and essential hypertensives. Int J Clin Pharmacol Res 1984;4(1):61−9.

[141] Elliott HL, McLean K, Sumner DJ, Donnelly RJ, Reid JL. Clinical evaluation of endralazine (BO22708), a new vasodilator, in essential hypertension. Clin Exp Hypertens, Part A 1982;4(8):1409−18.

[142] Elliott HL, Meredith PA, Reid JL. Pharmacodynamic and pharmacokinetic studies with the vasodilator endralazine. J Hypertens 1984;2(Suppl. 3):551−4.

[143] Bogers WA, Meems L. Endralazine, a new peripheral vasodilator. Evaluation of safety and efficacy over a 3 year period. Eur J Clin Pharmacol 1983;24(3):301−5.

[144] Zacest R, Reece PA. Endralazine and sexual arousal. Lancet 1983;1(8335):1221.

[145] Schenker E. Pyrido [4,3-c]pyridazines, DE 2221808; 1972.

[146] Schenker E. 3-Hydrazino cycloalkyl[c]pyridazines as antihypertensive agents, US 4478837 A; 1984.

[147] Schenker E. 3-Hydrazinopyridazine derivatives, CH 561186; 1975.

[148] Schenker E. 3-Hydrazinopyridazine derivatives, CH 565797; 1975.

[149] Schenker E, Salzmann R. Antihypertensive action of bicyclic 3-hydrazinopyridazines. Arzneimitt-Forsch 1979;29(12):1835−43.

[150] Kallos P, Kallos-Deffner L. Experimental and clinical investigations of the action of thephorin. Int Arch Allergy Applied Immunol 1950;1:189−216.

[151] Witek Jr TJ, Canestrari DA, Miller RD, Yang JY, Riker DK. The effects of phenindamine tartrate on sleepiness and psychomotor performance. J Allergy Clin Immunol 1992;90(6 Pt 1):953−61.

[152] Lehmann G. Pharmacological properties of a new antihistaminic, 2-methyl-9-phenyl-2,3,4,9-tetrahydro-1-pyridindene (tephorin) and derivatives. J Pharmacol Exp Therapeut 1948;92:249−59.

[153] Schallek W, Young B. Quinidinelike activity of thephorin. J Pharmacol Exp Therapeut 1952;105:291−8.

[154] Plati JT, Wenner W. Heterocyclic amines, US 2470108; 1949.

[155] Plati JT, Wenner W. Pyridindenes, US 2546652; 1951.

Chapter 9

Piperidine-Based Spiro-Fused Biheterocycles

Piperidine-based spiro-fused biheterocycles are represented on the pharmaceutical market with six drugs. Three of them are derivatives of 1,3,8-triazaspiro[4.5]decan-4-one: spiperone (**9.1.1**), fluspirilene (**9.1.2**), and mosapramine (**9.1.3**) are typical antipsychotics, and three other drugs with diverse spirocyclic structure include bronchodilator fenspiride (**9.1.22**), a new neuroleptic agent clospirazine (**9.1.29**), and an investigational sedative and anxiolytic drug pazinaclone (**9.1.36**) (Fig. 9.1).

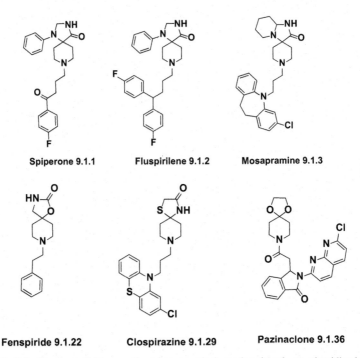

Spiperone 9.1.1   Fluspirilene 9.1.2   Mosapramine 9.1.3

Fenspiride 9.1.22   Clospirazine 9.1.29   Pazinaclone 9.1.36

FIGURE 9.1 Drugs available on the pharmaceutical market based on piperidine-based spiro-fused biheterocycles.

Piperidine-Based Drug Discovery. DOI: http://dx.doi.org/10.1016/B978-0-12-805157-3.00009-0

9.1 SPIPERONE (8015)

Spiperone (**9.1.1**) (Spiropitan) is a first generation typical antipsychotic, a spiro butyrophenone analog similar to haloperidol and other related compounds, which has been recommended in the treatment of schizophrenia.

Spiperone acts as a potent dopamine D2, serotonin 5-HT1A, and serotonin 5-HT2A antagonist. Although spiperone also binds at 5-HT2C receptors, it is one of the very few agents that display some ($\sim$1000-fold) binding selectivity for 5-HT2A versus 5-HT2C receptors.

Spiperone is a widely used pharmacological tool. [3*H*]Spiperone was extensively used as a radioligand for labeling 5-HT3−5-HT7 receptors [1−3].

The synthesis of spiperone (**9.1.1**) was carried out starting with simultaneous exothermic reaction of 1-benzylpiperidin-4-one (**9.1.4**) with aniline (**9.1.5**) and hydrogen cyanide, which took place after adding potassium cyanide solution in the mixture of (**9.1.4**) and (**9.1.5**) in glacial acetic acid to give 1-benzyl-4-(phenylamino)piperidine-4-carbonitrile (**9.1.6**). The nitrile function of obtained compound was hydrolyzed to carbamoyl group on heating with 90% sulfuric acid for 10 minutes at 70°C giving 1-benzyl-4-(phenylamino)piperidine-4-carboxamide (**9.1.7**). The last was heated at 200°C in formamide containing 98% sulfuric acid for three hours, which resulted in cyclizing to a spiro compound: 8-benzyl-1-phenyl-1,3,8-triazaspiro[4.5]decan-4-one (**9.1.8**). Prepared product was debenzylated by hydrogenation on palladium-on-charcoal catalyst in ethanol giving key compound: 1-phenyl-1,3,8-triazaspiro[4.5]decan-4-one (**9.1.9**). Alkylation of the last with 4-chloro-1-(4-fluorophenyl)butan-1-one (**9.1.10**) occurred in standard conditions − reflux in methyl isobutyl ketone in the presence of sodium carbonate and a few crystals of potassium iodide resulted in formation of spiperone (**9.1.1**) [3−6] (Scheme 9.1).

Fluspirilene (493)

Fluspirilene (**9.1.2**) (Imap) is a dopamine D2 receptor antagonist, which is a long-acting neuroleptic useful in the maintenance therapy of schizophrenic patients. Besides its actions on D2 receptors it also displays Ca^{2+} channel blocking activity.

Fluspirilene demonstrated marked efficacy and was generally well tolerated. Although occasional side effects were observed, such as parkinsonian symptoms, dystonia, akathisia, they are rare and do not differ significantly from other antipsychotics [7−21].

Fluspirilene (**9.1.2**) was synthesized practically in similar manner as spiperone (**9.1.1**) via condensation of 1-benzylpiperidin-4-one (**9.1.4**) with aniline (**9.1.5**) and hydrogen cyanide, which causes simultaneous introduction of the nitrile and secondary amino groups in the piperidyl ring at the 4-position (**9.1.6**), followed by conversion of nitrile function is to the amide

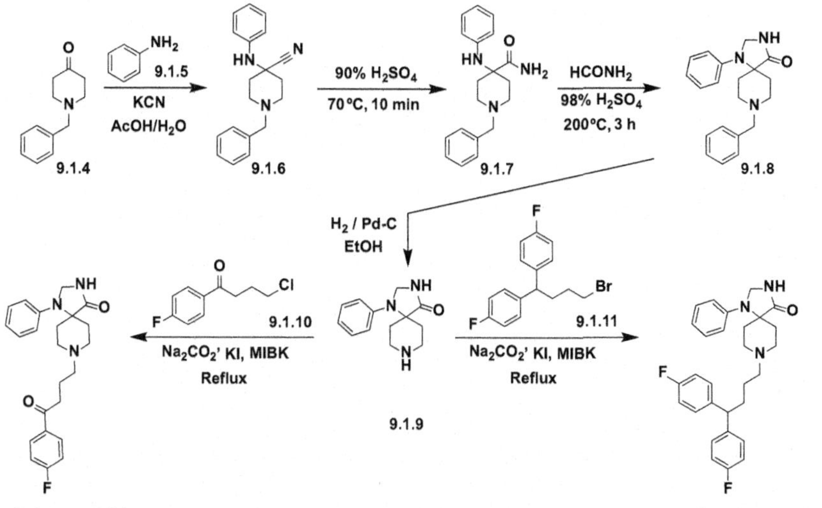

Spiperone 9.1.1   Fluspirilene 9.1.2

SCHEME 9.1   Synthesis of spiperone.

(**9.1.7**), subsequent cylization to (**9.1.8**) effected by treatment with formamide, deprotection debenzylation by means of hydrogen on palladium-on-charcoal catalyst yielded the key compound (**9.1.9**). Alkylation of the last with 1,1-bis(4-fluorophenyl)-4-bromobutane (**9.1.11**) on reflux in methyl isobutyl ketone in presence of sodium carbonate and potassium iodide resulted in formation of fluspirilene (**9.1.2**) [22,23] (Scheme 9.1).

Mosapramine (101)

Mosapramine (**9.1.3**) (Cremin) is a first generation atypical antipsychotic approved in Japan and available in a number of countries worldwide for the treatment of schizophrenia. It is a potent dopamine antagonist with high affinity to the D2, D3, and D4 receptors, and with moderate affinity for the 5-HT2 receptors.

Pharmacological profiles of mosapramine was associated with higher risks of extrapyramidal symptoms, hyperprolactinemia and drowsiness, but it displays generally low toxicity than the other antipsychotics for schizophrenia treatment. This drug may be beneficial for the improvement of so-called positive symptoms such as hallucinations, delusions, and racing thoughts [24−28].

Synthesis of mosapramine (**9.1.3**) also started from 1-benzylpiperidin-4-one (**9.1.4**), which was converted to 1-benzyl-4-(cyclohexylamino)piperidine-4-carbonitrile (**9.1.13**) in Strecker−Tiemann reaction conditions using as secondary amine piperidine hydrochloride (**9.1.12**) and as a source of cyanide ions potassium cyanide. The nitrile group in prepared aminonitrile

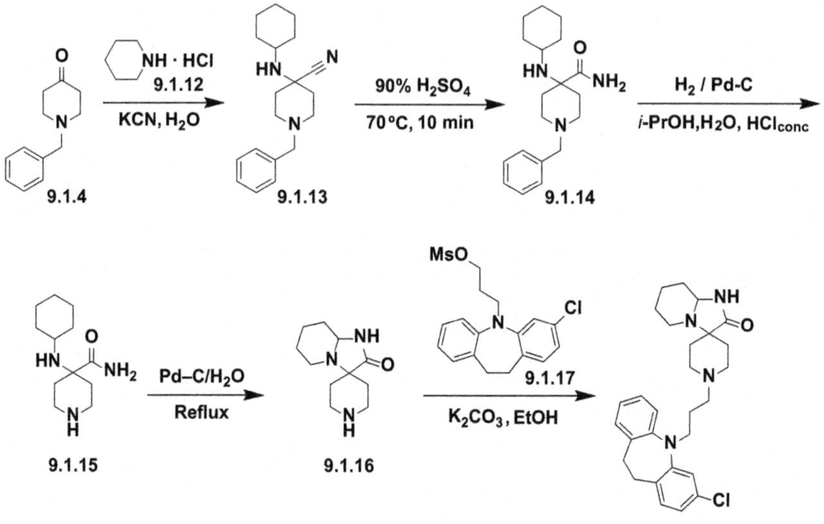

SCHEME 9.2 Synthesis of mosapramine.

(**9.1.13**) easily hydrolyzed to carbamoyl group on short-term heating with 90% sulfuric acid giving compound (**9.1.14**). The last was debenzylated on a palladium-on-charcoal catalyst in a mixture of *i*-propanol, water and concentric hydrochloric acid, which gave 4-(cyclohexylamino)piperidine-4-carboxamide (**9.1.15**) [29]. An interesting reaction of intramolecular condensation took place obtained with carboxamide (**9.1.15**) on boiling in water under reflux with palladium-on-charcoal catalyst giving hexahydro-2*H*-spiro[imidazo[1,2-*a*]pyridine-3,4'-piperidin]-2-one (**9.1.16**). The last was alkylated with 3-(3-chloro-10,11-dihydro-5*H*-dibenzo[*b,f*]azepin-5-yl)propyl methanesulfonate (**9.1.17**) in ethanol in presence of potassium carbonate to give desired mosapramine (**9.1.3**) [30] (Scheme 9.2).

Mosapramine has an asymmetric carbon atom (8a) in its imidazopyridine ring. The enantiomers of this agent were synthesized to compare their biological activities. The key intermediates, (*R*)-(−)- and (*S*)-(+)- hexahydro-2*H*-spiro[imidazo[1,2-*a*]pyridine-3,4'-piperidin]-2-one (**9.1.16**) were prepared by optical resolution of the corresponding racemic compound and further were alkylated with (**9.1.17**) to afford (*R*)-(−)- and (*S*)-(+)- mosapramine. There were few differences in the examined biological activities of the two enantiomers [31].

Fenspiride (188)

Fenspiride (**9.1.22**) (Eurespal) is a bronchodilator, a noncorticosteroid antiinflammatory drug, used as a part of complex therapy of bronchial asthma, in the treatment of certain respiratory diseases like rhinopharyngitis, laryngitis, tracheobronchitis, otitis and sinusitis, whooping cough, influenza. It has been described to have beneficial effects in patients with chronic obstructive pulmonary disease, although the mechanism of its action is not well known. It inhibits mucus secretion and reduces the release of tachykinins at a prejunctional level by its antimuscarinic action. It also may be an antagonist at the α adrenoceptor and H1 histamine receptors. The most common side effects are nausea, stomachache, and vomiting [32−43].

Two approaches were disclosed for the preparation of fenspiride (**9.1.22**). According to first of them 1-phenethylpiperidin-4-one (**9.1.18**) was converted to 4-hydroxy-1-phenethylpiperidine-4-carbonitrile (**9.1.19**) via adding solution of sodium cyanide to cooled solution of (**9.1.18**) in 15% hydrochloric acid [44].

Obtained product was hydrogenated in tetrahydrofuran by means of sodium aluminum hydride to give aminoethanol derivative (**9.1.20**), which on reflux in diethyl carbonate (**9.1.21**) in the presence of sodium methylate gave desired fenspiride (**9.1.22**) [45].

The second approach is based on sequence of series of reactions (**9.1.18** → **9.1.23** → **9.1.24** → **9.1.25** → **9.1.22**) started with Reformatskii reaction and finishing with a Curtius rearrangement. The synthesis started with a Reformatskii reaction of 1-phenethylpiperidin-4-one (**9.1.18**) with ethyl 2-bromoacetate and zinc dust in benzene which gave ethyl 2-(4-hydroxy-1-phenethylpiperidin-4-yl)acetate (**9.1.23**). The last was converted to hydrazide (**9.1.24**) on heating at 50−60°C in benzene with hydrazine hydrate. Obtained hydrazide was subjected to a Curtius rearrangement on reaction with nitrous acid ($NaNO_2$, HCl). Hydrazide (**9.1.24**) was converted to corresponding azide, which in reaction conditions was rearranged to intermediate isocyanate (**9.1.25**) and then after intramolecular cyclization converted to fenspiride (**9.1.22**) [46] (Scheme 9.3).

Some modifications of presented approaches were proposed [47,48].

Clospirazine (7)

Clospirazine (**9.1.29**) (Diceplon) is a contradictory compound. Its status as a drug is not clear. It has a potent sympatholytic action and was reported as a new neuroleptic agent with a characteristic spectrum of pharmacological activities, which differs from those of standard chloropromazine series compounds and has low toxicity [49−52]. It was also shown that clospirazine inhibits platelet type B monoamine oxidase [53].

At the same time it was reported that although its acute toxicity was lower than that of chloropromazine it caused pathological changes in livers and kidneys of rats [54] and in pregnant animals, resulting in a delay in fetus development [55]. Decades later it was shown that clospirazine displays

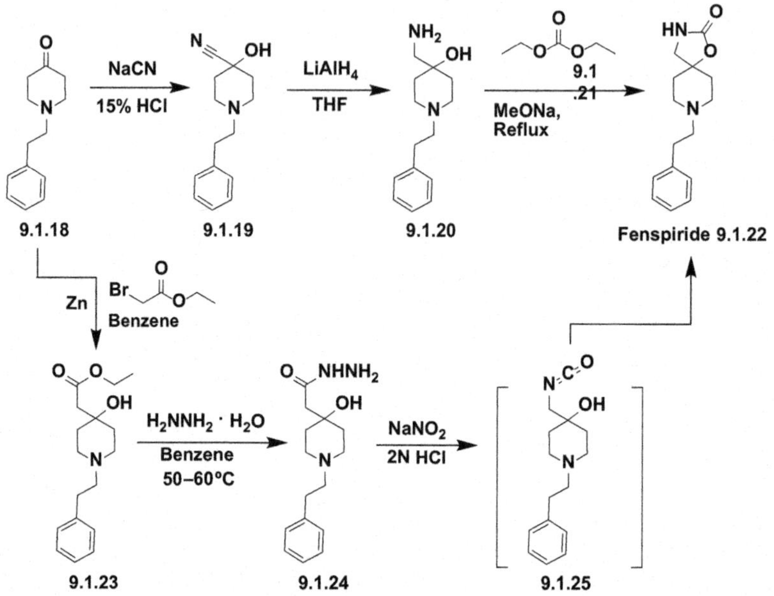

SCHEME 9.3 Synthesis of fenspiride.

extensive antitumor activity in pancreatic cancer via impairing ras-MAPK signaling [56,57].

Synthesis of clospirazine (**9.1.29**) is very simple and was disclosed in some patents [58,59].

In one of them 1-(3-(2-chloro-10H-phenothiazin-10-yl)propyl)piperidin-4-one (**9.1.26**) was reacted with thioglycolic acid (**9.1.27**) and ammonium carbonate in boiling benzene to give clospirazine (**9.1.29**) [58]. In a little bit other approach, the same starting ketone (**9.1.26**) was refluxed in benzene with thioglycolic acid amide (**9.1.28**) in presence of p-toluenesulfonic acid [59]. At last, 1-thia-4,8-diazaspiro[4.5]decan-3-one (**9.1.30**) was alkylated with 2-chloro-10-(3-chloropropyl)-10H-phenothiazine (**9.1.31**) in boiling ethanol in presence of potassium carbonate resulting in clospirazine (**9.1.29**) [58] (Scheme 9.4).

Pazinaclone (43)

Pazinaclone (**9.1.36**) (DN-2327) is an investigational sedative and anxiolytic drug, a nonbenzodiazepine compound that has a high affinity for benzodiazepine receptors. It acts as a partial agonist at γ-aminobutyric acid (GABA) benzodiazepine receptors. Pazinaclone's overall pharmacological profile is a very similar to the benzodiazepines including sedative and anxiolytic properties, but acting as a partial agonist pazinaclone reveal less amnestic effects. The acute behavioral effects and abuse liability, anticonvulsive effects of

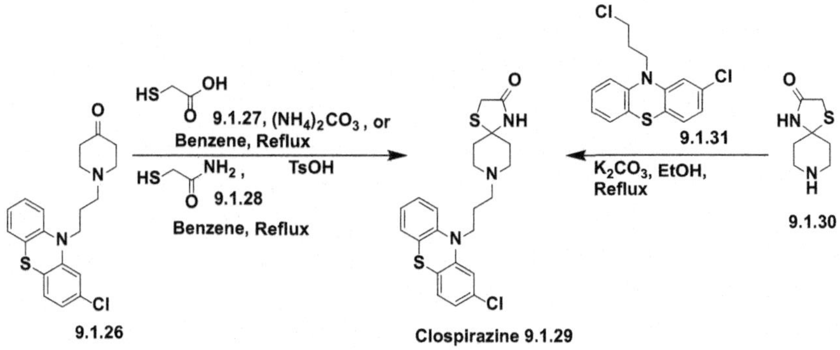

SCHEME 9.4   Synthesis of clospirazine.

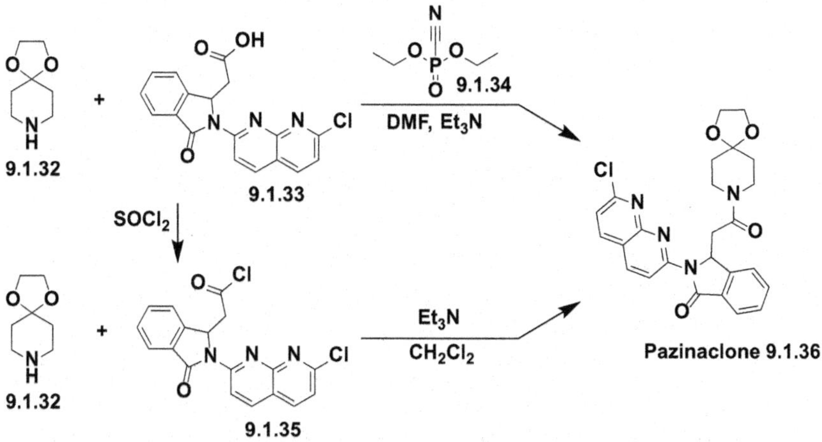

SCHEME 9.5   Synthesis of pazinaclone.

pazinaclone are comparable to those of the benzodiazepine anxiolytic alprazolam without relevant sedative properties, or signs of dependence. Pazinaclone showed a higher affinity for the GABA receptor in comparison to diazepam or flunitrazepam. Patients treated with pazinaclone reported more unwanted events, mostly dizziness and tiredness [60–68].

Pazinaclone (**9.1.36**) was synthesized starting from ethylene ketal of piperidin-4-one (**9.1.32**), which was acylated with specially prepared 2-(2-(7-chloro-1,8-naphthyridin-2-yl)-3-oxoisoindolin-1-yl)acetic acid (**9.1.33**) using the coupling diethyl cyanophosphonate (**9.1.34**) in dimethylformamide in the presence of triethylamine, or heretofore converting acid (**9.1.33**), to its chloride (**9.1.35**) on heating in thionyl chloride with further acylation in dichloroethane in the presence of trimethylamine, which gave final desired product (**9.1.36**) [69,70] (Scheme 9.5).

REFERENCES

[1] Chivers J, Jenner P, Marsden CD. Pharmacological characterization of binding sites identified in rat brain following in vivo administration of [3H]spiperone. Brit J Pharmacol 1987;90(3):467−78.

[2] Blackburn TP, Cox B, Thornber CW, Pearce RJ. Pharmacological studies in vivo with ICI 169,369, a chemically novel 5-HT2/5-HT1C receptor antagonist. Eur J Pharmacol 1990;180(2-3):229−37.

[3] Metwally KA, Dukat M, Egan CT, Smith C, DuPre A, Gauthier CB, et al. Spiperone: influence of spiro ring substituents on 5-HT2A serotonin receptor binding. J Med Chem 1998;41(25):5084−93.

[4] Janssen PAJ, Janssen C. 2,4,8-Triazaspiro[4.5]dec-2-enes, US 3155669; 1964.

[5] Janssen PAJ. 1-Oxo-2,4,8-triazaspiro[4.5]decanes; 1964.

[6] Janssen PAJ. 1-Benzyl-4-substituted piperidines, US 3161644; 1964.

[7] Janssen PAJ, Niemegeers CJE, Schellekens KHL, Lenaerts FM, Verbruggen FJ, Van Nueten JM, et al. Pharmacology of fluspirilene (R 6218) [8-[4,4-bis(p-fluorophenyl)butyl]-1-phenyl-1,3,8-triazaspiro[4,5]decan-4-one] a potent, long-acting, and injectable neuroleptic drug. Arzneimitt-Forsch 1970;20(11):1689−98.

[8] Immich H, Eckmann F, Neumann H, Schapperle O, Schwarz H, Tempel H. Joint clinical study of a long term neuroleptic drug fluspirilene. Arzneimitt -Forsch 1970;20(11):1699−701.

[9] Chouinard G, Annable L, Steinberg S. A controlled clinical trial of fluspirilene, a long-acting injectable neuroleptic, in schizophrenic patients with acute exacerbation. J Clin Psychopharm 1986;6(1):21−6.

[10] Gould RJ, Murphy KMM, Reynolds IJ, Snyder SH. Antischizophrenic drugs of the diphenylbutylpiperidine type act as calcium channel antagonists. Proc Natl Acad Sci USA 1983;80(16):5122−5.

[11] Abhijnhan A, Adams CE, David A, Ozbilen M. Depot fluspirilene for schizophrenia. Cochrane Database Syst Rev 2007;(1)):CD001718.

[12] Ciccarelli G, Nose F, Zuanazzi GF. Fluspirilene, a long-acting injectable neuroleptic. Acta Psychiatr Belg 1972;72(6):736−47.

[13] Wurthmann C, Klieser E, Lehmann E, Pester U. Test therapy in the treatment of generalized anxiety disorders with low dose fluspirilene. Prog Neuropsychopharmacol Biol Psychiatry 1995;19(6):1049−60.

[14] Quraishi S, David A. Depot fluspirilene for schizophrenia. Cochrane Database Syst Rev 2000;(2)):CD001718.

[15] Anonymous. Pimozide as maintenance treatment in chronic schizophrenia. Br J Clin Pract 1971;25(9):417−20.

[16] Villeneuve A, Dogan K, Lachance R, Proulx C. Controlled study of fluspirilene in chronic schizophrenia. Curr Therapeut Res 1970;12(12):819−27.

[17] Chouinard G, Ban TA, Lehmann HE, Ananth JV. Fluspirilene in the treatment of chronic schizophrenic patients. Curr Ther Res Clin Exp 1970;12(9):604−8.

[18] Onkenhout LA, Scheffer W, Amery W. Clinical study of an injectable long-acting neuroleptic agent: fluspirilene (R 6218). Psychiatr Neurol Neurochir 1970;73(4):285−91.

[19] van Epen JH. Experience with fluspirilene (R 6218), a long-acting neuroleptic. Psychiatr Neurol Neurochir 1970;73(4):277−84.

[20] Vereecken JL, Tanghe A. Fluspirilene and pipothiazine undecylenate, two long-acting injectable neuroleptics. A double-blind controlled trial in residual schizophrenia. Psychiatr Neurol Neurochir 1972;75(2):117−27.

[21] Soni SD. Fluspirilene in the treatment of non-hospitalized schizophrenic patients. Curr Med Res Opin 1977;4(9):645−9.

[22] Janssen C. Substituted 4-oxo-1,3,8-triazaspiro[4.5]decanes, BE 633914; 1963.

[23] Janssen PAJ. Substituted 1,3,8-triazaspiro[4.5]decanes, US 3238216; 1966.

[24] Sumiyoshi T, Suzuki K, Sakamoto H, Yamaguchi N, Mori H, Shiba K, et al. Atypicality of several antipsychotics on the basis of in vivo dopamine-D2 and serotonin-5HT2 receptor occupancy. Neuropsychopharmacology 1995;12(1):57−64.

[25] Takahashi N, Terao T, Oga T, Okada M. Comparison of risperidone and mosapramine addition to neuroleptic treatment in chronic schizophrenia. Neuropsychobiology 1999;39 (2):81−5.

[26] Fujimura M, Hashimoto K, Yamagami K. The effect of the antipsychotic drug mosapramine on the expression of fos protein in the rat brain; comparison with haloperidol, clozapine and risperidone. Life Sci 2000;67(23):2865−72.

[27] Kishi T, Matsunaga S, Matsuda Y, Iwata N. Iminodibenzyl class antipsychotics for schizophrenia: a systematic review and meta-analysis of carpipramine, clocapramine, and mosapramine. Neuropsychiatr Dis Treat 2014;10:2339−51.

[28] Kishi T, Matsuda Y, Matsunaga S, Iwata N. Aripiprazole for the management of schizophrenia in the Japanese population: a systematic review and meta-analysis of randomized controlled trials. Neuropsychiatr Dis Treat 2015;11:419−34.

[29] van de Westeringh C, van Daele P, Hermans B, van der Eycken C, Boey J, Janssen PAJ. 4-Substituted piperidines. I. Derivatives of 4-tertiaryamino-4-piperidinecarboxamides. J Med Chem 1964;7(5):619−23.

[30] Tashiro C, Horii I. Imidazopyridine-spiropiperidine compounds, US 4337260; 1982.

[31] Tashiro C, Setoguchi S, Fukuda T, Marubayashi N. Syntheses and biological activities of optical isomers of 3-chloro-5-[3-(2-oxo-1,2,3,5,6,7,8,8a-octahydroimidazo[1,2-a]pyridine-3-spiro-4′-piperidino)propyl]-10,11-dihydro-5H-dibenz[b,f]azepine (mosapramine) dihydrochloride. Chem Pharm Bull 1993;41(6):1074−8.

[32] LeDouarec JC, Duhault J, Laubie M. Pharmacological study of phenethyl-8-oxa-1-diaza-3,8-spiro(4,5)decanone-2 hydrochloride or fensperide (JP 428). Arzneimitt -Forsch 1969;19(8):1263−71.

[33] Evrard Y, Lhoste F, Advenier C, Duhault J. Fensperide and the respiratory tract: a new pharmacological approach. Semaine des Hopitaux 1986;62(19):1375−81.

[34] Salem H, Shemano I, Beiler JM, Orzechowski R, Hitchens JT, Clemente E. Fensperide. Nonsympathomimetic bronchodilator with antiallergic activity. Arch International Pharmacodyn Ther 1971;193(1):111−23.

[35] Shmelev EI, Kunicina Yu L. Comparison of fensperide with beclomethasone as adjunctive anti-inflammatory treatment in patients with chronic obstructive pulmonary disease. Clin Drug Invest 2006;26(3):151−9.

[36] Khawaja AM, Liu Y-C, Rogers DF. Effect of fensperide, a non-steroidal antiinflammatory agent, on neurogenic mucus secretion in ferret trachea in vitro. Pulm Pharmacol Therapeut 1999;12(6):363−8.

[37] De Castro CMMB, Nahori M, Dumarey CH, Vargaftig BB, Bachelet M. Fensperide: an anti-inflammatory drug with potential benefits in the treatment of endotoxemia. Eur J Pharmacol 1995;294(2):669−76.

[38] Bareggi SR, Guadagni L. Pharmacological research on a new antiinflammatory: fensperide. Atti della Accademia Medica Lombarda 1979;34(1-2-3-4):151−6.

[39] Baryshevskaia LA, Velikanov AK, Sedykh MI. Experience with fensperide in a rhinosurgeon's practice. Vestnik otorinolaringologii 2009;2:46−8.

[40] Butorov SI, Muntianu VI. The effectiveness of fensperide in the treatment of patients with chronic obstructive pulmonary disease. Klinicheskaia Meditsina 2007;85(5):43–7.

[41] Shmelev EI, Kunicina Yu L. Comparison of fensperide with beclomethasone as adjunctive anti-inflammatory treatment in patients with chronic obstructive pulmonary disease. Clin Drug Invest 2006;26(3):151–9.

[42] Shorokhova TD, Medvedeva IV, Lapik SV, Solov'eva OG, Gracheva EI, Iusupova RS. Effectiveness of fensperide in patients with chronic obstructive bronchitis. Klinicheskaia Meditsina 2001;79(8):55–7.

[43] Montes B, Catalan M, Roces A, Jeanniot JP, Honorato JM. Single-dose pharmacokinetics of fenspiride hydrochloride: phase I clinical Trial. Eur J Pharmacol 1993;45(2):169–72.

[44] Harper NJ, Fullerton SE. Ethynyl and styryl compounds of the prodine type. J Med Pharm Chem 1961;4:297–316.

[45] Regnier G, Douarec J, Le Canevari R. 1-oxa-2-oxo 3, 8-diaza spiro (4, 5) decanes, US 3399192 A; 1968.

[46] No Inventor data available. 1-Oxa-2-oxo-3,8-diazo-8-phenethylspiro[4,5]decane, BE 786631; 1972.

[47] Somanathan R, Rivero IA, Nunez GI, Hellberg LH. Convenient synthesis of 1-oxa-3,8-diazaspiro[4,5]decan-2-ones. Synth Commun 1994;24(10):1483–7.

[48] Regnier G, Canevari R, Le Douarec JC, Laubie M, Duhault J. 2-Oxazolidinones. II. Synthesis and pharmacological properties of 1,3,8-oxadiazaspiro [4,5] decan-2-one. Chim Therapeut 1969;4(3):185–94.

[49] Imamura H, Okada T, Matsui E, Kato Y. Studies on psychotropic drugs. VII: metabolic fate of APY-606, absorption, distribution, excretion and metabolism of tritiated APY-606. J Pharm Soc Jap 1970;90(6):813–17.

[50] Fryer IR. Chapter 1 In: Cain CK, editor. Antipsychotic and anti-anxiety agents in annual reports in medicinal chemistry (Engelhardt, E L), Volume 6. Elsevier; 1971

[51] Zirkle CL, Kaiser C. Chapter 1 In: Heinzelman RV, editor. Antipsychotic and Anti-anxiety Agents in Annual Reports in Medicinal Chemistry (Engelhardt, E L), Volume 7. Elsevier; 1972

[52] Nakanishi M, Okada T, Tsumagari T. Psychotropic drugs. 5. Pharmacological effects of 8-[3-(2-chloro-10-phenothiazinyl)propyl]-1-thia-4,8-diazaspiro[4,5]decan-3-one hydrochloride (APY-606). Yakugaku Zasshi 1970;90(7):800–7.

[53] Suzuki O, Seno H, Kumazawa T. In vitro inhibition of human platelet monoamine oxidase by phenothiazine derivatives. Life Sci 1988;42(21):2131–6.

[54] Takeuchi M, Namba T, Okada T, Moriguchi J. Psychotropic drugs. 8. Acute, subacute, and chronic toxicity of APY-606. Oyo Yakuri 1970;4(3):487–95.

[55] Hamada Y, Namba T, Okada T, Izaki K. Psychotropic drugs. 9. Teratological studies on the safety of APY-606. Oyo Yakuri 1970;4(3):497–504.

[56] Guo N, Liu Z, Zhao W, Wang E, Wang J. Small molecule APY606 displays extensive antitumor activity in pancreatic cancer via impairing ras-MAPK signaling. PLoS One 2016;11(5) e0155874/1-e0155874/17.

[57] Zhao W, Li D, Liu Z, Zheng X, Wang J, Wang E. Spiclomazine induces apoptosis associated with the suppression of cell viability, migration and invasion in pancreatic carcinoma cells. PLoS One 2013;8(6):e66362.

[58] Nakanishi M, Arimura K, Tsumagari T, Shiraki M. 3-Oxo-1-thia-4,8-diazaspiro[4,5] decanes, JP 45015979; 1970.

[59] Nakanishi M, Arimura K. Spirocyclic compounds, JP 49031994; 1974.

[60] Anonymous P. A 77000, DN 2327. Drugs R&D 1999;2(1):57–9.

[61] Evans SM, Foltin RW, Levin FR, Fischman MW. Behavioral and subjective effects of DN-2327 (pazinaclone) and alprazolam in normal volunteers. Behav Pharmacol 1995;6 (2):176–86.

[62] Mumford GK, Rush CR, Griffiths RR. Alprazolam and DN-2327 (pazinaclone) in humans: psychomotor, memory, subjective, and reinforcing effects. Exp Clin Psychopharmacol 1995;3(1):39–48.

[63] Suzuki M, Uchiumi M, Murasaki M. A comparative study of the psychological effects of DN-2327, a partial benzodiazepine agonist, and alprazolam. Psychopharmacology 1995;121(4):442–50.

[64] Ichida T, Hirouchi M, Mizutan H, Narihara R, Kuriyama K. Association and agonistic action of DN-2327, a novel isoindoline derivative, GABAB receptor in brain. Eur J Pharmacol 1994;267(1):43–7.

[65] Wada T, Fukuda N. Pharmacologic profile of a new anxiolytic, DN-2327: effect of Ro15-1788 and interaction with diazepam in rodents. Psychopharmacology 1991;103(3):314–22.

[66] Wada T, Nakajima R, Kurihara E, Narumi S, Masuoka Y, Goto G, et al. Pharmacologic characterization of a novel non-benzodiazepine selective anxiolytic, DN-2327. Jap J Pharmacol 1989;49(3):337–49.

[67] Wada T, Fukuda N. Effect of a new anxiolytic, DN-2327, on learning and memory in rats. Pharmacol Biochem Behav 1992;41(3):573–9.

[68] Linden M, Hadler D, Hofmann S. Randomized, double-blind, placebo-controlled trial of the efficacy and tolerability of a new isoindoline derivative (DN-2327) in generalized anxiety. Hum Psychopharmacol 1997;12(5):445–52.

[69] Goto G, Saji Y. Isoindolinone derivatives and their use, EP 174858; 1986.

[70] Goto G, Saji Y. Isoindolinone derivatives, production and use thereof, US 4778801; 1988.

Chapter 10

Classes of Piperidine-Based Drugs

To summarize the above-presented material, which was set out in a particular order, based on the type and attachment order of substituents coupled with piperidine ring, here the same compounds are shown in accordance with their pharmacological action. This order is different from the classical narration in pharmacology textbooks and based on the frequency of their appearance in corresponding classes of drugs.

Piperidine has been extensively utilized for the synthesis of a wide range of therapeutics, but more often, it is found in analgesics. The frequency of it appearance in different classes of drugs it looks like this:

1. Analgesics; 2. Antipsychotics; 3. Antihistamine drugs; 4. Anticholinergic drugs; 5. Local anesthetics; 6. Antithrombotic drugs; 7. Antiarrhythmic drugs; 8. Antihypertensive drugs; 9. Drugs for treating respiratory system diseases; 10. Antidepressants; 11. Antiparkinsonian drugs; 12. Hypolipidemic and antihyperlipidemic drugs; 13. Adrenergic (sympathomimetic) drugs; 14. Central nervous system stimulants; 15. Selective estrogen receptor modulators; 16. Antianginal drugs; 17. Drugs for treating protozoan infections; 18. Antimigraine Drugs; 19. Nootropic and neuroprotective drugs; 20. Antiemetics; 21. Antidiarrheal and prokinetic drugs; 22. Gastric antisecretory drugs; 23. Hypoglycemic drugs; 24. Drugs used in the treatment of rheumatoid arthritis; 25. Nicotinic cholinomimetics; 26. Immunosuppresant drugs.

10.1 PIPERIDINE-BASED ANALGESICS

The pain therapies that rely on drugs that have long been known to have analgesic properties are: nonsteroidal antiinflammatory drugs (NSAIDs) or cyclooxygenase inhibitors, opioid analgesics, and analgesic adjuvants, which include several classes of compounds.

The family of 4-phenylpiperidine analgesics are the first representatives of wholly synthetic opioid analgesics widely used in medicine from their inception to the present day. A series of these compounds was born in 1939 when pethidine (**5.1.91**) the prototype of entire family of this series. Hundreds of pethidine analogs were synthesized, some of them entered pharmaceutical market, some were withdrawn.

Piperidine-Based Drug Discovery. DOI: http://dx.doi.org/10.1016/B978-0-12-805157-3.00010-7
299

One of the most "crowded" classes of piperidine-based analgesics is commonly called 4-phenylpiperidines, which, in turn, can be divided into derivatives of 4-s, phenylpiperidine-4-carboxylic acids, 4-phenylpiperidin-4-yl propionates and 1-(4-phenylpiperidin-4-yl)propan-1-ones.

4-Phenylpiperidines

4-Phenylpiperidine-4-Carboxylic Acid Derivatives

Pethidine (**5.3.5**) was the first of them. It is used as an analgesic for the relief of moderate to severe pain including obstetric analgesia. Phenoperidine (**5.3.27**) is another opioid analgesic used in general anesthesia. Anileridine (**5.3.28**) is a compound where *N*-methyl group of pethidine is replaced by an *N*-aminophenethyl group, which increases its analgesic activity. Benzethidine (**5.3.29**) is another 4-phenylpiperidine derivative that is related to pethidine. Etoxeridine (**5.3.30**) was never commercialized and is not currently used in medicine. Furethidine (**5.3.31**) is another 4-phenylpiperidine derivative that is related to pethidine is not currently used in medicine. Piminodine (**5.3.32**), an analog of pethidine, was used in medicine for obstetric analgesia and in dental procedures briefly during the 1960s and 1970s. Morpheridine (**5.3.33**) is a strong analgesic which is approximately four times more potent than pethidine, but it is not currently used in medicine (Fig. 10.1).

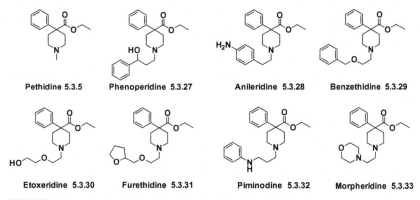

Pethidine 5.3.5 Phenoperidine 5.3.27 Anileridine 5.3.28 Benzethidine 5.3.29

Etoxeridine 5.3.30 Furethidine 5.3.31 Piminodine 5.3.32 Morpheridine 5.3.33

FIGURE 10.1 Analgesic 4-phenylpiperidine-4-carboxylic acid derivatives.

4-Phenylpiperidin-4-yl Propionates

Several 4-phenylpiperidin-4-yl propionates chemically closely related to pethidine have pharmacological action resembling that of pethidine. As a rule, they are more rapid in onset and have shorter duration action. This type of analgesic is available in a number of countries worldwide. Among them are desmethylprodine (**5.2.36**), prodine (**5.2.37**) (Nisentil), allylprodine (**5.2.38**), and trimeperidine (**5.2.39**) (promedol). Prodine is no longer commonly used in the United States because of the risk of serious complications. They have been used in obstetrics, as a preoperative medication, for minor surgical procedures, and for dental procedures (Fig. 10.2).

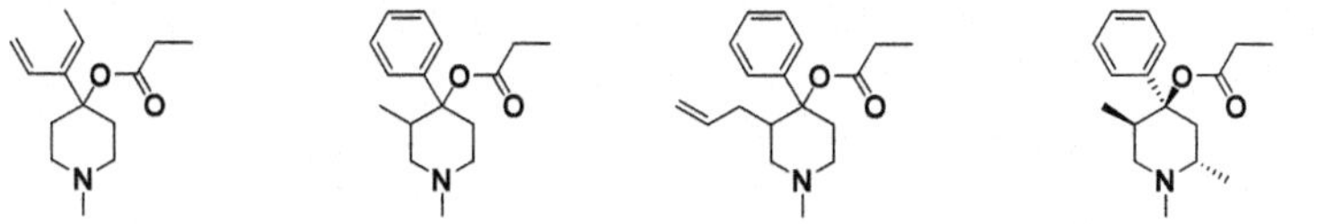

Desmethylprodine 5.2.36   **Prodine 5.2.37**   **Allylprodine 5.2.38**   **Trimeperidine 5.2.39**

FIGURE 10.2   Analgesic 4-phenylpiperidin-4-yl propionates.

1-(4-Phenylpiperidin-4-yl)Propan-1-Ones

Ketobemidone (**5.4.5**) (Ketorax) is a powerful opioid analgesic that was first synthesized in 1942. Its effectiveness against pain is in the same range as morphine, and it also has some noncompetitive *N*-methyl-D-aspartate (NMDA) antagonists properties. Ketobemidone is a medicine available in a number of countries worldwide, but it was not in clinical use in the United States since 1970 (Fig. 10.3).

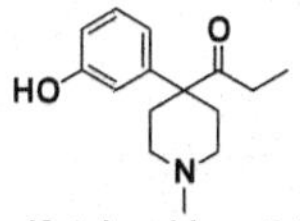

Ketobemidone 5.4.5

FIGURE 10.3   Analgesic 1-(4-phenylpiperidin-4-yl)propan-1-ones.

4-Anilidopiperidines

The 4-anilidopiperidines are the most potent class of opioid analgesics known to date. The prototype of this class, fentanyl (**5.7.1**), is about 300 times more potent than morphine in mice and rats, compared to 50−100 times in humans [320−330]. A very large number of fentanyl analogs − mefentanyl (**5.7.2**), phenaridine (**5.7.3**), ohmefentanyl (**5.7.4**), ocfentanil (**5.7.5**), brifentanil (**5.7.6**), mirfentanil (**5.7.7**), carfentanil (**5.7.8**), lofentanil (**5.7.9**), thiophentanil (**5.7.10**), remifentanil (**5.1.11**), sufentanyl (**5.7.12**), alfentanyl (**5.7.13**), and trefentanil (**5.7.14**) − have been created and achieved success in the pharmaceutical market (Fig. 10.4).

Miscellanous Analgesic Piperidine Derivatives

Piritramide (**5.9.5**) is one of an eponymous two-member class of opioids in clinical use with the other being bezitramide (**7.1.7**). As an analgesic it is weaker than morphine, but with more rapid-onset of analgesia. Piritramide is

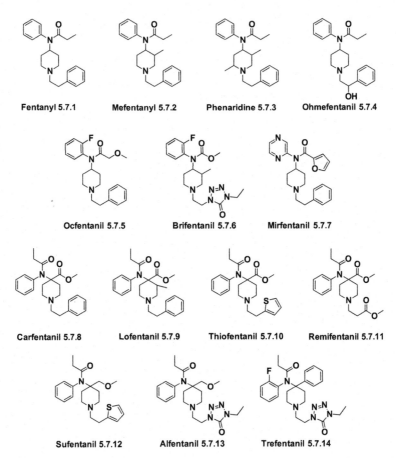

FIGURE 10.4 4-Anilidopiperidines.

marketed in certain European countries for postoperative analgesia and analgosedation but is not in clinical use in the United States. Bezitramide (**7.1.7**) (Burgodin) is an effective, long-acting opioid analgesic that causes analgesia, euphoria, and sedation. It was introduced for the treatment of severe, chronic pain in Europe, but it has never been marketed in the United States. Benzpiperylon (**7.1.43**) and piperylon (**7.1.44**) were proposed for the pharmaceutical market at the beginning of 1960s. They have good analgetic and antipyretic properties, very low toxicity, but did not get wide application and dissemination on the pharmaceutical market. Alvimopan (**5.10.9**) (Entereg) is a selective, peripherally acting μ-opioid receptor antagonist, with no central nervous system activity. It is used for shortening recovery time in patients who have had certain bowel surgeries (Fig. 10.5).

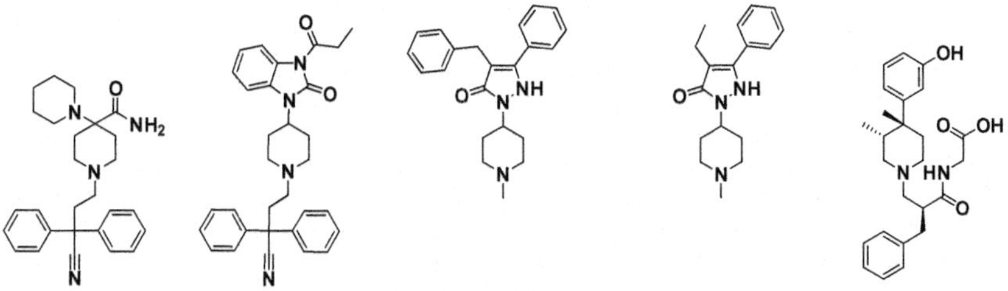

Piritramide 5.9.5   Bezitramide 7.1.7   Benzpiperylon 7.1.43   Piperylone 7.1.44   Alvimopan 5.10.9

FIGURE 10.5   Miscellanous analgesic piperidine derivatives.

10.2   PIPERIDINE-BASED ANTIPSYCHOTICS

Antipsychotics, also termed neuroleptics or major tranquilizers, are a group of drugs that are used to treat serious mental health conditions such as schizophrenia and delusional disorder, psychosis as well as other emotional and mental conditions. Antipsychotics fall into eight chemical groups including phenothiazines, thioxanthenes, dibenzoxazepines, dihydroindoles, benzamides, butyrophenones, diphenylbutylpiperidines and benzisoxazoles, which are considered typical or first-generation antipsychotics and were the first type of antipsychotics available. They have been prescribed since the 1950s and were the most prescribed treatments for psychosis for a long time. Antipsychotics help to control the symptoms of psychosis as well as less serious mental health conditions such as bipolar and mood disorder that may develop into later psychosis. Antipsychotics reduce or increase the effect of neurotransmitters in the brain to regulate levels.

Butyrophenones

Antipsychotics of butyrophenone 4-phenylpiperidin-4-ol derivatives series are represented by haloperidol (**5.2.7**), bromperidol (**5.2.14**), moperone (**5.2.15**), trifluperidol (**5.2.20**). Haloperidol, discovered in 1958, was for more than 50 years the most commonly used and widely prescribed antipsychotic. On the WHO Model List of Essential Drugs, haloperidol (**5.2.7**) was considered indispensable for treating psychiatric emergency situations. Bromperidol (**5.2.14**), a medicine available in a number of countries worldwide is another potent and long-acting neuroleptic, has been used as an antipsychotic. Moperone (**5.2.15**), a typical antipsychotic used in the treatment of schizophrenia, was discontinued. Trifluperidol (**5.2.20**) is another first-generation antipsychotic that has properties similar to those of haloperidol. It is considerably more potent drug, and causes relatively more severe side effects, especially tardive dyskinesia, a disorder resulting in involuntary, repetitive body movements (Fig. 10.6).

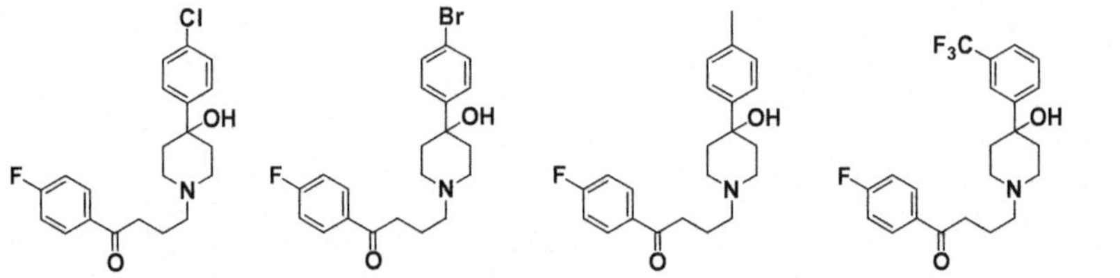

FIGURE 10.6 Piperidine-based antipsychotics—butyrophenones.

Another butyrophenone series of antipsychotics piperidine compounds depict as 1-butyrophenone substituted (piperidin-4-yl)-benzoimidazol-2-one (thione) derivatives represented by droperidol (**7.1.1**), an effective medication used in emergency medicine practice for control of a variety of uses, such as psychosis/agitation, and benperidol (**7.1.2**) a compound with general properties similar to those of haloperidol (**5.1.52**) and used for the treatment of schizophrenia and of aberrant sexual behavior in several European countries, and timiperone (**7.1.3**), a similar compound but one that contains a thiourea fragment in the benzoimidazole part of a molecule instead of urea. It was approved in Japan for the treatment of schizophrenia (Fig. 10.7).

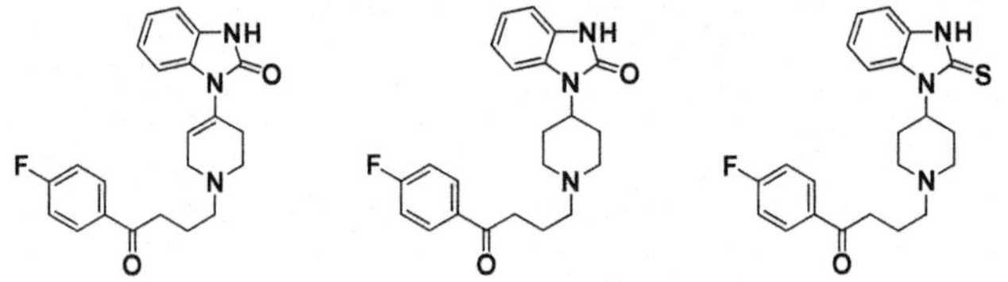

FIGURE 10.7 Piperidine-based bntipsychotics—butyrophenones.

Two other butyrophenone series representatives are Pipamperone (**5.9.4**) and Spiperone (**9.1.1**).

Pipamperone (**5.9.4**), introduced in 1961, is one of the early butyrophenone derivatives whose pharmacological and clinical profile was distinct from haloperidol and all other known antipsychotic drugs. Pipamperone, a typical antipsychotic, is used in the treatment of acute states of agitation and aggression in the acute and chronic psychotic states (schizophrenia and chronic nonschizophrenic delusions: paranoid delusions, chronic hallucinatory psychoses).

Spiperone (**9.1.1**) is a spiro-fused bicyclic 1-phenylimidazolidin-4-one and piperidine derivative with butyrophenone substituent in the first position of piperidine ring with typical antipsychotic properties. It has been recommended in the treatment of schizophrenia. Spiperone is a widely used also as a pharmacological tool that acts as a potent dopamine D2, serotonin 5-HT1A, and serotonin 5-HT2A antagonist (Fig. 10.8).

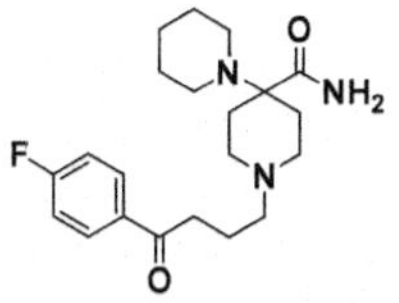

Pipamperone 5.9.4 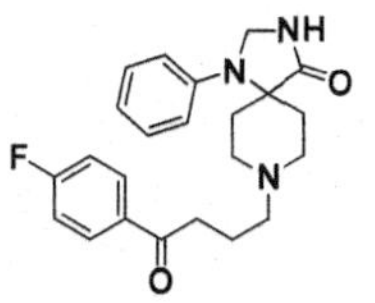   Spiperone 9.1.1

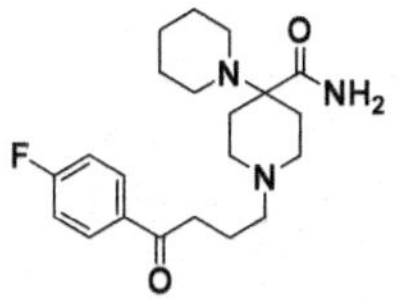

Pipamperone 5.1.275 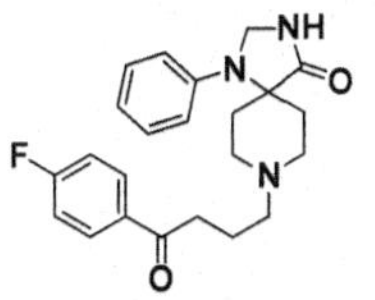   Spiperone 9.1.1

FIGURE 10.8 Piperidine-based bntipsychotics—butyrophenones.

Diphenylbutylpiperidines

Diphenylbutylpiperidines are presented with three drugs: penfluridol (**5.2.26**), pimozide (**7.1.4**) and Fluspirilene (**9.1.2**)

Penfluridol (**5.2.26**) is a highly potent, long-acting, oral antipsychotic compound, which was discovered at the same time as the discovery of the butyrophenone series of antipsychotic medications (haloperidol, bromoperidol etc.) and are used primarily to treat positive symptoms including the experiences of perceptual abnormalities (hallucinations) and fixed, false, irrational beliefs (delusions).

Pimozide (**7.1.4**) is an antipsychotic agent used to reduce uncontrolled movements or outbursts of words/sounds caused by Tourette syndrome in patients failed to respond satisfactorily to standard treatment.

Fluspirilene (**9.1.2**) is a long-acting antipsychotic agent used for the treatment of psychoses including chronic schizophrenia (Fig. 10.9).

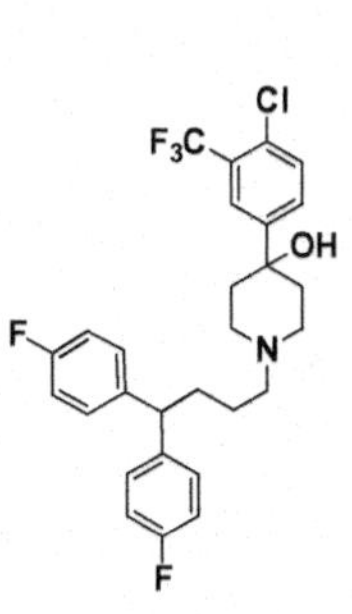

Penfluridol 5.2.26 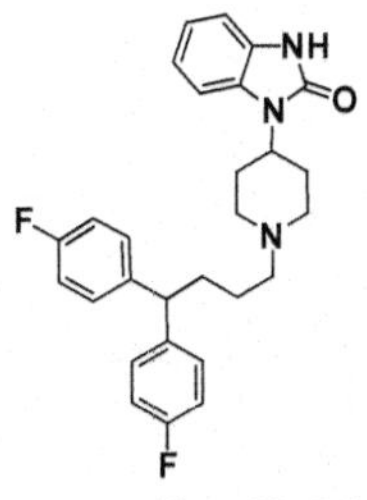   Pimozide 7.1.4 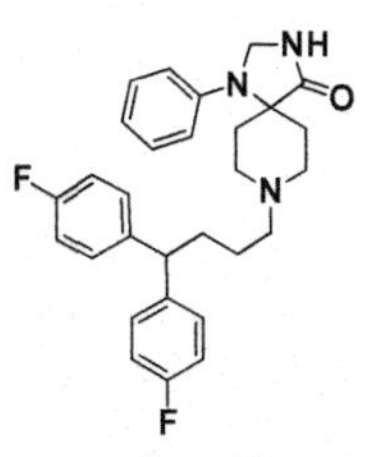   Fluspirilene 9.1.2

FIGURE 10.9 Piperidine-based antipsychotics—diphenylbutylpiperidines.

Benzisoxazoles

The benzisoxazoles series of antidepressants is also represented with three drugs: risperidone (**7.2.1**), paliperidone (**7.2.2**), and iloperidone (**7.2.3**).

Risperidone (**7.2.1**) is the first second-generation antipsychotic, used to treat schizophrenia and symptoms of bipolar disorder (manic depression). It is also used in irritability with autism.

Paliperidone (**7.2.2**) is used to treat schizophrenia and schizoaffective disorder in adults and adolescents as an adjunct to mood stabilizers and/or antidepressant therapy.

Iloperidone (**7.2.3**) is used for treating symptoms of psychotic (mental) disorders such as schizophrenia and certain mental/mood disorders (Fig. 10.10).

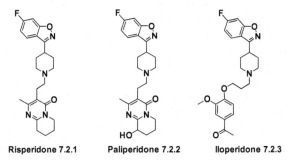

Risperidone 7.2.1   Paliperidone 7.2.2   Iloperidone 7.2.3

FIGURE 10.10 Piperidine-based antipsychotics—benzisoxazoles.

Phenothiazines

Antidepressant medication with phenothiazine-piperidine coupling is represented by thioridazine (**3.1.73**).

Thioridazine (**3.1.73**), a typical antipsychotic of the phenothiazine series, was widely used in the treatment of schizophrenia and psychosis, but was withdrawn worldwide in 2005 because it caused severe cardiac arrhythmias (Fig. 10.11).

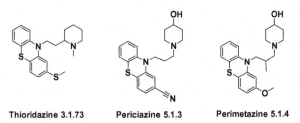

Thioridazine 3.1.73   Periciazine 5.1.3   Perimetazine 5.1.4

FIGURE 10.11 Piperidine-based antipsychotics—phenothiazines.

Periciazine (**5.1.3**), (Neuleptil) is another old drug of the piperidine—piperidine coupling. It has been shown to reduce pathologic arousal and affective tension in some psychotic patients, while the symptoms of abnormal mental integration are relatively unaffected. It is a sedative with weak antipsychotic properties. It is used as an adjunctive medication in some psychotic patients, for the control of residual prevailing hostility, impulsiveness and aggressiveness. Periciazine has been utilized also for the treatment of cannabis dependence. A close analog of periciazine (**5.1.3**) is perimetazine (**5.1.4**) (Fig. 10.11).

Dibenzazepines

This subchapter will present chemically related compounds that represent a combination of piperidine and the dibenzazepine derivatives.

Carpipramine (**5.9.7**) is available in both Japan and France. It is a drug classified as second-generation antipsychotic for the treatment of psychoses.

Clocapramine (**5.9.9**) is approved in Japan for the treatment of schizophrenia and was introduced into medicinal practice as a successor to carpipramine (**5.9.7**).

Mosapramine (**9.1.3**) is an antipsychotic for the treatment of schizophrenia with respect to actions on negative symptoms in schizophrenia and presumably with a lower tendency to induce neurological side effects (Fig. 10.12).

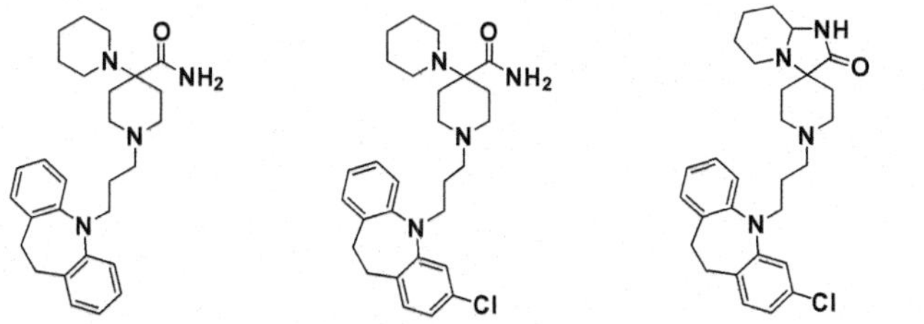

Carpipramine 5.9.7   Clocarpramine 5.9.9   Mosapramine 9.1.3

FIGURE 10.12   Piperidine-based antipsychotics—dibenzazepines.

Tetrahydro-γ-Carbolines and 3-(Piperidin-4-yl)-1*H*-Indoles

Sertindole (**7.2.42**) 3-(piperidin-4-yl)-1*H*-indole derivative is a second-generation atypical antipsychotic that is thought to give a lower incidence of extrapyramidal side effects than typical antipsychotic drugs.

Gevotroline (**8.1.40**) and carvotroline (**8.1.42**) have not yet entered the category of the vogue pharmaceuticals. Gevotroline and carvotroline are atypical antipsychotics under development for the treatment of schizophrenia (Fig. 10.13).

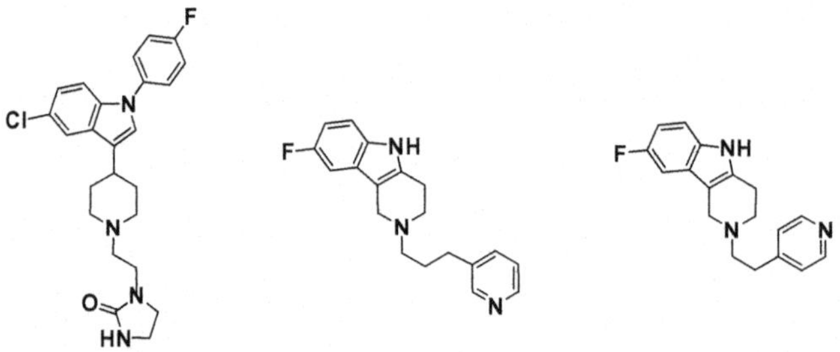

Sertindole 7.2.42   Gevotroline 8.1.40   Carvotroline 8.1.42

FIGURE 10.13 Piperidine-based antipsychotics—sertindole, gevotroline, and carvotroline.

10.3 PIPERIDINE-BASED ANTIHISTAMINE DRUGS

Drugs that antagonize the actions of histamine are referred to as antihistamines. Histamine is released by the immune system during allergic reactions causing itching, swelling and congestion, stimulating gastric secretion, constriction of bronchial smooth muscle, and dilating blood vessels that cause a fall in blood pressure. It is involved also in the regulation of basic body functions including energy and endocrine homeostasis, cognition, and memory, the cycle of sleeping, and waking.

Piperidin-4-Ylidene Diaryl[a,d][7]Annulenes

Cyproheptadine (**6.1.4**) is a first-generation antihistamine with additional anticholinergic, antiserotonergic, and local anesthetic properties. It is used to relieve allergy symptoms such as watery eyes, runny nose, itching eyes/nose, sneezing, hives, and itching.

Azatadine (**6.1.2**) also is a first-generation antihistamine and is used to treat sneezing, runny nose, itching, watery eyes, hives, rashes, itching, and other symptoms of allergies and the common cold. Loratadine (**6.1.3**) is a second-generation H_1-antihistamine with significantly lower probability of occurrence of adverse drug reactions, such as sedation, while still providing effective relief of allergic conditions that treats symptoms such as itching, runny nose, watery eyes, and sneezing from "hay fever" and other allergies. Desloratadine (**6.1.4**) is an active metabolite of loratadine (**6.1.3**) similar in safety and effectiveness. Ketotifen (**6.1.7**) is an antihistamine used in its oral form, it is used to prevent asthma attacks and in its ophthalmic form, it is used to treat allergic conjunctivitis, or the itchy red eyes caused by allergies. Alcaftadine (**6.1.8**) is an antihistamine medication used to prevent itching of the eyes due to allergies (Fig. 10.14).

4-(Benzhydryloxy)Piperidines

Diphenylpyraline (**5.1.6**) is a first-generation antihistaminic agent that antagonizes most of the pharmacological effects of histamine, including urticaria

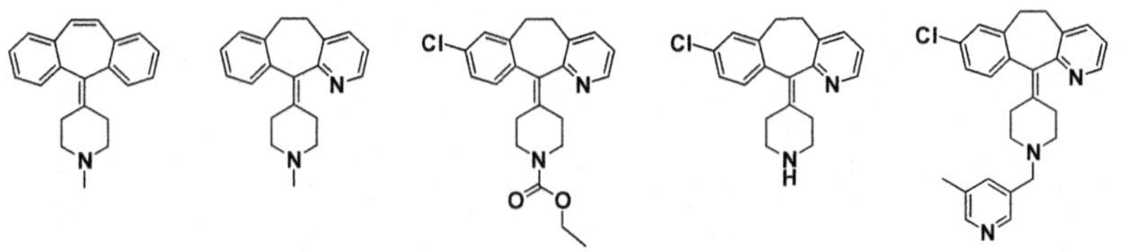

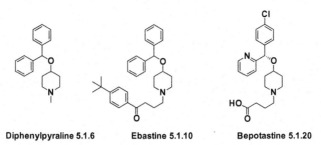

Cyproheptadine 6.1.1   Azatadine 6.1.2   Loratadine 6.1.3   Desloratadine 6.1.4   Rupatadine 6.1.5

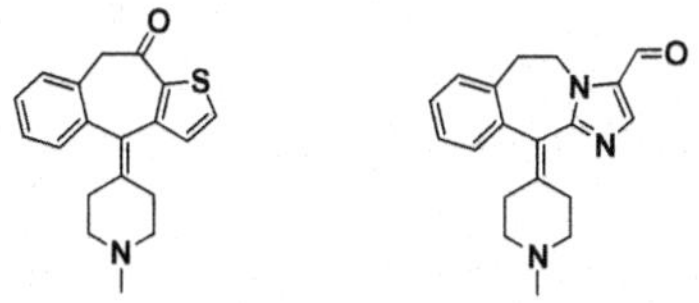

Ketotifen 6.1.7   Alcaftadine 6.1.8

FIGURE 10.14   Piperidin-4-ylidene diaryl[*a,d*][7]annulenes.

and pruritus. Diphenylpyraline may exhibit anticholinergic actions (as do most of the antihistamines). It prevents, but does not reverse, responses mediated by histamine alone. It is marketed in for the treatment of allergic conditions and pruritic skin disorders.

Ebastine (**5.1.10**) structurally related to diphenylpyraline (**5.1.6**) is a nonsedating second-generation long-acting, antihistamine that is indicated mainly for allergic rhinitis and chronic idiopathic urticaria.

Bepotastine (**5.1.20**) is also a second-generation histamine H1-receptor antagonist that is indicated mainly for allergic rhinitis and chronic idiopathic urticarial. It also suppresses some allergic inflammatory processes (Fig. 10.15).

Diphenylpyraline 5.1.6   Ebastine 5.1.10   Bepotastine 5.1.20

FIGURE 10.15   4-(Benzhydryloxy)piperidines.

Miscellanous Antihistamine Piperidine Derivatives

Bamipine (**5.8.4**) is a classical antihistaminic, a drug acting as an H1-antihistamine with anticholinergic properties used as an antipruritic ointment for the symptomatic relief of allergic conditions such as urticaria and in pruritic skin disorders.

Thenalidine (**5.8.5**) another antihistaminics used as an antipruritic but was withdrawn from the market due to a risk of neutropenia.

Astemizole (**5.6.59**) is a long-acting, nonsedating H1-receptor antagonist that prevents sneezing, runny nose, itching and watering of the eyes, and other allergic symptoms, used to treat allergies, urticaria, and other allergic inflammatory conditions. It was withdrawn from the US market in 1999.

Mizolastine (**5.8.12**) is a second-generation, long-acting H1-antihistamine used to relieve the symptoms of seasonal allergic rhinitis, urticarial, and other allergic reactions causing irritation of the eyes and nose.

Mebhydrolin (**8.1.34**) (Diazolin) somewhat out of fashion antihistamine, used for symptomatic relief of allergic symptoms such as nasal allergies and allergic dermatosis. It is not available in the United States, but it is in use in other countries.

Levocabastine (**5.3.41**) is a second-generation selective H1-antihistamine compound, used to treat eye symptoms of allergic conditions, such as inflammation, itching, watering, and burning (Fig. 10.16).

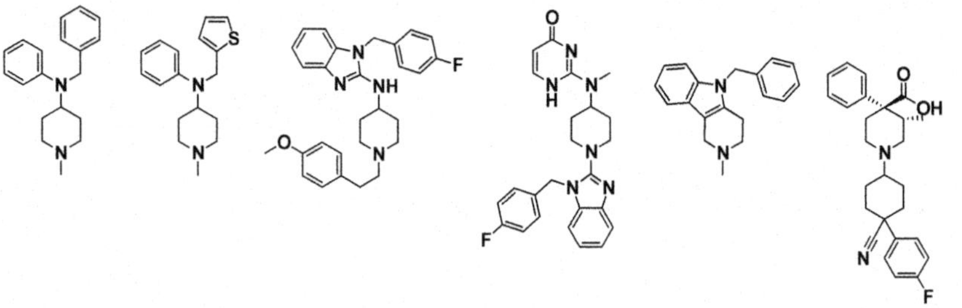

| Bamipine | Thenalidine | Astemizole | Mizolastine | Mebhydrolin | Levocabastine |
| 5.8.4 | 5.8.5 | 5.6.59 | 5.8.12 | 8.1.34 | 5.3.41 |

FIGURE 10.16 Miscellanous antihistamine piperidine derivatives.

10.4 PIPERIDINE-BASED ANTICHOLINERGIC DRUGS

Anticholinergic drugs reduce the effect of the neurotransmitter acetylcholine, which is involved in many major functions in the central and peripheral nervous systems and are commonly used in critical care medicine.

They are classified into two categories according to the receptors that they act on: antinicotinic agents and antimuscarinic agents.

Antinicotinic agents prevent the transmission of signals from motor nerves to neuromuscular structures of the skeletal muscle and are in therapeutic use in surgical procedures as muscle relaxants, either as adjuvants or as premedication drugs for anesthesia.

Antimuscarinic drugs are generally used in psychiatry to treat the extrapyramidal side effects of antipsychotic medications.

Muscarinic Receptor Antagonists

Piperidolate (**4.1.28**) (Dactiran) is a muscarinic receptor antagonist (anticholinergic). It inhibits the action of acetylcholine at postganglionic cholinergic sites and used in the symptomatic treatment of smooth muscle pain spasm associated with gastrointestinal disorders, dysmenorrhea, premature labor, threatened miscarriage.

Eucatropine (**5.1.36**), an anticholinergic and antimuscarinic, produces mydriasis applied topically in the eye producing prompt mydriasis free from anesthetic action, pain, corneal irritation. The action of eucatropine hydrochloride closely parallels that of atropine, although it is much less potent than the latter.

Propiverine (**5.1.41**) (Detrunorm) is an antimuscarinic agent used to treat a number of urinary symptoms and bladder problems such as frequency, urgency, and incontinence.

Pentapiperide (**5.1.45**) is an anticholinergic with antisecretory and antimotility activity on the upper gastrointestinal tract for adjunctive therapy in relieving the symptoms of peptic ulcer.

Mepenzolate (**4.1.32**) is a specific muscarinic receptor antagonist that decreases stomach acid production and bowel contraction. It is indicated for use as adjunctive therapy in the treatment of peptic ulcer.

Pipenzolate (**4.1.33**) is a muscarinic receptor antagonist with anticholinergic and antispasmodic action. Used for the treatment of gastrointestinal spasms, flatulent dispepsis, and other conditions (Fig. 10.17).

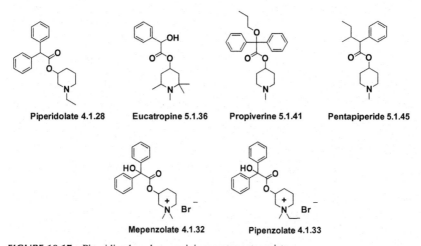

Piperidolate 4.1.28   Eucatropine 5.1.36   Propiverine 5.1.41   Pentapiperide 5.1.45

Mepenzolate 4.1.32   Pipenzolate 4.1.33

FIGURE 10.17   Piperidine-based muscarinic receptor antagonists.

Trihexyphenidyl (**2.1.1**) is a centrally acting muscarinic antagonist used to treat symptoms of Parkinson's disease or involuntary movements due to the side effects of certain psychiatric drugs (antipsychotics such as chlor-promazine/haloperidol) and as an antispasmodic.

Biperiden (**2.1.2**) is another centrally active anticholinergic drug. It is a muscarinic antagonist that has effects in both the central and peripheral nervous systems. It has been used in the treatment of parkinsonism, especially of the postencephalitic, arteriosclerotic, and idiopathic types and side effects (e.g., involuntary movements) caused by certain medicines. It is on the WHO Model List of Essential Drugs.

Cycrimine (**2.1.4**) (Pagitane) is another central anticholinergic. It is an antimuscarinic drug used to reduce levels of acetylcholine to return a balance with dopamine in the treatment and management of Parkinson's disease.

Diphenidol (**2.2.1**) is a muscarinic antagonist that exerts an anticho-linergic effect due to interactions with M1, M2, M3, and M4. It is employed as an antiemetic and antivertigo agent (Fig. 10.18).

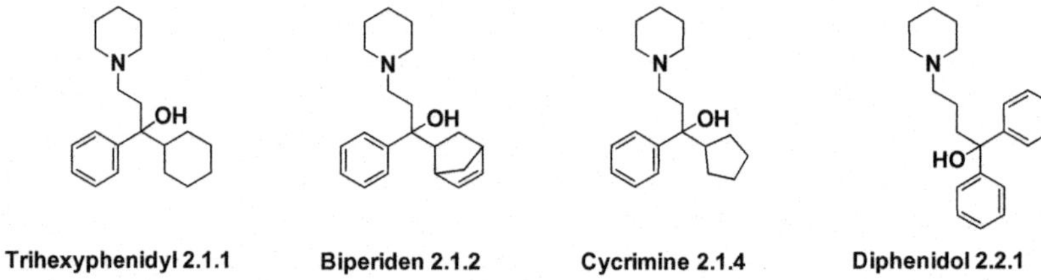

Trihexyphenidyl 2.1.1   Biperiden 2.1.2   Cycrimine 2.1.4   Diphenidol 2.2.1

FIGURE 10.18   Piperidine-based muscarinic receptor antagonists.

Flavoxate (**2.3.30**) is an anticholinergic with antimuscarinic action. It has direct muscle relaxant activity by counteracting smooth muscle spasms of the urinary tract and can be used to treat bladder spasms, for symptomatic relief of dysuria, urgency, nocturia, suprapubic pain, frequency and incontinence.

Pipoxolan (**1.4.2**) is an antimuscarinic medication used as an adjunct in the symptomatic treatment of musculoskeletal conditions associated with painful muscle spasm indicated as adjunct therapy in gastrointestinal disorders characterized by smooth muscle spasm such as: gastrointestinal colic, biliary colic, gastrointestinal spasm and hypermotility, spastic constipation, delayed relaxation of lower uterine segment, and dysmenorrhea.

Methixene (**4.1.3**) is an antimuscarinic that has antihistamine properties. It is used as an antispasmodic in the treatment of functional bowel hypermotility and spasm, as an antiparkinsonian agent and for drug-induced extrapyramidal disorders in the treatment of functional bowel hypermotility and spasm.

Fenpiverinium bromide (**2.6.1**) is an antimuscarinic agent, antispasmodic medication used for the treatment of pain and spasms of smooth muscles and other conditions.

Timepidium (**4.1.15**) is a muscarinic receptor antagonist. It is an anticholinergic agent with antispasmodic effect used for the symptomatic treatment of visceral spasms (Fig. 10.19).

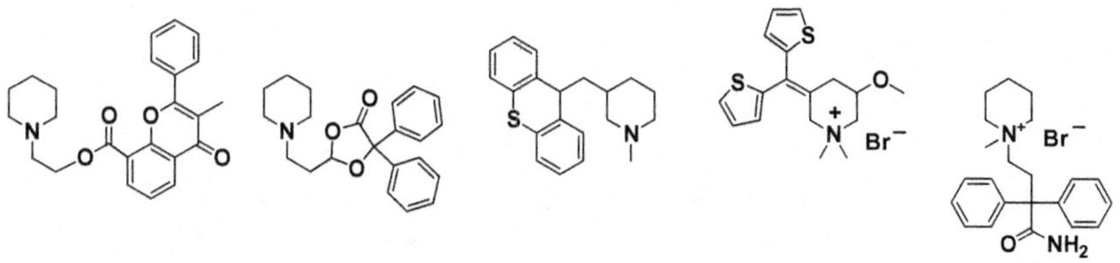

Flavoxate 2.3.30   Pipoxolan 2.4.2   Methixene 4.1.3   Timepidium 4.1.15   Fenpiverinium bromide 2.6.1

FIGURE 10.19 Piperidine-based muscarinic receptor antagonists.

10.5 PIPERIDINE-BASED LOCAL ANESTHETICS

Local anesthetics are drugs that cause reversible local anesthesia and a loss of nociception. When they are used on specific nerve pathways, effects such as analgesia (loss of pain sensation) and paralysis (loss of muscle power) can be achieved.

Local anesthetics are medications used for the purpose of temporary and reversible elimination of painful feelings in specific areas of the body by blocking nerve transmission to cause local loss of feeling and preventing muscle activity in the process without affecting consciousness.

The first synthetic local anesthetic drug, procaine, appeared in clinical practice in 1905. Later on there were thousands of compounds with analogous properties. However, only about 10−12 of them are used in practice, and anesthesia is still waiting an ideal agent. In 1947, lidocaine was introduced followed by bupivacaine in 1963.

From the chemical point of view, general anesthetics can be classified as anilides of *N,N*-dialkyl substituted α-aminoacids (aminoamide-type local anesthetics) or as esters of *p*-aminobenzoic acid and dialkylaminoalkanols (amino ester- type local anesthetics).

Aminoamide-Type Local Anesthetics

Mepivacaine (**3.1.31**) is an efficacious and useful intermediate-acting local anesthetic related to lidocaine. It has rapid-onset of action and indicated for infiltration, nerve block, and epidural anesthesia in dentistry.

Ropivacaine (**3.1.37**) is another long-acting aminoamide-type local anesthetic with both anesthetic and analgesic effects devoid of neuro and cardio toxicity and with rapid-onset of sensory block and prolonged postoperative analgesia. Ropivacaine is a drug that is injected before and during various surgical procedures or during labor and delivery causing loss of feeling and pain relief during surgical procedures, labor, delivery, or caesarean section.

Bupivacaine (**3.1.41**) is a widely used and long-acting local anesthetic prescribed for local or regional anesthesia or analgesia and indicated for the production of local or regional anesthesia or analgesia for surgery, dental and oral surgery procedures, diagnostic and therapeutic procedures, and for obstetrical procedures. It blocks initiation and transmission of nerve impulses at the site of application by stabilizing the neuronal membrane.

Levobupivacaine (**3.1.48**), the pure *S* (−)-enantiomer of bupivacaine, emerged as a safer alternative for regional anesthesia than its racemic parent with a clinical profile closely resembling that of bupivacaine. Levobupivacaine and bupivacaine produce comparable surgical sensory block with similar adverse side effects, and equal labor pain control with comparable maternal and fetal outcomes. The advantage is less motor block with levobupivacaine. Probably for pharmacoeconomic considerations levobupivacaine has not entirely replaced bupivacaine in clinical practice (Fig. 10.20).

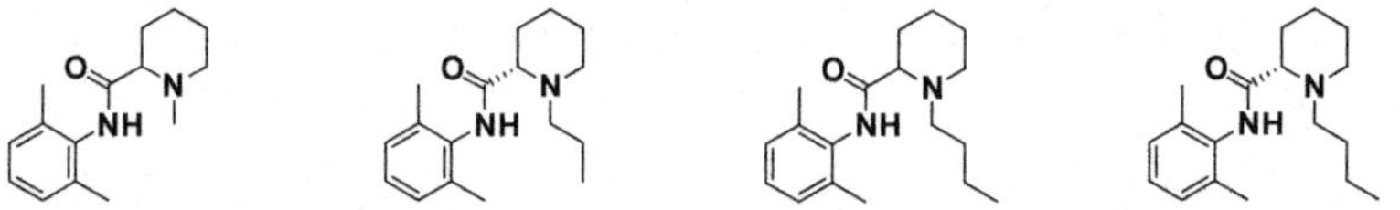

Mepivacaine 3.1.31   Ropivacaine 3.1.37   Bupivacaine 3.1.41   Levobupivacaine 3.1.48

FIGURE 10.20   Piperidine-based aminoamide-type local anesthetics.

Amino Ester–Type Local Anesthetics

Proparacaine (**2.3.31**) is a topical anesthetic drug of the amino ester group used as a local or spinal anesthetic for surface anesthesia, infiltration anesthesia, or regional nerve block.

Diperodon (**2.4.1**) is a local anesthetic formally belonging to the amino ester group with analgesic, antipyretic, antiinflammatory properties used to treat skin abrasions, irritations, and pruritus and used intrarectally for relief of pain from hemorrhoids, soreness of the rectal area, inflammation of the rectum, and anal itching (Fig. 10.21).

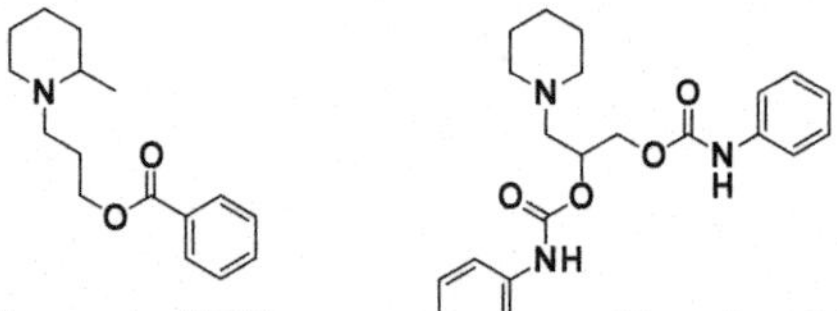

Piperocaine 2.3.31   Diperodone 2.4.1

FIGURE 10.21   Piperidine-based amino ester–type local anesthetics.

Dyclonine (**2.5.1**) is a β-amino-ketone, representing a unique local anesthetic agent structurally distinct from the two major classes of local anesthetic agents (aminoesters and aminoamides). This anesthetic very effectively produces topical airway anesthesia for laryngoscopy and awake endotracheal intubation. Dyclonine is also indicated for temporary relief following occasional mouth and throat symptoms including minor irritation, pain, sore mouth, and sore throat (Fig. 10.22).

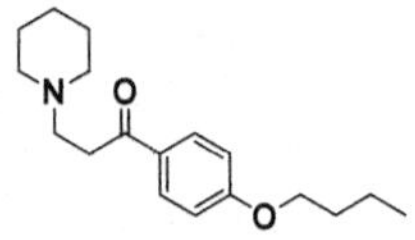

Dyclonine 2.5.1

FIGURE 10.22   Piperidine-based β-amino-ketone−type local anesthetics.

10.6   PIPERIDINE-BASED ANTITHROMBOTIC DRUGS

Thrombosis is the process of formation of thrombus in circulation from constituents of flowing blood. At the same time hemostatic plugs are the blood clots formed in healthy individuals at the site of bleeding stopping the escape of blood and plasma, whereas thrombi developing in the unruptured blood vessels may be harmful. Blood clots can develop in the arterial or venous circulation. By knowing the nature of thrombus, effective therapy could be chosen, i.e., arterial thrombus should be treated by antiplatelet agents and venous thrombosis should be treated by anticoagulants mainly. These drugs prevent thrombus extension, recurrence, and embolic complications, but they do not dissolve the already-formed clot. For the lysis of already formed thrombus fibrinolytic drugs are used.

Antiplatelet Drugs

Antiplatelet drugs decrease the ability of blood clots to form and are used in the prevention of thrombotic cerebrovascular or cardiovascular disease.

Series of pideridine-based antiplatelet drugs used to prevent blood clot formation, consequently thrombosis, include ticlopidine (**8.1.1**), clopidogrel (**8.1.2**), and prasugrel (**8.1.3**).

Ticlopidine (**8.1.1**) (Ticlid) is an effective inhibitor of platelet aggregation. It is used for preventing strokes in patients who have a history of stroke or transient ischemic attacks. It also used for preventing blood clots in stents placed in the arteries of the heart. Ticlopidine is a prodrug that is metabolized to an active form. It can cause life-threatening hematological adverse reactions, including neutropenia/agranulocytosis, thrombotic thrombocytopenic purpura, and aplastic anemia.

Clopidogrel (**8.1.2**) (Plavix) is another platelet aggregation inhibitor structurally and pharmacologically similar to ticlopidine. It is used to inhibit blood clots in a variety of conditions such as peripheral vascular disease, coronary artery disease, and cerebrovascular disease, and used to prevent heart attack or stroke.

Prasugrel (**8.1.3**) (Effient) is a third-generation thienopyridine and intended for use by patients with heart disease (recent heart attack, unstable angina) or those who undergo a certain heart procedures to prevent other serious vessel problems. Prasugrel may cause serious or life-threatening bleeding (Fig. 10.23).

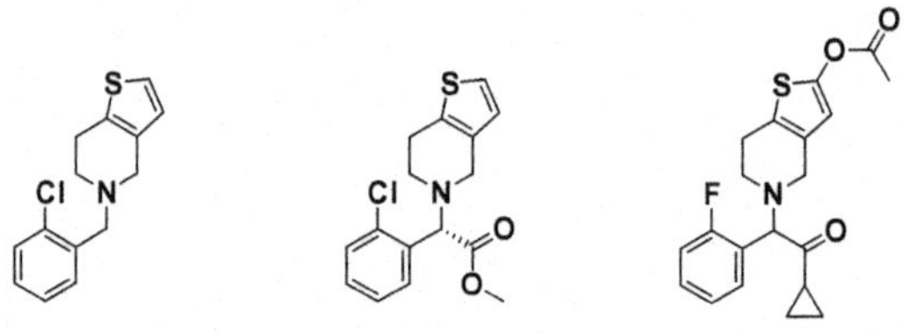

Ticlopidine 8.1.1   Clopidogrel 8.1.2   Prasugrel 8.1.3

FIGURE 10.23   Piperidine-based antiplatelet drugs.

Glycoprotein IIb/IIIa inhibitors are concidered another class of antiplatelet agents.

Lamifiban (**5.1.31**), an intravenously administered, selective, reversible, nonpeptide glycoprotein IIb/IIIa receptor antagonist, is a member of the aforementioned class of compounds that inhibit platelet aggregation and thrombus formation by preventing the binding of fibrinogen to platelets. A large phase III trial showed that the incidence of clinical events was significantly lower after lamifiban (with or without heparin) and aspirin therapy than after standard heparin and aspirin therapy. The most common adverse events associated with lamifiban were bleeding complications, which were increased by the concomitant administration of heparin (Fig. 10.24).

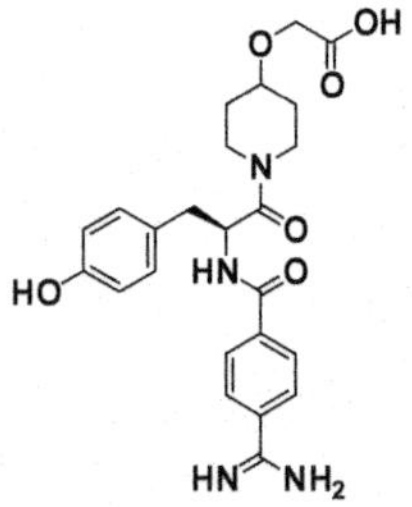

Lamifiban 5.1.31

FIGURE 10.24   Lamifiban.

Anticoagulant Drugs

Argatroban (**3.1.111**) is an anticoagulant, a type of "blood thinner," and is a highly selective direct thrombin inhibitor intended to both normalize platelet count in patients with heparin-induced thrombocytopenia (low platelet levels) and prevent the formation of thrombi. Common side effects of include nausea, vomiting, diarrhea, stomach pain, fever, headache, and back pain (Fig. 10.25).

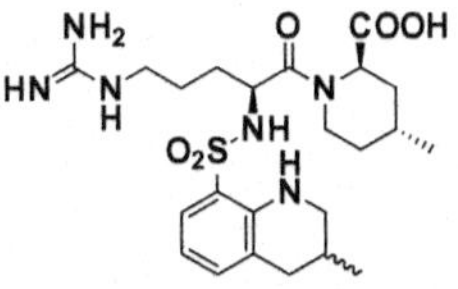

Argatroban 3.1.111

FIGURE 10.25   Argatroban.

10.7   PIPERIDINE-BASED ANTIARRHYTHMIC DRUGS

Arrhythmias, irregular heartbeating, are classified as tachycardia: if the heartbeating is too fast, bradycardia; if it is too slow, fibrillation; if it is too irregular, premature contraction, if it is too early.

There are hundreds of different types of cardiac arrhythmias. The normal rhythm of the heart (sinus rhythm), can be disturbed through failure of automaticity, or through overactivity, or premature atrial or ventricular contractions or fibrillation.

Antiarrhythmic drugs help return the heart to its normal sinus rhythm and rate and maintain it after it has been achieved, stabilize the heart muscle tissue.

According to existing classification antiarrhythmic drugs can be subdivided into five main classes.

The first class represents sodium channel blockers that in turn are subdivided to three, based on specific kinetics characteristics upon their effect on the length of the action potential duration (association/dissociation) within the group.

Two piperidine-derivative drugs represented in the series of compounds are included into Class Ic antiarrhythmics (sodium channel blockers with slow association/dissociation kinetics). These compounds are flecainide (**3.1.54**) and encainide (**3.1.62**).

Flecainide (**3.1.54**) is an antiarrhythmic drug used to treat certain types of serious irregular heartbeat such as persistent ventricular tachycardia and paroxysmal supraventricular tachycardia. It is also used to prevent certain types of irregular heartbeat from returning (such as atrial fibrillation). Serious side effect are: slow heart rate, weak pulse, fainting, slow breathing, dizziness, fainting, and fast or pounding heartbeat.

Encainide (**3.1.62**) was used as an antiarrhythmic drug in cases of ventricular arrhythmia and ventricular tachycardia but is no longer used because of its frequent proarrhythmic side effects.

Lorcainide (**5.7.53**) is another Class Ic antiarrhythmic agent that has been used for managing the prolonged duration of action, and with a desirable pharmacokinetic it is reported to be highly efficient for restoring normal heart rhythm in patients with premature ventricular contractions, ventricular tachycardia, and Wolff-Parkinson-White syndrome. However, an intriguing lorcainide trial story concerning a controlled trial of the antiarrhythmic drug showed that the drug is likely to be lethal. Nevertheless, it continues to be used in practice and in further clinical trials.

Pirmenol (**2.2.2**) is a new long-acting investigational Class Ia antiarrhythmic drug. The therapeutic response, lack of toxicity, and relatively long half-life indicate that pirmenol is a promising antiarrhythmic agent (Fig. 10.26).

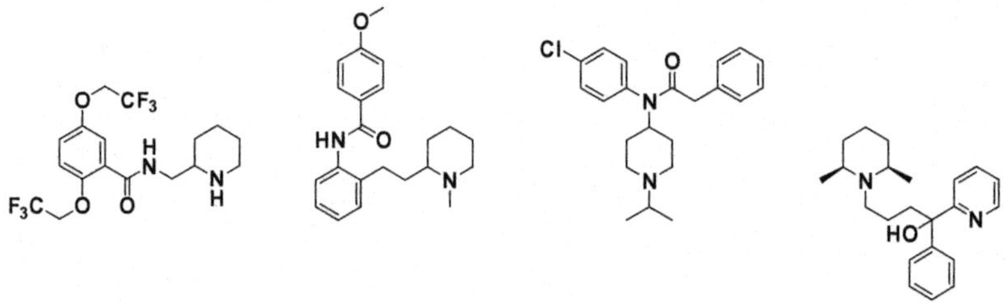

Flecainide 3.1.54   Encainide 3.1.62   Lorcainide 5.7.53   Pirmenol 2.2.2

FIGURE 10.26   Piperidine-based antiarrhythmic drugs.

10.8   PIPERIDINE-BASED ANTIHYPERTENSIVE DRUGS

Antihypertensive drugs are medications that help lower blood pressure. There are a variety of classes of antihypertensive medications and they include a number of different classes of drugs such as diuretics, which help the body get rid of excess sodium ions and water thereby control blood pressure:

- Beta-blockers, which reduce the heart rate, the heart's workload, and the heart's output of blood, which lowers blood pressure;
- Calcium channel blockers, which prevent calcium from entering the smooth muscle cells of the heart and arteries relax and open up narrowed blood vessels, reduces the heart rate and lowers blood pressure;
- Angiotensin-converting enzyme inhibitors, which help the body produce less angiotensin, which helps the blood vessels relax and open up, which, in turn, lowers blood pressure;
- Angiotensin II receptor blockers, which block the receptors so the angiotensin fails to constrict the blood vessels with the result that they stay open and blood pressure is reduced;
- Alpha blockers, which reduce the arteries' resistance, relaxing the muscle tone of the vascular walls;
- Vasodilators, which cause the muscle in the walls of the blood vessels to relax, allowing the vessel to dilate, which leads to better blood flow.

Four piperidine derivatives are representatives of different classes of antihypertensive drugs:

- Benidipine (**4.1.39**) (Coniel) is a long-acting orally active antihypertensive agent with several unique properties. It is triple L-, T-, and N- type calcium channel blocker for the treatment of hypertension. The vasorelaxant effect of benidipine is attributable to high affinity dihydropyridine binding sites (i.e., the binding site in Ca^{2+} channels).
- Ketanserin (**5.4.10**) (Sufrexal) is classified as a potent, orally very effective antihypertensive drug. It is a 5-HT2 receptor antagonist, which also possesses weak α1-adrenoceptor antagonistic activity, which may contribute to the acute blood pressure lowering effects explaining its prolonged antihypertensive mechanism of action in patients with essential hypertension. Ketanserin has a paradoxical effect in wound healing. It stimulates wound closure but inhibits hypertrophic scar formation.
- Endralazine (**8.1.52**), a novel direct-acting peripheral vasodilator, is chemically and pharmacologically related to hydralazine and is used as an antihypertensive drug. The side effects reported to occur during oral endralazine therapy are not essentially different from those of hydralazine.
- Indoramin (**5.6.4**) (Doralese) is a postsynaptic adrenergic antagonist (α1-selective blocker) and at the same time an antagonist of histamine H1 and 5-HT receptors with no reflex tachycardia and direct myocardial depression action. It is a discontinued piperidine antiadrenergic drug used to treat hypertension, benign prostatic hypertrophy, Raynaud's phenomenon (Fig. 10.27).

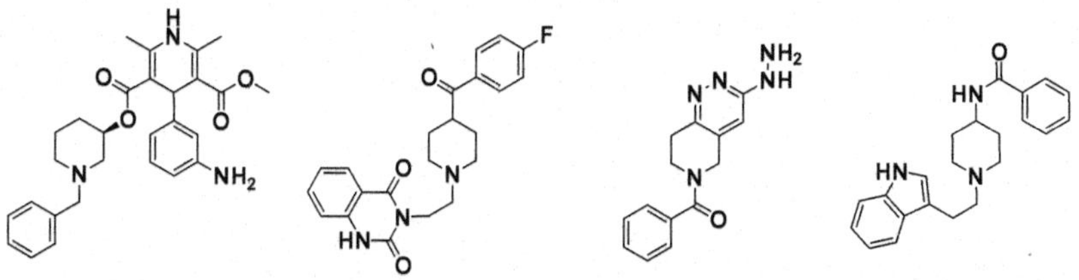

Benidipine 4.1.39   Ketanserin 5.4.10   Endralazine 8.1.52   Indoramin 5.6.4

FIGURE 10.27 Piperidine-based antihypertensive drugs.

10.9 PIPERIDINE-BASED DRUGS FOR TREATING RESPIRATORY SYSTEM DISEASES

Cough Suppressants

Cloperastine (**2.3.1**) is a cough suppressant that has been shown to possess dual activity. It is a drug used in the treatment of acute and chronic cough having a dual mechanism of action at the central bulbar cough center and at peripheral receptors in the tracheobronchial tree showing also mild bronchorelaxant and antihistaminic properties.

Benproperine (Cofrel) (**2.3.2**) is a racemate comprising equimolar amounts of (*R*)- and (*S*)-benproperine. The use of either enantiomer does not show any advantage over the racemate with regard to their antitussive effect. It is used as a cough suppressant in the treatment of acute and chronic bronchitis, cough due to varieties of causes.

Pipazethate (Pipazetate, Theratuss, Lenopect) (**2.3.3**) is a centrally acting antitussive drug that also has bronchodilatory and local anesthetic activities and has been introduced as a cough remedy. It suppresses irritative and spasmodic cough by inhibiting the exitability of the cough center and of the peripheral neural receptors in the respiratory passage. It has a bronchodilator effect, which reduces the increased resistance to expiration during paroxysms of cough. It is devoid of addictive liability and does not depress respiration (Fig. 10.28).

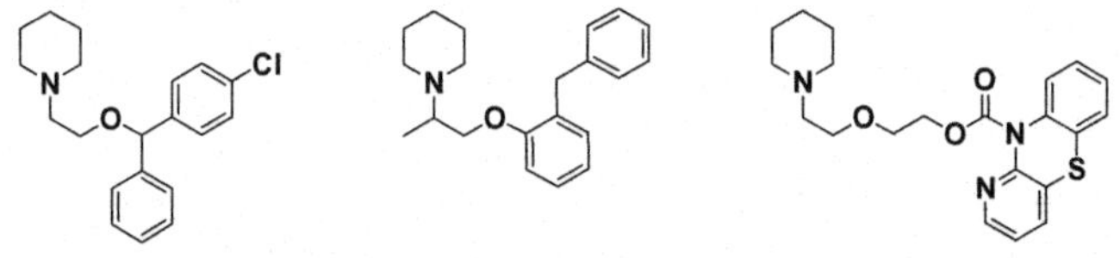

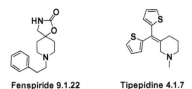

Cloperastine 2.3.1   Benproperine 2.3.2   Pipazetate 2.3.3

FIGURE 10.28   Piperidine-based cough suppressants.

Fenspiride (**9.1.22**) (Eurespal) is another antitussive, bronchodilator, a non corticosteroid antiinflammatory drug, used as a part of complex therapy of bronchial asthma, in the treatment of certain respiratory diseases like rhino-pharyngitis, laryngitis, tracheobronchitis, otitis and sinusitis, whooping cough, influenza, as well as for maintenance treatment of asthma (Fig. 10.29).

Fenspiride 9.1.22   Tipepidine 4.1.7

FIGURE 10.29   Piperidine-based cough suppressants.

Tipepidine (**4.1.7**) has been widely used solely as an antitussive and expectorant in Japan since 1959 in the management of cough. Possible side effects, especially in overdose, may include drowsiness, vertigo, delirium, disorientation, loss of consciousness (Fig. 10.29).

10.10   PIPERIDINE-BASED ANTIDEPRESSANTS

Antidepressants are drugs used to prevent or treat depression. The available antidepressant drugs include the selective serotonin reuptake inhibitors (SSRIs), norepinephrine-dopamine reuptake inhibitors (NDRIs), monoamine oxidase inhibitors (MAOIs), tricyclic antidepressant, tetracyclic antidepressants, and others. The overall clinical effect is increased mood and decreased anxiety.

Selective Serotonin Reuptake Inhibitors

Paroxetine (**4.1.49**) (Paxil) belongs to a class of SSRIs. As with other antidepressant agents, several weeks of therapy may be required before a clinical effect is seen. Paroxetine may be used to treat major depressive disorder, panic disorder with or without agoraphobia, obsessive-compulsive disorder, social anxiety disorder, generalized anxiety disorder, posttraumatic stress disorder and premenstrual dysphoric disorder. Side effects of weight gain, sexual dysfunction, sedation, and constipation are mentioned.

Femoxetine (**4.1.57**) is another SSRI and possesses a similar pharmacological activity as paroxetine. The overall clinical effect of femoxetine is increased mood and decreased anxiety. Side effects include dry mouth, nausea, dizziness, drowsiness, sexual dysfunction, and headache. Fluoxetine is the most anorexic-stimulating SSRI (Fig. 10.30).

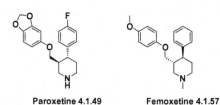

Paroxetine 4.1.49   **Femoxetine 4.1.57**

FIGURE 10.30   Piperidine-based selective serotonin reuptake inhibitors.

Norepinephrine-Dopamine Reuptake Inhibitors

Pipradrol (**3.1.17**) is an antidepressant in a group of NDRIs. Unlike most antidepressants, NDRIs are generally considered stimulants and have a unique set of side effects such as loss of appetite, weight loss, headache, dry mouth, skin rash, sweating, ringing in the ears, shakiness and nervousness, fast heartbeat, nore frequent urination, etc. Once used as a treatment for obesity and dementia, pipradrol is now regulated as a Schedule IV compound in the United States and is no longer widely used in most countries (Fig. 10.31).

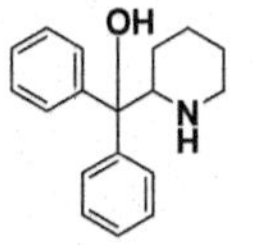

Pipradrol 3.1.17

FIGURE 10.31   Piperidine-based norepinephrine-dopamine reuptake inhibitors.

10.11 PIPERIDINE-BASED ANTIPARKINSONIAN DRUGS

Trihexyphenidyl (**2.1.1**) is classified as an antiparkinsonian agent. It is a centrally acting muscarinic antagonist used as an antispasmodic for treatment of parkinsonian disorders or drug-induced extrapyramidal movement disorders. It can cause side effects such as an allergic reaction, fever, fast or irregular heartbeats, anxiety, hallucinations, confusion, agitation, hyperactivity, or loss of consciousness, seizures, rash.

Pridinol (Nonplesin) (**2.1.3**) is another antiparkinsonian drug. The drug is also used to produce muscle relaxation in the treatment of muscle spasm and immobility associated with strains, sprains, and injuries of the back, and the muscle spasms that can occur in multiple sclerosis (Fig. 10.32).

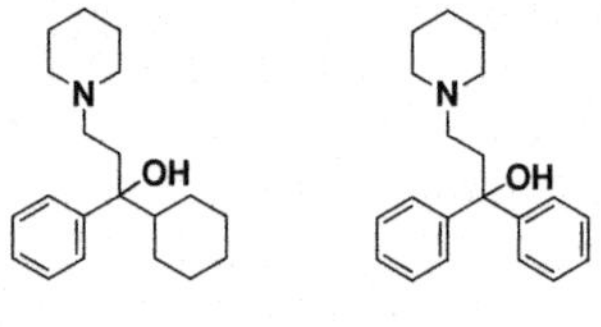

Trihexyphenidyl 2.1.1   **Pridinol 2.1.3**

FIGURE 10.32   Piperidine-based antiparkinsonian drugs.

10.12 PIPERIDINE-BASED HYPOLIPIDEMIC AND ANTIHYPERLIPIDEMIC DRUGS

Hypolipidemic and antihyperlipidemic drugs are agents that decrease lipid concentrations in blood. Lomitapide (**5.6.32**) (Juxtapid) is a new, oral lipid-lowering agent for the treatment of familial hypercholesterolemia. It works inhibiting microsomal triglyceride transfer protein. It is the first in a new class of lipid-lowering agents to improve lipoproteins (total cholesterol, low-density lipoprotein, and nonhigh-density lipoprotein cholesterol and (apo B) in patients. Common side effects may include vomiting, indigestion, stomach pain, diarrhea, constipation, and weight loss (Fig. 10.33).

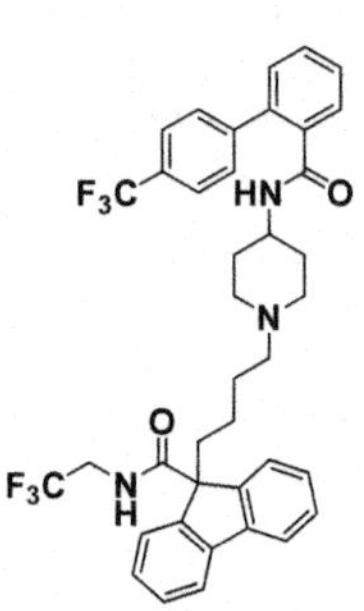

Lomitapide 5.6.32

FIGURE 10.33   Lomitapide.

10.13 PIPERIDINE-BASED ADRENERGIC (SYMPATHOMIMETIC) DRUGS

Adrenergic drugs stimulate the sympathetic (adrenergic) nervous system — the part of the autonomic nervous system that regulates involuntary reactions to stress. It stimulates the heartbeat, sweating, breathing rate, and other stress-related body processes. Sympathomimetics are catecholamines or analogs of catecholamines that can be divided into two classes: α-agonists and β-agonists. A third class of sympathomimetics that affects norepinephrine storage, release and uptake by sympathetic nerves are used as research tools.

Rimiterol (**3.1.82**) is a third-generation short-acting selective β2-adrenoreceptor agonist which rapidly produces effective bronchodilatation. It is used in the management of bronchial asthma. It relaxes the smooth muscle in the lungs and dilates or expands the bronchi in lungs and makes breathing easy as possible. Side effects such as fine tremor of skeletal muscle, palpitations, muscle cramps, tachycardia, nervous tension, headache, and peripheral vasodilatation have been observed (Fig. 10.34).

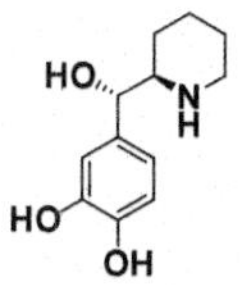

Rimiterol 3.1.82

FIGURE 10.34 Rimiterol.

10.14 PIPERIDINE-BASED CENTRAL NERVOUS SYSTEM STIMULANTS

Central nervous system stimulants speed up mental and physical processes and can be useful in the treatment of certain medical conditions such as narcolepsy and attention-deficit disorders. These medications have many unpleasant side effects and there is a high potential for addiction.

Methylphenidate (**3.1.6**) (Ritalin) is a commonly prescribed central nervous system stimulant. It is used most commonly in the treatment of attention-deficit disorders in children and for narcolepsy. Common side effects include: addiction, nervousness, anxiety and irritability, insomnia, decreased appetite, headache, nausea, dizziness (Fig. 10.35).

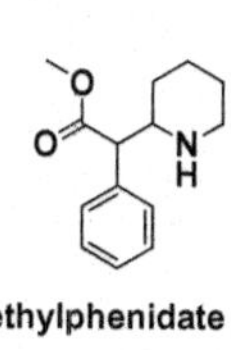

Methylphenidate 3.1.6

FIGURE 10.35 Methylphenidate.

10.15 PIPERIDINE-BASED SELECTIVE ESTROGEN RECEPTOR MODULATORS

Raloxifene (Evista) (**1.3.4**) is a second-generation selective estrogen receptor modulator (SERM) that functions as an estrogen antagonist on breast and uterine tissues, and as an estrogen agonist on bone. Raloxifene is used for treating and preventing osteoporosis and as a remedy to reduce the risk of invasive breast cancer. Common side effects of Evista include: hot flashes, increased sweating, headache, dizziness, spinning sensation, leg cramps or leg pain, joint pain, nausea, and vomiting (Fig. 10.36).

Raloxifene 2.3.4

FIGURE 10.36   Raloxifene.

10.16 PIPERIDINE-BASED ANTIANGINAL DRUGS

Antiangina drugs are medicines that relieve the symptoms of angina pectoris (chest pain due to an inadequate supply of oxygen to the heart muscle).

Perhexiline (**3.1.11**), an antianginal drug and a coronary vasodilator, reduces the frequency of moderate-to-severe attacks of angina pectoris due to coronary artery disease in patients who have not responded to other conventional therapy. Recently, calcium channel antagonist properties of this drug have been found.

Perhexiline therapy provides symptomatic relief in the majority of patients with minimal side effects on careful therapeutic level monitoring for dose titration. Transient side effects such as anorexia, weight loss, and lethargy, have been observed (Fig. 10.37).

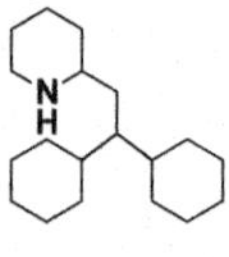

Perhexiline 3.1.11

FIGURE 10.37   Perhexiline.

10.17 PIPERIDINE-BASED DRUGS FOR TREATING PROTOZOAN INFECTIONS

Mefloquine (**3.1.27**) is a very potent blood schizontocide for both malaria prophylaxis and for acute treatment of falciparum malaria with very few side effects (Fig. 10.38).

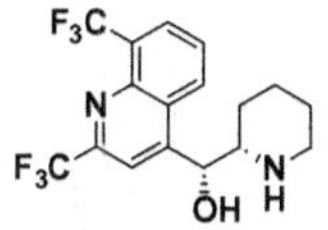

Mefloquine 3.1.27

FIGURE 10.38 Mefloquine.

10.18 PIPERIDINE-BASED ANTIMIGRAINE DRUGS

Antimigraine drugs are medicines used to prevent or reduce the severity of migraine headache. Migraines are different from other headaches because they occur with symptoms such as nausea, vomiting, or sensitivity to light.

Antimigraine drugs or drugs that prevent migraine or cure migraine include follows: triptans, β-blockers, anticonvulsants, methysergid, calcium channel blockers, antidepressants, clonidine (α-blocker), and pizotifen and its analogs, NSAIDs.

Naratriptan (**7.2.22**) (Amerge) is a selective 5-HT(1B/1D) receptor agonist that belongs to a family of tryptamine based drugs (triptans) that are indicated for the acute treatment of migraine headaches. Naratriptan is not intended for the prophylactic therapy of migraine. Common side effects may include: dizziness, drowsiness, feeling weak or tired, flushing, nausea (Fig. 10.39).

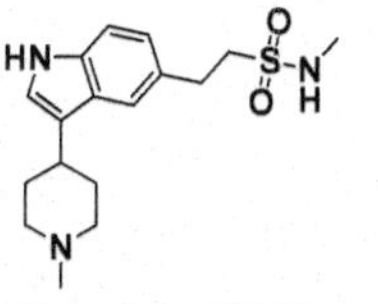

Naratriptan 7.2.22 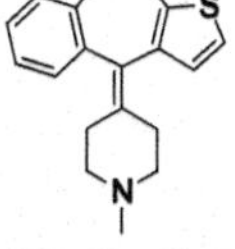   Pizotifen 6.1.7

FIGURE 10.39 Piperidine-based antimigraine drugs.

Pizotifen (**6.1.6**) (Sandomigran) is an antimigraine drug, acting principally as an antagonist of serotonin 5-HT2A and 5-HT2C receptors used in the preventive treatment of migraine and eating disorders. Possible side effects include dizziness, drowsiness, and weight gain (Fig. 10.39).

10.19 PIPERIDINE-BASED NOOTROPIC AND NEUROPROTECTIVE DRUGS

Sabeluzole (**5.8.22**) has shown positive effects on memory and has been proposed as a medication for the treatment of patients suffering from chronic neuro-degenerative diseases such as dementia of the Alzheimer type or Alzheimer's disease, amyotrophic lateral sclerosis, dementia associated with Parkinson's disease, and other central nervous system diseases that are characterized by progressive dementia. Except for cognitive-enhancing, it was also described to have antiischemic, antiepileptic properties.

Lubeluzole (**5.8.26**) a novel neuroprotective compound closely related to sabeluzole (Fig. 10.40).

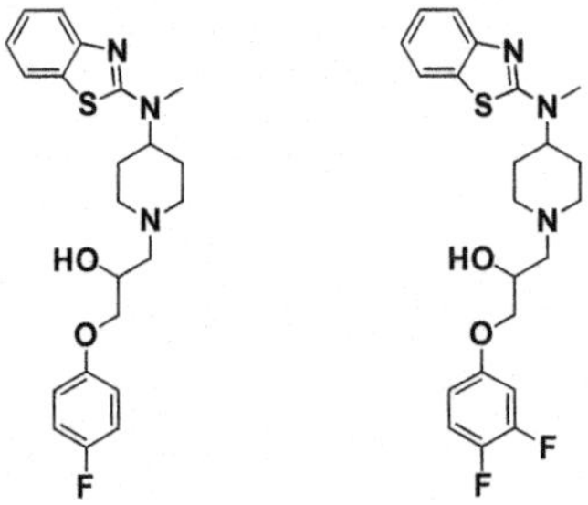

Sabeluzole 5.8.22   **Lubeluzole 5.8.26**

FIGURE 10.40   Piperidine-based nootropic and neuroprotective drugs.

10.20   PIPERIDINE-BASED ANTIEMETICS

Drugs that are useful in the suppression of nausea or vomiting.

The number of antiemetcs on the pharmaceutical market seems to be around 80. Four of them arpe piperidine derivatives.

Domperidone (**7.1.6**) (Motilium) is a peripherally selective D2-like receptor antagonist. It speeds gastrointestinal peristalsis, causes prolactin release, and is used as an antiemetic. Domperiodone provides relief of such symptoms as anorexia, nausea, vomiting, abdominal pain, early satiety, bloating, distension in patients with symptoms of diabetic gastropathy. It also provided short-term relief of symptoms in patients with dyspepsia or gastroesophageal reflux, and prevented nausea and vomiting. Possible side effects are headache, dizziness, dry mouth, nervousness, flushing, or irritability.

Metopimazine (**5.5.4**) (Vogalene) is a dopamine receptor antagonist that is used for the prevention and treatment of patients who consider nausea and vomiting as severe adverse events to chemotherapy.

Pipamazine (**5.5.6**) (Mornidine) was formerly used as another antiemetic, but was eventually withdrawn from the market in 1969, after reports of hepatotoxicity and hypotension.

Clebopride (**5.6.11**) (Cleboril) is another selective D2 receptors antagonist with antiemetic and prokinetic properties and is used as an antiemetic for chemotherapy-induced nausea and vomiting. Possible side effects may include restlessness, drowsiness, diarrhea, hypotension, hypertension, dizziness, headache, depression.

Cinitapride (**5.6.17**), a close structural analog of clebopride, has been marketed in Spain (Cidine) and Mexico (Pemix) as a prescribed antiemetic medication for vomiting (Fig. 10.41).

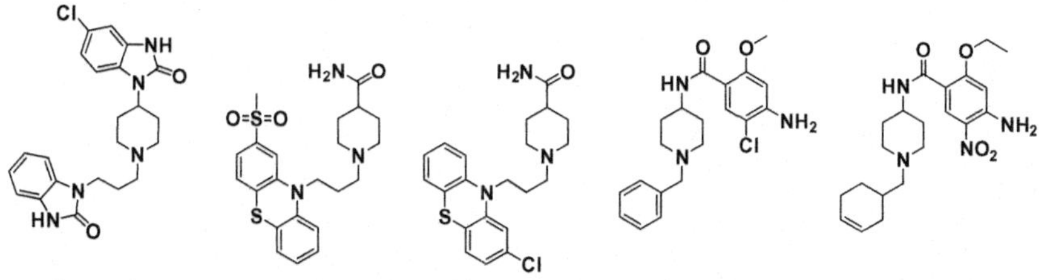

Domperidone	Metopimazine	Pipamazine	Clebopride	Cinitapride
7.1.6	5.5.4	5.5.6	5.6.11	5.6.17

FIGURE 10.41   Piperidine-based antiemetics.

10.21 PIPERIDINE-BASED ANTIDIARRHEAL AND PROKINETIC DRUGS

Antidiarrheals

Antidiarrheal drugs are fiber-forming agents that relieve the symptoms of diarrhea. The most effective drugs are opioid derivatives, which slow intestine motility to permit greater time for the absorbtion of water and electrolytes.

Loperamide (**5.2.35**) (Imodium) is a long-acting synthetic antidiarrheal; it is peripherally acting μ-opioid receptor agonist that is not absorbed from the gut and has no effect on the adrenergic system or central nervous system but may antagonize histamine and interfere with acetylcholine. This decreases the number of bowel movements and makes the stool less watery. Loperamide is also used to reduce the amount of discharge in patients who have undergone an ileostomy. It is also used to treat on-going diarrhea in people with inflammatory bowel disease. Loperamide treats only the symptoms, not the cause of the diarrhea (e.g., infection). Common side effects include dizziness, drowsiness, vomiting, and constipation.

Diphenoxylate (**5.3.35**) (Diocalm) is a close analog of loperamide and is commonly used as effective adjunctive therapy in the management of diarrhea in numerous settings of inflammatory bowel disease.

Zaldaride (**7.1.5**) is a novel, potent, and selective inhibitor of calmodulin that plays an important role in the regulation of wide variety of cellular processes and decreases the severity and duration of traveler's diarrhea. Mechanism of action is not clear. Zaldaride has been shown to be an effective antidiarrheal and decreases the duration of diarrhea (Fig. 10.42).

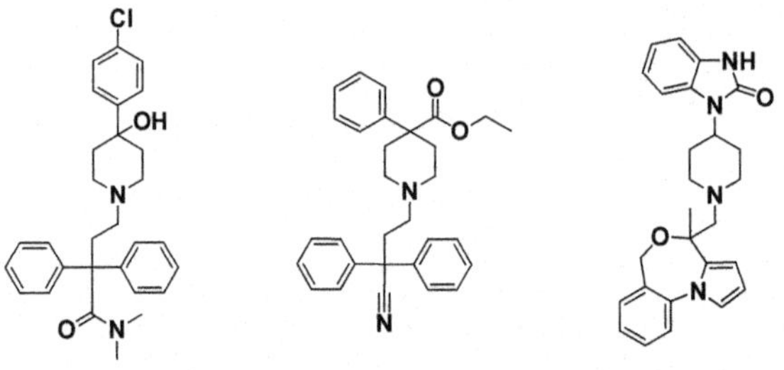

Loperamide 5.2.35   **Diphenoxylate 5.3.35**   **Zaldaride 7.1.5**

FIGURE 10.42 Piperidine-based antidiarrheals.

Prokinetics

Prokinetic (promotility) drugs formed the mainstay of treatment for dysmotilities (chronic intestinal pseudo-obstruction, slow-transit constipation, and gastroparesis).

Cisapride (**5.6.45**) (Propulsid), an oral prokinetic agent belonging to the pharmacotherapeutic group of propulsives that acts as a serotonin 5-HT4 agonist, was the most prescribed promotility agent used for gastroesophageal reflux disease, chronic intestinal pseudo-obstruction, slow-transit constipation and gastroparesis. It also provides for symptomatic relief of nocturnal heartburn due to gastroesophageal reflux disease in adults. It increases lower esophageal sphincter tone, accelerates gastric emptying, and increases small-bowel motility. In many countries cisapride has been withdrawn due to reports about long QT syndrome generated by cisapride, which predisposes to arrhythmias.

Prucalopride (**5.6.23**) (Resolor) is a drug acting as a selective, high affinity 5-HT4 receptor agonist that targets the impaired motility associated with chronic constipation, thus normalizing bowel movements. The drug has also been tested for the treatment of chronic intestinal pseudo-obstruction (Fig. 10.43).

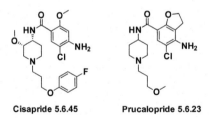

Cisapride 5.6.45   **Prucalopride 5.6.23**

FIGURE 10.43 Piperidine-based prokinetics.

10.22 PIPERIDINE-BASED GASTRIC ANTISECRETORY DRUGS

These drugs reduces the normal rate of acid secretion into the stomach. Such drugs include antimuscarinic drugs, H 2-receptor antagonists, and proton-pump inhibitors and the cytoprotective agents (antacids, sucralfate) that are used in the treatment of gastrointestinal disorders like gastritis and are considered gastric antisecretory drugs.

Troxipide (**4.1.60**) (Aplace) is a gastro protective agent with antiulcer, antiinflammatory and mucus secreting properties, which neither inhibits acid secretion nor has an acid-neutralizing activity but has been clinically proven to heal gastritis and gastric ulcers. Troxipide has shown to inhibit neutrophil mediated inflammation and oxidative stress in addition to improving the gastric mucus composition and output. Common side effects are headache and constipation (Fig. 10.44).

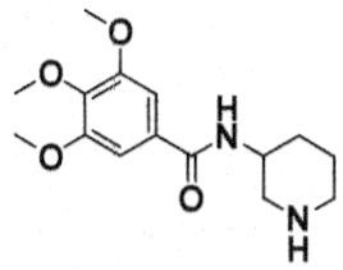

Troxipide 4.1.60

FIGURE 10.44   Troxipide.

10.23   PIPERIDINE-BASED HYPOGLYCEMIC DRUGS

Hypoglycemic or antihyperglycemic drugs lower glucose levels in the blood. They are commonly used in the treatment of diabetes mellitus and can be classified as sulfonylureas, biguanides, thiazolidinediones, a-glucosidase inhibitors, and glitins.

Linagliptin (**4.1.70**) (Trajenta), a representative of gliptins, is an oral, highly selective inhibitor of dipeptidyl peptidase-4 (DPP-4) and can be used to treat diabetes mellitus type 2. It is the first agent of its class to be eliminated predominantly via a nonrenal route (Fig. 10.45).

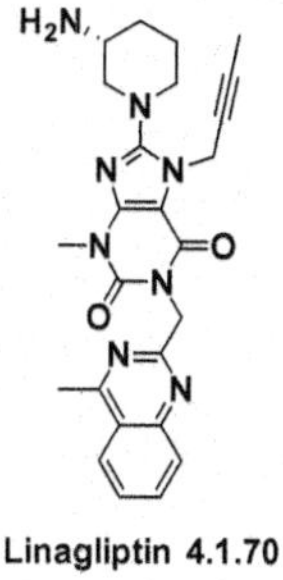

Linagliptin 4.1.70

FIGURE 10.45   Linagliptin.

10.24 PIPERIDINE-BASED DRUGS USED IN THE TREATMENT OF RHEUMATOID ARTHRITIS

Different drugs used in the treatment of rheumatoid arthritis. Some are used primarily to ease the symptoms of disease, others are used to slow or stop the course of the disease and to inhibit structural damage.

Tofacitinib (**4.1.83**) (Xeljanz) is an oral Janus kinase inhibitor used for the treatment of moderate-to-severe rheumatoid arthritis.

Besides rheumatoid arthritis, tofacitinib has also been studied in clinical trials for the prevention of organ transplant rejection, and is currently under investigation for the treatment of psoriasis. Known adverse effects include nausea and headache as well as more serious immunologic and hematological adverse effects (Fig. 10.46).

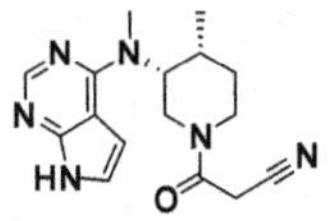

Tofacitinib 4.1.83

FIGURE 10.46   Tofacitinib.

10.25 PIPERIDINE-BASED NICOTINIC CHOLINOMIMETICS

These compounds mimic the effects of acetylcholine at nicotinic receptors. Lobeline (**3.1.90**) acts on nicotinic cholinergic receptors and has been proposed for a variety of therapeutic uses including respiratory disorders, peripheral vascular disorders, insomnia, and smoking cessation. It has been proposed for treatment of other drug dependencies; however, there is only modest evidence for efficacy.

Lobeline inhibits also the function of vesicular monoamine and dopamine transporters and diminishes the behavioral effects of nicotine and amphetamines. Lobeline binds to μ-opiate receptors, blocking the effects of opiate receptor agonists (Fig. 10.47).

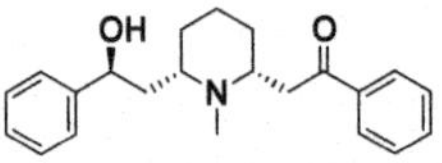

Lobeline 3.1.90

FIGURE 10.47   Lobeline.

10.26 PIPERIDINE-BASED IMMUNOSUPPRESANT DRUGS

Immunosuppressant drugs are a class of drugs that suppress, or reduce, the strength of the body's immune system.

The series of ascomycin (**3.1.112**) derivatives – pimecrolimus (**3.1.113**), tacrolimus (**3.1.114**) as well as sirolimus (**3.1.115**) and its derivatives everolimus (**3.1.116**) and temsirolimus (**3.1.117**) formally – can be considered 1,2-disubstituted piperidines and are immunosuppressive drugs whose main use is after organ transplant to reduce the activity of the patient's immune system and so the risk of organ rejection. It is also used in a topical preparation in the treatment of severe atopic dermatitis, severe refractory uveitis after bone marrow transplants, and the skin condition vitiligo (Fig. 10.48).

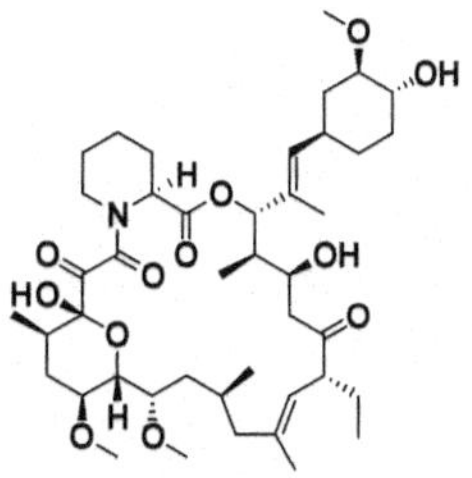

Ascomycin 3.1.112 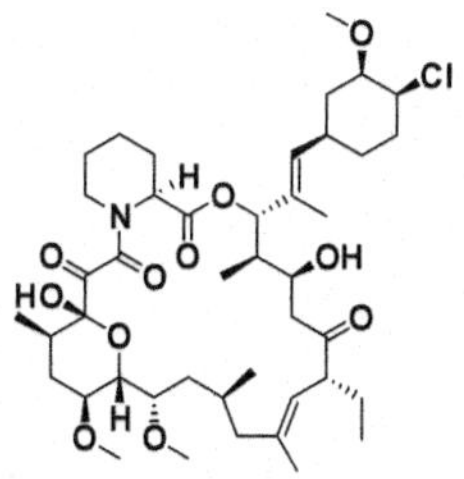   Pimecrolimus 3.1.113 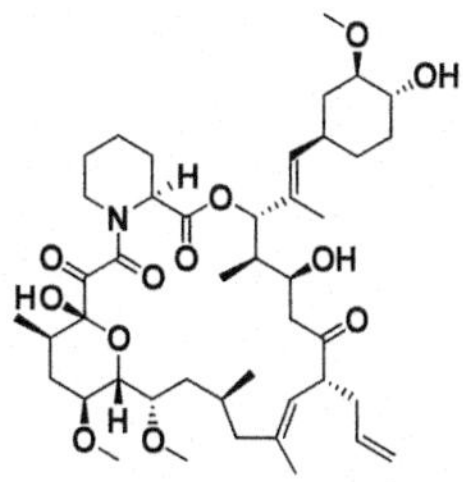   Tacrolimus 3.1.114

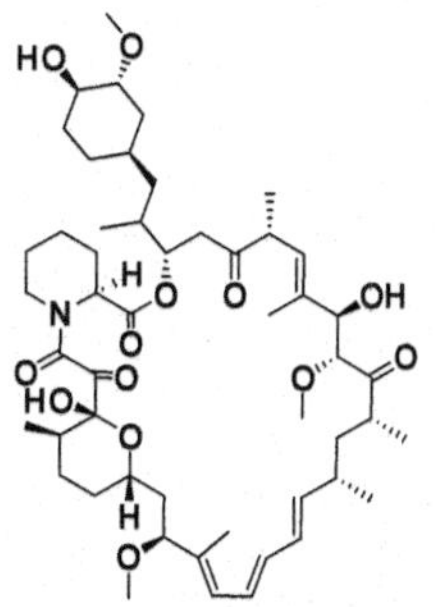

Sirolimus 3.1.115 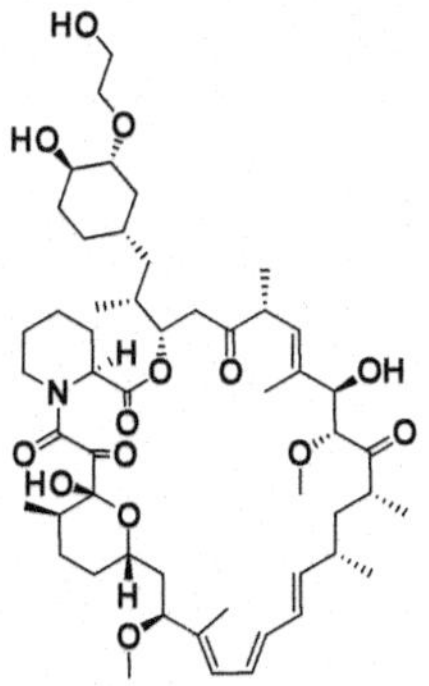   Everolimus 3.1.116 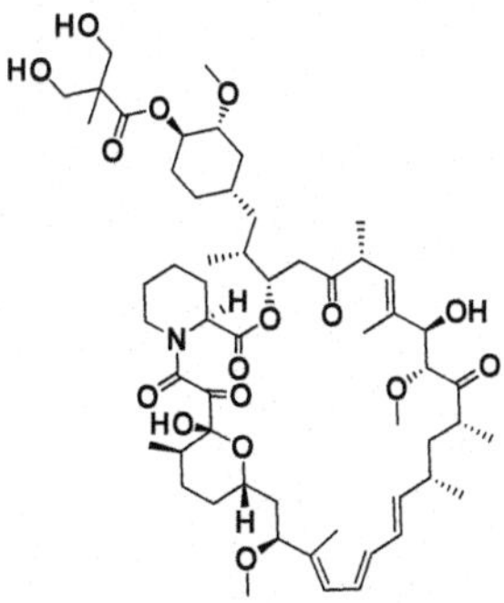   Temsirolimus 3.1.117

FIGURE 10.48   Piperidine-based immunosuppressants.

10.27 CONCLUSION

Careful contemplation of the structures of drugs existing on the pharmaceutical market makes more or less evident that the "language" by means of which drugs are transferring information to receptors could be based on an "alphabet" in which interval defined for each "letter," its average "width" is very close to that determined by distance between two functionalities in drugs chemically represented as derivatives of β- or γ-aminoesters, β- or γ-diamines, aryl- or hetarylethyl- or propylamines, or aminoacids. This same observation applies to the structure of all biogenic amines, catecholamines — dopamine, noradrenaline, adrenaline, histamine, which mediate an infinite set of a mammalian's functioning.

A variety of substituents (side chains) attached on that β- (-C-C-) or γ-(-C-C-C-) moieties express the meaning of the particular "letter," its "sound value," and represent together a definite sequence of two-, three-, sometimes, four "letter" "order-word molecules" of a biologically active entity.

Perhaps the distance between the nitrogen atom and the fourth or third position in a piperidine ring represents certain optimum interval between "letters" in the substrat-receptor communication "language" "alphabet," which explains the prevalence of the piperidine ring in the drug's world.

From this extremely simplificated view on things, it becomes clear why the piperidine ring, which by itself does not have any pharmacological value, represents an optimal carrier for different pharmacophores — "letters" bearing specific information and explains why the piperidine ring is the most frequently occurring framework in biologically active compounds for transmitting messages of very diverse profile.

This observation fits perfectly with Lipinski's rule of five or rule of three and points towards novel research opportunities for piperidine-based drug discovery and development.

Index—Trade Names

Page numbers followed by "*f*" refer to figures.

Index—Substance Classes

Page numbers followed by "*f*" refer to figures.

Printed in the United States
By Bookmasters